Mastering ArcGIS

Second Edition

Maribeth Price
*South Dakota School of
Mines and Technology*

McGraw Hill **Higher Education**

Boston   Burr Ridge, IL   Dubuque, IA   Madison, WI   New York   San Francisco   St. Louis
Bangkok   Bogotá   Caracas   Kuala Lumpur   Lisbon   London   Madrid   Mexico City
Milan   Montreal   New Delhi   Santiago   Seoul   Singapore   Sydney   Taipei   Toronto

Higher Education

MASTERING ARCGIS, SECOND EDITION

1 2 3 4 5 6 7 8 9 0 QPD/QPD 0 9 8 7 6 5

ISBN 0–07–298417–1

Publisher: *Margaret J. Kemp*
Senior Sponsoring Editor: *Daryl Bruflodt*
Developmental Editor: *Debra A. Henricks*
Associate Marketing Manager: *Todd L. Turner*
Lead Project Manager: *Joyce M. Berendes*
Senior Production Supervisor: *Kara Kudronowicz*
Lead Media Project Manager: *Judi David*
Senior Media Technology Producer: *Jeffry Schmitt*
Designer: *Laurie B. Janssen*
Cover Designer: *Rokusek Design*
(USE) Cover Image: *Maribeth H. Price, with special thanks to ESRI, Inc.*
Compositor: *Lachina Publishing Services*
Typeface: *10/12 Times Roman*
Printer: *Quebecor World Dubuque, IA*

Library of Congress Cataloging-in-Publication Data

Price, Maribeth Hughett, 1963–
  Mastering ArcGIS / Maribeth H. Price. — 2nd ed.
    p.   cm.
  ISBN 0–07–98417–1 (hard copy : alk. paper)
  1. ArcGIS. 2. Geographic information systems. I. Title.

G70.212P74   2006
910'.285—dc22   2005011621
                                           CIP

www.mhhe.com

Table of Contents

Skills Reference List

Chapter 9 Layouts, cont.

Chapter 10 Geocoding

Chapter 11 Basic Editing

Chapter 12 More Editing

Chapter 13 Geodatabases

Chapter 14 Network Analysis

Chapter 15 Raster Analysis

Preface

Welcome to Mastering ArcGIS, a detailed primer on learning to use the ArcGIS™ software by ESRI®, Inc. This book is designed to offer everything you need to master the basic elements of GIS.

> **Notice:** ArcGIS™, ArcView™, ArcMap™, ArcCatalog™, ArcInfo™, ArcInfo Workstation™, ArcEditor™, and the other program names used in this text are registered trademarks of ESRI, Inc. The software names and the screen shots used in the text are reproduced by permission. For ease of reading only, the ™ symbol has been omitted from the names; however, no infringement or denial of the rights of ESRI® is thereby intended or condoned by the author.

Previous experience

This book assumes that the reader is comfortable using Windows™ to carry out basic tasks such as copying files, moving directories, opening documents, exploring folders, and editing text and word processing documents. Previous experience with maps and map data is also helpful. No previous GIS experience or training is needed to use this book.

Elements of the package

This learning system includes a textbook and a CD-ROM, including:

➢ Fifteen chapters on the most important capabilities of ArcGIS.

➢ Comprehensive tutorials in every chapter to learn the skills. Each step is demonstrated in a video clip.

➢ A set of exercises, map documents, and data for practicing skills independently.

➢ Reference sections on skills with video clips demonstrating each one.

Philosophy

This text reflects the author's personal philosophies and prejudices developed from ten years of teaching GIS at an engineering school.

➢ **GIS is best learned by doing it, not by studying it.** The laboratory is THE critical component of the course, and theory is best introduced sparingly and integrated with experience. Hence this book is heavy on experience and light on theory. Instructors who love teaching theory will probably find it necessary to supplement this book with one of the fine theoretical textbooks available.

➢ **Independent work and projects are critical to learning GIS.** This book includes a wealth of exercises in which the student must find solutions independently without a cookbook recipe of steps. A wise instructor will also require students to choose and carry out an independent research project.

> ➤ **GIS analysis should be taught before GIS data development.** Analysis is more interesting and easier to learn for the beginner. Creating general exercises for data development is also difficult because it generally requires additional software, which varies from organization to organization and never seems to work as seamlessly as one would hope. It also often requires using aspects of GIS not covered in the first semester, such as using a few steps of GRID to byteswap an image. The author has found that student projects go much better when students are encouraged to work mainly with data already in GIS format, and leave data management for later.

This book assumes that the student has access to ArcView 9.0 or ArcGIS Desktop 9.0 or higher. Many important GIS functions require the more expensive ArcEditor or ArcInfo licenses, but they are not covered in this text. The Spatial Analyst extension is required to do Chapter 15.

Chapter sequence

The book contains an introduction and 15 chapters. Each chapter includes roughly one week's work for a typical three-credit semester course, although Chapter 15 is best covered over two weeks. This book intentionally contains somewhat more material than the average GIS class can cover during a single semester, in recognition of the fact that instructors vary in their consideration of which topics are most important. In the author's introductory GIS course, we work steadily on the material in the book for 11-12 weeks, and the remainder of the semester is devoted to project work. The remaining chapters provide helpful information for student projects or material for continued study after the course is over.

An introductory chapter describes GIS and gives some examples of how it is used. It also provides an overview of GIS project management and how to develop a project and proposal. The remaining chapters can be divided into five basic sections:

I. Basic Skills: Chapters 1-5 introduce basic skills on working with GIS data, including learning the ArcCatalog and ArcGIS interface, creating maps, working with tables, and generating map layouts and reports.

II. GIS Analysis: Chapters 6-8 introduce different analysis techniques, including queries, spatial joins, and map overlay. Chapter 9 shows how to create map layouts.

III. Data Entry: Chapters 10-12 teach geocoding and editing. Chapter 11 provides the critical skills needed by most people to do basic data entry; Chapter 12 can be considered optional and covers more advanced ways to manipulate features.

IV. Advanced Topics: Chapters 13-15 introduce advanced data management and analysis, including working with geodatabases and using networks. Chapter 15 provides an introduction to raster analysis for those with the Spatial Analyst extension.

The first five chapters are best followed in order to ground the student in critical skills. After completing these basic skills, some variation in chapter order is fine. Those preferring a more project-oriented approach, for example, might wish to use the following order: Chapters 1-5 on basic skills, Chapter 11 or 11-12 for editing, Chapters 6-8, 10, 14 and possibly 15 for studying analysis, and Chapter 9 for presenting the results. Minor adjustments to the exercise problems may be needed if the chapter order is varied.

Chapter layout

Each chapter is organized into the following sections:

- ➢ **Concepts:** provides basic background material for understanding the principles and techniques involved in using ArcGIS. A set of review questions follows the concepts section.

- ➢ **Tutorial:** contains a step-by-step tutorial demonstrating the concepts and skills learned in the chapter. The tutorials begin with detailed instructions, which gradually become more general as mastery is built. Every step in the tutorial is demonstrated by accompanying video clips.

- ➢ **Exercises:** presents a series of problems to build skill in identifying the appropriate techniques and applying them without step-by-step help. Through these exercises the student builds an independent mastery of GIS processes. The problems are graduated in difficulty from the first to the last, and end with a challenge problem. Answers or solution methods are included for selected exercises.

- ➢ **Skills Reference:** provides step-by-step instructions for carrying out the most frequent and important tasks being learned. This material is similar to the program documentation but is organized by topic and demonstrated with video clips.

The CD-ROM contains all the necessary data and documents to follow the tutorials and complete the exercises.

Instructors should use judgment in assigning exercises, as the typical class would be stretched to complete all the exercises in every chapter. Assign about half of them as a starting point, and adjust if necessary. A very good student can complete the entire set in 3-5 hours, most would need 6-8 hours, and a few would require 10 or more hours. Students with extensive computer experience generally find the material easier than those who make only limited use of computers.

Using this text

In working through this book, the following sequence of steps is suggested:

- ➢ READ through the Concepts section to get familiar with the principles and techniques.

- ➢ SKIM the Skills Reference section to find out what techniques will be learned. You may wish to view the videos at this time, or simply use them later for reference as needed.

- ➢ ANSWER the Chapter Review Questions to test your comprehension of the material.

- ➢ WORK the Tutorial section for a step-by-step tutorial and explanation of key techniques.

- ➢ REREAD the Concepts section to reinforce the ideas.

- ➢ PRACTICE by doing the Exercises.

Using the tutorial

The tutorial provides step-by-step practice and introduces details on how to perform specific steps. Doing the tutorial is extremely important to the learning process. Students should be encouraged to think about the steps as they are performed, and not just race to get to the end.

It is extremely important to follow the directions carefully. Skipping a step or doing it incorrectly may result in a later step not working properly. Saving often will make it easier to go back and correct a mistake in order to continue on. Once in a while a step will not work due to differences between computer systems or software versions. Having an experienced user nearby to identify the problem can really help. If one isn't available, however, just skip the step and move on without it.

Using the videos

The CD distributed with the book contains two types of videos. The Tutorial Videos demonstrate each step of the tutorial. They are numbered in the text for easy reference. The Skills Reference Videos show how to perform generic tasks, such as deleting a file or changing symbols on a map.

The videos are intended as a supplement and as an alternate learning strategy for those who find following a series of written steps cumbersome. It would be extremely tedious to watch all of them. Instead, use them in the following situations:

> When the student does not completely understand the written instructions or cannot find the correct menu or button.

> When a step cannot be made to work properly.

> When a reminder is needed to do a previously learned skill in order to complete a step.

> Whenever a student finds that watching the videos enhances learning.

The CD contains an index file with hyperlinks to each clip. As you work through a chapter, keep the index on the screen, and click the appropriate link to see a video. The tutorial clips are distinguished by numbered steps, the Skills Reference Videos by their headings.

Installing the CD-ROM components

The CD-ROM contains one folder, MGIS. Inside the folder is a folder containing the video clips, a video index document (in Microsoft® Word or HTML format), and a self-installing archive of the tutorial data. See the README file on the CD for any last-minute changes to the installation instructions below.

To view the videos:

Insert the CD in the drive and use Windows Explorer to open the D:\MGIS folder (if your CD-ROM drive is designated by a different letter, substitute the correct letter for D). Double-click the VidIndex.doc or VidIndex.htm file to open it. Click on a blue video hyperlink to start a video. For best resolution, size the video window as large as possible.

To install the training data:

The mgisdata.exe archive contains the documents and data needed to do the tutorials and exercises. The student must copy this folder to his own hard drive. If more than one person on the computer is using this book, then each person should make her own copy of the data in a separate folder. The installed data requires approximately 100 MB of disk space. It is put on the CD as a self-extracting zip file. To install the data for the exercises, follow these instructions:

➢ Place the book CD in your computer's CD-ROM drive and wait for the CD splash screen to appear. Indicate your acceptance of the license agreement.

➢ Click on the link to Download Data Sets. If a dialog window appears asking whether to Run or Save the data, choose Run.

➢ When a dialog box appears asking whether to Open or Save the data, choose Open. Don't choose Save because it will only copy the data archive instead of extracting it.

➢ Click the button with ellipses to set the folder to extract the data to (see (a) in the figure). The data will be placed in a folder called mgisdata in whatever location you choose. In other words, if you select C:\temp as the target folder, then the data will be placed in C:\temp\mgisdata.

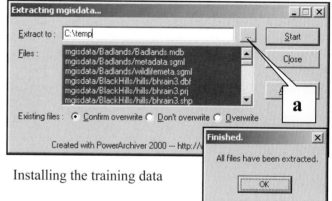

Installing the training data

➢ Click Start to start installing the data. It may take several minutes. Wait until you see the Finished window announcing that all the files have been extracted, then click OK.

System requirements

To use the tutorials and do the exercises in this book, the student must have access to a computer with the following characteristics:

Hardware:

Pentium™-class computer with 800 MHz processor or better and 256 MB RAM

Suitable sound/graphics card (nearly all systems have one)

Software:

Windows NT™ or Windows 2000™ or Windows XP™ operating system

A Web browser, such as Netscape or Internet Explorer, or Microsoft® Word

Microsoft® MediaPlayer™ 7.0 or higher (can be downloaded free from Microsoft®)

ArcView™ 9.0 or higher, or ArcGIS™ Desktop 9.0.x or higher

For assistance in acquiring or installing these components, contact your system administrator, hardware/software provider, or local computer store.

Acknowledgements

I would like to thank many people who made this book possible. Governor Janklow of South Dakota funded a three-month summer project in 2000 that got the book started, as part of his Teaching with Technology program. Many students in my GIS classes between 2000 and 2005 tested the text and exercises and helped immensely in making sure the tutorials were clear and worked correctly. Reviewers Michael Harrison, Ralph Hitz, Olga Medvedkov, James W. Merchant, Raymond L. Sanders, Jr., Yifei Sun, Fahui Wang, and Judy Sneller provided detailed and helpful comments, and the book is better than it would have been without their efforts. I also thank the reviewers who provided valuable advice for the second edition, including Richard Aspinall, Joe Grengs, Tom Carlson, Susan K. Langley, Henrietta Loustsen, Xun Shi, Richard Lisichenko, John Harmon, Michael Emch, Jim Sloan, Sharolyn Anderson, Talbot Brooks, Qihao Weng, Jeanne Halls, and Mark Leipnik.

ESRI, Inc. was prompt and generous in its granting of permission to use the screenshots, data, and other materials throughout the text. I also thank George Sielstad, Eddie Childers, Mark Rumble, Tom Junti, and Patsy Horton for their generous donations of data. I am grateful to Dale Nickels and Steve Bauer for their long-term computer lab administration, without which I could not have taught GIS courses or developed this book. I thank Linda Heindel for organizing student feedback and assisting with the initial round of edits on the first draft. I thank editors Tom Lyons and Lisa Bruflodt of McGraw-Hill for their unfailing encouragement and enthusiasm about the book as it took shape, as well as their excellent feedback. I thank the McGraw-Hill team working on the second edition, especially Debra Henricks. I am grateful to Daryl Pope who first started me in GIS, and John Suppe who encouraged me to return to graduate school and continue doing GIS on a fascinating study of Venus. And certainly not least, I thank Curtis Price and my daughters Ginny and Madeleine for their understanding and support during the many, many hours I spent creating this book.

Introduction

What Is GIS?

Objectives

- ➤ Developing a basic understanding of what GIS is, its operations, and its uses

- ➤ Getting familiar with GIS project management

- ➤ Learning to plan a GIS project and write a project proposal

Concepts

What is GIS?

GIS stands for Geographic Information System. In practical terms, a GIS is a set of computer tools that allows people to work with data that are tied to a particular location on the earth. Although many people think of a GIS as a computer mapping system, its functions are broader and more sophisticated than that. A GIS is a database that is specially designed to work with map data.

For example, consider the accounting department of the local telephone company. They maintain a large computer database of their customers, in which they store the name, address, phone number, type of service, and billing information for each customer. This information is only incidentally tied to where customers live; they can carry out all of the important functions (billing, for example) without needing to know where each house is. Of course, they need to have addresses for mailing bills, but it is really the post office that worries about where the houses actually ARE. This type of information is called **aspatial data,** meaning that it is not tied, or only incidentally tied, to a location on the earth's surface.

In the service department, however, they need to work with **spatial data** to provide the telephone services. When hundreds of people call in after a power outage, they must analyze the distribution of the calls with respect to the telephone lines and isolate the location where the outage occurred. When a construction company starts work on a street, workers must be informed of the precise location of buried telephone cables. If a developer builds a new neighborhood, the company must be able to determine the best place to tie into the existing line network so that the services are most efficiently distributed off the main trunk lines. When technicians prepare lists of house calls for the day, they need to plan the order of visits to minimize the amount of driving time. In these tasks, location is a critical aspect of the job, and the information is spatial.

In this example, two types of software are used. The accounting department uses special software called a database management system, or DBMS, which is optimized to work with large volumes of aspatial data. The service department needs access to a database that is optimized for working with spatial data, a Geographic Information System. Because these two types of software are related, they often work together, and they may access the same information. However, they do different things with the data.

A GIS is built from a collection of hardware and software components.

> **A computer hardware platform.** Due to the intensive nature of spatial data storage and processing, GIS was once limited primarily to large mainframe computers or expensive workstations. Today it can run on a typical desktop personal computer.

> **GIS software.** Many companies offer software for GIS applications, and they vary widely in cost, ease of use, and level of functionality. All of them should offer at least some minimal set of functions as described in the next few paragraphs. In this book we study one particular package that is very powerful and widely used, but others are available and may be just as suitable for certain applications.

> **Data storage.** Some projects use only the hard drive of the GIS computer. Other projects may require more sophisticated solutions, if very large volumes of data are being stored, or multiple users need access to the same data sets. Backup capability is also an issue when expensive and irreplaceable data are being used. Compact disk writers and/or tape systems are highly useful for backing up and sharing data.

> **Data input hardware.** Many GIS projects require sophisticated data entry tools. Digitizer tablets enable the shapes on a paper map to be entered as features in a GIS data file. Scanners create digital images of paper maps. An Internet connection provides easy access to large volumes of GIS data. High-speed connections are preferred, as GIS data sets are often inconveniently large to download by slower methods.

> **Information output hardware.** A quality color printer capable of letter-size prints provides the minimum desirable output capability for a GIS system. Printers that can handle map-size output (36" × 48") will be required for many projects.

> **GIS data.** Data comes from a variety of sources and in a plethora of formats. Gathering data, assessing its accuracy, and maintaining it usually constitute the longest and most expensive part of a GIS project.

> **GIS personnel.** A system of computer and hardware is useless without trained and knowledgeable people to run it. The contribution of professional training to successful implementation of a GIS is often overlooked.

GIS software varies widely in functionality, but any system claiming to be a GIS should provide the following functions at a minimum:

> Data entry from a variety of sources, including digitizing, scanning, text files, and the most common spatial data formats. Ways to export information to other programs should also be provided.

> Data management tools, including building data sets, editing spatial features and their attributes, and managing coordinate systems and projections.

> Thematic mapping (displaying data in map form), including symbolizing map features in different ways and combining map layers for display.

> Data analysis functions for exploring and testing spatial relationships in and between map layers.

➤ Map layout functions for creating soft and hard copy maps, including titles, scale bars, north arrows, and other map elements.

GIS systems are put to many uses, but the common goal of all, and the strength of GIS, lies in providing the means to collect, manage, and analyze data to produce information for better decision making. This book is a practical guide to understanding and using a particular Geographic Information System called ArcGIS. Using this book, you can learn what a GIS is and what it does, and how to apply its capabilities to solve real-world problems.

A history of GIS

Geographic Information Systems have grown from a long history of cartography begun in the lost mists of time by early tribesmen who made sketches on hides or formed crude models of clay as aids to hunting for food or making war. Ptolemy, an astronomer and geographer from the second century B.C., created one of the earliest known atlases, a collection of world, regional, and local maps and advice on how to draw them, that remained essentially unknown to Europeans until the 15th century. Translated into Latin, it became the core of Western geography, influencing cartographic giants such as Gerhard Mercator, who published his famous world map in 1569. The 17th and 18th centuries saw many important developments in cartography, including the measurement of a degree of longitude by Jean Picard in 1669, the discovery that the earth flattens toward the poles, and the adoption of the Prime Meridian that passes through Greenwich, England. In 1859, French photographer and balloonist Gaspard Felix Tournachon founded the art of remote sensing by carrying large format cameras into the sky. In an oft-cited early example of spatial analysis, Dr. John Snow mapped cholera deaths in central London in September, 1854, and thereby was able to locate the source of the outbreak--a contaminated well. However, until the 20th century, cartography remained an art and a science carried out by laborious calculation and hand drawing.

As with many other endeavors, the development of computers inspired cartographers to see what these new machines might do. The early systems developed by these groups, crude and slow by today's standards, nevertheless laid the groundwork for modern GIS. Dr. Roger Tomlinson, head of an Ottawa group of consulting cartographers, has been called the "Father of GIS" for his promotion of the idea to use computers for mapping, and his vision and effort in developing the Canada Geographic Information System (CGIS), an early GIS birthed in the mid-1960's. Another pioneering group, the Harvard Laboratory for Computer Graphics and Spatial Analysis, was founded in the mid-1960's by Howard Fisher. He and his colleagues developed a number of early programs between 1966 and 1975, including SYMAP, CALFORM, SYMVU, GRID, Polyvrt, and ODYSSEY. Other notable developers included professors Nystuen, Tobler, Bunge, and Berry from the Department of Geography at Washington University during 1958-1961. In 1970, the U.S. Bureau of the Census produced its first geocoded census and developed the early DIME data format based on the CGIS and POLYVRT data representations. DIME files were widely distributed and were later refined into the TIGER format. These efforts had a pronounced effect on the development of data models for storing and distributing geographic information.

In 1969, Laura and Jack Dangermond founded the Environmental Systems Research Institute (ESRI), which pioneered the powerful idea of linking spatial representation of features with attributes in a table, the core idea that revolutionized the industry and launched the development of Arc/Info, a program whose descendents have captured about 90% of today's GIS market. Other vendors are still active in developing GIS systems, which include packages MAPINFO, MGA from Intergraph, and IDRISI from Clark University.

What can a GIS do?

GIS works with many different applications: land use planning, environmental management, sociological analysis, business marketing, and more. Any endeavor that uses spatial data can benefit from a GIS. For example, researchers at the U.S. Department of Agriculture, Rocky Mountain Research Station in Rapid City are conducting a study of elk habitat in the Black Hills of South Dakota and Wyoming. They have placed radio transmitter collars on about 70 elk bulls and cows (Fig. I.1c). Using the collars and a handheld antenna, they track the animals and obtain their locations. Several thousand locations have been collected (Fig. I.1a), and the scientists study the characteristics of the habitat where elk spend time.

The elk locations are entered into a GIS system for record keeping and analysis. Each location becomes a point with attached information, including the animal ID number, and date and time of the sighting. Information about vegetation, slope, aspect, elevation, water availability, and other site factors can be derived by overlaying the points on other data layers, allowing the biologists to compare the characteristics of sites utilized by the elk. Figure I.1b shows a map of distances to major roads in the central part of the study area. The elk locations tend to cluster in the darker roadless areas, and statistical analysis demonstrates this observation empirically. The next section describes a GIS model to evaluate elk habitat.

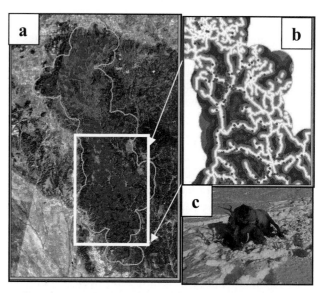

Fig. I.1. Analyzing elk habitat use. (a) Elk locations and study area. (b) Locations on a map of distance to nearest road. (c) Collaring an elk.

Modeling elk habitat with GIS

The Habitat Capability model (HABCAP), developed by the same researchers at the Rocky Mountain Research Station, offers an example of a GIS application that uses many classic spatial analysis functions. The model was designed to help the Forest Service evaluate different timber-cutting plans in terms of their effects on wildlife such as elk. It is presented here, slightly simplified, to demonstrate some types of analysis a GIS can perform. (Users interested in learning more about or using the actual HABCAP model may visit http://www.fs.fed.us/rm/rapidcity/.)

The HABCAP model focuses on three key factors shown by research to influence the suitability of habitat for elk: forage, cover, and roads. Forage meets the food requirements of the animals. Cover provides places to hide from predators as well as thermal protection during periods of extreme heat or cold. Generally these two factors are somewhat exclusive: dense tree stands good for cover do not allow growth of grasses, and the grass-filled meadows provide poor cover. Thus the *juxtaposition* of forage and cover also become important: meadows full of tasty grass will not be utilized by the elk if they are too far from the protection of the trees. Finally, traffic stresses the elk, and studies show that they tend to avoid areas close to roads. Primary roads with heavy traffic have a greater impact than lightly used roads.

The vegetation and road data needed by the model are routinely collected and maintained by the U.S. Forest Service as GIS data **layers**. The roads layer contains line features representing each road segment, each of which is also linked to a row in a table indicating whether it is a primary, secondary, or primitive road (Fig. I.2a). The vegetation layer contains polygons, or enclosed areas representing stands of vegetation with similar characteristics. Each polygon is linked to a table which stores information about it, including the dominant cover type (such as aspen or ponderosa pine), and the structural stage, a variable representing the size and density of the trees (Fig I.2b). The COVSS field in the table is a combination of those two factors, with the first three letters representing the cover type (TAA = aspen), and the rest the structural stage (0, 1, 2, 3A, etc.) The white areas show private land where no vegetation data are available.

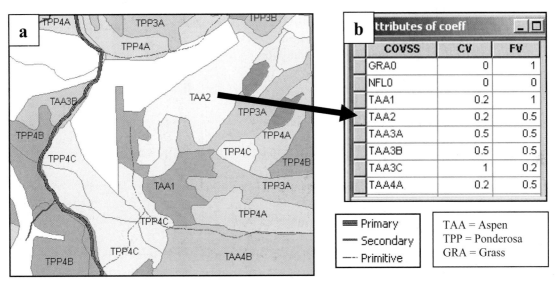

Fig. I.2. Vegetation data for the model. (a) Polygons with a combined cover type – structural stage code, with roads. (b) A table linking each vegetation code to the forage (FV) and cover (CV) coefficients.

The model uses three coefficients to measure habitat suitability: forage value (FV), cover value (CV), and proximity value (PV). Each variable rates a particular stand of vegetation based on its quality as forage, cover, and whether the two exist close together. A value of zero indicates low value in one of the categories, a value of 1 indicates the highest value. The overall habitat suitability (HS) of each polygon is calculated using the formula

$$HS = (3 * FV * CV * PV) / 5$$

The value 3 is used to weight the forage variable as being more important than the others. The HS factor thus specifies the quality of the polygon as elk habitat on a scale of zero to one. The model proceeds by determining the values of FV, CV, and PV for each polygon. After HS is found for each polygon, the values are halved in areas close to roads.

Finding FV and CV is quite straightforward. Each combination of cover type/structural stage is assigned an FV and CV based on its characteristics. In Figure I.2b, a polygon with GRA = Grass and a structural stage of 0 (first line in the table) describes an open meadow, and thus rates an FV value of 1 and a CV value of 0. The TAA4A value indicates a stand of large aspen with relatively open canopy, providing both food and cover, and is given FV = 0.5 and CV = 0.5. This step demonstrates the basic GIS function of **reclassification**: using one set of values to derive a second set more directly useful to the problem at hand.

Finding PV is more complex and requires the kind of spatial analysis that a GIS is designed to provide. Basically, the ideal habitat (PV = 1) has both cover and forage available within a short distance of each other. As the separation increases, the habitat quality goes down, and PV should decrease.

First, every polygon is reclassified again based on its CV and FV values. A polygon with CV < 0.5 and FV ≥ 0.5, for example, becomes an "F" polygon, indicating it has primarily forage and little cover. A "C" polygon contains good cover but little forage. A "B" polygon contains some of both (Fig I. 3a).

Following this step a **dissolve** operation is used to remove extraneous boundaries between polygons with the same designation (Fig. I.3b). Next a **query** is performed, which extracts the C and F polygons and places them in a new layer by themselves, leaving the B polygons out.

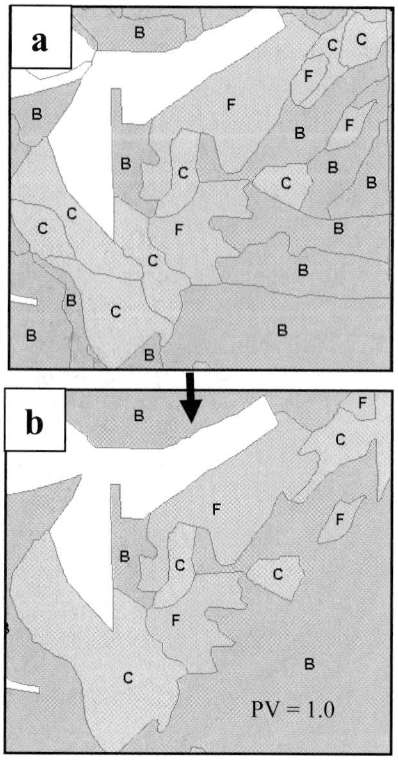

As one examines Figure I.3, some important observations emerge. A B polygon has PV = 1 because both cover and forage are found inside the polygon. Parts of an F polygon that are close to a B or a C polygon are also high quality, but the quality decreases with distance from the boundary. Notice that as a result of the dissolve, the boundary of an F polygon must be next to either a C or B polygon, indicating that in either case, cover is available just outside the boundary. For a C polygon, its boundary represents the location of the closest forage. Thus, the boundaries of both C and F polygons represent the closest available source of the missing factor inside the polygon. The areas closest to the boundary can be given a PV = 1, and areas further away should receive a lower PV.

In a GIS this function is performed using **buffers**, which are polygons created within a certain distance of a feature. (Usually buffers extend *outside* the polygon; in this case the buffers are created *inside* it.) Three buffers are constructed delineating the areas within 100, 200, and 300 meters of the boundaries, and the polygons are given a PV value of 1, 0.5, and 0.2 respectively (Fig. I.4). Areas further than 300 meters are given PV = 0.

Afterwards, the layer containing the buffers must be recombined with the original vegetation polygons. This process uses a GIS function called **union**, which combines the layers to create all possible polygons (Fig. I.5a). The result is a single layer containing the original B, C, and F

Fig. I.3. A dissolve

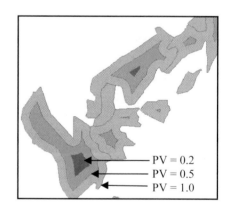

Fig. I.4. Buffering the C and F polygons

polygons as well as the buffers. Each resulting polygon has a value for all three variables, CV, FV, and PV, and the final habitat suitability rating, HS, is calculated from the variables using the equation HS = (3*FV * CV * PV) / 5.

The final step in the model creates buffers around the roads. Primary roads with high traffic are given 300 meter buffers, secondary roads receive 60 meter buffers, and primitive roads are ignored. Another union operation combines the road buffers with the habitat suitability layer, and a final calculation is performed to reduce the HS values inside the road buffers by 50%, yielding the final map of elk habitat suitability (Fig. I.5b).

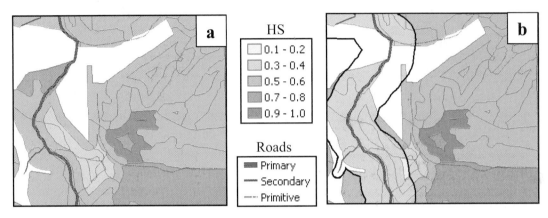

Fig. I.5. Habitat suitability (HS) calculated from the model. (a) Calculated from PV, FV, and CV. (b) Reduced by 50% in areas close to the roads.

The HABCAP model uses only a few of the many functions available in a GIS system, but it serves to demonstrate the power of linking spatial data in a map with tabular data in a table and using a variety of tools to explore the relationships between them. You will discover many more capabilities and applications of GIS as you work your way through this text.

GIS project management

A GIS project may be a small effort spanning a couple of days by a single person, or it can be an ongoing concern of a large organization with dozens of people participating. Large or small, however, projects often follow the generalized model shown in Figure I.6. A project usually begins with an assessment of needs. What specific issues must be studied? What kind of information is needed to support decision making? What functions must the GIS perform? How long will the project last? Who will be using the data? What funding is available for start-up and long-term support?

Without a realistic idea of what the system must accomplish, it is nearly impossible to design it efficiently. Users may find that some critical data are absent, or that resources have been wasted acquiring data that no one ever uses. In a short-term project the needs are generally clear-cut. A long-term organizational system will find that its needs evolve over time, requiring periodic reassessment. A well-designed system will adapt easily to future modification. The creator of a haphazard system may be constantly redoing previous work when changes arise.

In studies seeking specific answers to scientific or managerial questions, a methodology or **model** must be chosen. Models convert the raw data of the project into useful information using a well-defined series of steps and assumptions. Creating a landslides hazards map provides a simple example. One might define a model such that if an area has a steep slope and consists of a shale rock unit, then it should be rated as hazardous. The raw data layers of geology and slope can thus be used to create the hazard map. A more complex model might also take into account the dip (bedding angle) of the shale units. Models can be very simple as in this example, or more complex, like the elk model presented earlier.

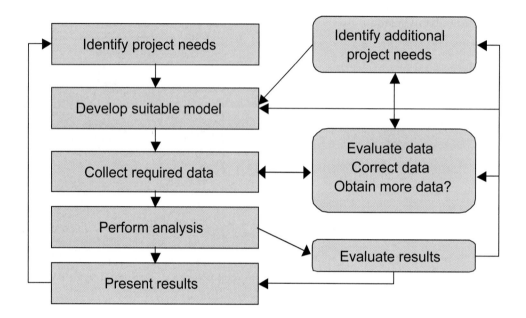

Fig. I.6. Generalized flow diagram representing steps in a GIS project

Once needs are known and any appropriate models have been designed, data collection can begin. GIS data are stored as layers, with each layer representing one type of information, such as roads or soil types. The needs dictate which data layers are required and how accurate they must be. A source for each data layer must be found. In some cases data can be obtained free from other organizations. General base layers such as elevation, roads, streams, political boundaries, and demographics are freely available from government sources (although the accuracy and level of detail are not always what one might wish). More specific data must often be developed in-house. For example, a utility company would need to develop its own layers showing electric lines and substations—no one else would be likely to have it.

The spatial detail and accuracy of the data must be evaluated to ensure that it is able to meet the needs of the project. For example, an engineering firm creating the site plan for a shopping mall could download elevation contours of the site for free. However, standard 10-foot contours cannot provide the detailed surface information needed by the engineers. Instead, the firm might contract with a surveying firm to measure contours at half-foot intervals.

After the data are assembled, the analysis can begin, based on the model chosen. During the analysis phase it is not unusual to encounter problems which might require making changes to the model and/or data. Thus the steps of model development, data collection, and analysis often become iterative, as experience gained is used to refine the process. The final result must be checked carefully against reality in order to recognize any shortcomings and provide guidelines for improving future work.

Finally, no project is complete until the results have been presented, perhaps informally to a supervisor, perhaps published in a scientific journal, or perhaps as the focus of a heated public meeting. The form of the presentation influences the format and style of the presentation, which can take the form of maps, reports, presentation slides, etc.

Project case study: A thorny issue

A project recently commissioned by Badlands National Park (Fig. I.7) can serve as an example of a GIS project in action. Canada thistle (*Cirsium arvense*) is a highly invasive, noxious weed that is widespread throughout the park. Proposed control measures included spraying with Tordon, a broad-based herbicide. However, the herbicide also kills native plants, and it has a long residence time in soils, leading to potential groundwater contamination. The park approached the author with a proposal to use GIS analysis to better locate areas of infestation, thereby minimizing the amount of pesticide used.

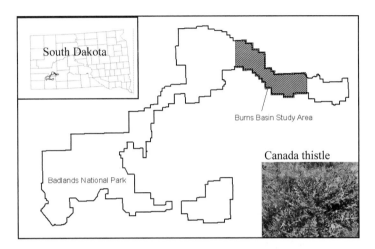

Fig. I.7. Burns Basin study area in Badlands National Park

Step 1: Identify project needs

Work began by identifying the project objectives in order to write the proposal. The objectives centered on a specific research question: Can the occurrence of Canada thistle be predicted from knowledge of environmental factors at known thistle localities? The objectives included:

> ➤ Correlating known thistle occurrence to environmental parameters such as soil type
> ➤ Developing a model to predict the occurrence of thistle
> ➤ Testing the model predictions using additional field surveys
> ➤ Finishing predictions/testing prior to the anticipated field spraying operation in June

Step 2: Develop model

For this project a **multivariate regression** model was chosen. The familiar **linear regression** model calculates a best-fit line through a cloud of points in an *x-y* graph (Fig. I.8), and the equation of the line $y = mx + b$ can then be used to predict values of y (dependent variable) for any proposed value of x (independent variable). A multivariate regression applies this concept to multiple independent variables at once. In GIS, each independent variable is represented by a map layer. Since thistle presence is a true/false variable, a variation of the multivariate regression model, **logistic regression,** must be used.

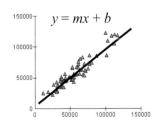

Fig. I.8. Linear regression

Analysis of the literature on Canada thistle revealed that the presence of water and anthropogenic disturbance are significant factors in thistle infestations. Likely other factors included soil types, water availability, vegetation communities, the prior existence of thistle, and prevailing winds.

The author and the park prepared a joint research proposal and received funding from the National Park Service Geographic Information Systems Funding Program (Midwest Regional Office). The project had a $10,000 budget and a six-month duration. The project proposal is included at the end of this chapter. The short time frame and small budget dictated that the project focus on data that were easily available.

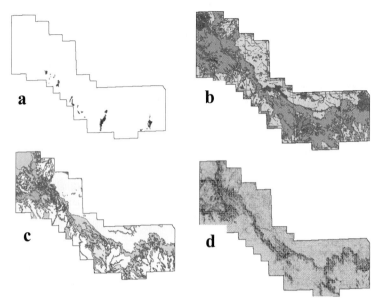

Fig. I.9. Some data layers used for the thistle project: (a) Known thistle localities, (b) Soils map, (c) Geology map, and (d) Slope.

Step 3: Collect data

The park provided most of the data layers, including geology, soil types, vegetation communities, slope, aspect, water bodies, and roads. The author also included vegetation and soil moisture indices derived from Landsat satellite imagery. The park also had a layer of known thistle localities plotted during a horseback survey in 1998. Figure I.9 shows some of the data layers included in the project.

Reliable layers for hiking trail usage, wind direction, and wetlands would have been a welcome addition, but were dismissed for reasons of time and economy. The scope and budget of a project often require decisions of this sort, at the risk of leaving out critical information. Some projects have more clearly defined layers that are critical to the project. In this one the important factors were not entirely known, so the approach adopted was to include everything easily available that might have an influence.

Data collection for the project included locating all of the layers and bringing them together. Additional data management tasks included merging separate files for each quadrangle into single map layers and removing information outside the study area (clipping). Analysis functions were used to create slope and aspect maps from the original digital elevation models and to calculate indices from the satellite images. When all the data processing was complete, the analysis could start.

Step 4: Perform analysis

As work began, a need to revise the proposed model quickly surfaced due to two difficulties. First, the logistic regression requires input locations representing the absence of thistles as well as their presence. A no-thistle data layer was constructed by assuming that the survey found all the thistle areas and by randomly choosing other points outside those areas which were assumed to be thistle free. Second, logistic regression works for numerical data only. Geology, soil type, and vegetation are non-numeric quantities and could not be included in the regression. We brought in an additional model based on a binomial statistical analysis of the abundance of thistle relative to the area fraction of each variable. The two models would be run independently and the results would be combined to produce the final result.

This part of the project demonstrates that the initial planning often cannot foresee all eventualities, and that additional data or a revised model may be needed to complete a project. Thus, the procedure shown in Figure I.6 may not necessarily follow a linear progression but may include cycles of reevaluating the data or model, finding additional data, or making corrections to the procedures.

The logistic regression calculated correlations between the numeric variables and the thistle locations/nonlocations, and produced a map showing the probability of finding thistle at any location on the map. The results are shown in Figure I.10a. The white areas represent a low probability of finding thistle and the brown areas represent a high probability.

In the binomial model, the number of thistle locations falling inside a particular map unit (e.g., a soil type) was compared with the number of locations that would be expected if the locations were random. (If random, the expected number is proportional to the area fraction of each map unit.) The comparisons were converted to Z-scores, in which a positive value greater than 2 indicates an affinity of the site to thistles, and a negative score less than –2 indicated an avoidance of the unit by thistles. Values between –2 and 2 were neutral with respect to thistles. The analysis was applied separately to the soil, geology, and vegetation layers. The results of the three layers were added together to determine a combined Z-score. Figure I.10b shows the model predictions. The blue areas indicate a low probability of thistles, the orange and brown areas predict a high probability of thistles.

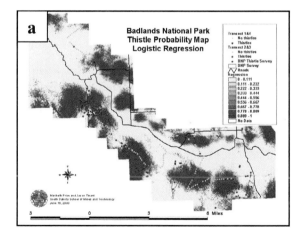
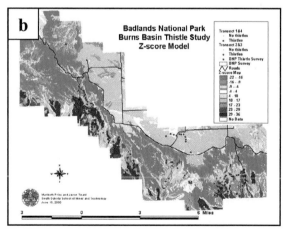

Fig. I.10. Thistle study results: (a) Probability map from logistic regression, showing likelihood of thistle infestation, (b) Combined Z-score map showing areas thistles avoid (blue) and prefer (browns).

Step 5: Evaluate results

Once the analysis is complete, the results must be evaluated for validity. In this case we chose to hike transects across the park, stopping at regular intervals to record the presence or absence of thistles (systematic sampling). The GIS helped in defining the transects, using criteria that each transect should cross both high and low thistle probabilities, should begin and end near a road, and should not cross any impassibly steep areas.

Figure I.11 shows portions of two transects. The red points indicate the presence of thistles, and the black points were thistle free. Note that virtually all of the red points occur in high probability areas, yet many black points occur there as well. After statistical evaluation of the results, the study concluded that the model could locate areas where thistles were able to grow, but could not well identify where thistles were actually growing at any given time.

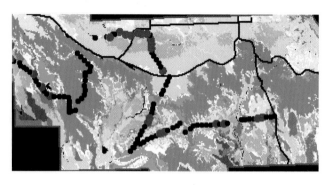

Fig. I.11. Portions of field survey plots on the model prediction map. Black points are thistle free; red points have thistles.

The final phase of the study included identifying future areas for improvement. Using current infestations and prevailing winds and hiking trails to identify areas at risk of future infestation could be very helpful. Also, the application of the model to the entire park could have provided useful information. Had more funding been available, the study could have improved the model, initiating another cycle as shown in the project flow diagram (Fig. I.6).

Step 6: Present results

The results were presented at several conferences, and a final report was submitted to the park in time to plan for the spraying (which unfortunately was cancelled for other reasons).

Types of GIS projects

The Badlands thistle study represents a **project-oriented GIS,** in which the scale and time frame of the project are well defined and the project life is finite. At the other end of the spectrum lies the **enterprise GIS,** designed for long-term support of critical functions within an organization. A city planning department, for example, is continually collecting and updating information as new parcels are developed, zoning changes, owners and tax rates fluctuate, and regulations require more and more documentation. An enterprise GIS undergoes many project cycles, and data gathering and analysis are ongoing concerns, not finite tasks. They also can encounter special issues, including sharing of data by multiple users, keeping track of updates, protecting privacy, and managing large volumes of data. However, the tasks of defining needs, collecting and evaluating data, performing analysis, and presenting results are just as applicable to the large system, even though they proceed in a more circular and ongoing fashion.

Planning a GIS project

Learning GIS by doing exercises from a book can take you only so far. The best way to learn is to develop and carry out an actual GIS project, ideally under the supervision of a more experienced user who can help solve the inevitable difficulties encountered in the real world with real data. After completing the first eight chapters of this book you will know enough to begin planning a project. If you work for an organization, then a topic or a problem may suggest itself immediately, and most of the data may already be present in-house. If not, choose an interesting topic and begin checking if appropriate data are available. For a first project it is wise to choose a problem for which most of the data needed are already available in GIS format.

Preparing a project proposal is an excellent way to develop a project. In many cases a proposal is required to obtain funding or approval for work to begin. Even if not required, however, the exercise of producing a proposal helps focus the work and provides immediate feedback on the feasibility of the project. Steps toward building a proposal include the following (in practice, the first four steps often occur in overlapping cycles until a suitable project is found):

➢ Decide on a general topic of interest.
➢ Formulate a specific question to be answered or a product to be made.
➢ Decide on an approach or model to use.
➢ Determine the data needed for the project, and confirm each layer's availability.
➢ Write the proposal according to established guidelines.

Step 1: Find a general topic

Deciding on a general topic can be difficult, as it involves matching your interests with suitable data. Many projects sound terrific until it becomes clear that the needed data are impossibly difficult to obtain. For example, almost every year one of the author's students wants to do a project analyzing the spatial distribution of mines in the Black Hills with respect to geology, faults, and soils. Although digital soil and geology maps are available, they are not sufficiently detailed to produce any worthwhile results. Mine inventories exist for the hills, but they are scattered over multiple publications with sometimes sketchy information, and it would require a great effort to pull together a useful data set. Although the topic might make a suitable Ph.D. dissertation, it is clearly too much work for a class project!

Since it is easy to develop a topic only to find that the data resources are inadequate, working backwards from the data is often helpful. Examine the data sources available and use them as a springboard to the imagination. Peruse your organization's data. Search the Internet, especially government sites such as the U.S. Geological Survey or the Environmental Protection Agency. Talk to colleagues. Visit the local GIS expert and ask for ideas and data holdings.

Step 2: Formulate a specific question

Once a topic is identified for which data are probably available, the project must be formulated as a specific research question. This step ensures that the project will yield a useful result. Too often, projects are conducted by gathering a lot of information and then analyzing it to see if anything interesting turns up. All too often, nothing significant emerges and the supervisor or instructor is less than impressed. For example, one might propose to "study thistles in Badlands National Park." In the first place, the topic is not even a question. Second, it is too broad, giving no guidance as to the data layers required or the analysis to be performed.

A good research question is focused, limited, and answerable. Look again at the research question for the thistle case study: "Can the occurrence of Canada thistle be predicted from a knowledge of environmental factors at known thistle localities?" The question has an answer: yes, no, or maybe. It narrows down the data requirements to thistle locations and environmental factors, enabling one to start tracking down layers to see what is available. It is unlikely to go off on irrelevant tangents such as the palatability of thistles to wildlife.

Projects may alternatively be structured around a specific product rather than a research question, and some fall more naturally into this category. For example, an agency could attempt to develop a 1:250,000 earthquake hazards map for Southern California.

Step 3: Develop the methodology

After asking the question, a methodology for determining the answer (or creating the product) must be defined. The major steps involved should be clearly articulated. In some cases, such as the thistle project, a formal model is implemented (logistic regression). Often, however, a sequence of analysis steps constitutes the methodology. Using the earthquake hazards map example, the model would include defining the important independent variables (fault proximity, fault activity, population density, soil acceleration potential) which will determine the dependent variable (earthquake hazard rating). If any independent variables must be derived from available base maps, such as calculating soil acceleration potential from standard maps of soil texture, then procedures must be outlined and equations defined. The goal is to develop a sequence of steps which clearly indicates the data layers needed and defines how they will be manipulated to give the final result.

Step 4: Find the data

Before the project can be considered viable, you must ensure that all of the data layers required by the model are readily available, are adequate to the task, and are affordable within the resource limitations of the project (both money and time). Failure to ascertain these conditions can derail the project later on, sometimes fatally, if a critical data set turns out to be unobtainable. Find all the data sources before committing to the project!

Step 5: Write the proposal

Once steps 1-4 have been done, the proposal nearly writes itself. If the proposal is a formal document being submitted to an agency (or instructor) for funding or approval, be sure to obtain the relevant proposal guidelines and follow them carefully, *especially the page limits*. If you are doing the proposal as an exercise, follow the fairly standard outline below. Use the exact headings shown in boldface to organize the proposal.

Introduction. Give a brief overview of the general problem you are addressing and why it is important. State the overall goals of the project and transition gracefully to a statement of the research question near the end.

Objectives. State the specific objectives to be accomplished or products to be created. A bulleted list is often easiest to read. Keep it short. One to four objectives are ample for most projects.

Methodology. Explain the rationale behind the methodology and list the major steps. Standard procedures can be glossed over or ignored, but unique aspects (such as how to calculate the soil acceleration from the county soil maps) should be clearly defined.

Data sources. List the data layers required and acknowledge the sources (e.g., the USGS website or the colleague who gave you the data). Sometimes placing this section before the Methods section makes more sense to the reader.

Work plan. List the major tasks and the estimated completion dates.

Budget (if applicable). List the itemized budget items and amounts, and total funds and/or in-kind contributions requested.

The length of the proposal depends both on the complexity of the project and the guidelines set forth by a funding agency. A simple class project can be succinctly described in one to two pages as a general rule. Extended or complex projects require additional description.

Once the project proposal is complete, bring it to a GIS professional to review, if one is available. He or she can identify any problems with methodology or data that might impair the success of the project, or suggest alternate approaches not previously considered. When all is in order, start the project and have fun with it.

Example of a GIS Project Proposal

Prediction of thistle-infested areas in Badlands National Park using a GIS model

Dr. Maribeth Price
South Dakota School of Mines and Technology
Rapid City, SD 57701
November 18, 1999

Introduction

Canada thistle *(Cirsium arvense)* is a highly invasive, noxious weed that is widespread throughout Badlands National Park (BADL). The widespread occurrence of this causes potential problems such as displacing native plants, harboring predators in black-footed ferret reintroduction sites, and causing economic damage to adjacent agricultural lands. The Canada thistle population exceeds the park's current chemical, mechanical and biological control programs. Thus, an intensive herbicide treatment program has been proposed.

Tordon™, a picloramine compound manufactured by Dow Chemical, has been shown to be effective in thistle control. However, the herbicide also kills native plants and has a long residence time in soils, creating a significant potential for groundwater contamination. Limiting the herbicide application to infested areas reduces environmental risk and also provides significant cost savings. Unfortunately, the rugged nature of Badlands National Park makes field identification of thistle-infested areas both cost and time prohibitive.

An alternative approach includes developing a GIS model to predict the occurrence of thistle in the park based on environmental factors of known infestations, such as soil type, slope, and water availability. Such a model might assist BADL in targeting the herbicide applications to areas of highest need, as well as providing information on thistle habitat which could help in locating additional infestations throughout the park.

Objectives

This project will ascertain whether a GIS model based on environmental variables at known thistle localities can be used to predict the occurrence of thistle throughout the park. The specific objectives include:

➢ Correlating known thistle occurrence to environmental parameters such as soil type
➢ Developing a model to predict the occurrence of thistle
➢ Testing the model predictions using additional field surveys
➢ Finishing predictions/testing prior to an anticipated field spraying operation in June

Methodology

The GIS analysis will use standard multivariate regression techniques to identify potential spatial correlations between approximately 50 field-mapped thistle patches and various terrain and vegetation parameters, and to develop a predictive model for thistle infestation. In addition to site factors, studies of Canada thistle suggest that infestation is affected by water availability and anthropogenic disturbances. Parameters tested will include soil type, geologic unit, vegetation community, slope, aspect, distance to water, distance to roads, and Landsat-derived Normalized Difference Vegetation Index (NDVI) and Normalized Difference Moisture Index (NDMI). The final product will include a map showing the predicted thistle density and distribution for the 2000 growing season.

After the start of the growing season, and prior to spraying, the map will be field checked to develop an accuracy estimate for the regression model. Systematic sampling sites along several transects will be located using the BADL Trimble GPS unit, and the presence/absence of thistle will be measured and compared to the model predictions. The transects will be chosen such that they adequately represent all levels of thistle probability developed by the model, they start and end near roads, and they do not cross areas of impassibly steep slopes.

Data sources

The South Dakota School of Mines and the Badlands National Park possess all of the data, software, and equipment needed to conduct the analysis.

1998 Horseback Survey of thistle infestations (shapefile)
1:24,000 vegetation communities map (coverage)
1:24,000 USGS Geologic Map and 1:24,000 SSURGO soil maps (coverages)
1:24,000 USGS Digital Elevation Model quadrangles (grids)
1:24,000 USGS Digital Line Graph roads, streams, and water bodies (coverages)
May 03, 1998 and November 17, 1998 Landsat TM images

Work plan

Jan-Feb 2000	Compile data from SDSMT and BADL sources.
	Merge quads into single layers.
	Clip layers to study area boundary in Burns Basin.
	Calculate NDVI and NDMI from Landsat images.
Mar-Apr 2000	Perform multivariate regression analysis to produce predictive map.
	Create metadata for final layers.
	Choose field transects and sampling interval.
May 2000	Field check predictive map.
June 2000	Analyze results and write final report for BADL.

Budget

Salary and Benefits	
Faculty, 3 weeks	$3900
Graduate student, 5 weeks	$2600
Undergraduate, 1 week for field work	$350
Travel	
6 trips to BADL by car, ~700 miles @ $0.27	$200
Ropes, stakes, misc. field equipment	$50
Overhead	
43% of salaries	$2945
Total requested	$10,045

Chapter 1. Introducing ArcGIS

Mastering the Concepts

Objectives

➢ Understanding the architecture of the ArcGIS program

➢ Becoming familiar with the types of data files used in ArcGIS

➢ Learning how to explore data files using ArcCatalog

Concepts

ArcGIS overview

ArcGIS is a GIS developed and sold by Environmental Systems Research Institute, Inc. (ESRI). It has a long history and has been through many versions and changes. Originally developed for large mainframe computers, in the last 10 years it has metamorphosed from a system based on typed commands to a full-featured graphical user interface (GUI), which makes it much easier to use. Because of the size and complexity of the program (actually a suite of programs), and because users have come to depend on certain aspects of the software, much of the code is carried forward and included in the new versions. Knowing this background helps a student of ArcGIS understand the nature of the ArcGIS system, and helps explain some of its odd features and characteristics.

For example, the software originally used sets of files called coverages to store the geographic data. These files were developed using a database called INFO, which was state-of-the-art at the time, but appears primitive today. ESRI has now developed a new data model, the geodatabase, based on current database technology and with exciting new features. However, rewriting the millions of lines of code developed to process and to analyze coverages takes time. Thus today the current version of ArcGIS can use both the old and new data models; however, some important functions are only available for the old

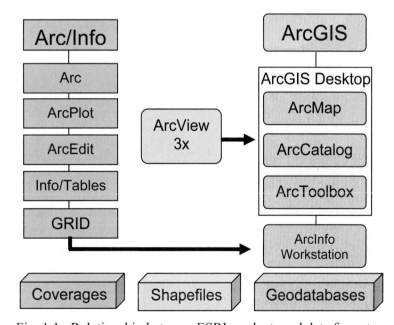

Fig. 1.1. Relationship between ESRI products and data formats

coverages. Other functions only operate on the new type of files. Thus for the time being, one must understand and be able to work with both the old and the new.

The older core of the ArcGIS system was called Arc/Info and included a basic set of programs: Arc, ArcEdit, and ArcPlot (Fig. 1.1), which utilized the coverage data model. Also included was a version of the INFO database system called Tables, and a programming language called Arc Macro Language, or AML. Optional programs could also be purchased to extend the functionality, including GRID, TIN, and COGO. All of these programs were command based, meaning that the user typed commands into a window to make the program work.

The difficulty of learning Arc/Info prompted ESRI to create a piece of software called ArcView, which was based on a GUI and designed to be easy to use. However, it was not as powerful as Arc/Info. ArcView was designed primarily to view and analyze spatial data, rather than create it. ArcView also used a simpler data model, called the shapefile, although it could read coverages and convert them to shapefiles. Beginners in GIS often learned ArcView first, and then began learning Arc/Info as their needs and abilities advanced. Advanced Arc/Info users, however, would use ArcView when they could, because it was so much easier to use, especially for creating maps.

ArcGIS, released in 2001, is a synthesis of the powerful Arc/Info system with the easy-to-use interface of ArcView, updated to use the latest advances in desktop computing and database technology. It contains two basic programs, collectively referred to as **ArcGIS Desktop** (Fig. 1.2).

➢ **ArcMap** provides the means to display, analyze, and edit spatial data and data tables. Similar in appearance to its ArcView predecessor, it nevertheless contains powerful new functionality.

➢ **ArcCatalog** is a tool for viewing and managing spatial data files It resembles Microsoft Windows Explorer, but it is specially designed to work with GIS data. It should **always** be used to delete, copy, rename, or move spatial data files.

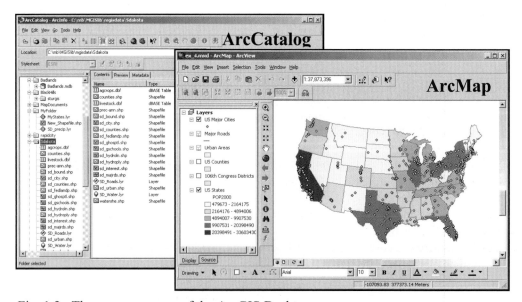

Fig. 1.2. The core programs of the ArcGIS Desktop

In addition, ArcGIS Desktop contains **ArcToolbox**, a collection of tools and functions that work in ArcCatalog and ArcMap, such as converting between data formats, managing map projections, and performing analysis. Users may create and add their own tools or scripts for special or often-used tasks. The ESRI website at www.esri.com has a large library of scripts and tools that can be downloaded to extend the ArcGIS functionality.

Finally, the original Arc/Info command-line software can still be accessed in the additional module called **Workstation Arc,** which is still used by many organizations that may be tied to the older coverage model for various reasons, such as having a large number of specialized programs written in the older AML programming language.

The ArcGIS system also provides different levels of functionality that all use the same basic interface. Users can save money by buying only the functions they need. These levels include the following:

> ➢ **ArcView** provides all of the basic mapping, editing, and analysis functions for shapefiles and geodatabases and is the level of functionality most users will require on a regular basis. It includes ArcMap and ArcCatalog, and a subset of ArcToolbox functions.

> ➢ **ArcEditor** includes all the functions of ArcView but adds editing capabilities needed to work with the advanced aspects of the geodatabase, such as topology and network editing. Additional functions reside in ArcToolbox with this level.

> ➢ **ArcInfo** provides access to the full functionality of the ArcGIS Desktop tools, and the full version of ArcToolbox. In addition, it includes the original core Arc/Info software, now called Workstation Arc.

This book focuses almost exclusively on the functions available with an ArcView license, although it mentions some of the additional capabilities as appropriate. Users can read the software documentation to learn more about the advanced topics.

The ESRI system of GIS programs, then, is a fairly complex set of tools with a long history, designed to work with a number of different data formats, also with a long history. We turn now to a discussion of how spatial and aspatial data are stored in the computer.

Introduction to raster and vector data models

The literature and concepts regarding data models are extensive and complex. This section provides a brief overview to get users started.

Many different data formats have been invented to encode data for use with GIS programs, however most of them follow one of two basic approaches: the **raster model** or the **vector model.** In either approach, the critical task includes representing the information at a point in space using x and y coordinate values (and sometimes z for height). The x and y coordinates are the spatial data. The information being represented, such as a soil type or a chemical analysis of a well, is called the attribute data. Raster and vector data models both store spatial and attribute data, but they do it in different ways.

Both data systems are **georeferenced**, meaning that the information is tied to a specific location on the earth's surface. One can use a variety of different coordinate systems for georeferencing, as we will see in Chapter 3. As long as the coordinate systems match we can display any two spatial data sets together, such as roads and houses, and have them appear in the correct spatial relationship to one another.

The raster model

A raster data system has the benefit of simplicity. A set of spatial data, such as a land use map, is represented as a series of small squares, called cells or pixels (Fig. 1.3). Each pixel has a numeric code indicating the land use, and the raster is stored as an array of numbers. To display it, a different color is assigned to each code value.

A raster data set is laid out as a series of rows and columns. Each pixel has an "address" indicated by its position in the array, such as row = 3 and column = 6.

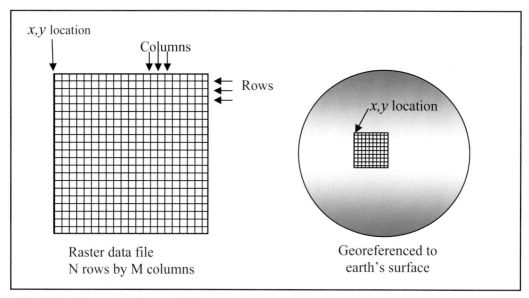

Fig. 1.3. The raster data model uses an array of values to represent a map. The raster is tied to a real-world location using the *x-y* coordinates of the upper left corner.

Georeferencing a map in an *x-y* coordinate system requires four numbers: an *x-y* location for one pixel in the raster data set, and the size of the pixel in the *x* and *y* directions. Usually the upper-left corner is chosen as the known location, and the *x* and *y* pixel dimensions are the same, so that the pixels are square. From these four numbers, it is possible to calculate the coordinates of every other pixel based on its row and column position. In this sense, the georeferencing of the pixels in a raster data set is implicit—one need not store the *x-y* location of every pixel.

The *x* and *y* dimensions of each pixel define the **resolution** of the raster data. The higher the resolution, the more precisely the data can be represented. Consider the 90-meter resolution roads raster in Figure 1.4. Since the raster cell dimensions are 90 meters, the roads are represented as much wider than they actually are, and they appear blocky rather than forming smooth curves. A 10-meter resolution raster could represent the roads more accurately; however, the file size would increase by 9 × 9, or 81 times!

Two styles of raster data can be stored (Fig. 1.4). A **discrete** raster has relatively few possible values that tend to repeat themselves in adjacent cells. Categorical data, which falls into a few named categories, is typically discrete also. Roads are discrete, with a value representing the location of the road segment and its type (primary or secondary). Land use is both categorical and discrete, with relatively large patches of adjacent cells sharing the same land use code. The codes are drawn from a list of defined values. When a land use boundary is crossed, the code changes abruptly to a new one with no intermediate values. A **continuous** raster data set is one with a large selection of numeric values that can range smoothly from one location to another. A digital elevation model, or DEM is an example of continuous data—cells are unlikely to have the same elevation value as their neighbors, and thousands of different values are possible in a single file. Different strategies are used to display discrete versus categorical data.

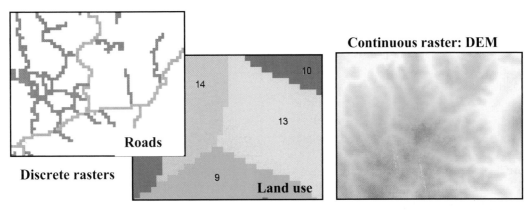

Fig. 1.4. Discrete rasters store categorical data such as land use or road types. Continuous rasters store data which vary smoothly over a surface, such as a digital elevation model (DEM) or rainfall.

The vector model

Vector data uses a series of *x-y* locations to store information. Three basic vector objects exist: points, lines, and polygons. These objects are called **features.** Point features are used to represent objects that have no dimensions, such as a well, or a sampling locality. Line features represent objects in one dimension, such as a road or a utility line. Polygons are used to represent areas, such as a parcel or a state.

In all cases, the features are represented using one or more *x-y* coordinate locations (Fig. 1.5). A point consists of a single *x-y* coordinate. A line includes two or more coordinates—the endpoints of the line are termed **nodes** and each of the intermediate points is called a **vertex.** A polygon is a group of **vertices** that define a closed area.

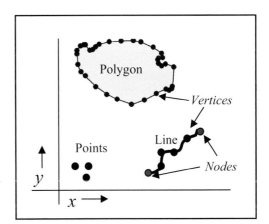

Fig. 1.5. The vector data model uses a series of *x-y* locations to represent points, lines, and polygon areas.

To some extent, the type of object used to represent features depends on the scale of the map. A large river would most likely be represented as a line on a map of the United States, because at that scale it is too small for its width to encompass any significant area on the map. If one is viewing a typical USGS topographic map, however, the river can encompass a larger area and might be represented as a polygon.

In the GIS view of the world, like features are grouped together into single data sets, which are often called **layers.** Roads and rivers are different types of objects and would be stored in separate layers. The picture of ArcMap in Figure 1.2 has six different data layers shown in the area on the left: Major Cities, Roads, Urban Areas, Congressional Districts, Counties, and States. Two of these layers, Cities and States, are currently drawn in the display.

Each feature in a vector has information about it, its attributes. The attributes are stored in a table, with a unique feature identification code (FID) linking the feature with its attributes (Fig. 1.6). Each feature corresponds to one and only one line (record) in the table. The attributes for a state might include its name, abbreviation, and population. When a state is highlighted on the map, its record in the table is also highlighted, and vice versa. It is this live link between the spatial and attribute information that gives the GIS system its power. It enables us, for example, to create a map in which the states are colored based on their populations (Fig. 1.7). This **thematic mapping** is only one example of how linked attributes can be used to analyze geographic information.

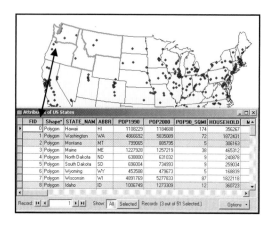

Fig. 1.6. Each state is represented by a spatial feature (polygon), which is linked to a record of information in the attribute table.

A group of like objects stored together is called a **feature class** (Fig. 1.7). A feature class can only contain one kind of object—it can include point features, or line features, or polygon features, but never a combination. In addition, objects in feature classes share a table containing their attributes, and so must have like attributes. A river and a highway would not be found in the same feature class because their attributes are likely to be very different. The pavement type would not be a logical attribute for a river!

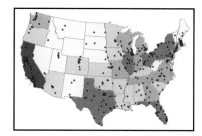

Fig. 1.7. A states feature class and a cities feature class

Some data formats, such as shapefiles, only contain one feature class. Others, such as coverages and **feature datasets** in geodatabases, can contain multiple feature classes that are in some way related to each other. For example, a feature dataset called Transportation in a geodatabase might contain the feature classes Roads, Traffic Lights, Railroads, and Canals.

Two basic vector models exist, **topological models** and **spaghetti models.** A spaghetti model simply stores features of the file as independent objects, unrelated to each other. A topological data model also stores information about how the features are spatially related, such as whether they are adjacent or connected. ArcGIS uses both kinds of models. A roads shapefile is a spaghetti model; features are stored as simple strings of *x-y* coordinates. A Transportation feature dataset in a geodatabase may contain topological rules for its feature classes, such as that roads must always connect to other roads, or that a traffic light may only occur at a road intersection.

Data files in ArcGIS

ArcGIS can read a variety of different file formats. As indicated before, many of these come from older versions of the software. Some can come from other programs such as image processing packages and CAD systems. Each of these file formats is described in some detail in the paragraphs that follow. The icon illustrations show how these data sets appear in ArcCatalog.

Shapefiles

Shapefiles are vector data files developed for the early version of ArcView and have been carried over into ArcGIS. They do not store topology, and they can contain only one feature class. They may contain points, lines, or polygons, and their attribute data are stored as dBase files (extension .dbf). A shapefile is actually a collection of files on the disk, with a common name but different extensions (e.g., roads.shp, roads.dbf, roads.shx). Shapefile features have green icons, with different symbols for point, line, and polygon shapefiles.

Coverages

A coverage is the vector data format developed for Arc/Info and is the oldest of the data formats. Coverages are topological data sets and usually contain multiple feature classes. Coverages are also composed of multiple files on the disk and even spread data among multiple folders. A coverage data set includes a folder containing several data files with an .adf (arc data file) extension. In addition, more files are stored in a folder called info that must be in the same directory. A folder containing one or more coverages is called a workspace, and it includes the info folder as well as folders for each coverage. All of the spatial and attribute information for coverages are stored in INFO format data files. Coverages have yellow icons.

Geodatabases

Geodatabases represent an entirely new model for storing spatial information that far exceeds the capabilities of coverages or shapefiles. They can contain multiple feature classes, including tables not linked with spatial data. They are stored as single database files, and they require an underlying database system to operate. They can store topological relationships between features and feature classes. They can also store rules on how the features in the geodatabase can behave and what attribute values they can have; for example, a traffic light must always be associated with a street intersection, or a telephone pole must be made of either wood or metal. These rules make it easier to model the behavior of features and reduce errors associated with data entry. Geodatabases have grey icons. Chapter 13 describes this model and its capabilities in more detail.

Two types of geodatabases are used by ArcGIS. Personal geodatabases are based on Microsoft Access database technology. A version of Access, called Jet, comes with ArcGIS and allows the user to create and manage personal geodatabases. Users may also access organizational databases through Database Connections.

Database connections

Some organizations maintain relational database management systems (RDBMS), such as Oracle or SQLServer, intended to provide sophisticated analysis and management of large data volumes accessed by multiple users. For example, at least three groups of users might need to edit parcel records in a large city database: the tax department, the city surveyors, and the deeds office. Instead of maintaining three sets of data, an RDBMS allows users to "check out" certain portions of data for editing and merges the changes back into the central database. ArcGIS can establish a database connection with such a system, allowing users to view and manipulate the data in ArcMap or ArcCatalog instead of the database program itself.

Layer files

A layer file does not contain spatial data. Instead it references a spatial data file and stores information on how it is to be displayed. A layer file can be created to store a set of symbols for displaying a particular dataset, and it can be used over and over. For example, if a city always uses the same symbols to show land use data, then a layer file can be created so that the same symbols can be loaded and used each time a new map is made. Several layers can be combined to form a group layer. Layers have diamond-shaped icons and can refer to raster or vector data.

Rasters

Rasters are arrays of numbers stored in binary format (base 2). Rasters consist of the data itself plus a header that gives information about the file, such as its number of rows and columns and its georeferencing information. This information may be stored in a separate file, or as the first part of the binary raster. Many different formats of rasters exist, and ArcGIS recognizes a great many of them. Rasters can be displayed in many different ways, but they cannot be analyzed unless they are converted to grids. A list of supported raster formats can be found in the ArcMap Help under the index heading "rasters, formats."

Tables

Tables can exist as separate data objects that are unassociated with a spatial data set. These are called standalone tables. They may be stored in dBase format (.dbf) or as comma-delimited text files. INFO files may also be created, stored, and accessed as standalone tables. Info table icons have a yellow strip at the top; dBase table icons have a green strip.

Grids

A grid is a raster format developed by ESRI and used with ArcView, Arc/Info GRID, and ArcGIS software. Grids can be displayed in a variety of ways, and they can also be manipulated and analyzed by the ESRI software Spatial Analyst or Grid.

TINs

TINs are Triangulated Irregular Networks that store surface information, such as elevation, using a set of nodes and triangles. TINs are used to display and analyze 3D surface information, create contour maps, and perform other functions.

CAD drawings

Data sets created by CAD programs can be read by ArcGIS, although they cannot be edited or analyzed unless they are converted to shapefiles or geodatabases. A CAD file may contain multiple feature classes, which correspond to the layers of the drawing, and can be opened separately and viewed just like feature classes in a coverage or a geodatabase. One can also access CAD drawings, which portray all the features in the CAD file with preset symbols. In a drawing, the feature classes are not accessible individually. CAD layers have blue icons.

🖳 Internet servers

Many organizations now make data available over the Internet. Users can connect to these data sources and download information for their work. To connect, you need to know the URL of the service, such as www.geographynetwork.com (Fig. 1.8).

Two types of services are offered. An **image service** allows people to display the information and print out a map from it, but they cannot change how it is displayed or make a copy of the data. A **feature service** allows people to download the data, view it, and save the features as a shapefile for later use. One of the largest Internet servers is called the Geography Network. Developed and maintained by ESRI, it offers a wide selection of information.

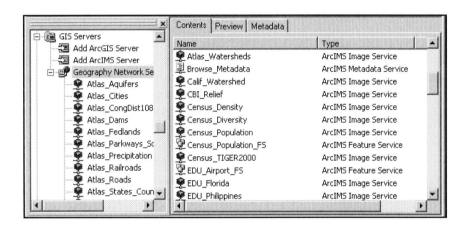

Fig. 1.8. The Geography Network offers many types of map data to use in GIS projects.

Properties of spatial data files

Shapefiles

Shapefiles are among the simplest spatial data sets. A shapefile contains only one feature class: points or lines or polygons, never a mixture. Each feature is independent and unrelated to any other features. The attributes associated with each feature are stored in a dBase file. Shapefiles can, however, store multifeatures, which are single features made of multiple objects. For example, the state of Hawaii requires multiple polygons to represent each island, but it can be stored as a multifeature so that it has only one record in the attribute table.

Although a shapefile appears as one icon in ArcCatalog, it is actually composed of multiple data files, which can be seen individually in Windows Explorer (Fig. 1.9).

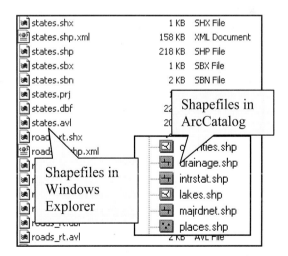

Fig. 1.9. Shapefiles are groups of files but appear as single entries in ArcCatalog.

Note that the states shapefile has eight different files associated with it. The .shp file stores the coordinate data, the .dbf file stores the attribute data, and the .shx file stores a spatial index which speeds drawing and analysis. These first three files are required for every shapefile to function properly. Additional files may also be present: the .prj file stores projection information, the .avl file is a stored legend, and the .xml file contains metadata. Similar files are present for the roads_rt shapefile. Note that to copy a shapefile to a new location all of these files must be moved together. ArcCatalog takes care of this automatically, but Windows Explorer does not.

In a shapefile attribute table, the first two columns of data are reserved for storing the feature identification code (FID) and the coordinate geometry (SHAPE) field. These fields are created and maintained by ArcGIS and should never be modified by the user. All other fields are added by the user and can be modified without difficulty.

Shapefiles also make use of a notion called coincident geometry. Two adjacent polygons in a shapefile share a common boundary (Fig. 1.10). The boundary is stored twice, once for each feature, yet measures can be taken to ensure that the boundary has exactly the same vertices for each polygon, so that the boundaries coincide exactly.

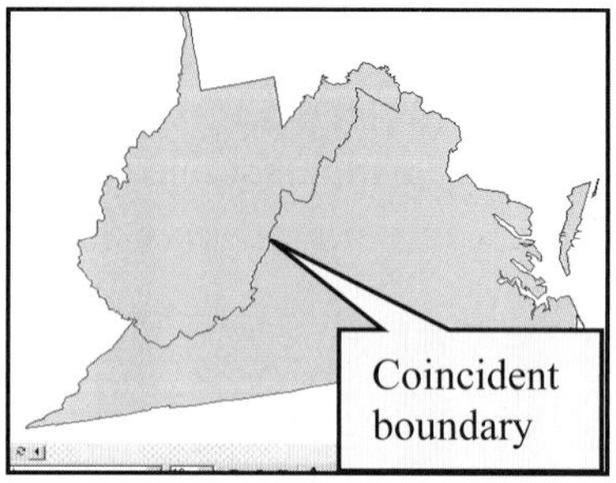

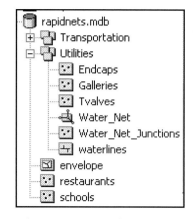

Fig. 1.10. A coincident boundary gets stored twice but is exactly the same for both features.

Geodatabases

A geodatabase is stored as a single file with an .mdb extension, using Microsoft Access database technology. Yet it can contain multiple feature classes, geometric networks, tables, rasters, and other objects such as rules (Fig. 1.11). Spatial data can be created as coverages or shapefiles and loaded into the geodatabase as feature classes.

Feature classes may exist as individual elements in a geodatabase (as do the restaurants or schools), or they may be grouped into **feature datasets.** A feature dataset contains a collection of related feature classes with the same coordinate system. The Utilities feature dataset in Figure 1.11 contains several feature classes related to a water network, such as waterlines and T-valves. A feature dataset can store complex associations between feature classes, such as networks or topology.

Fig. 1.11. A geodatabase containing two feature datasets and several feature classes.

A **network** consists of interconnected features along which flow can be initialized and studied, such as water in streams, electricity through wires, or traffic on road systems. The Utilities dataset in Figure 1.11 contains a network named Water_Net constructed from its feature classes. Chapter 14 describes some special analysis functions that can be used with networks.

Feature datasets may also contain **planar topology**, which tracks spatial relationships within or between layers. To create topology, the user specifies rules describing the required and permitted relationships. For example, in an ideal data set, counties should not have gaps between them and should not extend even a tiny bit outside their state. Such errors are commonly introduced during creation and editing of data; geodatabase topology assists in finding and correcting them. Editing with topology requires an ArcEditor or ArcInfo license.

Finally, geodatabases may contain rules that assist in the entering and validation of attribute data. Called domains, these rules specify which values or range of values may be entered in a particular field; a percent field, for example, should only contain numbers between 0 and 100. Other features of geodatabases that facilitate editing are discussed in Chapter 13.

Coverages

A coverage is a complex data format, and it is worthwhile to take some time to understand how it works. Coverages contain multiple feature classes, and often some feature classes are combined to create new feature classes. For example, a polygon feature class requires a point feature class to form polygon labels and a line, or arc, feature class to form the boundaries of the polygons (Fig. 1.12). From these two feature classes, the polygon feature class is created. As another example, a region is a single feature made of multiple polygons, such as the state of Hawaii which includes several islands. To build a region feature class in a coverage requires the presence of label, arc, and polygon feature classes.

Feature classes in coverages include the spatial coordinates as well as the attribute information. Attributes are stored in INFO tables and have special names: a polygon attribute table (.PAT), arc attribute table (.AAT), region attribute table (.RAT), etc. These feature tables also contain fields created and maintained by the GIS software, which should never be modified. The critical fields present vary depending on the feature type, but all feature classes have a *cover#* or *cover_* field and a *cover-id* field, where *cover* is the name of the coverage. (For example, a coverage named roads would have fields called roads# and roads-id.) The cover# field is analogous to the FID in a shapefile or geodatabase. The cover-id is a numeric identification code that can be modified by the user.

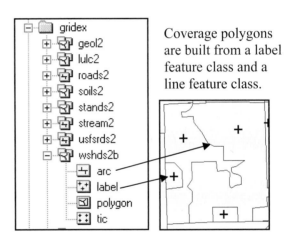

Coverage polygons are built from a label feature class and a line feature class.

Fig. 1.12. Coverages usually contain multiple feature classes.

Coverages also store topology, or the spatial relationships between features. For example, a polygon is composed of individual arcs and a label point (Fig. 1.12). Adjacent polygons share the same arc, so it need be stored only once. The polygon table keeps track of which arcs and labels belong to each polygon, and the arc attribute table stores which polygons are on either side of an arc.

VERY IMPORTANT TIP: Do not use Windows to copy or delete coverages, shapefiles, and geodatabases. These data sets may span multiple files and folders, and they might not be copied or deleted correctly. Always use ArcCatalog to delete or copy spatial data sets to prevent problems.

Introduction to metadata

Metadata contains data about data. It includes important information such as who created it, where it came from, what coordinate system it uses, what the fields in the attribute tables mean, and more (Fig. 1.13). Without appropriate metadata, a data set can be useless.

The content and format of metadata is established by the Federal Geographic Data Committee, and metadata which follows these standards is referred to as FGDC-compliant. ArcCatalog provides tools which make it easy to create and update metadata and which ensure that when the data set is copied or moved, its metadata goes with it.

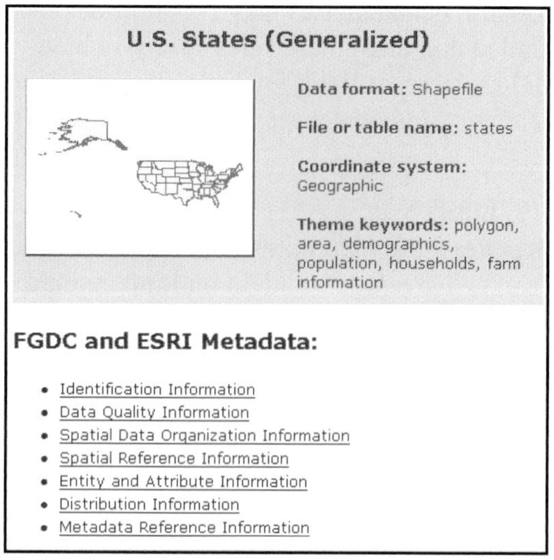

U.S. States (Generalized)

Data format: Shapefile

File or table name: states

Coordinate system: Geographic

Theme keywords: polygon, area, demographics, population, households, farm information

FGDC and ESRI Metadata:

- Identification Information
- Data Quality Information
- Spatial Data Organization Information
- Spatial Reference Information
- Entity and Attribute Information
- Distribution Information
- Metadata Reference Information

Fig. 1.13. Metadata contains important information about layers so that other people may use it more effectively.

A fully completed metadata set has hundreds of pieces of information and can be quite time-consuming to develop. However, much of the metadata is entered by ArcGIS itself based on information in the data set. Other required information, such as the name and address of the organization creating the data, can be copied and reused between data sets.

Finally, partial metadata is better than no metadata at all. Even if you don't have access to all the information required by the metadata, filling out the information you do have is helpful to your organization and to any others sharing the data. As a data user, once you have had the frustration of dealing with an undocumented data set, you will come to appreciate the value of metadata!

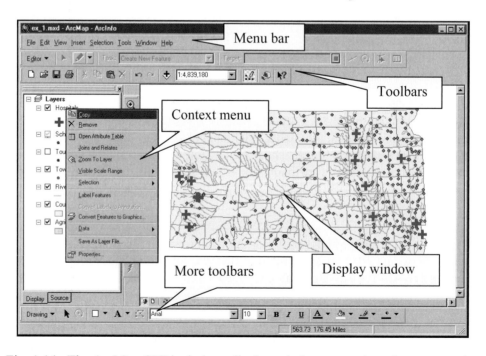

Fig. 1.14. The ArcMap GUI includes a display window, menu bars for commands, toolbars to provide functions, and context menus for added flexibility.

Overview of the ArcGIS interface

The graphical user interface, or GUI, used by ArcGIS may look complicated at first, but there are many similarities between the three programs ArcCatalog, ArcMap, and ArcToolbox that make them easier to learn. The interface has also been designed to pack as many commands and functions as possible into a small space on the screen (Fig. 1.14). Moreover, it is easy to customize the GUI to provide the look and functionality desired. Let's look at some of the unique features of the GUI by examining the ArcMap window. ArcCatalog has a similar layout.

The menu bar

The menu bar is the top row of the GUI; it contains drop-down menus organized by function (Fig. 1.15). You can use the Customize function to add additional commands, or even your own commands, to the menus.

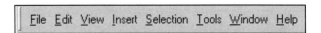

Fig. 1.15. A menu bar

The toolbars

ArcGIS has many toolbars, organized by function (Fig. 1.16). These toolbars can be torn out of their locations and moved to a different spot in the GUI, or even off the window entirely. You can also add your own tools to the toolbars using the Customize function.

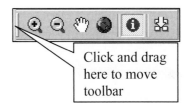

Fig. 1.16. A toolbar

Context menus

Context menus pop up everywhere in ArcGIS. A context menu appears when you right-click or left-click a certain object, and the menu that appears will depend on which object was clicked and which mouse button was used (Fig. 1.17). For example, right-clicking a symbol in the legend allows you to select a color, but left-clicking it lets you choose a different symbol. Right-clicking the layer name gives yet another menu. Context menus are a great way to pack a lot of commands into the GUI without cluttering it.

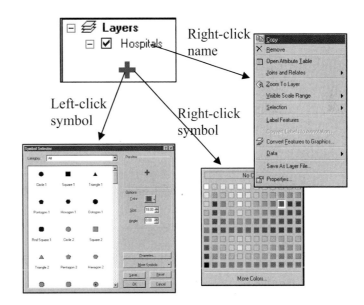

Fig. 1.17. Context menus are sensitive to what is being clicked and which mouse button is used.

Object properties

ArcGIS is an object-based program, meaning that nearly everything in it, from a data set to a map to a button, is considered an object. The program manipulates these objects and keeps track of their properties. For example, to open a table, the program sends an Open command to the table object. Objects also have properties that define what they are and how they behave. Often accomplishing things in ArcGIS involves setting properties of various objects. For example, to draw a data layer using certain symbol colors, you set the Symbol Properties of the data layer. Or to add fields to a shapefile, you modify its Fields Properties.

Properties of data layers are accessed by right-clicking on the layer and choosing Properties from the context menu. The Layer Properties window has multiple tabs that control different properties (Fig. 1.18). As you work through this text, you will learn the properties of all sorts of objects and how to set and modify them.

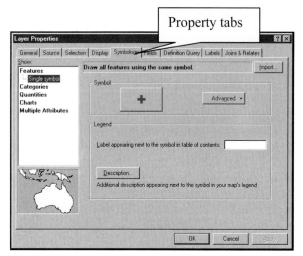

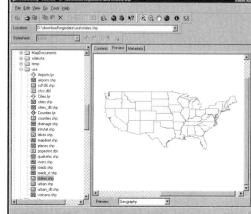

Fig. 1.18. The Layer Properties window has tabs for viewing and modifying layer properties, such as its source data and how it is displayed.

Fig. 1.19. ArcCatalog lets you create, explore, and manage GIS data sets.

About ArcCatalog

ArcCatalog (Fig. 1.19) is like having a version of Windows Explorer that is smart about spatial data and helps manage it. ArcCatalog can view data layers, set their properties, and manage data sets and files. ArcCatalog is also used to create new data sets and convert between data formats.

It should be emphasized that ArcCatalog should nearly always be used to copy, delete, rename, and modify spatial data sets. It knows the data formats and requirements and can ensure that these functions are carried out properly. Using Windows Explorer to manage GIS data files may cause damage to the data and render it unusable.

Viewing data files

ArcCatalog gives users many ways to view and get information about data. The left window in Figure 1.19 shows the directory tree information. The content window on the right has three main tabs to provide different views of the data: **Contents, Preview,** and **Metadata** (Fig. 1.20).

The **Contents** tab shows what is inside a folder or a multifeature data set.

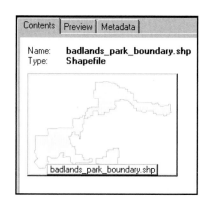

Fig. 1.20. View tabs in ArcCatalog include the file Contents, a data Preview, and Metadata Viewer/Editor.

➤ If a folder is clicked in the tree window, then the contents of the folder are displayed on the right. Buttons in the main button bar of ArcCatalog also switch among the display modes: large icons, small icons, details, or thumbnails.

➤ If a single data object such as a shapefile is highlighted, then the Contents window will show information about the object. If a thumbnail, or a small snapshot of what it looks like, has been created for the data set, then the thumbnail will be displayed as shown in Figure 1.20.

The **Preview** tab shows what the data set contains. You can switch between viewing the spatial features and viewing the attribute table associated with the features by clicking on the drop-down box at the bottom of the Contents window to set it to Geography or Table.

➤ In Preview mode, you can use the Zoom and Pan tools to zoom in and out, pan, and return to the full extent of the data set (Fig. 1.21).

➤ You can use the Identify tool to get information about a feature in the map.

➤ You can use the Thumbnail tool to create a thumbnail for the data set. If you are zoomed in when creating the thumbnail, it will create it according to how the map currently appears.

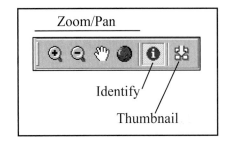

Fig. 1.21. Buttons for previewing data layers

The **Metadata** tab allows the user to view, create, edit, update, and import/export metadata for a data layer (Fig. 1.22).

➤ The drop-down Stylesheet tab allows a choice among several formats for displaying the metadata.

➤ The Metadata editing tools allow users to easily create and update metadata. ArcCatalog itself fills out many of the required fields automatically.

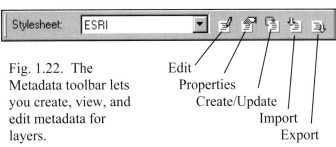

Fig. 1.22. The Metadata toolbar lets you create, view, and edit metadata for layers.

Edit
Properties
Create/Update
Import
Export

About ArcToolbox

The ArcToolbox contains an assortment of functions, or tools, for managing and analyzing data. The commands are implemented as easy-to-use menus. The tools are organized into a hierarchical system of toolboxes containing related tools (Fig. 1.23). If the user has purchased optional program extensions to ArcGIS, such as Spatial Analyst, that functionality will appear as additional toolboxes. Users may also create their own toolboxes inside ArcToolbox and fill them with frequently used tools, or even create new tools to put in them.

The functions available in the toolbox depend on the software license obtained. Users holding only an ArcView license will find mainly functions to perform common analysis tasks, to convert between basic data formats, and to work with map projections. This includes about 30 functions. The ArcEditor license adds a few functions associated with editing. The ArcInfo Toolbox is packed with capabilities.

ArcToolbox is a powerful part of the ArcGIS "geoprocessing environment," a new concept implemented in Version 9 that aims to provide seamless analysis and management functions that work across the entire ArcGIS system. Tools in ArcToolbox can be run from either ArcMap or ArcCatalog, and most are designed to work with all spatial data types, including shapefiles, coverages, and geodatabase feature classes.

Fig. 1.23. ArcToolbox

The geoprocessing environment includes an application called ModelBuilder, which allows users to string tools together with input and output layers to create more complex tools. These models can streamline processing when several analysis steps are always repeated in the same order. The new models can be saved in the toolbox and run as needed. Users can also write scripts, or programs which string together analysis steps. Like models, scripts can be used over and over to perform the same series of functions on different data layers. Tools, models, and scripts can all exist in ArcToolbox. Advanced users will want to learn more about the geoprocessing environment, models, and scripting, because all three help streamline GIS work and can add flexibility and power to the user's repertoire of GIS tricks.

The geoprocessing environment utilizes environment settings that control many aspects of how tools work. For example, users can set a default working directory where all outputs are placed, or specify that resulting layers occupy only a restricted geographic area of interest. The default settings provide reasonable service for all the exercises in this book. Users interested in advanced geoprocessing will need to learn more about these settings prior to changing the defaults.

Summary

➢ A Geographic Information System is designed as a database system that uses both spatial and aspatial data in order to answer questions about where things are and how they are related. It has many functions, including creating data, making maps, and analyzing relationships.

➢ GIS software by ESRI, Inc. has a long history with several major transformations along the way. The current version of ArcGIS Desktop employs a menu-based interface, with optional access to the older command-line functionality of Arc/Info. The Desktop consists of three programs, ArcMap, ArcCatalog, and ArcToolbox.

➢ GIS data are stored in one of two basic formats: raster and vector.

➢ Raster data employs arrays of values representing conditions on the ground within a small square called a pixel. The array is georeferenced to a ground location using a single x-y point.

➢ Vector data uses sequences of x-y coordinates to store point locations, lines, and polygon area features. Every feature is linked to an attribute table containing information about the feature.

➢ ArcGIS uses a variety of data formats old and new, including shapefiles, coverages, geodatabases, grids, images, TINs, and CAD drawings.

➢ Metadata files store information about GIS data layers to help people understand and use them properly. Metadata can be created in ArcCatalog, and the files are automatically copied and updated along with the data sets.

➢ The ArcGIS Desktop interface uses menus, toolbars, and context menus to facilitate the display, management, and analysis of data. Use these features to access and modify objects such as data layers, tables, and reports.

➢ ArcCatalog contains many functions for creating data, exploring files, and managing GIS data. It also provides tools for viewing and editing metadata.

➢ ArcToolbox contains functions for processing, managing, and analyzing GIS data. Users may customize it by building models or writing scripts to repeat often-used sequences.

VERY IMPORTANT TIP: Do not use Windows to copy or delete coverages, shapefiles, and geodatabases. These data sets may span multiple files and folders and might not be copied or deleted correctly. Always use ArcCatalog to delete or copy spatial data sets to prevent problems.

IMPORTANT TIP: Although spaces are permitted in names of files and folders, they can cause problems for some GIS functions. It is recommended NEVER to use spaces when naming files and folders that will contain GIS data, nor to let spaces appear in any folders above them.

Chapter Review Questions

You may need to consult the Skills section at the end of the chapter to answer some of these questions.

1. What feature most distinguishes a Geographic Information System (GIS) from a Database Management System (DBMS)?

2. Explain the relationship between Arc/Info and ArcGIS Desktop.

3. Explain how a raster is georeferenced.

4. If each of the following data were stored as rasters, state which ones would be discrete and which ones would be continuous: rainfall, soil type, voting districts, temperature, slope, and vegetation type.

5. Imagine you are looking at a database which contains 50 states, 500 cities, and 100 rivers. How many feature classes are there? How many features? How many attribute tables? How many total records in all the attribute tables?

6. Of the three vector data formats, shapefiles, coverages, and geodatabases, which ones can store multiple feature classes? Which one is the most recent? Which one is a single file? Which ones can store points, lines, and polygons?

7. What does it mean to connect to a folder in ArcCatalog? What is the benefit of this feature?

8. What functions can be performed on a table while previewing it in ArcCatalog?

9. Can the Search function in ArcCatalog do more than locate files by finding their names? What other kinds of search can it do?

10. What is metadata and why is it important?

Mastering the Skills

Teaching Tutorial

Preparing to begin

Each step of the tutorials in this book is illustrated by a video clip on the book's CD. You can view these clips whenever you want a demonstration of one of the steps in the tutorial. To view videos, do the following.

→ Place the book's CD in the computer's CD-ROM drive. Wait for the splash screen to appear.

→ Click the button to accept the license agreement. The main window appears. (Fig. 1.24)

→ If needed, size the document window to a narrow strip on the left side of the screen, so that you can see all the numbers and titles clearly. Put your ArcMap window on the right, so you can see it also.

→ In the Chapter 1 section, click on the number of the tutorial video you want to see. Windows MediaPlayer will start playing the clip. If asked whether you want to open the clip in Windows Explorer, say NO.

→ Size the play window as large as possible for best resolution.

 → When the video finishes, click the Minimize button in the upper right corner of the MediaPlayer window, to get it out of the way.

→ The headings under the Skills section contain links to performing different skills introduced in the chapter. Use these videos as a reference if you have forgotten how to do something.

→ Before starting the tutorial, make sure that you have installed the mgisdata folder from the CD to the computer hard drive. See the Preface for detailed instructions.

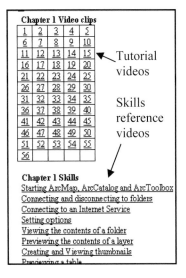

Fig. 1.24. The Video Index file provides links to demonstration videos.

ArcGIS demonstration

The following examples provide step-by-step instructions for doing basic tasks and solving basic problems in ArcGIS. The steps you need to do are highlighted with an arrow →; follow them carefully. Click on the video number in the VideoIndex to view a demonstration of the steps.

→ Start ArcMap.

1→ Click the button next to "An Existing Map" and click Browse for Maps in the box below. Click OK.

1→ Navigate to your mgisdata directory and open the MapDocuments folder. Click on the ex_1.mxd document and click Open.

1➜ Choose Save As from the File menu and save a copy of the map document under a new name, such as "ex_1mine.mxd". Then save periodically as you work.

TIP: In these tutorials, values you must enter are enclosed in quotes. Do NOT type the quotes unless the instructions specifically direct you to type them.

You will see a map of South Dakota in the display area on the right, and a list of data layers in the Table of Contents area on the left. First let's explore some of the data layers.

2➜ Locate the Identify tool on one of the menu bars and click on it. The Identify window appears.

2➜ Place the tool on top of one of the towns and click. The town will briefly flash on the screen, and the attributes of the town are displayed in the Identify window.

2➜ Close the Identify window by clicking the X in the upper right corner.

TIP: You must click exactly on the center of the point to identify the town. If you are a little off, you may get county or river information instead. Try clicking the town again.

Toolbars in ArcMap can be moved and docked at different locations, even outside the program.

3➜ Locate the toolbar with the Identify button again. At its top or left, find a faint grey line. This is its handle.

3➜ Click the line and drag the menu out of the ArcMap window. Then click it and drag it to a spot with the other menu bars at the top of the window. Leave it wherever you like.

The Find tool is a handy way to seek a feature. It will look for features having a specified string across all the layers in the map or an individual layer. It can also search all fields or confine the search to a single field.

4➜ Click the Find tool (Fig. 1.25).

4➜ Make sure the Features tab is clicked so it is active.

4➜ Enter the name "Pennington".

4➜ Choose Counties as the layer to search.

4➜ Click Find.

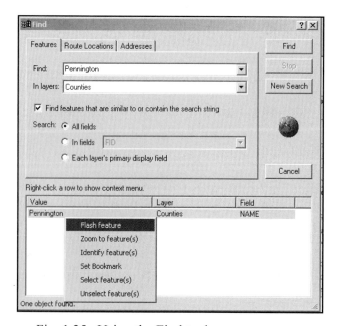

Fig. 1.25. Using the Find tool

4➡ When the feature has been found, choose it from the list and right-click on its name. Choose Flash Feature and watch for it to Flash on the screen (Fig. 1.25). (Move the Find window off the western part of South Dakota so you can see.)

4➡ Next choose Select Feature and notice that the feature you chose is highlighted in the map with a blue line. When you have seen it, right-click the name again and choose Unselect Feature.

4➡ Experiment with finding a couple more features. When done, click the Cancel box to close the Find window.

Now let's use the measuring tool.

 5➡ Locate and click on the Measure tool.

5➡ Click on the upper left corner of the state and then click on the upper right corner.

5➡ Double-click to end the line.

5➡ Read the distance from the lower left corner of the ArcMap window.

1. How far is it across the northern border of South Dakota? _____

6➡ Choose the square county in the upper left corner of the state, and click on each corner in turn with the Measure tool. Notice that the Measure tool reports the total length as well as the last segment.

6➡ When you get back to the first corner, double-click to end the line and read the perimeter of the county.

2. What is the perimeter of the county? _____

3. What is the name of the county? (**Hint:** Use Identify) _____

Map Tips are little flags that pop up and give the name of a feature when the cursor is held over it. They are useful for identifying features quickly. The Lakes layer currently has the Map Tip function turned on. Let's see how they work.

7➡ Use the cursor to hover over a river or lake until the name of the feature is displayed. You may need to gently move the cursor around to find the right spot. Try several different rivers to get the hang of it.

Now let's learn how to turn the Map Tips on and off.

7➡ Right-click the Hospitals layer name and choose Properties from the menu.

7➡ Click the Fields tab and verify that the Primary display field is set to NAME. This is the field shown in the Map Tips.

7➡ Click the Display tab and check the box next to the words Show Map Tips.

7➡ Click OK to close the Properties box and keep the changes just made.

7➡ Turn the Hospitals layer on by clicking the check box next to its name.

7➡ Use the cursor to hover over a hospital marker until a small yellow box appears with the name of the hospital. Find out the names of some other hospitals.

> **TIP:** If the Show Map Tips text is dimmed in the Display Properties menu, then no spatial index currently exists for the data layer. Spatial indices must be created in ArcCatalog before Map Tips can be used.

Now we will experiment with the zooming buttons.

8➜ Uncheck the Towns and Hospitals boxes in the Table of Contents so they are not being displayed.

8➜ Check the Tourist Spots layer to display it.

 8➜ Click the Zoom In tool. Place the cursor at the upper left corner of Harding County and click and hold. Continue holding the mouse button down and drag a box around the rest of the county. When finished with the box, let go of the mouse button.

> **TIP:** If you do not like the area you chose, or if you made a mistake, click the Previous Extent button to return to the original extent. You can use the Next Extent button to move back the other way.

8➜ Click once in the corner of the county and notice that the view zooms in and places the point you clicked at the center. The Zoom Out tool works the same way.

 8➜ Click on the Pan tool, then click and drag inside the display window to move the map around. When you release the mouse button, the map redraws.

 8➜ Click the Zoom Out Center button a few times to zoom out. Notice that the center stays put.

8➜ Click the Zoom In Center button a few times to zoom in.

8➜ Click the Full Extent button to view the total area of all the layers in the Table of Contents.

Bookmarks provide a handy way to zoom into extents that you use frequently, or that you want others to be able to find easily.

9➜ Click on the View menu and choose Bookmarks > Black Hills from the menu. The scene zooms into the Black Hills region of western South Dakota.

9➜ Click the Full Extent button to return to all of South Dakota again.

It is easy to create and manage bookmarks.

 10➜ Use the Zoom In tool to draw a box around the intersection of the two interstates in eastern South Dakota (Fig. 1.26).

10➜ Click on the View menu and choose Bookmarks > Create.

10➜ Type "Sioux Falls" in the box to name the bookmark.

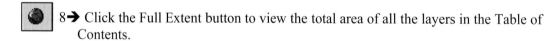

Fig. 1.26.
Sioux Falls

10➔ Return to the Full Extent, and then use the new bookmark to zoom into Sioux Falls again.

TIP: Use View > Bookmarks > Manage to examine and remove bookmarks.

Scale Range is a layer property that allows additional control for displaying features. You can specify a range of map scales at which the map layer is drawn, to avoid cluttering maps with too much information. Try it with the Schools layer.

11➔ Return to the Full Extent.

11➔ Turn on the Schools layer. Notice how some areas are nearly obscured by schools because there are so many.

11➔ Right-click the Schools layer name and choose Properties from the menu.

11➔ Click the General tab if it is not already displayed.

11➔ Fill the button next to "Don't show layer when zoomed:" and enter the value "500000" in the box for "out beyond." Notice that the current scale is approximately 1:4 million in the scale box on the main toolbar (Fig. 1.27).

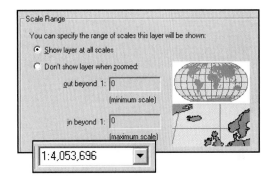

Fig. 1.27. Setting the display scale

11➔ Click OK, and watch the schools disappear from the map. Note that the check box next to the layer name is now dimmed.

11➔ Use the bookmark to zoom into Sioux Falls again, and the schools should appear. If not, then use the Zoom tool to zoom more into the urban area shown in blue. Keep zooming until the schools appear.

ArcMap provides many options for displaying features using different symbols. In this map we see two basic ones: displaying every feature in the layer with one symbol (the rivers, for example), and displaying features based on an attribute (the roads). We will now learn how to modify these symbols.

12➔ In the Table of Contents, right-click the Schools symbol (not the name). A menu of colors will appear. Choose a bright red color for the schools.

The Symbol Selector can modify the type, size, and color of each symbol.

12➔ Click on the Schools symbol in the Table of Contents. The Symbol Selector window appears (Fig. 1.28).

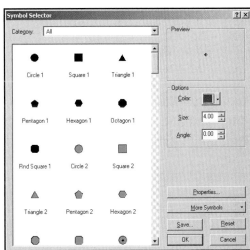

Fig. 1.28. The Symbol Selector window

12➔ Scroll down the window showing the symbols until you find the School 2 symbol. Click on it to select it. Then click on the color patch to select a red color for the school, and set the size to 22 pts. Click OK when finished.

Line and polygon symbols can also be modified using these two methods. Next we will learn how to assign symbols based on an attribute.

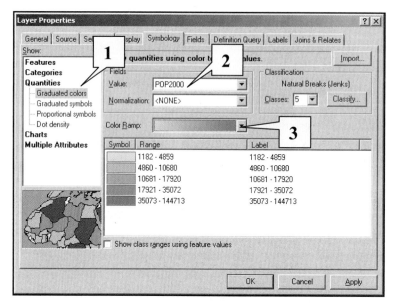

Fig. 1.29. The Symbology tab of the Layer Properties window

13➔ Click on the Full Extent button to zoom to the state. Turn Towns back on.

13➔ Right-click on the Counties layer name and choose Properties from the menu.

13➔ Click the Symbology tab.

13➔ In the box on the left (#1), click Quantities > Graduated Colors (Fig. 1.29).

13➔ Use the drop-down Value box to choose the field POP2000 (#2).

13➔ Use the Color Ramp drop-down box to choose a light green color ramp (#3). Click OK to produce the map (Fig. 1.30).

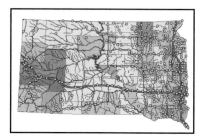

Fig. 1.30. The map after step 13

Now we are going to use selection to extract a certain subset of records to look at, based on their values in a field of the attribute table. This action is called a **query.** Once records are selected you can examine their distribution on the map, get statistics about them, export them to a new data file, and many other operations.

14➔ Right-click on the Counties layer and choose Open Attribute Table from the context menu which pops up.

14➜ Examine the attribute fields in the table. Use the scroll bar at the bottom to scroll right to see the fields that are out of the display area. Notice that these fields contain mostly demographic information.

14➜ Click on the Options button in the table and choose Select By Attributes.

In the dialog box you will enter an expression to find all counties with a population greater than 20,000 people. It will look like this "POP1990" > 20000. (When entering expressions, you must include quotes.)

14➜ Double-click on "POP1990" from the list of fields. Check to make sure the field name appears in the box below.

14➜ Click once on the > sign.

14➜ Type "20000" from the keyboard. Click Apply.

14➜ Close the Select By Attributes box by clicking on the X in the upper right corner.

14➜ Move the table window away from the map by clicking on its title bar and dragging it to one side, so you can see the map clearly.

Examine the map, noticing that the counties with population greater than 20,000 are now outlined in blue and that their associated records in the table are also highlighted in blue. (You may need to scroll down the table to find a highlighted county.)

14➜ Click the Selected button on the table so that only the selected records are displayed.

14➜ When done examining the selected records, click the All button to show all of them again.

You can use the Statistics command to calculate basic statistics about the selected records (or about all of the records if none of them are selected).

15➜ In the table, right-click on the name of the POP1990 field and choose Statistics from the context menu.

15➜ Examine the statistics and the frequency diagram. Note that only the selected records are used to calculate the statistics.

4. How many counties are currently selected? _____

5. What is their total population? _____

15➜ Use the drop-down box to choose another field and view its statistics.

15➜ Close the Statistics box.

15➜ Click Options and choose Clear Selection from the menu.

15➜ Close the attribute table.

➜ Exit ArcMap by choosing File > Exit or by clicking the X box in the upper right corner of the ArcMap window. Save the changes to the map document when prompted.

Exploring data with ArcCatalog

ArcCatalog (and ArcMap) access data through **connections,** which are links to folders containing GIS data. By default, the main computer hard drive will always show as a connection (C:\).

→ Start ArcCatalog.

16→ Examine the folder tree on the left side and find the default connection, C:\.

16→ Click the plus sign next to it to expand the contents of the drive and see the subfolders.

Although you can navigate through folders to find any data on C:\ from the default connection, you can also establish connections to subfolders, thus creating handy shortcuts to frequently used data. You may already have one shortcut to the mgisdata folder, or it may be absent. If no mgisdata connection exists, then add one.

16→ Look for a connection to the mgisdata folder as shown by the red oval in Figure 1.31. The first part of the name may be different, depending on where the data were installed (such as D:\student\mgisdata).

16→ If the connection is already there, go on to step 17.

 16→ To add the connection, click the Connect to Folder button.

16→ Navigate to the directory containing the mgisdata, and click on the mgisdata *folder* to select it. Do not select any of the subfolders, just the mgisdata folder.

16→ Click OK to add the connection.

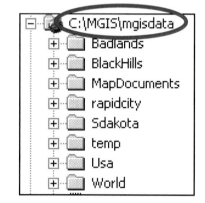

Fig. 1.31. A connection to the mgisdata folder in ArcCatalog

Adding connections is critical for accessing data not on the main hard drive, such as on a second disk drive, a network drive, or a CD-ROM drive. Connections are saved between sessions, and they work in ArcMap and ArcToolbox as well. Once you add a connection, it will be present until you delete it. Connections tend to build up over time, so once in a while go through them and delete ones no longer being used.

 17→ Locate the connection to the mgisdata folder just added. Click to select it and then click the Disconnect button. It disappears from the list.

 17→ Click the Connect button, navigate to the mgisdata folder again, and add the connection again.

Now let's examine some features of ArcCatalog. First, you can use the Options in ArcCatalog to control a number of convenient features.

18→ Adjust the folder tree by clicking the plus signs until you can see the mgisdata\MapDocuments folder, and click it to highlight it.

18→ Click the Contents tab in the ArcCatalog main window.

18→ Click Tools > Options on the main menu bar.

18➜ Click the General tab.

The General tab can control which types of files and services appear in the Catalog. By default, all are shown. You can also choose to hide or show file extensions, such as .shp or .mxd.

TIP: Another option is to show a special icon for folders containing GIS data. This option may be handy at times, but it does slow down the performance of ArcCatalog.

18➜ Uncheck the box next to Hide File Extensions.

18➜ Click OK to close the Options menu and apply the changes. Notice that the map documents now appear with an .mxd extension.

19➜ Open the Options menu again and click the File Types tab. Examine the option to add additional non-GIS file types for display, such as Word documents.

19➜ Click the Contents tab. By default, ArcCatalog shows only the name and type of file. Click to add the file size to the list. Click OK and examine the documents to see the change.

TIP: You can click the edge of a contents column and drag it to increase or decrease the column width.

We will examine some of the other options as we go along. For now, let's practice viewing files in ArcCatalog.

20➜ Adjust the folder tree so that you can see the contents of the mgisdata\Rapidcity folder (Fig. 1.32).

20➜ Examine the many different icons in this folder. These are only a small sample of the many types of spatial and attribute data that ArcCatalog can use.

20➜ Click the plus sign next to the citybnd layer. The expanded list shows each of the feature classes of the coverage. The shapefiles do not expand because they can only have one feature class. Recall that a polygon coverage would have a minimum of three feature classes that make up the polygons: arcs, labels, and polygons. Other features such as tics and regions may also be present.

20➜ Expand the TM_24sep98MS raster layer to see the seven color bands that make up this image. Each band shows a different range of light wavelengths as measured by the Landsat satellite.

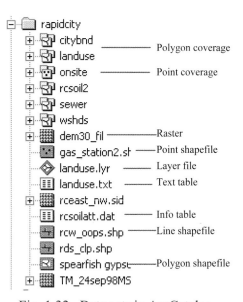

Fig. 1.32. Data sets in ArcCatalog

20➔ Click one of the feature layers in this directory, and notice the information that appears in the main window when the Contents tab is clicked. A coverage lists all the feature classes. A shapefile has a name and a thumbnail picture, if one has been created. You will learn to create thumbnails in a few minutes.

20➔ Expand the Badlands folder in the mgisdata folder, and then expand the geodatabase Badlands (Fig. 1.33). Expand the Infrastructure **feature dataset** to see the **feature classes** (including one line feature class and two point feature classes).

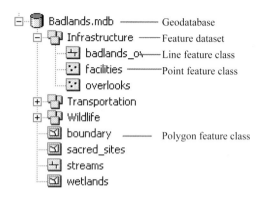

Fig. 1.33. A geodatabase in ArcCatalog

6. How many shapefiles are there in the folder mgisdata\BlackHills\Sturgis? _____ How many coverages? _____ How many .dbf files? _____ How many info files? _____ (**Hint:** Info files have a yellow border like coverages, .dbf files have a green border like shapefiles.)

The Contents tab provides some useful information and is also the fastest option when exploring in ArcCatalog. It also has several ways to display the contents of folders: as large icons, as a list with small icons, as a detailed list, and as thumbnails. The view style is controlled using these buttons (Fig. 1.34).

Fig. 1.34. Content view styles

21➔ Click the Rapidcity folder to select it, and then click each of the view buttons in turn. When finished, set it back to List.

We will now look at two slower options with more information: the Preview tab and the Metadata tab. The Preview tab shows a quick sketch of the data layer or a look at its tables.

22➔ Select the rds_clp shapefile in the mgisdata\Rapidcity folder, and note the appearance of the main window, showing a thumbnail of the data set.

22➔ Click the Preview tab. The view changes to display the roads using a random symbol.

22➔ Experiment with the Zoom and Pan tools to examine areas of the roads more closely. These buttons work the same as the ones in ArcMap.

Zoom/Pan tools

22➔ Click the Identify button and click on a road to get information about it. Close the Identify window when finished.

23➔ Click the Create Thumbnail button to create a snapshot of the current view. Nothing apparently happens, until….

23➔ Click the Contents tab. This time the thumbnail appears with the data set name and information.

23➔ Click the Rapidcity folder to select it. Click the view tab to the thumbnail option. All of the data layers appear with thumbnails, if they have been created, or icons.

23➔ Return to Preview mode and make several more thumbnails, then examine them again using the Contents tab.

Not only can you preview a data set's features, but you can also look at its attribute table, change the appearance of the table, and even add and delete fields.

24➔ Click on the landuse data set, make sure the Preview tab is clicked, and choose Table from the drop-down menu that currently reads Geography. The landuse table appears.

24➔ Scroll to the right through the end of the table, noting all of the fields.

7. How many records (rows) are there in this table? _____

24➔ Hold the cursor over the right edge of the FID field until it turns into a double arrow bar. Click and drag the edge to the left or right to reduce or increase the width of the column.

24➔ Click the Options button in the table window (if necessary, expand the window to the right to see it better).

Note the menu options. Find searches the table for particular text, as has already been demonstrated. Add Field adds a new field to the table, and Export saves a copy of the table as a .dbf file. Other options are also available.

25➔ Right-click the field name of AREA to display a context menu. Choose Sort Ascending or Sort Descending to sort the field.

8. What is the land use code of the largest polygon in this data set? _____ Of the smallest one? _____

25➔ Right-click the AREA field and choose Statistics to see basic statistics and a frequency diagram of the values. Close the Statistics menu when done looking.

25➔ Right-click the LU_CODE field and choose Freeze/Unfreeze. This places the field to the left of the table and keeps it there as you scroll to the right.

TIP: More than one field can be frozen at a time. Unfreezing allows the field to scroll again, although it remains on the left side until ArcCatalog closes.

You can even add or delete fields from the table in Preview mode.

26➔ Click the gas_station2 shapefile to select it, and examine the fields.

26➔ Click the Options menu and choose Add Field.

26➔ Type "NOTES" for the field name.

26➔ Use the drop-down box to set the field type to Text.

26➔ Enter a length of "25" for the field. (For numeric fields don't change any of the field properties until you are more experienced.) Click OK.

26➔ Scroll to the right to see the new field added to the end of the table. Data are added to fields in ArcMap.

26➔ Right-click the new NOTES field and choose Delete. Choose Yes in the warning box.

You can also search for information in a field or in the entire table.

27➔ Click the Options button and choose Find.

27➔ Enter "Sinclair" as the text to search for.

27➔ Examine the find options available, and then click the Find Next button.

27➔ Note the first Sinclair station is found. Close the Find box.

TIP: Except for the Add/Delete field options, none of these operations changes the data that are stored on the disk. You can sort, freeze, and do statistics without making a single change to the actual data.

Layer properties

Every layer has various properties that can be viewed and set in ArcCatalog. We will examine some of these properties now to start getting familiar with them. The succeeding chapters will cover how to work with the properties.

28➔ Find the Rapidcity folder in your mgisdata directory in ArcCatalog.

28➔ Right-click the gas_station2 shapefile and choose Properties from the context menu.

28➔ Click the General tab. There is not much to set here.

28➔ Click the Fields tab. You can view, add, and delete fields here too, and will learn to do so in Chapter 5.

28➔ Click the Index tab.

Shapefiles can have two types of indices. An attribute index can be created for individual fields and enhances performance when searching or querying that field. A spatial index decreases the time needed to draw and query the layer, and is required for certain operations, such as Map Tips.

Shapefiles, coverages, rasters, and geodatabases all have different properties. Let's look at the properties for coverages.

29➔ Close the Shapefile Properties window.

29➔ Right-click the citybnd coverage and choose Properties from the menu.

29➔ Click the General tab.

Recall that coverages are topological data sets, meaning that they store spatial relationships between features. The General tab shows and updates the topology of the feature classes in a coverage. Notice the Build and Clean buttons, which are used to generate and update topology.

9. Which feature classes are present in this coverage? _____

29➔ Examine the Projection tab. This tab lists the coordinate system, which you will learn about in Chapter 3.

29➔ Examine the Tics and Extent tab. Tics are locations with known real-world *x-y* coordinates, used to register a paper map on a digitizer and allow it to be converted to a digital map with a real-world coordinate system. The Extent shows the range of *x-y* coordinates present in the data set.

29➔ Examine the Tolerances tab. These tolerances are used during editing and when updating topology.

29➔ Click OK to close the Coverage Properties box.

Next we will examine properties of feature classes in a geodatabase. These are similar to shapefile properties, except that two additional tabs for Subtypes and Relationships are added. Subtypes are rules used to help validate data entry, and Relationships establish links between tables. Note that the feature datasets and the geodatabase itself have different properties than the feature classes inside it.

30➔ Expand the Badlands geodatabase in mgisdata\Badlands until the boundary feature class is visible.

30➔ Right-click the boundary feature class and choose Properties. Examine each of the tabs in turn. Close the Properties box when finished.

30➔ Right-click the Transportation feature dataset and notice it has only a General tab.

30➔ Right-click the Badlands.mdb geodatabase and examine its properties. Notice that it only has two tabs, General and Domains.

30➔ Finally, examine the properties of one of the rasters and one of the tables in the mgisdata\Rapidcity folder.

Next we will briefly examine some ways to create and delete data files.

31➔ Right-click the mgisdata folder and choose New > Folder from the context menu.

31➔ When the folder appears in the tree, type in the name "MyFolder" and press the Enter key.

IMPORTANT TIP: Although spaces are permitted in names of files and folders, they can cause problems for some GIS functions. It is recommended NEVER to use spaces when naming files and folders that will contain GIS data, nor to let spaces appear in any folders above them.

31➔ Right-click MyFolder and choose New > Shapefile.

31➔ Give the shapefile the name "myshape". Use the drop-down box to set the Feature Type to Polygon.

31➔ Click the Edit button to assign a coordinate system.

31➔ Choose Select in the Spatial Reference Properties box. Navigate through the folders to select Geographic > North America > North American Datum 1983.prj. Click Add and then OK.

31➔ Click OK once more to create the shapefile. See it appear in MyFolder.

As another example, let's create a layer file. A layer file references another spatial data set and stores information on how it is displayed. The layer file can be displayed in ArcCatalog to give a more detailed picture of the data. It can also be added to ArcMap to easily draw the features a specific way, or it can be used to specify symbols for another data layer.

32➜ Right-click MyFolder and choose New > Layer from the menu.

32➜ Type in "MyStates" for the layer name.

32➜ Click the Browse button in the Create New Layer window and navigate to the mgisdata\usa folder. Click the states.shp file to select it and click Add.

32➜ Keep the check in the Create Thumbnail box. Also place a check in the Relative Pathname box. This ensures that as long as the files retain the same relationship to each other in their respective folders, they will continue to work even if the entire directory structure is moved to another disk, or transferred to another person. Click OK.

33➜ Click the Preview tab.

33➜ Right-click the MyStates layer and choose Properties. Click the Symbology tab.

33➜ If you want to, try using the example in the tutorial you did earlier to create a Quantities: Graduated Color map of the states using POP2000 as the Value field.

33➜ After you click OK, the symbols for the layer may not update in the Preview window. Click the Contents tab, and then click the Preview tab again. The new symbols should appear now.

33➜ Zoom in to the lower 48 states. Click the Thumbnail button to create a new thumbnail for the layer.

33➜ Click the Contents tab to view the new thumbnail (Fig. 1.35).

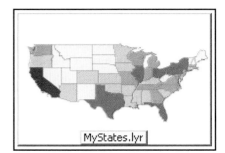

Fig. 1.35. Your thumbnail after step 33

Viewing metadata

We rely on metadata to provide information when data sets are sold or shared, or even when we come back to a data set months later and cannot remember the details about it. However, one must make a commitment to building and maintaining metadata, because it does take time and vigilance. Fortunately, ArcMap has the tools to make it easier.

34➜ In ArcCatalog, navigate to the mgisdata\usa folder and click on the states data set to select it.

34➜ Make sure that the Stylesheet drop-down box on the Metadata toolbar says FGDC ESRI.

34➜ Click the Metadata tab and examine the information.

34➜ Click on the green text **Abstract.** The abstract information folds up and is no longer visible. Click **Abstract** again to display the text once more.

34➜ Scroll down and click **Publication Information.** It tells who created the data and who published it.

34➜ Click the Spatial tab at the top of the metadata. It switches to another "page" that contains information about the coordinate system, extent, and accuracy of the data.

34➜ Click the Attributes tab to see a list of the fields in the attribute table, their definitions, and their descriptions.

34➜ Click the AREA attribute and read the information about it. Notice that the description indicates the units are square miles. Had the creator of this data set not entered this information into the metadata, the user might have no idea whether the areas represented square miles or square kilometers.

Already you are beginning to see how incredibly useful these data are. They are so important that a federal organization, called the Federal Geographic Data Committee (FGDC), has compiled standard rules about what kind of information goes into metadata, and how it is organized and stored. This allows everyone to define and manage metadata more easily, find the information they are looking for, and use software tools to create and maintain the data. These rules comprise the FGDC Metadata Standard. The FGDC standard format is a text file with a very specific layout so that different programs can find the information they need.

35➜ Click the Stylesheet drop-down tab in the metadata toolbar and change it from FGDC ESRI to FGDC Classic.

35➜ Scroll down to examine the information, and then return to the top of the document.

Metadata contains seven main sections shown as hyperlinks to the information in the document. The metadata standard has hundreds of "fields" of information, organized into groups and subgroups for easy access. The field names are shown in italics, and the information in plain text.

The ESRI stylesheet uses exactly the same information but presents it in a different way, or style. (Hence the name stylesheet). The information itself is stored in a document using a formatting language called XML, or Extensible Markup Language. It is very similar to the HTML used to create Web page documents, but has additional commands. Several other stylesheets can be used in ArcCatalog to view the metadata—but it is all coming out of the same .xml file.

36➜ Change the stylesheet to Xml to see what this file looks like. Then set it back to the FGDC ESRI stylesheet.

ArcCatalog uses the information already stored in the data sets to fill out certain fields in the metadata, such as the coordinate system, the spatial extent, and the fields. If these characteristics change, ArcCatalog will automatically update the metadata. By default, metadata is created, or updated, every time the user looks at the metadata sheet. The Metadata options control whether update occurs automatically or manually.

36➜ Choose Tools > Options from the main menu bar and click the Metadata tab. Read through the options, but do not change them at this time. Click Cancel when done.

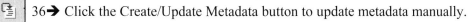 36➜ Click the Create/Update Metadata button to update metadata manually.

The rest of the metadata fields must be entered by the user. Metadata is a complex topic and requires training beyond the scope of this book to fully understand it and create it. For detailed information and copies of the metadata standard, refer to the FGDC website for geospatial metadata standards: www.fgdc.gov/metadata/contstan.html.

This is the end of the tutorial. However, you may do the following optional exercises on Internet map servers and searching if desired.

➜ If stopping here, close ArcCatalog.

Working with Internet map services (optional)

You can utilize the Internet to locate, view, and copy data as well. Many organizations maintain server sites that provide valuable data. Some of these sites are free. One major site developed by ESRI, Inc. is the Geography Network. It offers two main types of services: image service and feature service. A feature service allows people to download and save the information as shapefiles. An image service allows people to display and query the data, but they cannot save the features.

37➔ Click the Contents tab in the ArcCatalog window.

37➔ Scroll down to the bottom of the folder tree and find the entry labeled GIS Servers.

37➔ Expand it if necessary, and double-click the icon labeled Geography Network Services. It may have a little red X on the icon, indicating it is not currently active.

37➔ Expand the plus box of Geography Network Services to view its contents.

The red X disappears and a list of services appears as icons. These service entries are similar to other ArcCatalog data sets, and they can be viewed with the Contents and Preview buttons just like other data sets. The only difference is that the information is transmitted via the Internet. Preview works slowly with Internet data, so it is best to keep ArcCatalog in Contents mode until you want to see a specific layer.

38➔ Locate the FEMA_Flood service in the list and click it to highlight it.

38➔ Click the Preview tab. Zoom in to the United States for a closer look.

38➔ Click the ArcMap icon to start the program, and choose to begin with a new, empty map. Position the ArcMap window so you can see it when in ArcCatalog.

38➔ Click and drag the FEMA image service into the ArcMap window. Switch to ArcMap to view the FEMA service.

39➔ Notice the many layers added to the Table of Contents in the ArcMap window. Many of these have special scale ranges and will not appear until you zoom in.

39➔ Zoom in to the state of New Jersey. Several cities should appear now.

39➔ Zoom very closely in to the Philadelphia area (draw a small box around just the yellow city symbol).

39➔ The FEMA flood zones of the Delaware River should now be visible, color-coded by type in shades of green. Expand the FEMA service if needed to view the legend.

Now let's save a copy of some data from a feature service.

40➔ Still in ArcMap, right-click the FEMA_Flood service at the top of the Table of Contents and choose Remove.

40➔ Switch back to ArcCatalog. Locate the EPA Hazards FS feature service and click the plus sign to expand it.

40➔ Preview the Superfund Sites layer. Then preview its table.

40➔ Right-click the Superfund Sites layer and choose Export. Save it to the mgisdata\usa directory as supersites.shp. Now it can be accessed at any time.

41➔ Switch to ArcMap and use the Add Data button to add the supersites and states shapefiles to the map. If a warning about the coordinate system appears, click Yes to continue.

41➔ Right-click the supersites layer and choose Properties from the menu.

41➔ Click the Display tab and turn on Map Tips.

41➔ Click the Fields tab and set the Primary Display field to HUD_EPA_2 (the site name). Click OK.

41➔ Zoom into South Dakota.

10. How many superfund sites are there in South Dakota? _____ List their names:

_____

➔ Explore more of the Geography Network services, if you have time.

42➔ When finished, close ArcMap and do not save your changes.

42➔ Switch back to ArcCatalog and locate the Geography Network Services entry in the folder tree. Right-click it and choose Disconnect.

Searching for data (optional)

ArcCatalog has a nice search tool for finding data with certain characteristics or in a certain area. Although we will explore searching only our mgisdata folder, this tool can also search major file servers and Internet data providers.

 43➔ Click the Search button on the ArcCatalog main toolbar.

43➔ Click the Name & location tab, if necessary, and click the Browse button to specify the disk to search.

43➔ Navigate to and select your mgisdata folder and click Add.

43➔ Click the Geography tab. Check the box to use Geographic location in the search.

43➔ Use the Map drop-down box to set the map to Other. Navigate to the mgisdata\usa folder and choose states.shp.

43➔ Zoom into South Dakota.

 43➔ Click the Draw Box button and click and drag a box around the state of South Dakota. Notice that the latitude-longitude coordinates update as the cursor moves.

43➔ Fill the button to find data entirely within the location.

43➔ Click Find Now and wait until the search is complete (the magnifying glass in the window stops moving).

> **TIP:** The search will be saved as another Catalog entry named "My Search" by default. You can also specify a different name and create several different searches.

44➔ Close the Search box.

44➔ Look in the folder tree and locate the MySearch folder near the bottom under the Search Results entry. Click to expand MySearch. All the data sets found will be listed as links to the original data source.

TIP: The Search option uses the file metadata, and may fail to find files if the metadata are missing or incomplete.

Use the Preview button to view these search results just as you can any other data set. You can also click and drag a search result to ArcMap to begin working with it.

44➔ Click one of the search results and click the Preview button.

44➔ Click the ArcMap icon to launch ArcMap. Choose to start with a new, empty map.

44➔ Position the ArcMap window so that a portion of it is visible outside the ArcCatalog window.

44➔ Click on one of the search results data sets and drag it on top of either white window in the ArcMap GUI.

TIP: If the ArcMap window is not visible, drag the data set to the ArcMap icon on the computer Start menu and wait a moment, still holding down the mouse button. The ArcMap window will appear on top of the other windows, and you can then finish dragging the data set to it.

This is the end of the optional tutorial section.

➔ Close ArcCatalog and ArcMap.

Exercises

These exercises are all intended to be done in ArcCatalog.

1. How many feature datasets are there in the Badlands geodatabase in the mgisdata\Badlands folder? List their names. *3 infrastructure, transportation, wildlife*

   How many total feature classes does it have? *17*

   How many each of point, line, and polygon feature classes does it have?
   point = 2   line = 5   polygon = 10

2. What is the coordinate system of the utmzones shapefile in the mgisdata\world folder? *horizontal c.s. = "GCS_ Geographic"*

3. What four data types are present in the country shapefile attribute table in the mgisdata\world folder?

4. What is the positional accuracy of the majrdnet shapefile in the mgisdata\usa folder?

5. What is the largest lake in the United States? What is its area?

6. Which state has a county named Itawamba?

7. What is the minimum, maximum, and average number of nozzles in Rapid City gas stations?

8. How many rasters does the mgisdata\Rapidcity folder contain? List them.

9. How many rows and columns does the Landsat image L7_Aug20_W84 have? What is the cell size?

10. What is the name of the westernmost sacred site in the Badlands? *Vision Questing Area*

Challenge Problem

Create a layer file in MyFolder for the precip-ann shapefile in the mgisdata\Sdakota folder. Call it "SD_precip". Symbolize the polygons by the total amount of rain (attribute RANGE) using a yellow-to-blue color scheme. Place a copy of the ArcCatalog window showing your layer file and map in an answer document.

TIP: Pressing the keys Alt-PrntScrn together will capture a copy of the active window to the Clipboard. You can then paste it into a Word or PowerPoint document.

Skills Reference

NOTE: All skills in this chapter are used with ArcCatalog, rather than ArcMap.

Starting ArcMap or ArcCatalog

1. Look on the computer desktop for an icon named ArcMap or ArcCatalog (Fig. 1.36). Double-click it to start the program.

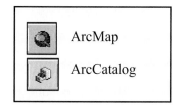

Fig. 1.36. Program icons

2. If no icon is present on the desktop, click the Start button on the computer's menu bar. Navigate to Programs > ArcGIS and choose the name of the program desired.

3. From ArcCatalog, launch ArcMap by clicking the ArcMap button in the menu bar.

4. From ArcMap, launch ArcCatalog by clicking the appropriate icon in the menu bar.

Starting ArcToolbox

ArcToolbox is a dockable window that sits inside ArcCatalog or ArcMap.

1. Click on the ArcToolbox icon in either ArcCatalog or ArcMap to open the window.

Connecting and disconnecting from folders

In order to access data files from ArcCatalog, you must set up a connection to the appropriate disk or folder. This saves time when frequently accessing data in a subfolder deep below the top. You can set up connections either to a drive letter such as D:\ or to a subfolder in the drive. Connections can be deleted when they are no longer in use.

1. To **connect** to a folder, click on the Folder Connect button in ArcCatalog.

2. Navigate down the directory tree to the folder to connect to.

3. Highlight the folder (or drive letter) and click OK (Fig. 1.37).

4. To **disconnect** from a folder, click the connection in ArcCatalog to highlight it, and then click the Disconnect Folder button.

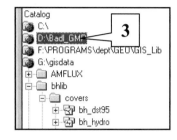

Fig. 1.37. Folder connection

Connecting to an Internet service

Using ArcCatalog, you can connect to and access data from Internet Map Servers. You must know the URL of the service, such as www.geographynetwork.com. Secure services also require a login and password. ArcGIS Servers are similar to ArcIMS servers, but may serve data either over a local network or over the Internet.

1. In the left side of ArcCatalog, scroll to the bottom and expand the GIS Servers entry. Double-click the Add ArcIMS Server entry.

2. Type in the URL of the service in the top box.

3. If the server requires a user login and password, enter these at the bottom of the window, and click OK.

Setting options

You can control the way the ArcCatalog performs various actions, displays information, and so on using this dialog box. You can also set defaults for the way tables, images, and other features are displayed.

1. Click Tools in the ArcCatalog menu bar and choose Options.

2. Click the appropriate tab to set the options.

3. When done setting options, click OK.

Viewing the contents of a folder

1. Click on the folder in the tree window.

2. Click on the Contents tab in the content window.

3. Choose one of the display options from the toolbar: Large icons, List, Details, or Thumbnails (Fig. 1.38).

Fig. 1.38. View icons

Creating and viewing thumbnails

1. Make sure that the layer is highlighted in the tree window and the Preview tab is clicked.

2. If desired, use the Zoom and Pan buttons to modify the appearance of the layer.

3. Click the Thumbnail button in the Zoom/Pan menu.

4. To view all the thumbnails in a folder, click on the folder, make sure the Contents tab is clicked, and choose the Thumbnail display option.

5. To view the thumbnail for a single layer, click on the layer in the tree window to highlight it and make sure the Contents tab is clicked.

Previewing a layer

1. Click on the data layer to highlight it in the tree window.

2. Click the Preview tab (Fig. 1.39).

3. Choose Geography from the drop-down menu at the bottom of the Preview window to preview the spatial data, or Table to preview the table.

4. Use the Zoom, Pan, Full Extent, or Identify buttons to explore the geography preview.

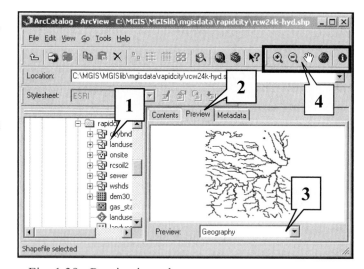

Fig. 1.39. Previewing a layer

Previewing a table

ArcCatalog has many tools to investigate the contents of a table. The table can be an attribute table of a layer or a standalone table.

1. Click on a data layer or table in the tree window, and make sure that the Preview tab is clicked.

2. Choose Table from the drop-down box at the bottom of the Preview window.

Sorting the table

3. Right-click the field name and choose Sort Ascending or Sort Descending from the context menu.

Getting statistics on a field

4. To get statistics for a numeric field, right-click on the field name and choose Statistics.

Freezing/unfreezing columns

5. To hold a field at the left side of the table while you scroll to the right, right-click the field and choose Freeze/Unfreeze from the context menu.

6. The field will move to the left edge of the table and remain there as you scroll to the right. More than one field can be frozen at a time.

Finding text in a field

7. To find text or values in a particular field, click on the field name to highlight the field (Fig. 1.40).

8. Shift-click to add additional highlighted fields to search, if desired.

9. Click the Options button in the lower right of the table, and choose Find from the menu. (If needed, enlarge the ArcCatalog window to the right to find the Options button.)

10. Type in the text to find, modify the search settings if desired, and click Find Next.

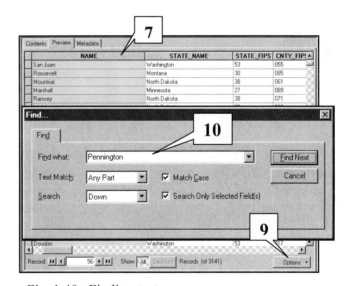

Fig. 1.40. Finding text

Adding/deleting fields from Preview mode

11. To add a field, click on the Options button and choose Add Field. Enter the table name and field type. For more information on adding fields to tables, see Chapter 5.

12. To delete a field, right-click on the field name and choose Delete Field. This action cannot be undone.

Using layer properties

Data sets, including layers and tables, have properties that can be modified in ArcCatalog. The types of properties will vary with the type of data set.

1. To access the layer/table properties, right-click on the layer/table name in the tree window and choose Properties. Double-clicking the layer/table also opens its Properties.

2. Click the tab containing the properties to change.

   Shapefile tabs include: General, Fields, and Indexes.

   Geodatabase tabs for feature classes include: General, Fields, Indexes, Subtypes, and Relationships.

   Coverage tabs include: General, Projection, Tics and Extent, and Tolerances.

   Subsequent chapters will describe some of these properties and how to set them.

Searching for data

The search engine in ArcCatalog locates data sets on a particular disk or server that cover the geographic area specified. The results are written to a catalog entry and can be easily dragged into ArcMap or copied to a different directory.

> **TIP:** The Search option uses the file metadata, and it may fail to find files if the metadata are missing or incomplete.

1. Click the Search button in ArcCatalog.

2. Click the **Name & location** tab (Fig. 1.41).

3. To search for particular types of files, choose them from the Type list. Use Ctrl-click to select more than one. To search for all types, don't select any, or click Clear.

4. Choose Catalog to search only the connected folders, or Disk to search the entire disk.

5. Set the disk or other location to search.

6. Click the **Geography** tab.

7. If you know the map coordinates to search, type these directly into the boxes.

8. OR, to locate an area to search on a map, use the Map drop-down box at the bottom of the Geography tab to choose a display map to search on. Use one of the default maps, or select a different one from the disk by entering <Other>.

9. OR, use the buttons to zoom into the appropriate region and draw a box around the target area.

10. OR, select a place name from the list at the top of the tab. The list will depend on the map chosen in step 8. Thus, to search for a county, choose US Counties as the map layer.

11. Choose to search for data entirely within your location, or overlapping your location.

12. To search for data for specific dates, click the **Date** tab. Fill out the information.

13. You can also search for explicit fields and values in the metadata by clicking the **Advanced** tab. This option is only recommended for experienced users. However, you can use it to find data produced by certain agencies or having certain keywords. There are many ways to refine a search.

14. In the Geography tab, type in a name to save the search under, or use the default, My Search.

15. When ready, click Find Now. Wait, as the search may take some time to complete. The status bar at the bottom of the window shows the folders being searched. Click the Stop button at any time to quit the search.

16. When the search is complete, click on My Search (or whatever you named it) in the ArcCatalog tree window to expand it and view the results.

17. The Search routine places links to the original data inside the My Search entry. You can preview the geography and tables of these links just as you would other files and drag and drop them into ArcMap.

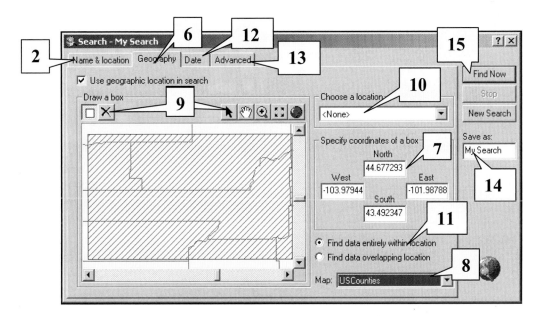

Fig. 1.41. Searching for geographic data

Drag and drop files to ArcMap

After searching through data in ArcCatalog, you can easily drag them into ArcMap for use.

1. Position the ArcCatalog and ArcMap windows so that both are visible.

2. Click the file in ArcCatalog that you want to open in ArcMap. Hold down the mouse button and drag it into the ArcMap window (you must be in the map window or the Table of Contents—not on a menu bar). Release the mouse click to drop the file in.

> **TIP:** If ArcMap is not visible on the screen, drag the file onto the ArcMap icon on the computer's Start Menu bar. Hold it there for a moment until ArcMap opens, and then drag the file into ArcMap.

Viewing metadata

1. Click on the file you wish to view and click the Metadata tab.

2. Optionally, you can change the way the metadata look by choosing a different stylesheet from the drop-down menu. They are all based on the same metadata file, written in XML; they just present the information differently.

Creating and deleting files

Create new shapefiles, geodatabases, and coverages in ArcCatalog.

1. Click on the folder to contain the new file.

2. Choose File > New > and the type of file to create (Fig. 1.42).

3. Follow the instructions in the menu for creating the file. Different file types require different parameters.

4. To delete a file, right-click it and choose Delete. Click Yes to confirm that it should be deleted.

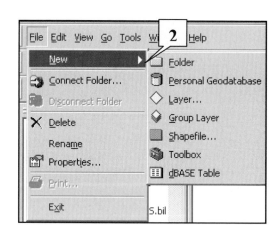

Fig. 1.42. Creating a file

Chapter 2. Working with ArcMap

Mastering the Concepts

Objectives

> ➢ Learning how to manage and use ArcMap windows and menus

> ➢ Adding feature datasets and tables to a map document

> ➢ Drawing features using different symbols

> ➢ Getting information about features using Find and Identify

> ➢ Understanding and using map scale when displaying features

> ➢ Labeling features as simple text, dynamic labels, and annotation

Concepts

Map documents

ArcMap works with map documents, just as Microsoft Word works with Word documents. A map document is a collection of different spatial data layers and tables, along with instructions for how the layers will be displayed. Just as text on a page has properties such as its font, size, and style, map features have properties that control the symbol, color, and style with which they are drawn. Tables have properties that specify which fields are shown, how many decimal places are included, and so on. The map document keeps track of all these layers and their properties, so that when it is opened again, the map appears exactly as it was when it was last saved. Even the size of the windows and the locations of the toolbars are stored when saving the document.

Map documents organize the data using **data frames.** A data frame is a window in the document that contains groups of related layers that are drawn together. Maps can contain multiple data frames. For example, a document could contain a data frame with a detailed map of South Dakota, and another frame showing the location of South Dakota in the United States (Fig. 2.1). Both frames are included when printing the map.

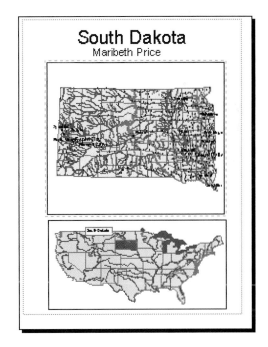

Fig. 2.1. A map document with two data frames, one of South Dakota and one of the United States

Unlike the pictures placed in this book document, which are completely saved within the document itself, a map document does not store the features and tables of the data sets within the map document. Instead, the map stores the name and disk location of the spatial data—its **source.** Changes to the map document do not affect the source data except for specific actions such as adding fields or editing shapes. Changing display symbols, sorting tables—these actions do not affect the source data at all.

It is important to realize that, because the source data are stored outside the map document, changes to source data made in one map document affect all map documents using the same data set. For example, imagine that Jim creates a map of the United States, saves it, and goes to lunch. During lunch, Peggy opens her map, which uses the same states shapefile, and begins to practice editing. She deletes the states of California and Texas, and then accidentally saves her changes. When Jim comes back from lunch and opens his map again to continue working, he finds that California and Texas are missing.

Most organizations take steps to protect their spatial data sets from being accidentally changed, precisely because the mistake of one can affect the work of many. Often the source files are write-protected so that only specific users are authorized to make changes. Others like Peggy, who want to practice editing the data, must make a copy of the original to work with. Before making any permanent changes to source files, stop and carefully consider who owns and uses the data, and who else might be affected by changes. If in doubt, create a copy to work on.

The storage of the map document and the source data in separate locations also has implications for sharing map documents. Giving a colleague a copy of a map document, by e-mail for example, only works if the recipient has access to exactly the same data in the same disk location as the original user. Map documents keep track of the source files by storing the name and location of each file as a **pathname.**

Map documents and pathnames

A **pathname** is the list of folders that one must traverse to reach a particular file on the disk. A pathname begins with the drive letter and lists successive folders, each one beneath the previous, separated by backslashes. The pathname c:\mgisdata\usa\states.shp thus refers to the shapefile states.shp which resides in the usa folder, which in turn resides in the mgisdata folder, which is found on drive C:\ (Fig. 2.2).

The pathname c:\mgisdata\Usa\States.shp is termed an absolute pathname because it starts at the drive letter and proceeds downward to the file. When searching for a file using an absolute pathname, the search always begins at the drive level.

A relative pathname is used to indicate that a search should be made starting at the current folder location, rather than from the drive letter. A double dot indicates movement up one folder. Thus the pathname ..\Usa\States.shp indicates a search up one folder and then down into the Usa folder and thus to the States shapefile. It provides an alternate route from the MapDocuments folder to States.

Map documents can store either absolute or relative pathnames, and the choice is made by the user creating the map document. By default a map document stores absolute pathnames. Imagine that the map document ex_1.mxd in Figure 2.2 refers to the States file in the Usa folder. When the map document is opened in ArcMap, it searches starting at the C:\ drive down through C:\mgisdata\Usa\States.shp in order to locate and draw the states in the map.

However, what if the user decides to move the mgisdata folder to another location on a network drive designated with the letter F:\? Upon opening the ex_1.mxd map document in the new location, it searches down the C:\ path as before but can no longer locate the mgisdata folder or its contents. The links to the spatial data are broken, and in ArcMap the States layer will appear with a red exclamation point (Fig. 2.3). The user can manually locate the data to fix the links, but this can be time-consuming if the map document contains many layers.

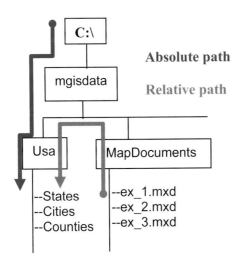

Fig. 2.2. Pathnames

Fig. 2.3. The red exclamation point indicates that the pathname link is broken, probably because the data file has been moved or deleted.

A better solution is to direct the map document to store relative pathnames. In this case the States pathname is stored as ..\Usa\States.shp, and it no longer matters whether the mgisdata folder exists on the C:\ drive, the F:\ drive, or even under C:\students\project\mystuff. As long as the relative positions of the MapDocuments and Usa folders remain the same, the map document can always locate the data it is looking for.

The choice of absolute or relative pathname depends on the situation. If the map documents will always refer to a central data server folder that all users on a network have access to from many different computers, and the users want to transfer map documents from computer to computer, then absolute pathnames will more reliably locate the data layers. On the other hand, if a user plans to move the map documents and data together as a unit, then relative pathnames work better. The map documents on the CD that comes with this book store relative pathnames, so that the links to the data layers continue to work no matter where you install the mgisdata folder.

VERY IMPORTANT TIP: Pathnames, including the names of map documents and spatial data sets, should not contain spaces or unusual characters such as #,@,&,*, etc. Use the underscore character _ to create spaces in names if needed. Although ArcCatalog and ArcMap allow spaces in names, the spaces create problems for certain functions. Thus, it is wise to avoid them entirely.

ArcMap windows and menus

The ArcMap Graphical User Interface (GUI) packs a lot of functionality into a deceptively simple space. This section introduces the primary features and functions of the interface, including the menus, toolbars, and layout of the ArcMap main window.

The menu bar

The menu bar usually lies at the top of the window and includes groups of commands in several categories (Fig. 2.4). The Customize function allows the user to add commands to the menu bar. Many functions can also be accessed using buttons in the toolbars. The menu bar can be torn off its location and moved to another site in the GUI, or even off the GUI entirely.

The toolbars

The toolbars contain buttons that perform actions, or buttons that convert the mouse cursor into different types of tools (Fig. 2.4). For example, the Add Data button launches a dialog box to add layers to the map, and the Identify button converts the cursor into a tool for getting information about individual features in the map (Fig. 2.5). Toolbars are organized into groups of related tasks, such as the Zoom/Pan toolbar, the Editing toolbar, the Layout toolbar, etc.

Fig. 2.4. A menu bar (top) and a toolbar (bottom)

Like the menu bar, toolbars can be torn from their locations and docked somewhere else, including off the GUI. Buttons can be added to existing toolbars, and the user can create entirely new toolbars for favorite sets of commands.

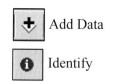

Fig. 2.5. Some buttons

Context menus

A context menu appears when the user *right*-clicks certain objects, such as layer names, symbols, field headings, and more (in contrast to the more usual *left*-click for most operations). The menus that appear will depend on the object clicked, and even the situation at hand. To determine whether an object has a context menu, simply right-click it to see if one appears.

The Table of Contents

The Table of Contents window lists the data layers and tables that have been added to the map (Fig. 2.6). It also provides access to several functions.

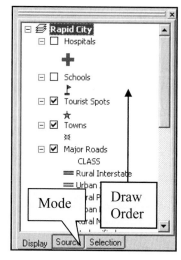

Fig. 2.6. The Table of Contents

➤ The window has three modes: **Display**, **Source**, and **Selection.** In Display mode, only the layers are listed. In Source mode, the layers are organized into their home directories, and additional non-displaying objects such as tables are listed as well. The Selection tab shows the selection status of each layer, i.e., the number of features that have been selected using a query, as discussed in Chapter 6. Clicking the tabs at the lower right corner of the Table of Contents determines the current mode.

➤ Checking and unchecking the boxes by each layer turns the layer display on and off.

➤ In Display mode, you can control the order in which layers are drawn by clicking the layer and dragging it to a new position in the list. The map is drawn from the bottom up. When adding layers, ArcMap automatically puts polygon layers at the bottom, then line layers, then point layers. The order can be important. If a polygon layer appears above a point layer in the list, then the points will be obscured by the polygons being drawn on top of them.

➤ The Table of Contents shows the symbols being used to draw each layer, and also provides access to the menus and tools that manage the symbols. Clicking and right-clicking a layer symbol launch different ways to modify it.

➤ Double-clicking a data layer name opens its Layer Properties window.

➤ Right-clicking a layer name in the Table of Contents produces a context menu which provides access to many functions and properties associated with the layer.

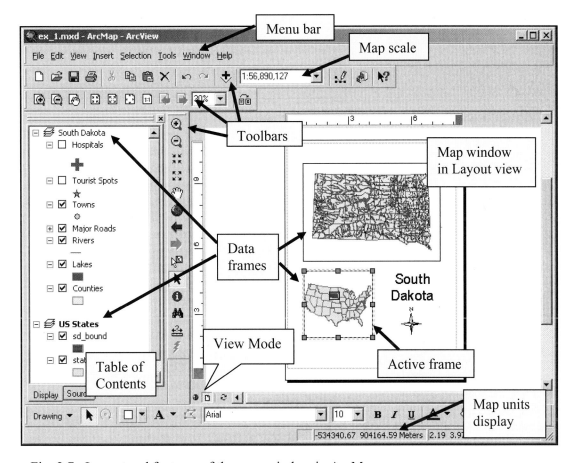

Fig. 2.7. Layout and features of the map window in ArcMap

The map window

The map window is the main display area (Fig. 2.7). Data in the window are organized into groups called **data frames,** which are boxes containing views of the data in the group. The map window has two view modes: Data view mode and Layout view mode. In Data view, only a single data frame is displayed. Layout view is a picture of what the map document would look like if printed, and it shows all of the data frames on the page, along with titles, scale bars, legends, and any other map information that has been placed on the page. The ArcMap interface in Figure 2.7 shows a map with two data frames in Layout view.

The Help system

ArcGIS includes extensive help files with a wealth of important information. Help is requested from the Help entry on the main menu bar. On the left are three tabs allowing the user to search for information, and the right shows the current entry on display (Fig. 2.8). Four search methods are available. The Contents tab shows an organized outline of material, much like a library of books. To view the contents of a book, click the plus sign to expand the entry. The Index tab contains a wealth of frequently used entries. Typing a word in the box on top causes the window to jump through the index to the matching word. The Search tab allows the user to enter a word or phrase and search the entire Help text. Finally, the Favorites tab can be used to save entries that are frequently consulted.

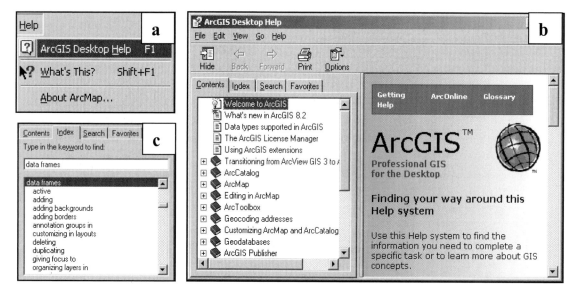

Fig. 2.8. ArcGIS Desktop help: (a) getting help, (b) the Help window showing the Contents tab, (c) the Index tab

Data frames

Data frames contain layers and tables to display and analyze together and are used to help organize the map. A map document can have multiple data frames. The map document in Figure 2.7 has two: a frame called South Dakota and another called US States. The US States frame is outlined with blue squares because it is currently the **active frame.** All commands and actions, such as Adding Data, occur in the active frame. When the map window is in Data view, only the active frame is visible. In Layout view, all the frames are visible.

Data frames have properties, just like other objects in ArcMap. Frame properties include the coordinate system, size and position, borders, rules for labeling features in the frame, and others. The chapters ahead explain more about data frame properties.

The coordinate system of a data frame defaults to the coordinate system of the first data set loaded into it. The coordinate system can also be set explicitly. ArcMap practices on-the-fly projection, meaning that every data layer in the frame is automatically reprojected to the frame coordinate system so that they will all display together.

Data layer properties

Data layers, including shapefiles, geodatabases, and coverages, and even tables, have properties that control how they are stored and displayed. Some properties can be accessed in ArcCatalog, some in ArcMap, and some in both.

Layer properties are accessed by right-clicking the name of the layer and choosing Properties from the context menu. The Properties dialog box has many tabs, each of which provides the interface for setting layer properties. For example, the General tab specifies the name of the layer, whether it is currently visible, and at what range of scales it is displayed. Many of these options can also be set elsewhere. The Symbology tab provides detailed control of the colors, sizes, and shapes of symbols being used to draw the layer.

Working with symbols and styles

Symbols come in three types: marker symbols for point data, line symbols for line data, and fill symbols for polygon data (Fig. 2.9). Each type has its own basic properties to set.

Marker symbols have a symbol, a size, a color, and an angle. Line symbols have a symbol, a color, and a thickness. Polygon symbols have a fill style, a fill color, an outline thickness, and an outline color. The fill can be a solid color, or it can be patterns such as stripes, or even pictures made from bitmap files.

The Symbol Selector allows the user to select an existing symbol, use it as it is, or modify its properties to suit specific needs, such as changing its color (Fig. 2.9). For example, the Expressway symbol could be given a different color and line width.

Many symbols come with ArcMap in addition to the basic ones that appear in the Symbol Selector. The user can load additional symbol sets and modify them to suit specific applications, or create entirely new symbols and symbol sets.

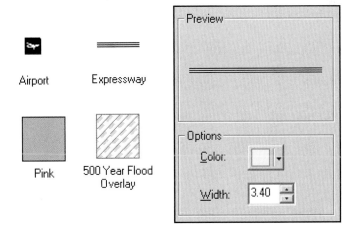

Fig. 2.9. Examples of symbols: a marker for airports, a line symbol for expressways, a pink solid fill symbol, and a 500 Year Flood patterned fill symbol. Once a symbol is chosen in the Symbol Selector, you can modify its properties.

Map scale concepts

What is map scale?

Map scale is a measure of the size at which features in a map are represented. The scale is expressed as a fraction, or ratio, of the size of objects on the page to the size of the object on the ground in real life. Because it is expressed as a ratio, it is valid for any unit of measure. So for a common U.S. Geological Survey topographic map which has a scale of 1:24,000, one inch on the map represents 24,000 inches on the ground (or one meter represents 24,000 meters on the ground, and so on). Imagine a map made to the scale of 1:100,000. You can use the map scale and a ruler to determine the true distance of any feature on the map, such as the width of a lake (Fig. 2.10). Measure the lake with a ruler, and then set up a proportion such that the map scale equals the measured width over the actual width (x). Then solve for x. Keep in mind that the actual width and the measured width will have the same units. You can convert these units if need be.

Often people or publications refer to large-scale maps and small-scale maps. A large-scale map is one in which the *ratio* is large (i.e., the denominator is small). Thus a 1:24,000 scale map is larger scale than a 1:100,000 scale map. Large-scale maps show a relatively small area, such as a quadrangle, while small-scale maps show bigger areas such as states or countries.

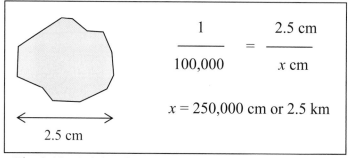

Fig. 2.10. Solving for the actual size of a lake

Fig. 2.11. The scale box in ArcMap shows the current scale of the map on the screen.

Scale with GIS data

When data are stored in shapefiles or other spatial data formats, they do not have a scale, technically speaking, because only the coordinates are stored. They acquire a scale once they are drawn on the screen or on a piece of paper. ArcMap has a scale box on the main toolbar that reports the current scale of the map on the screen (Fig. 2.11), and it updates every time the zoom scale changes. Likewise, you can determine the scale of a map to be printed or set a specific scale to print to.

However, the concept of scale does apply to GIS data, in the sense that most data layers have an intrinsic scale at which they were created. The map in Figure 2.12 shows congressional districts in pink and the state outlines in thick black lines. Notice how the state boundaries are more angular and less detailed than the districts because they were digitized at a smaller scale. Thus, although it is possible to take small-scale data and zoom in to large scales, the accuracy and detail of the data will

Fig. 2.12. These two layers showing Massachusetts originated from maps with different scales.

suffer. The original scale of a map or data set is thus an important attribute, and it should be included when creating metadata for layers.

One should exercise caution in using data at scales very different from the original. When selecting or creating data, it is also important to use data that are appropriately scaled for the analysis. Zooming into a data set may give a false impression that the data are more precise than they actually are. A pipeline digitized from a 1:100,000 scale map, for example, has an uncertainty of about 170 feet in its actual location due just to the thickness of the line on the paper. Displaying the pipeline on a city map at 1:60,000 might look fine, but zooming in to 1:2000 would not help at all should you desire to locate the pipeline by digging.

From looking at Figure 2.12, one might conclude that it is desirable to always obtain and use data at the largest possible scale. However, large-scale data require many more data points per unit area, increasing data storage space and slowing the drawing of layers. Every application has an optimal scale, and little is gained by using information at a higher scale than needed.

Reference scale

Data frames have a property called the **reference scale,** which stipulates the scale for which the layers are designed to be displayed (Fig. 2.13a). When placing symbols and labels on the map, you naturally give the symbols their most aesthetically pleasing size relative to the current map scale. However, because symbols normally remain the same size as the scale changes, at some zoom levels they appear too large, while at others they appear too small (Fig. 2.13b).

a

The reference scale

Fig. 2.13. (a) The reference scale sets the proportional size of the features and their text. (b) If the reference scale is not set, the symbols will always appear the same size. (c) If the reference scale is set, then zooming in or out changes the size of the text and symbols in proportion to the features.

If the reference scale is set, then the symbols will get larger or smaller as the zoom scale changes, just as the features do (Fig. 2.13c). You can set the reference scale, clear it, and automatically zoom to it.

b

Zoom in—reference scale not set

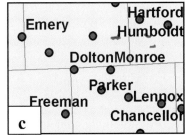

c

Zoom in—reference scale set

TIP: Be sure to understand the difference between the terms *map scale*, *reference scale*, and *visible scale range*. Map scale determines the ratio of features on the map to features on the ground. The reference scale controls the sizing of symbols and text when zooming from one map scale to another. The visible scale range determines the range of scales at which a layer will appear.

Labeling concepts

Labeling involves taking an attribute of a feature, such as its name, and placing it on the map close by. ArcMap provides three basic types of labeling. You can place **interactive labels** one at a time, or automatically label many features using **dynamic labeling,** or take precise control of labeling by using **annotation.**

Interactive labels

Interactive labels are constructed by the user one at a time, and they can be used to label map features or provide titles and supporting text for a map layout. One labeling method involves placing text interactively on the map in one of three ways: as a plain text graphic, as a callout box, or splined along a line. A callout places text in a box with a pointer to indicate the feature of interest. Splining causes labels to follow along a feature such as a road or creek. Alternately, you can label a feature using the value from a field in its attribute table.

More complex text can be constructed using the wrapped text tools. The text is placed inside a shape defined by the user, and rudimentary control of margins and columns is provided, as well as background and border options. If the wrapped text box changes shape, the text inside is automatically reformatted to fit.

All interactive labels remain on the map as graphic elements, unless you decide to select and delete them.

Dynamic labels

Dynamic labels are created from an attribute and are actively managed by ArcMap as you turn layers on and off, zoom in and out, etc. Every time you redraw the map, ArcMap redraws the labels. It optimizes the display by omitting overlapping labels. Thus the number of labels displayed can vary, depending on the zoom scale and label size.

Creating dynamic labels involves specifying the label field in the Layer Properties menu, setting options for how they appear, and turning them on (Fig. 2.14). Choices include the following:

> Setting options that control the size, font, and color of the labels.

> Setting priorities for where they are placed.

> Adopting guidelines to resolve overlap conflicts between labels.

> Specifying different classes of labels. For example, large cities could have a bigger label and take priority over smaller cities.

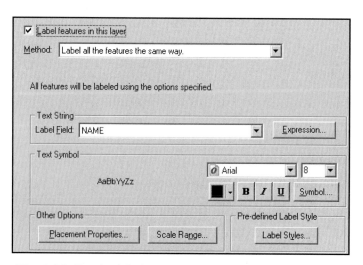

Fig. 2.14. Set the labeling properties of dynamic labels using the Labels tab in the Layer Properties window.

> Using the data frame properties to control which layers have priority over other layers. For example, in case of a conflict, states should always be labeled before cities.

> Specifying that labels be displayed only at a range of scales, just as for layers.

The **Placement Properties** help to manage the overlap and optimization of the labels to give the best possible result. Points, lines, and polygons have different placement options.

Point label overlaps are resolved by trying to offset the label from the point, while putting as many labels as possible in the preferred location.

Line labels can be placed several ways: they can be placed horizontally or vertically at the middle or ends of the line. You can also spline text along a linear feature, placing it inside, above, or below the line (Fig. 2.15).

Polygon labels may be forced to be horizontal, or the software can try to place them at the same angle as the polygon's longest boundaries to maximize the fit (Fig. 2.15). In addition, the user can request to omit labels that would extend beyond the borders of the polygon.

Duplicate label options are used in situations such as one street composed of many blocks. You can control whether a label is placed for every block, or only one per street.

Fig. 2.15. Line labels may be splined along features such as these creeks. Polygon labels may be horizontal or straight with the long edges.

Finally, the **Conflict Detection** options can handle priorities for different layers. **Label weights** control which layers take priority; for example, states might have a higher priority than cities and be preferentially labeled in case of a conflict between the two. **Feature weights** specify the layers whose features cannot be overlapped by labels. For example, labels from cities should probably not obliterate markers for Superfund sites. Finally, label **buffers** prevent new labels from encroaching on the edges of existing labels.

Because the actual labels placed vary, depending on scale and symbol size, printing maps based on dynamic labels can bring surprises. What shows on the screen does not necessarily appear the same way on the map. The Print Preview option shows the printed map and labels as they will appear. For many purposes a few extra or missing labels do not matter. If they do, however, converting labels to annotation provides greater control over label placement.

Annotation

Annotation provides precise control of each label. Ordinary dynamic labels can be converted to annotation. The labels that fit on the map are automatically placed, and the overlapping labels are directed to an overflow window. From this window you can work with the labels individually to place them exactly as needed. Annotation always appears in the same place and at the same size regardless of printing or zooming changes. Annotation may be stored three ways.

> You can store it in the map as simple text.

> You can store it as a feature class in a geodatabase and use it in other map documents.

➢ You can store it in the same geodatabase as the features linked to it—of course the features must then be in a geodatabase and not in a shapefile or coverage. Feature-linked annotation is automatically created, moved, and deleted as the features are edited, but it is only available to ArcEditor or ArcInfo users, not to ArcView users.

Summary

➢ ArcMap provides a fully customizable graphical user interface (GUI) composed of menu bars, toolbars, windows, and context menus.

➢ Map documents store references to layers and information about them, such as how to draw them. The actual data exists outside the document and may be used by many people.

➢ Map documents may store either absolute or relative pathnames for data files. Absolute pathnames work better when the data will always be in the same place for every user. Relative pathnames are best if the data will be moved about.

➢ The Table of Contents lists the available data layers and provides control for the order of drawing, and through context menus it gives access to properties and commands associated with each layer.

➢ Data frames contain layers to be drawn together, and they also have properties including the coordinate system and the reference scale.

➢ Data layers have multiple properties covering issues such as the source data, the projection, display, labeling, symbols, and more.

➢ Map scale is the ratio of size in the map to size on the ground. The original scale of a geographic data set determines the scale of its usefulness.

➢ The display scale controls the range of scales at which a layer appears. The reference scale of a data frame can be used to control the scaling of symbols and labels.

➢ Labels can be placed dynamically or as permanent annotation. Annotation may be stored as text in the map, as a feature class in a geodatabase, or as feature-linked annotation in the same geodatabase as the features.

VERY IMPORTANT TIP: Pathnames, including the names of map documents and spatial data sets, should not contain spaces or unusual characters such as #, @, &, *, etc. Use the underscore character _ to create spaces in names if needed. Although ArcCatalog and ArcMap allow spaces in names, the spaces create problems for certain functions. Thus, it is wise to avoid them entirely.

TIP: Be sure to understand the difference between the terms *map scale*, *reference scale*, and *visible scale range*. Map scale determines the ratio of features on the map to features on the ground. The reference scale controls the sizing of symbols and text when zooming from one map scale to another. The visible scale range determines the range of scales at which a layer will appear.

Chapter Review Questions

You may need to consult the Skills Reference section to answer some of these questions.

1. Describe the difference between a data layer and a data frame.

2. What is a map document, and what types of information are stored in it?

3. How does ArcMap decide which layers to draw first?

4. Mary creates a map document and e-mails it to a colleague in another state. He calls and tells her the document opens but no map appears. Explain what is wrong.

5. What does it mean if a layer's check box is dimmed?

6. Why can different labels appear each time you zoom or pan a layer with labels?

7. What is the difference between the scale range and the reference scale?

8. You measure a football field (100 yards long) on a detailed map and find that it is 0.5 inches long. What is the scale of the map?

9. Describe the difference between dynamic labeling and annotation.

10. What characters should be avoided when naming GIS files, folders, and map documents?

Mastering the Skills

Teaching Tutorial

The following examples provide step-by-step instructions for doing basic tasks and solving basic problems in ArcGIS. The steps you need to do are highlighted with an arrow ➔; follow them carefully. Click on the video number in the VideoIndex to view a demonstration of the steps.

➔ Start ArcMap, if necessary. Navigate to the mgisdata\MapDocuments folder and open the map document ex_2.mxd.

➔ Choose Save As from the File menu and save your own copy of the map document under a new name such as ex_2mine.mxd. Save it frequently as you work.

TIP: The names of map documents, spatial data sets, and folders should not contain spaces or unusual characters such as #, @ , &, *, etc. Use the underscore character _ to create spaces in names if needed. Although ArcCatalog and ArcMap allow spaces in names, the spaces can create problems for certain advanced functions. Thus, it is wise to avoid them entirely.

Working with the map display

First, let's briefly review some of the techniques for zooming and panning that we learned in Chapter 1.

  1➔ Use the Zoom In tool to draw a box around the lower part of the large lake in central South Dakota (Fig. 2.16).

1➔ Hover the cursor over the lake for a moment until its name appears.

Fig. 2.16. Zoom to this lake

1. What is the name of this lake?

_____

2. What is the function called that causes the lake name to appear? _____

  1➔ Use the Pan tool to drag the two small towns near the bottom part of the lake to the center of the view.

  1➔ Use the Zoom In Center tool to zoom in. Keep clicking the tool until the schools appear on the map.

3. At approximately what scale do the schools appear? _____

1➔ Click the Zoom Out Centered ONCE to zoom back out.

4. What is the scale now? _____ Make a guess what the minimum scale for the school display is set to. _____

2➜ Right-click the Schools layer in the Table of Contents and choose Properties from the context menu.

2➜ Click the General tab and read the minimum display scale.

5. What is the minimum display scale of the Schools layer? _____

2➜ Click the button to Show Layer at All Scales and click OK.

2➜ Use the Full Extent button to zoom to all of South Dakota again.

Note that the schools are now displayed for the entire state.

TIP: You can also set/clear the range scale by right-clicking the layer name in the Table of Contents and choosing Visible Scale Range.

3➜ Use the Previous Extent button to return to the close-up of the Pierre area.

Now notice a curious thing: north of the two towns (Pierre and Ft. Pierre), two schools show up in the middle of Lake Oahe. This seems very odd, unless they mean schools of fish.

3➜ Click on the Identify button and click on each school in turn.

6. What are the schools' names? _____

Each school's name is followed by the term (historical), so probably these schools were built and used prior to the dam that created Lake Oahe.

4➜ Close the Identify Results menu.

4➜ Use the Forward Extent button to return to the display of the state.   ➡

4➜ Click the Find button. Fill out the dialog box to search for "Meadowbrook" in the NAME field of the Schools layer.

4➜ When it finds the school, right-click on the school name and choose Zoom to Feature(s).

4➜ Right-click the school name again and choose Flash Feature to locate it. You may need to move the Find dialog box to see the feature flash. Repeat if needed.

7. To what urban area does Meadowbrook School belong? _____

5➜ Close the Find dialog box.

5➜ Choose View > Bookmarks > Rapid City from the main menu bar.

5➜ Click the Measure tool and use it to determine the length of Rapid Creek inside the urban area of Rapid City (inside the blue boundary).

8. What is the length of Rapid Creek inside Rapid City, in miles? _____

Suppose that instead of miles, you wanted to know the distance in kilometers. Instead of converting the units, simply change the measurement units of the data frame and measure it again.

6➔ Double-click the name of the Layers data frame in the Table of Contents (or right-click and choose Properties).

6➔ Click the General tab, and set the Display units to kilometers in the drop-down list.

6➔ Click OK to close the window, and measure the creek length again.

9. What is the length of Rapid Creek inside Rapid City, in kilometers? _____

Using data frames

Now we will learn some things about data frames. First we will examine the map in Layout view, rename the existing data frame, and add a new one.

7➔ Click the Layout view icon in the lower left of the display area to switch from Data view to Layout view.

7➔ Click on the Layers data frame name, then click again and type in a new name for it, "Rapid City".

7➔ Click the minus sign in the box next to the data frame name to collapse the list of data layers in it.

7➔ Choose Insert > Data Frame from the main menu bar.

7➔ Rename the new frame "South Dakota".

The new data frame appears with a set size in the middle of the page. It is highlighted with blue handles, indicating it is the active frame. The active frame is also shown in boldface type in the Table of Contents. Next, resize both frames so they do not overlap each other on the page.

8➔ Place the cursor inside the South Dakota frame, then click and drag it to the lower right, slightly off the page.

8➔ Click once inside the Rapid City frame to make it the active frame. Place the cursor over the lower left or right corner, and click and drag to resize the frame to occupy only the top half of the page.

8➔ Move and resize the South Dakota frame until the page appears similar to the one shown in Figure 2.17.

9➔ Make sure that the South Dakota frame is the active frame.

9➔ Click the Add Data button. Navigate to your mgisdata\Sdakota directory and choose the sd_bound.shp file. Watch it appear in the data frame.

9➔ Switch back to Data view by clicking the world icon (Fig. 2.18), or choose View > Data view in the main menu bar.

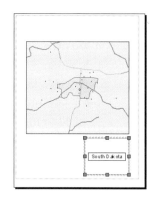

Fig. 2.17. Layout view

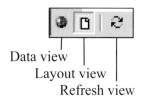

Data view
Layout view
Refresh view

Fig. 2.18. View buttons

You will use this data frame to create a location map for Rapid City. First, however, notice that the state appears elongated in the east-west direction. This elongation provides evidence that the data layer is stored in geographic data units, that is, latitude-longitude. Let us examine the coordinate system in a little more detail.

> 10➔ Right-click the sd_bound layer name in the Table of Contents and choose Properties, then click the Source tab to see the projection information.

10. What is the coordinate system of this data set? _____

By default, the data frame adopts the coordinate system of the first data set loaded into it. Any other data sets loaded now would be drawn in this same coordinate system, no matter what other coordinate system they may have.

> 10➔ Look at the map units display at the bottom of the ArcMap window and verify that the units are shown in degrees of longitude and latitude.

99°19'44.11"W 44°25'23.86"N

The map units displayed have also defaulted to the units of the first coordinate system—degrees. These display units can be changed, as you did earlier from miles to kilometers when measuring Rapid Creek.

Although data are commonly stored in a geographic coordinate system (GCS) with units of degrees, the distortion inherent in a GCS makes it a poor choice for a map, and professional cartographers avoid using GCS in maps for this reason. We will now set the map projection of the frame. Afterward, the GCS units are projected on the fly to whatever coordinate system is chosen for the data frame.

> 11➔ Close the Layer Properties window.
> 11➔ Right-click the South Dakota frame name and choose Properties from the context menu.
> 11➔ Click the Coordinate System tab, and verify that the frame coordinate system is the same as the sd_bound coordinate system.
> 11➔ In the box showing the coordinate system tree, click the following links to find the coordinate system desired: Predefined > Projected Coordinate Systems > State Plane > NAD 1983 > NAD_1983_StatePlane_South_Dakota_North_FIPS_4001. Click OK.

Now that the data frame has this coordinate system, any other data layer loaded into the frame will be reprojected to match it. Next, you will set up an **extent rectangle** to show the location of Rapid City in this frame. This trick is another data frame property.

> 12➔ Open the South Dakota frame Properties window, and choose the Extent Rectangles tab.
> 12➔ Click the Rapid City frame in the list to the left, and click the double arrow to place it in the list to the right.
> 12➔ Click the Frame button and choose the border by clicking on the drop-down list next to the sample. Select the 4.0 Point line symbol. Click OK in both boxes.
> 12➔ Use the Zoom In tool to zoom in as far as possible to the state without cropping any of it.

12➔ Switch to Layout view to verify that the border is shown. Extent rectangles only appear in Layout view.

Labeling features interactively

We need to add a few labels to this map. We can do this easily using the text tool in the Drawing toolbar.

13➔ Make sure that South Dakota is the active frame and return to Data view.

 13➔ Click the New Text tool on the Drawing toolbar. The black triangle indicates that more versions of this tool are accessible by clicking the drop-down arrow.

13➔ Click inside the state, then type the name "South Dakota". Press Enter.

Notice the blue dashed line that appears around the text, indicating that it is currently selected and will be affected by changes in its font or other properties.

13➔ Use the drawing toolbar buttons to set the font size to 28 and make the text Bold. Click and drag the text to center it inside the state.

TIP: To delete text, click the black arrow on the Drawing toolbar, click the text to select it, and press the Delete key. You can also use the black arrow to draw a box around many text boxes, or hold down the Ctrl key to select multiple text boxes for deletion.

14➔ Use Add Data to add the shapefile sd_urban.shp from the mgisdata\Sdakota folder to the South Dakota frame. If necessary, use the Identify button to determine which spot is Rapid City.

14➔ Click the drop-down arrow on the Text tool and choose the Callout text button. Click on Rapid City and drag slightly to the lower right to create the label.

14➔ Type in the name "Rapid City" and press Enter. The new label is selected. Adjust its location and pointer end if necessary (Fig. 2.19). Also adjust the South Dakota label if needed so that both can be clearly seen. (Make sure the Select Elements tool is selected on the drawing toolbar to move the labels.)

14➔ Turn off the sd_urban layer by clicking its check box.

14➔ Switch to Layout view to see the changes.

Fig. 2.19. Labeling the state

 14➔ Use the Layout Zoom tool to zoom into the South Dakota frame.

Notice that the South Dakota label has run over the edge of the state, and the Rapid City label is so enlarged that the pointer is not even visible. This happens because the font sizes remain the same with respect to the screen, but the state became smaller when you switched to Layout view.

Setting the frame's reference scale will fix the label sizes so they remain the same relative to the state size.

15➜ Go back to Data view.

15➜ Right-click the South Dakota frame and choose Reference Scale > Set Reference Scale from the menu.

15➜ Switch to Layout view again. This time the labels are scaled down along with the state.

The multi-line text tools are useful for placing paragraphs on a map document. Several different shapes can be chosen, including rectangles, circles, and polygons. The size and shape of the text boxes can be adjusted, and the text inside will automatically be wrapped to fit.

16➜ Return to Data View. Click the New Rectangle Text button on the Drawing toolbar and click and drag to create a rectangle in the NW corner of South Dakota, then release the mouse button. A dotted box with "Text" inside it will appear.

16➜ Double-click on the "Text" to open the Properties box. Enter a short paragraph in the box, saying whatever you like, without pressing the Enter key.

16➜ Click the Frame tab and choose a border for the text.

16➜ Click the Columns and Margins tab. Leave the margins as is, but set the number of columns to 2. Click OK to close the text Properties box.

16➜ Adjust the rectangle size by clicking and dragging its corners. Change the font or other properties of the text using the Drawing toolbar or by double-clicking on the text again to edit its properties.

16➜ While the text box is still selected, press the Delete key to get rid of it.

Next we will change symbols and create labels for the schools in Rapid City.

17➜ Return to Data view. It will show the South Dakota frame since that was the most recent active view.

17➜ To switch frames while in Data view, right-click the Rapid City frame name and choose Activate from the context menu.

17➜ Click the plus sign next to the Rapid City frame to expand its contents again.

17➜ Click the Schools symbol to launch the Symbol Selector. Choose the symbol School 2, click on the color box and select a red color from the palette, and change the symbol size to 16 pts. Click OK.

First we will demonstrate using the labeling tool to label features individually. Later we will create dynamic labels.

18➜ Choose the Label tool from the text tool set. This tool uses the display field set in the layer properties to label the features clicked. The Label Tool Options window appears.

18➜ Keep the default options to find the best label placement and to use the label style set for the layer.

18➜ Click inside the blue area and watch the Rapid City label appear. Use the drawing toolbar to change its font size to 14 and move it to a better location, if needed.

TIP: To delete a newly added label, press the Delete key while the label is still selected.

18➔ Click somewhere on the orange background (Pennington County).

No label appears when you click the county. It is likely that the "best label placement" for the label is the middle point of the county, which does not show up in this view. Hence no label appears. Adjust the label placement options to solve this problem.

19➔ Change the labeling option to "Place label at position clicked." (If you have already closed the Label Tool Options menu, click the Label tool again to open it.)

19➔ Click the orange background again, and the county label will appear. Adjust its font to 24 pts. and move it to a better location if desired.

19➔ Click on a section of Rapid Creek to label it. Adjust the size, font, or location if needed.

19➔ Click on the major highway running east-west near the top of the map.

Notice that when you click the highway, only an empty label box appears. You might need to set the display field for the layer.

20➔ Close the Label Tool Options window.

20➔ Open the Major Roads layer properties and click the Labels tab. The Label field is currently set to DESCRIP. Close the Properties window.

20➔ Use the Identify tool to click the highway now. Notice that its DESCRIP field is blank, which explains the labeling problem. SIGN1 looks like a better field for labeling. Close the Identify Results window.

20➔ Open the Label tab in the Properties window again, and change the Label field to SIGN1. Click OK to close the Properties window.

20➔ Click the Label tool again, and click on the highway to create a new label.

TIP: If you still have trouble making a label appear, be sure to click exactly on top of the highway symbol. The Map Tips option is turned on for this layer, so it may help to wait until the tip appears, and then click to place the label.

20➔ In the Label Tool Options menu, change the second option to Choose a Style.

20➔ Click the Interstate Highway symbol from the list that appears. Label the highway again.

20➔ Choose the U.S. Route symbol and then label the smaller road heading south out of town.

20➔ Close the Label Tool Options window.

TIP: Simple labels, text boxes, and graphics are collectively termed Elements. To Select All Elements or Unselect All Elements, use the functions in the Edit pull-down menu on the main menu bar.

Creating dynamic labels and annotation

These labels we have created are simple graphics placed on the screen. Now you will learn how to create dynamic labels for features based on one of their attribute fields and convert them to annotation.

21➔ Right-click the Schools layer and choose Properties from the context menu, then click the Labels tab.

21➔ Check the Label features box. Verify that NAME is the Label Field. Click OK.

Notice that some of the schools may not have labels because they overlapped with names of other schools. We can make this fact even more obvious by increasing the size of the text.

22➔ Open the Layer Properties for Schools again, and make sure the Labels tab is showing. Increase the size of the text to 12 pt. and make the style Bold. Click OK.

22➔ Zoom into the northeast part of the city between Rapid Creek and the interstate. Notice how more school labels appear now that we have more space.

22➔ Use the button to return to the previous zoom extent.

Now let's experiment with some label placement options.

23➔ Open the Layer Properties for Schools to the Labels tab and click the Placement Properties button.

23➔ Click the Placement tab, if necessary. Notice the graphic indicating the placement preferences.

23➔ Click Change Location and find the Prefer Top Center, all allowed icon (Fig. 2.20). Click it to select it and click OK.

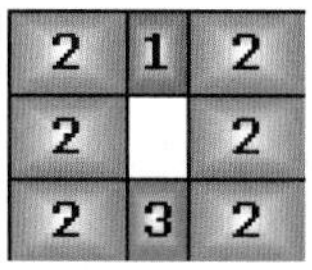

Prefer Top Center, all
allowed

Fig. 2.20.

The matrix in Figure 2.20 represents a set of rules for placing the labels. The white space in the center represents the point itself. The most preferred label location is the top center, designated with a 1. The boxes labeled 2 are second in priority, and the 3's show the least desirable location. A zero in a box would indicate that labels must never be placed there.

23➔ Click the Conflict Detection tab, and set the buffer distance to 0.25 (units are fraction of the label height). This option will suppress labels within this distance of each other for a cleaner-looking map. Click OK.

23➔ Change the font size to 10. Click OK. See the changes.

TIP: Turn labels on and off for a layer by right-clicking the layer name in the Table of Contents and choosing the Label Features option. If the menu choice is checked, the labels are on, and choosing it will turn them off. If unchecked, choosing it will turn them on.

The Labeling toolbar offers several convenient functions for working with labels. One function displays labels that are being suppressed because they overlap other labels.

24➔ Right-click the gray menu area to see the other menus available, and choose the Labeling toolbar.

24➜ Click on the View Unplaced Labels button to show the unplaced labels, which appear red. (If you don't see any unplaced labels, click the Fixed Zoom Out tool once or twice until they appear.)

24➜ Choose Labeling > Options from the Labeling menu and set the unplaced labels color to something other than red. Click OK.

24➜ Click the View Unplaced Labels button again to hide the unplaced labels. If you zoomed out to see them, return to the original extent.

Before we convert the labels to annotation for the finishing touches, let's label Rapid Creek again with a symbol that angles along the creek.

25➜ Select and delete the previous label(s) for Rapid Creek.

25➜ Open the Layer Properties for the Rivers layer and click the Labels tab.

25➜ Check the box to turn on the labels.

25➜ Change the font symbol to dark blue italic. Leave the other options as they are.

25➜ Click the Placement Properties button and click the Placement tab.

25➜ Fill out the boxes as shown in Figure 2.21 to place labels parallel to the lines and above or below them, with a 10-meter offset, at the best location, and removing duplicate labels. Click OK and OK.

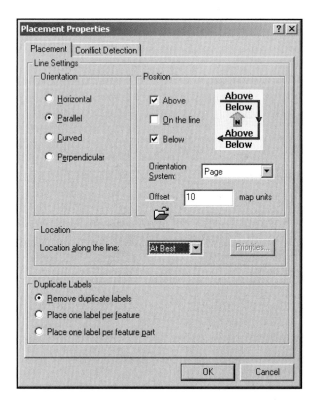

Fig. 2.21. Label placement options for lines

TIP: Turn labels on and off for a layer by right-clicking the layer name in the Table of Contents and choosing the Label Features option. If the menu choice is checked, the labels are on, and choosing it will turn them off. If it is unchecked, choosing it will turn them on.

Now we will convert the text to annotation for the final touches. In order to try out the entire process, the schools should have several unplaced labels.

26➜ In the Labeling menu, click the View Unplaced Labels button.

26➜ Click the Fixed Zoom Out button until at least two of the schools show as unplaced labels.

26➜ Right-click the Rapid City data frame and choose Convert Labels to Annotation.

26➜ Fill the button to store annotation in the map, since these are shapefiles rather than geodatabase layers.

Annotation is created for an entire data frame rather than for single layers. Notice that both Schools and Rivers are listed for conversion, since both of them currently have dynamic labels. We only want to create annotation for Schools, and so we must turn off the labels for Rivers.

27➔ Cancel the Convert Labels to Annotation window, right-click the Rivers layer, and uncheck Label Features.

27➔ Open the Convert Labels to Annotation window again and fill the button to create annotation in the map.

27➔ Choose to create annotation only for the features in the current extent. We don't want to label every school in the state!

27➔ Keep the box checked to convert unplaced labels to unplaced annotation. These labels will be put in an overflow window for individual placement later.

27➔ Click Convert to create the annotation.

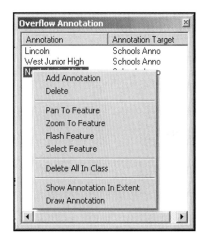

View the new annotation. It looks similar to the dynamic labels, but the labels are now individually adjustable. Notice that the unplaced labels were put in an overflow window (Fig. 2.22). We will now place these labels manually.

28➔ In the overflow window, right-click the first label and choose Flash Feature. Repeat if needed until you see the school flash. (You may need to move the overflow window to see the flash.)

28➔ Right-click the school name again and choose Add Annotation. The label is added to the map. The light blue dashed line around the name indicates that it is currently selected.

28➔ Adjust the newly added position to fit. Select and move other school names if needed.

Fig. 2.22. Manually placing overflow annotation

One benefit of annotation is customizing the label. Look at the long name for the South Dakota School of Mines and Technology. It would look better to call it by its nickname, SD Tech.

29➔ If the South Dakota School of Mines is in the overflow window, add it to the map.

29➔ Select the school name, then double-click inside the dashed box. A text dialog box will appear.

29➔ Replace the text with "SD Tech".

29➔ Click the Change Symbol button, and change the color of the text to blue, to distinguish the college from the K-12 schools. Click OK and OK.

29➔ Find the new label (it is far to the left). Put the cursor inside the label, then click and drag it to a good location near the school.

➔ Place the remaining overflow labels. To help fit the annotation, adjust the location of other schools and/or shorten the school names.

> **TIP:** If you are not sure which school is which, use the Identify tool to learn its name. Click on the black arrow in the Drawing toolbar to continue moving the annotation.

Now that these labels are annotation, they will be scaled up and down in size as the map scale changes.

30➜ Use the Zoom In tool to zoom into a section of the city with only a few schools. Although the school symbols remain small, the text becomes very large.

Next we will set the reference scale so that both symbols and text are scaled.

30➜ Use Previous Extent to return to normal.

30➜ Right-click the Rapid City frame name and choose Reference Scale > Set Reference Scale from the context menu.

30➜ Zoom in again, and notice how this time the symbols increase in size as well.

30➜ Return to the previous extent.

> **TIP:** To set an exact reference scale such as 1:24,000, right-click the data frame name, choose Properties, click the General tab, and type a specific reference scale in the appropriate box.

Some annotation properties can be modified, such as placing a scale range to show it only at certain scales. It can also be removed from the map if no longer needed.

31➜ Open the Rapid City data frame properties and choose the Annotation Groups tab.

31➜ Click the Schools Anno group and choose Properties.

31➜ Set the option to not show the annotation when zoomed out beyond 1:125,000. Click OK.

> **TIP:** To remove annotation, click the Annotation Group name to select it, and click the Remove Group button.

Annotation is saved by saving the map document. This time save it under a new name to use later when learning to create layouts.

31➜ Click OK to close the Frame Properties window, and click File > Save As from the main menu bar.

31➜ Navigate to your mgisdata\MapDocuments folder for the location, and name the document "ex2_schools". (It is not necessary to enter the .mxd extension.) Click Save.

Labeling classes with different label styles

As a final example, we will create labels for subsets of features in a layer. We will label large towns with large text and smaller towns with small text (Fig. 2.23). This technique requires creating two classes of labels, setting an expression which defines each class (population < or > 10000), and setting the symbol properties for each class.

Fig. 2.23. Town labels

➔ Open the original *ex_2.mxd* document.

32➔ Zoom into the greater Sioux Falls area as shown. Sioux Falls is in southeastern South Dakota at the intersection of Interstates I90 and I29.

32➔ Open the Labels tab of the Layer Properties for Towns.

32➔ Check the Label Features box.

32➔ Choose the method to Define classes.

32➔ The current class is Default. Uncheck the Label Features in this class box. Since we will be creating new classes, we don't want the default class displayed.

32➔ Click Add to add a new class.

32➔ Name the new class "Cities" and click OK.

The new class is created, and now all of the buttons and properties in this menu refer to the Cities class. Next we need to define the members of the class using a selection expression and set the symbol properties (Fig. 2.24).

33➔ Click the SQL Query box and enter the expression "POP100" >= 10000.

33➔ Click Verify to check the expression for errors. (If an expression is incorrect then no labels may appear.) Click OK and OK.

33➔ Check the box to Label Features in the Class.

33➔ Make sure AREANAME is the label field.

33➔ Click the Symbol button and choose Country 1 as the symbol for Cities. Click OK.

Now the Cities label class is completely defined. (You could edit other properties for label placement and scale range, but we won't do so at this time.) You can always go back to edit them later. We now create the other class, Villages.

34➔ Click the Add button and name the new class "Villages". Click OK.

34➔ Repeat the steps for the Villages class just as for Cities, but for the labels choose the symbol Capital and make it Bold, and enter the SQL expression "POP100"< 10000. Verify it and click OK and OK.

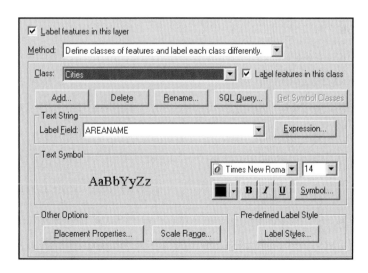

Fig. 2.24. Labeling properties window

34➔ When finished setting the properties for both classes, click OK to close the property sheet and enact the changes.

TIP: To go back and edit properties for a label class later, access the class by clicking the Class drop-down button in the property sheet and choose the class to edit.

Notice on the map now that Sioux Falls appears in large type and the other towns appear in small type. You can add as many label classes to a layer as desired. Like other dynamic labels, these class labels can be converted to annotation and will retain their different sizes in the conversion.

Using the Label Manager

A map with many labeled layers can be cumbersome to adjust if one must keep opening label properties for each layer. The Label Manager facilitates the control of labeled layers and makes it easier to modify the properties of multiple layers.

 35➔ Click the Label Manager button on the Labeling toolbar.

The Label Manager (Fig. 2.25) lists all the label classes from all the layers. Notice the two label classes you just added for the Towns layer: Cities and Villages. Imagine that you want to temporarily turn off the Villages labels.

> 35➔ Uncheck the Villages label class but leave Cities checked and click Apply. The village labels disappear.
>
> 35➔ Uncheck the Towns layer to turn off ALL labels and Apply. Then turn them back on.
>
> 35➔ Click the Villages label class in the Towns layer to highlight it. Notice the label properties appear in the window, similar to the way they appear in the Labels tab of the Layer Properties menu. You can use the Label Manager to edit the label properties of your label classes.
>
> 35➔ Change the Cities font color to green.
>
> 35➔ Click OK when you are done editing the labeling properties.

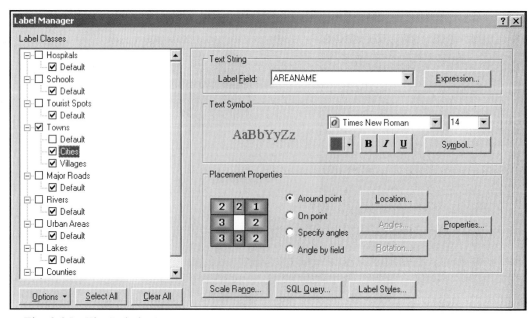

Fig. 2.25. The Label Manager

Using symbols and styles

Data layers may be represented with a variety of different symbols. The symbol properties of a layer control how it is displayed. To demonstrate, we will examine different ways of symbolizing town features. The towns of South Dakota in the map are currently drawn using a single symbol, a pink circle. First we will learn to change the characteristics of this symbol. Chapter 4 explains how to create and edit symbols.

36➜ Right-click the Towns layer and uncheck the Label Features option so the town labels are no longer being displayed.

36➜ Click the Full Extent button to view the entire state again.

36➜ Right-click the symbol for Towns in the Table of Contents. (Make sure to click the symbol, not the name.) Choose a different color from the palette.

36➜ Left-click the Towns symbol in the Table of Contents. This action launches the Symbol Selector, which has more options for changing the symbol.

36➜ Choose a different symbol from the palette. Experiment with changing its color and size. Then click OK to make the changes take effect.

➜ Examine the new symbols in the map. If they are unsatisfactory, open the Symbol Selector again and fiddle with the symbol some more.

ArcMap has many other symbol sets that come with it, and new symbols can be created as needed. Let's look at some of the other standard symbols available.

37➜ Click on the Towns symbol again to open the Symbol Selector.

37➜ Scroll down to the bottom of the symbol palette and note that the very last symbol is called Dam Lock.

37➜ Click on the More Symbols button.

The menu shows a long list of different **styles,** which are groups of related symbols used for a specific purpose. Notice that the first two, *yourname* and *ESRI,* are already checked, meaning that they are being displayed in the palette. You will now choose the Civic style and add it.

37➜ Click on the Civic symbol style, and notice the new symbols added to the bottom of the palette.

37➜ Scroll through this very long list, noticing that it is in alphabetical order.

37➜ Choose any symbol from the Civic list to display the towns.

Experiment with choosing different styles and symbols for line and polygon features.

➜ Click the symbol for the rivers and choose several different line styles from the default styles and/or add additional styles.

➜ Click the Counties layer symbol and choose several different shade styles.

This is the end of the Tutorial.

➜ Exit ArcMap. Do NOT save changes.

Exercises

Open the original ex_2.mxd map document and use it to answer the following questions.

TIP: To submit certain answers to your instructor, you will need to capture an image from the screen and place it in an answer document. Whenever a question has the statement **Capture** at the end, capture the answer by making sure the ArcMap window is active, and simultaneously press the Alt and PrintScrn keys on the keyboard. This action places the active window on the Clipboard, and it can then be pasted directly into a Word or Powerpoint document. You can also paste it first into the Paint program and cut and paste only the areas of the window that you want in your answer. If you need help, ask your instructor.

---

1. Find the Mount Rushmore Memorial in the Tourist Spots layer and zoom into it. Then answer the following questions.

   In which county is it?

   What is the name of the river running north of it?

   What is the name of the closest town?

   How far from the town is the memorial, in miles?

   What are the names of the two tourist spots to the immediate northwest of the memorial?

2. Create a road map of South Dakota with the following characteristics:

   ➢ Turn off the display for all layers except Roads and Counties.

   ➢ Make the outline of the counties medium grey so they are less prominent.

   ➢ Change the symbols of the roads to look like this legend. Use the "Expressway" symbol for the interstates and a thick single line for the others. Use a different color for each one.

CLASS	
≡	Rural Interstate
≡	Urban Interstate
—	Rural Principal Arterial
—	Urban Principal Arterial
—	Rural Minor Arterial

   Capture the map.

3. Locate Spink County and zoom into it so that the entire county is visible. Set up the map so that

   ➢ it displays the schools with purple school symbols, and

   ➢ the county outlines are thick black lines.

4. Next, label each of the towns in Spink County (do NOT include the ones outside Spink!). Use a boldface type, and be sure to adjust the labels for best readability. **Capture** your map.

5. Turn off the Schools layer. Create labels for the rivers in Spink County, using italic blue text that follows along the river. Also label the two lakes. **Capture** your map.

6. What is the driving distance across South Dakota along Interstate 90, in miles? In kilometers?

7. Set up labels for the towns in South Dakota such that cities with more than 25,000 people are labeled in 14 pt. bold text, towns with more than 5000 people are labeled with 10 pt. bold text, and hamlets with less than or equal to 5000 people are labeled with 8 pt. italic text. Set the scale range on the hamlets so they do not show when zoomed out beyond 1:2 million. **Capture** a map of the entire state showing the labels, and **capture** another one showing the southeast corner of the state with the hamlet labels.

TIP: To select the towns, set up a double expression that looks like "POP100" > 5000 AND "POP100" < 25000. The field name must be repeated for the expression to work correctly.

---

8. Add another data frame to the map that contains the lower 48 states with a light blue symbol, and rivers and lakes with a dark blue symbol. Set the coordinate system to USA Contiguous Equidistant Conic. Group the lakes and rivers as a single layer named Hydrology. Place South Dakota on the map in dark purple, and label it with a callout label.

9. Make a layout with the two frames. Put on the title "South Dakota" and put your name below it. Print the map to turn in.

10. Create a labeled school map of Sioux Falls, just as you did for Rapid City, using annotation to get the best label placement possible. Label it with your name in one corner. **Capture** your map.

Challenge Problem

Figure out how to make the frame borders in Exercise 9 look like the one shown, and change the backgrounds to a pale grey color to accent the map colors. Print the layout again.

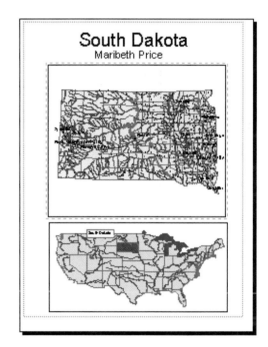

Skills Reference

Starting ArcMap

ArcMap can be started one of several ways.

1. Click the Start button on the Start Menu bar of the computer and navigate to Programs > ArcGIS > ArcMap, and click it to start. (The location of ArcGIS in the Start menu may differ. Consult your system administrator if you are having problems finding ArcMap.)

2. Find the ArcMap icon on the desktop and double-click it (Fig. 2.26).

3. Click the appropriate button in ArcCatalog (Fig. 2.26).

4. As ArcMap starts, it displays a dialog screen (Fig. 2.27). Choose to start with a new empty map, open a map template, or open an existing map.

5. To open existing maps, click "Browse for maps" to search for a map on the disk, or click one of the recently used maps shown in the list.

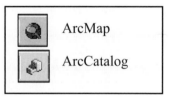

Fig. 2.26. Program icons

Fig. 2.27. ArcMap startup dialog

Managing toolbars

1. To move a toolbar to another location on the GUI, click and hold on its handle (a shadowed line on its left edge) and drag it to the desired position (Fig. 2.28). This is a docked toolbar.

Fig. 2.28. Handle

2. To move a toolbar off the GUI entirely, use the handle to drag it off the GUI. It then becomes a floating toolbar. It can be docked again by dragging it to a location on the GUI.

3. To open a toolbar that is not present, right-click on a toolbar or empty space and choose the desired toolbar from the alphabetical list.

4. To close a toolbar, click on the X box in its upper right corner, OR right-click on a toolbar or empty space and choose the toolbar to close from the list.

Adding data

1. Click the Add Data button on the main toolbar OR choose File > Add Data from the main menu bar.

2. Use the Add Data dialog box to navigate to the folder containing the data to open.

3. Click on the data layer to select it. Select multiple layers by holding down the Control or Shift key while clicking additional data sets.

4. Click Add.

TIP: If the disk or directory containing the data does not appear in the list, you probably need to add a connection to it using the Add Connection button, as described in Chapter 1.

Saving an ArcMap document

1. Click on the Save button, OR choose File > Save from the menu bar.

2. If the document has not previously been saved, a dialog box appears for the user to select a location and name for the map document. Navigate to the directory to save it to, and then enter a name for the document.

3. Choose File > Save As from the menu bar to save the map document under a new name, leaving any previously saved version unchanged. Specify the location and name as for step 2.

TIP: The names of map documents, spatial data sets, and folders should not contain spaces or unusual characters such as #, @, &, *, etc. Use the underscore character _ to create spaces in names if needed. Although ArcCatalog and ArcMap allow spaces in names, the spaces can create problems for certain advanced functions. Thus, it is wise to avoid them entirely.

Opening an ArcMap document

1. Click the Open button OR Choose File > Open from the main menu bar.

2. Use the dialog box to navigate to the directory containing the file.

3. Click on the file to select it and choose Open.

4. If the currently open map has unsaved changes, you will be prompted to save them before opening the new document. ArcMap can only open one map document at a time.

5. To create a new, empty map document, click the New Map File button, OR choose File > New from the main menu bar.

Controlling display with the Table of Contents

1. To **turn a layer on or off,** click inside the check box.

> **TIP:** If the check box is dimmed, then the current map scale is outside the display range set for the layer. Try zooming to another scale, or reset the scale range in the layer properties.

2. To **change the draw order,** click and hold on the layer name and drag it higher or lower in the list. A black bar indicates where the layer will be placed. Release the mouse button to finish.

> **TIP:** If the layers don't seem to go anywhere when you drag them, and the black line does not appear, check to make sure that the Display tab in the lower right corner of the Contents box is clicked. You cannot move layers when the Source tab is clicked.

3. **Hide the legend** for a data layer by clicking the minus sign in the box next to the layer. Click the plus sign to expand it again.

Grouping layers for display

Grouped layers can be turned on and off with one mouse click.

1. Click the first layer in the group to highlight it, and then click the other members of the group while holding down the Ctrl key. When you are done, all the group layers should be highlighted in blue.

2. Right-click a member of the group and choose Group (Fig. 2.29).

3. Click on the New Group Layer name once to highlight it, then click on it again to type in a name for it. Press Enter to finish.

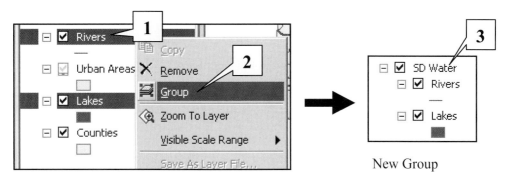

Fig. 2.29. Creating a group layer

Groups have joint properties such as setting scale ranges for the entire group at once. You can also add group members directly from the disk without needing to add them to the map document first. All layers in a group must be in the same data frame.

4. Right-click on the Group name and choose Properties from the context menu.

5. Click the General tab to set scale ranges for the group.

6. Click the Group tab and click Add to add more members to the group from the disk. Use Remove to remove members from the group. The Properties button is used to specify different properties for individual group members.

Using the Zoom/Pan tools and other zooming techniques

 1. To zoom into an area, click the Zoom In tool. Position the cursor to the upper left of the area to zoom to, then click and drag a box to encompass the area desired. Release the mouse button to finish.

 2. To zoom out, click the Zoom Out tool and click once at the center zoom point.

 3. To zoom in by a fixed amount while remaining centered within the current extent, click the Fixed Zoom In tool.

 4. To zoom out by a fixed amount while remaining centered within the current extent, click the Fixed Zoom Out tool.

 5. To move the map, click the Pan tool, click on the map, and drag it to the desired area.

 6. Click the Previous Extent button to return to the extent just before the current one.

 7. Click the Next Extent button to return to the extent after the current one (this button is dimmed unless the Previous Extent button has been clicked).

 8. Click the Full Extent button to view the full area of all the layers in the data frame.

9. To set a **bookmark,** zoom to the desired area and click View > Bookmarks > Create. Type in a name for the bookmark.

10. To zoom to an existing bookmark, click View > Bookmarks and choose the desired bookmark (Fig. 2.30).

11. To delete an existing bookmark, choose View > Bookmarks > Manage. Select the bookmark name by clicking on it, and click the Delete button.

12. To zoom to the reference scale, right-click the frame name and choose Zoom to Reference Scale from the menu.

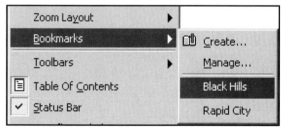

Fig. 2.30. Zooming to a bookmark

Identifying features

1. Click on the Identify tool.

2. Click on the feature to identify.

3. Clicking the identified feature in the dialog box will cause it to flash briefly (Fig. 2.31).

Sometimes you may have difficulty getting the right feature identified. Identify has some options to help: selecting the feature from the top layer, from the selectable layers, from all the layers, or from a requested layer.

4. In the Identify Results box, click the drop-down list at the top and select the option or layer desired.

5. If All Layers is the option selected, a list of features appears in the Results box, all of which were present at the point clicked. Click on a layer to view the attribute for that feature. Click on a different layer to see its attributes.

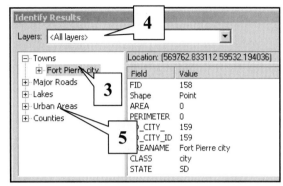

Fig. 2.31. Using Identify

Setting Map Tips

1. Right-click the name of the layer in the Table of Contents and choose Properties from the context menu.

2. Click the Fields tab and set the Primary display field for the text to appear in the tip.

3. Click the Display tab and check the box to Show Map Tips.

TIP: If the Show Map Tips box is dimmed, then you must create a spatial index for the layer using ArcCatalog.

Measuring features

1. Right-click the data frame name and choose Properties from the context menu.

2. Click the General tab, and set the Display units to the desired option (miles, kilometers, etc.). Click OK.

3. Click the Measure tool. Place the cursor at the starting location on the map and click.

4. Place the cursor at the next location and click again. Keep adding points as long as desired. The lengths will be displayed in the lower left corner of the GUI. Both the total length and the length of the last segment are shown.

5. Finish measuring by double-clicking on the last location.

Finding features by searching attributes

1. Click the Find tool and type the text to search for in the first box (Fig. 2.32).

2. Choose the layer or set of layers to be searched.

3. Check the box for an exact match or for features that are only similar to or contain the string.

4. Choose to search all fields, a specific field, or the primary display field.

5. Click Find to begin the search. Records that match the search will be listed at the bottom of the Find box.

6. Right-click one of the found records to display a context menu with options for flashing, zooming to, identifying, bookmarking, or selecting the found feature (Fig. 2.33).

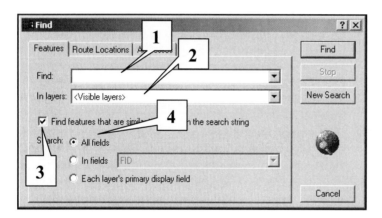

Fig. 2.32. Finding features

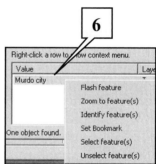

Fig. 2.33. The Find list

Setting data frame and layer properties

1. Right-click the frame or layer name and choose Properties from the context menu.

2. Click the tab containing the properties to set.

3. Enter the appropriate information and settings.

Defining symbols for a layer

You can set symbols a number of ways in ArcMap. Here are a few of the most common ways—all of these assume that all features in the layer are being drawn with the same symbol.

Changing the color of the current symbol

1. Right-click on the layer symbol in the Table of Contents, and choose a color from the palette that appears.

Changing properties of the current symbol

1. Click on the layer symbol in the Table of Contents to open the Symbol Selector window (Fig. 2.34).

2. Choose a symbol from the scroll box.

3. Modify the symbol's color, size, thickness, outline, or other attributes by setting the options provided. Make additional changes using the Properties button.

4. To load additional symbols in the scroll box, click the More Symbols button and choose from the list of categories.

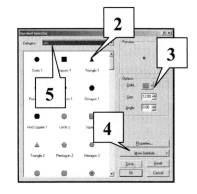

Fig. 2.34. Symbol Selector

5. Use the drop-down Category box to change which categories are currently visible in the box.

6. Click OK when finished modifying the symbol.

Labeling features interactively

1. Look for one of the labeling tools on the drawing toolbar. Click on the black arrow for a drop-down menu to choose one of the tools.

Adding text to the map

1. Click the Add Text tool.

2. Click the desired location on the map and enter the text. Press Enter when finished.

3. Newly entered text is always selected, as shown by the dashed blue box around it. At this point you can change its font, size, or color using the menus on the Drawing toolbar. Or use the cursor to click and drag it to a new location.

TIP: To delete text, click the black arrow on the Drawing toolbar, click the text to select it, and press the Delete key. You can also use the black arrow to draw a box around many text boxes, or hold down the Ctrl key to select multiple text boxes for deletion.

TIP: To delete all graphic elements including text, choose Edit > Select All Elements from the main menu bar, and press the Delete key. Be careful! This option also deletes any annotation saved in the map document!

Labeling a feature with its attribute

1. If needed, right-click the layer name, choose Properties, and click the Fields tab. Set the Primary Display Field to the desired attribute to appear in the label. Click OK.

2. Choose the Label tool from the labeling drop-down button. The Label Tool Options window will appear.

3. Choose whether to find the best placement (e.g. the center of a feature) or to place the label where you click. Also choose whether to use the symbol properties already set for the layer or to choose a new symbol style now. Click the X box to close the window.

4. Click on the feature to be labeled. ArcMap will label the topmost layer clicked, if more than one layer is present.

TIP: The text symbol is set for the layer using the Layer Properties. Modify the default symbols by right-clicking the layer name, choosing Properties, clicking the Labels tab, and setting the text symbol.

Splining text from an attribute along a line

Splining text makes it follow along a linear feature such as a road or stream.

1. Set the font size and style options desired using the Drawing toolbar.

2. Choose the Spline Text tool from the labeling drop-down button.

3. Click along the feature to define the line along which the text will appear. Double-click to end the line.

4. Type the text in the box and press Enter.

5. Use the Drawing toolbar to modify the position or font characteristics as needed.

Adding a callout label

A callout places text in a box with a pointer to the feature of interest.

1. Choose the Callout Text box tool from the labeling drop-down button.

2. Click on the feature to be labeled, hold, and drag to define the direction and length of the callout pointer.

3. Enter the text into the box and press Enter.

4. Click and drag on the text box to change its location, if desired. Click and drag on the blue dot to change the location of the pointer. Use the tools on the Drawing toolbar to modify the callout's font, style, size, etc.

Creating wrapped text boxes

1. Choose one of the wrapped text tools: the New Polygon text, the New Rectangle text, or the New Circle text.

2. For the circle or rectangle text tool, click and drag to draw the circle/box, releasing the mouse when it reaches the desired size and shape. For the polygon tool, click on each vertex to define the desired shape, and double-click when finished.

3. Click in the empty shape to open the text Properties box (Fig. 2.35). Type in the text to be displayed. Do not use the Enter key unless you wish to enforce a new line within the text.

4. Set the symbol, spacing, or other options if needed.

5. Use the other tabs to change the margins, columns, frame border, size, position, or area background of the text box as needed.

6. Click OK to place the text.

7. To modify the text later, double-click to open its properties box.

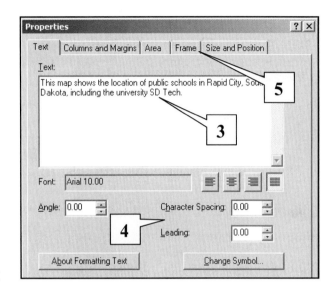

Fig. 2.35. Wrapped text properties window

Setting the reference scale

1. Use the Pan/Zoom or Bookmark tools to zoom to the desired scale.

2. Right-click the data frame name in the Table of Contents and choose Reference Scale > Set Reference Scale.

> **TIP:** To set an exact scale value such as 1:24,000, right-click the data frame name, open the data frame Properties, click the General tab, and type a specific reference scale in the appropriate box.

3. To remove the reference scale, right-click the data frame name in the Table of Contents and choose Reference Scale > Clear Reference Scale.

4. To zoom to the reference scale, right-click the data frame name and choose Reference Scale > Zoom to Reference Scale.

> **TIP:** Annotation, once created, retains its original reference scale set when it was created, even if the reference scale of the data frame is changed later.

Labeling features using dynamic labels

1. Right-click the layer name in the Table of Contents and choose Properties from the menu.

2. Click the Labels tab.

3. Check the Label Features box (Fig. 2.36).

4. Make sure the method is set to label all the features the same way.

5. Choose the Label Field. Click the Expression button to enter a VBA (Visual Basic for Applications) script.

6. Edit the font settings, or select a predefined text symbol by clicking the Symbol button and choosing a predefined symbol style.

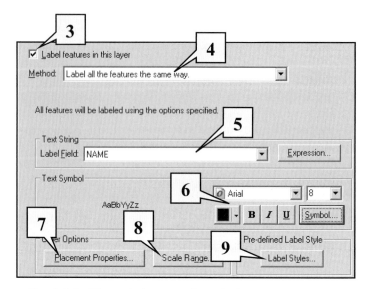

Fig. 2.36. The Label Properties window settings

7. For detailed control of label placement, click the Placement Properties button.

8. Set the scale range if desired, by using the label's scale range, or typing in new values. If the map scale is outside the specified range, the labels will not be drawn.

9. Select a label style, if desired. A label style includes BOTH a text symbol and predefined label placement options.

10. Click OK to place the labels.

TIP: Turn labels on and off for a layer by right-clicking the layer name in the Table of Contents and choosing the Label Features option. If the menu choice is checked, the labels are on, and choosing it will turn them off. If it is unchecked, choosing it will turn them on.

Converting dynamic labels to map annotation

These directions show how to create annotation stored as text graphics in a map. Consult the Help files for information on creating annotation as a geodatabase feature class.

1. Use the Layer Properties to create dynamic labels for all the desired layers. Take care in setting the properties, weights, and so on, because these will control the labels that appear.

2. Annotation will be created for all layers in the data frame with dynamic labels turned on. Turn off any dynamic labels for layers not to be converted.

3. Right-click the data frame name and choose Convert Labels to Annotation. The dialog box will appear (Fig. 2.37).

4. Examine which layers will be converted to ensure they are the desired ones.

5. Choose to create annotation in the map.

6. Check the box to Convert unplaced labels if you want to place overlapping labels interactively.

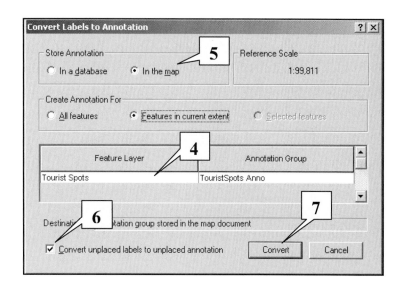

Fig. 2.37. Creating and placing annotation

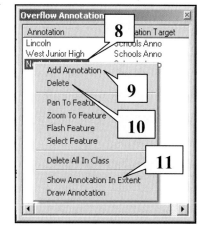

7. Click Convert to create the annotation.

Unplaced labels will be placed in an overflow window if you checked that option. The next step is to place these labels interactively.

8. Right-click a label in the overflow list and choose a method to locate it (flash, pan, zoom, etc.).

9. Choose Add Annotation to add the label to the map. Click and drag it to adjust its location if necessary.

10. If you decide not to place the label, delete it from the list using Delete.

11. Use Show Annotation In Extent to list only the labels in the current view. Place all of these before zooming to another location—it saves time.

Chapter 3. Coordinate Systems and Map Projections

Mastering the Concepts

Objectives

➢ Learning why and how coordinate systems are used in GIS

➢ Understanding the difference between geographic coordinates and projected coordinates

➢ Getting familiar with different types of map projections

➢ Understanding distortions caused by map projections

➢ Managing projections for images

Concepts

About map projections and GIS

A successful GIS system depends in large part on using projections correctly, and a person's skill in managing and converting projections can dictate the value of a database. Unfortunately, projections can also be somewhat daunting to those encountering them for the first time. You will probably need to review this chapter several times to become comfortable with these concepts. In the end, one learns best about projections by working with them.

In order to display and analyze maps on the screen, a GIS system uses **coordinate pairs** which specify the location and shape of a particular feature. A point is described by a single *x-y* coordinate pair located in space; a line is a series of *x-y* coordinates. A coordinate system is an agreed-upon range of coordinates used to portray the features.

Imagine that a surveyor is mapping an industrial site, using typical surveying instruments that measure angles, directions, and distances. He probably begins at a fence corner at the lower right of the complex and assigns that point the coordinates 0,0. This is the **origin** of the coordinate system (Fig. 3.1). The surveyor then determines that the southwest corner of building A is 175 meters east and 200 meters north of the origin, and he assigns that corner the coordinates (175, 200). The surveyor uses meters to record the distances and determine the coordinates, thereby

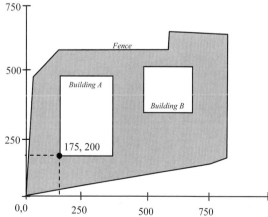

Fig. 3.1. An arbitrary coordinate system used for surveying a site

105

establishing meters as the **map units.** The combination of origin and map units becomes the **coordinate space** of this map. In this case the coordinate space is arbitrary, chosen for convenience.

Now the surveyor can use this coordinate space to tell other people to go to a particular location. He might tell his assistant to walk to (225, 398) and place a flag to mark the location for a new monitoring well. The assistant can do this, as long as he knows the coordinate space in use. Without knowing the coordinate space, however, the directions would be useless.

Now imagine that the surveyor's colleague is surveying another site across town. She also chose the corner of her site as the origin, but she is measuring her distances in feet. The two surveyors take their drawings back to the office and decide to create a map showing the location of both sites within the city, so they each transfer the coordinates into a GIS. When they plot the sites, however, the two different sites show up in exactly the same spot, even though in real life they are several miles apart. This problem occurs because both surveyors used an arbitrary (0,0) origin unrelated to the other site. Moreover, the woman's buildings look about three times as large as the man's buildings because she is measuring in feet instead of meters! To plot these together the two surveyors would need to define a single common origin and convert both sites to the same map units.

To avoid problems such as this, geographers and map creators usually employ standard **coordinate systems** whose characteristics are agreed upon beforehand. The surveyors could have used a global positioning system (GPS) instrument to establish the *x-y* coordinates of their origins in a common coordinate system known as Universal Transverse Mercator (UTM), and then entered their surveyed points relative to that base point. They then would have been spared the extra effort of converting the drawings into the same system.

Geographic coordinate systems

The most commonly used coordinate system on earth is based on spherical units, because the earth approximates a sphere (Fig. 3.2). The units are called **degrees.** The earth is divided into 360 degrees around the equator, which are called degrees of **longitude.** Usually longitude is expressed as degrees east or west of the **Prime Meridian,** which is the line along which the longitude equals 0 and passes through Greenwich, England. Longitudes thus fall in the range -180 to +180. **Latitudes** measure distance from the earth's equator, which has latitude equal to 0,

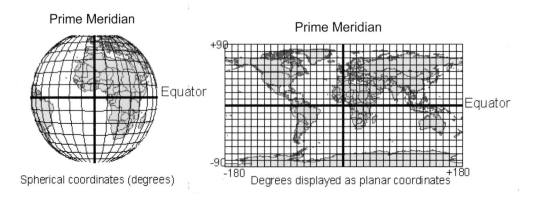

Fig. 3.2. The geographic coordinate system using degrees latitude and longitude

and latitudes range from -90 at the south pole to +90 at the north pole. This system of measurement is called a **geographic coordinate system,** or sometimes a **GCS.**

Although strictly speaking a GCS records locations on a sphere, in a GIS the coordinates are often drawn as though they represented a planar coordinate system on a piece of paper, as a matter of convenience (Fig. 3.2). However, distortions are introduced by this practice. Notice on the sphere that a latitude line at the equator has a distance equal to the circumference of the earth, a latitude line at the poles has a distance of zero, and that latitude lines in between have intermediate values. Yet in the planar display, these lines are all represented as the same length as the equator. Thus the further north or south an area is from the equator, the more it gets distorted.

Spheroids and datums

A geographic coordinate system is not quite the perfect system for measuring locations on the earth, however, because the earth is not a perfect sphere. It bulges at the equator due to its rotation. The bulge can be modeled mathematically using a **spheroid,** an oblong sphere with a major and minor axis. (The bulge in Figure 3.3 is highly exaggerated.) Latitude-longitude locations defined on a spheroid are thus more accurate than those defined on a sphere. The spheroid becomes a part of the GCS definition. (Many people use the term **ellipsoid** rather than spheroid, but since ESRI uses spheroid in the windows and manuals, we stick to that terminology here.)

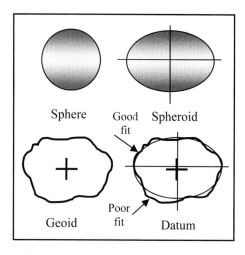

Many different spheroids have been used in the past because estimates of the earth's shape have varied over time. Clarke 1886 is a very common spheroid used in North America, but it is gradually being replaced by satellite-measured spheroids, such as WGS84, which are more accurate. Thus, when one uses a GCS to record locations, it is important to specify which spheroid is being used. The same location might have a slightly different latitude and longitude in GCS systems based on different spheroids, and a misalignment of features on the map could occur.

Fig. 3.3. Relationship between the spheroid, geoid, and datum

Even when a spheroid is used, however, the map accuracy will still not be perfect because of topography, as shown in Figure 3.3 (topography exaggerated). The actual shape of the earth is best represented by the **geoid,** which is the measurement of the earth's radius with respect to a datum which can be defined either locally or globally. This datum is often sea level, but sea level itself varies from place to place according to differences in gravitational force which arise from rotational, topographic, and compositional differences in the earth's mantle. The geoid can only be approximated mathematically because of local anomalies at various scales. To achieve a more accurate map, the spheroid is moved until the geoid at the area being mapped (arrow) corresponds best with the spheroid surface, as shown. The combination of the spheroid and the location where its center is placed is called a **datum.** The purpose of the datum is to create the best possible fit between the actual earth geoid and the spheroid used for mapping. In this example, the center of the spheroid is slightly offset from the center of the geoid in order to achieve the best fit. Some datums have this offset, others do not. A datum in which these two centers coincide is called an earth-centered datum.

Notice, however, that the improved fit at the arrow means that other areas of the globe, such as the southern hemisphere in this example, have a worse fit. Thus the best datum will change depending on one's location at the earth's surface. One should not use a North American Datum to map Europe, and vice versa. Also, for world maps, one would use a best-fit datum for the whole earth rather than a specialized datum for one part.

Several datums are commonly used in North America. The North American Datum of 1927, often abbreviated NAD 1927 or even NAD27, is based on the Clarke 1866 spheroid. The North American Datum of 1983, or NAD 1983, is based on the GRS_1980 spheroid. Another datum commonly used for satellite data is the World Geodetic Survey of 1984, or WGS 1984. Because NAD 1983 is relatively recent, one finds many maps that still use NAD27. However, the coordinates of a particular location can differ by up to several hundred meters between NAD27 and NAD83. Thus, one usually converts all the layers within a particular database into a common datum to avoid alignment problems.

Figure 3.4 shows a portion of Rapid City, South Dakota. Notice that the roads are displaced about 215 meters to the south relative to the image because the image coordinate system is based on NAD83, whereas the roads are in NAD27.

Datum conversions are not simple mathematical formulas but require localized estimates and fitting, and they are not exact. Several different methods may be used, and some methods are unique to particular datums and cannot be used for others. Thus, it is not desirable to convert map layers back and forth repeatedly from datum to datum. Usually one picks a datum to be used for a particular project and sticks with it, converting layers as needed. The choice of datum is usually a matter of convenience. If most layers to be placed in a database already use NAD27, it is easier to stick with the older datum. Alternatively, an organization may decide that all layers must use the NAD83 datum as a standard policy, in which case non-matching

Fig. 3.4. Offset of roads due to different datums

layers would require conversion. If highest accuracy is a concern, then most people choose NAD83. Most GPS and other data collection systems can be set to work with a variety of datums.

Map projections

Because the shape of the earth is close to a sphere, the most accurate way to map it is to use a spheroid. However, globes and spheroids are inconvenient objects for putting in a report or mounting on a wall. Therefore, a major effort of cartography includes converting the locations of features in three dimensions on a globe to locations in two dimensions on a piece of paper. This process is called **projection.**

You can imagine the projection process as a clear globe with the countries of the earth printed in black. If you place a light bulb at the center of the earth and wrap a piece of paper around the equator in the shape of a cylinder (Fig. 3.5), the country boundaries will be cast as shadows on the paper. Tracing them gives you a map which you can then unroll and lay flat. In cartography, however, projections are usually employed as mathematical formulas that convert degrees of latitude and longitude into planar *x-y* coordinates on the paper. During this conversion the degrees are usually converted into feet or meters, which become the new map units.

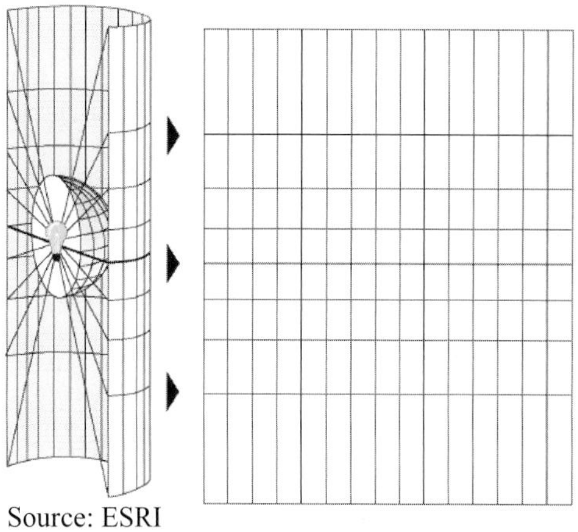

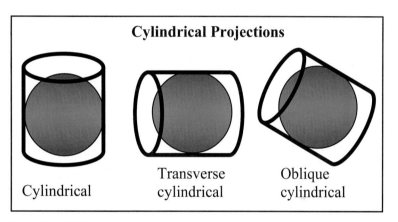

Source: ESRI

Fig. 3.5. Projecting a map from 3D to planar coordinates

Projections are grouped into three major classes, depending on how the paper is situated relative to the globe. A **cylindrical projection** uses a paper cylinder which lays tangent to (touches) the earth at the equator (Fig. 3.6). Rotating the paper sideways and making it tangent to the earth along a line of longitude produces a **transverse cylindrical projection.** An **oblique cylindrical projection** places the tangent at an angle. In each case, distortion is absent along the tangent and increases away from it.

Cylindrical Projections

Cylindrical

Transverse cylindrical

Oblique cylindrical

Fig. 3.6. Cylindrical projections

A **conic projection** is based on setting a paper cone over the sphere of the earth (Fig. 3.7). If the cone intersects the globe along one line of latitude, at which the paper is tangent to the globe, then the projection is a **tangent conic** projection. The line of tangency is called the standard parallel. One can also place the cone *through* the sphere so that it touches in two places—a **secant conic projection.** Such projections have two standard parallels.

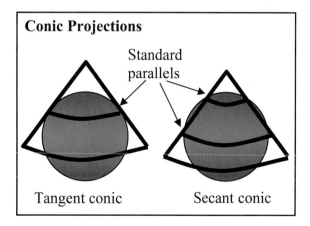

Conic Projections

Standard parallels

Tangent conic

Secant conic

Fig. 3.7. Conic projections use a cone of paper around the globe.

One can also place the paper flat against the globe at a single point, producing an **orthographic** projection (or **stereographic** projection) (Fig. 3.8). Typically these projections are used for displaying the earth's poles, and for that reason they are sometimes called polar projections.

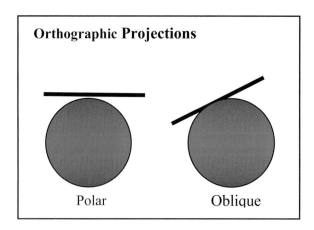

Fig. 3.8. Orthographic projections use a flat piece of paper against the globe.

Map projections typically have certain properties that are described using map **parameters** (Fig. 3.9). One parameter is the line of longitude which constitutes the **central meridian,** or $x = 0$ line on the map. The central meridian is chosen based on its position near the center of the map being created. A map of England would probably have a central meridian close to 0 degrees. A map of the United States would probably have a central meridian of about −100 degrees because that meridian is close to the center of the continental U.S. The **latitude of origin,** also sometimes called the **reference latitude,** is the $y = 0$ line. The equator is often used as the reference latitude. A conic map projection will also have one or more **standard parallels** in its list of parameters, the latitudes which define the lines of tangency or secancy for the cone. Maps may also have a **false northing** or **false easting** (Fig. 3.9). These are arbitrary numbers added to the x-y coordinates in order to translate the map to a new location in coordinate space. Most often the false easting and northing are used to ensure that all the coordinates of the map are positive.

Distortion in map projections

Although map projections conveniently convert spherical coordinates into planar coordinates, they inevitably introduce distortions in the map. Imagine a beach ball with countries drawn on it, which is then cut in half. You cannot lay the pieces flat without stretching, shrinking, or cutting different areas. Four properties of map features may be distorted by projections: **area, distance, shape,** and **direction.** Some map projections reduce or eliminate certain distortions at the expense of the others. For example, an equidistant map portrays distances correctly with minimal distortion of area, but shape and direction are distorted. A conformal map preserves shape at the expense of the other three.

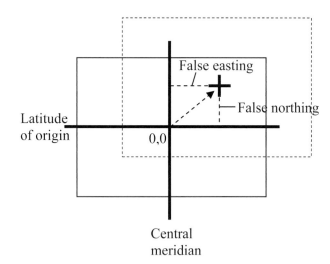

Fig. 3.9. Some parameters of map projections

Every map projection has its own distortions, but a few generalizations are useful. Cylindrical maps such as Mercator tend to preserve direction and shape but distort distance and area. Conic projections such as Lambert Conformal Conic tend to preserve distance and area at the expense of shape and direction. Orthographic projections tend to preserve distance and area but not direction or shape. The most commonly used projections and a summary of their properties can be found on the inside front cover of this text for quick reference.

The choice of projection depends on the task at hand. If the project includes measuring distances and areas, it is important to choose equidistant or equal-area map projections. If the maps are used for navigating by compass, maps which correctly display direction are important. ArcMap supports a wide variety of different map projections.

TIP: The ArcGIS online documentation includes a good reference book on map projections and coordinate systems, titled *Understanding Map Projections*. The ArcMap Help section also has some good material about projections.

A note on terminology

The preceding discussion uses some potentially confusing terms, because they have been applied in different ways by different people. For example, ESRI may use "coordinate system" to refer to a coordinate space with a given origin and defined units (such as a geographic coordinate system). ESRI now also uses the term to indicate a complete definition of a data layer's coordinate reference, including the GCS, the spheroid, the datum, and the projection if present. Although usually the meaning is clear in context, care must be taken in using these terms. The "projection" is understood as a specific mathematical transformation from a spherical to a planar coordinate system.

However, this definition conflicts with a common usage of the term "projection." Many people, and previous versions of ArcInfo, use "projection" to describe a coordinate system in the second sense above, that is, the complete definition of a layer's spatial reference. Many people still use the term in this way, even though it is somewhat misleading. A layer stored in spherical coordinates with latitude-longitude degrees was said, in the old terminology, to have a geographic "projection" despite the fact that no actual projection (mathematical conversion from 3D to 2D) had been applied to it. Users need to be aware of both of these definitions and be able to ascertain the correct meaning in a particular instance, which is usually clear in context.

Common projection systems

Two projections deserve special mention as they are commonly used in the United States: Universal Transverse Mercator (UTM) and State Plane. Neither of these is a single projection. Instead the terms UTM and State Plane refer to families of projections with similar characteristics. Users can choose the most appropriate member of the family based on the location of the map.

The UTM system

The UTM system is based on, as the name suggests, a transverse cylindrical projection. The line of tangency of the cylinder to the sphere falls along a line of longitude which becomes the central meridian (Fig. 3.10a). A zone is defined to include three degrees on each side of the central meridian between the 80S and 84N latitude lines. Since distortion is zero along the line of tangency and increases with distance from it, the distortion inside the zone is minimal. To map different locations, the cylinder is rotated about the sphere in 6 degree intervals, producing a total of 60 zones around the world. Each zone is indicated by a number. Figure 3.10b shows the UTM zones of North America.

Each UTM zone has its own central meridian and splits at the equator into a north and south component. In order to eliminate negative x-y coordinates, southern zones have a false northing of 10 million meters applied, and both northern and southern zones have a false easting of 500,000 meters. UTM is very convenient because users need only remember the zone number. Once the zone is specified, the other map parameters are set according to the standards for that zone.

UTM works best for small areas that lie completely within a single zone, since distortion increases away from the zone center. The zone used for a particular map depends on the location being mapped. The UTM system is used extensively for U.S. Geological Survey 1:100,000 and 1:24,000 scale publications, for county maps, for states within single zones, and for many project maps covering small areas.

If the region of interest lies on the boundary between two zones, the user should choose the zone containing the larger area. For example, South Dakota is split by a zone boundary. The western quarter is in UTM Zone 13, and the right three-quarters lies in UTM Zone 14 (Fig. 3.10b). UTM maps of the entire state would be better in UTM 14. However, the size and position of the state relative to the zones indicates that UTM is not the ideal projection for statewide maps. California, Oregon, and Washington are in the same fix, as are other states.

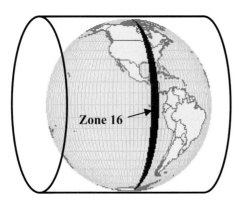

Fig. 3.10 (a) The UTM projection system has 60 north-south zones each 6 degrees wide.

Fig. 3.10 (b) UTM zones of the conterminous U.S.

The State Plane Coordinate System

The State Plane Coordinate System (SPCS) includes an assortment of projections. It was developed in the 1930s by the U.S. Coast and Geodetic Survey for large-scale mapping in the U.S. To maintain the desired level of accuracy, large states like California and Texas are broken into zones with different parameters used to minimize distortion in each zone. Like the UTM system, a State Plane coordinate system is uniquely identified by its name and zone.

The State Plane system uses three different projections: Lambert Conformal Conic, Transverse Mercator, and Oblique Mercator. North-south oriented states or zones use a Transverse Mercator projection, as it provides the least distortion. East-west oriented states or zones generally use a

Lambert Conformal Conic projection. A state may use more than one projection depending on the size and shape of its zones. Alaska, for example, has 10 zones and uses all three projections. Zones are identified formally using a FIPS code, but they are often informally referred to by zone number, e.g., California SPCS 27 Zone 7 (FIPS 0407).

The original SPCS was based on NAD 1927 and used units of feet. Improvements in surveying necessitated the revision of the SPCS in 1983. At that time it adopted the NAD 1983 datum and metric units. Certain zones were also revised and in some cases eliminated. Even though the commonly used units changed from feet to meters, users may specify either unit for both versions of SPCS.

Additional information about both the State Plane and UTM systems can be found in the ArcMap Help (index headings State Plane Coordinate System and UTM).

National Grids

Many countries possess a coordinate system set up, like State Plane, to minimize distortion of national maps. Examples include the New Zealand National Grid, or Germany Zone 5, or the Qatar National Grid. A variety of different projections, datums, spheroids, etc. are employed. ArcMap maintains a collection of these coordinate system definitions for use with international data.

Map projections in ArcMap

One benefit of using ArcMap includes being able to display data in a variety of projections, making it easy to choose the best projection for a particular task. However, it is important to understand how ArcMap works with projections. Every data set containing map features has a coordinate system which consists of a geographic coordinate system and a datum. Data may also be projected or unprojected.

Coordinate system definitions

If **unprojected,** the coordinates are stored in the GCS coordinates, decimal degrees of latitude and longitude. The layer's coordinate system definition would then include simply a GCS and a datum, such as GCS_North_American_1983. The stored x-y coordinates of a data set on the disk would then fall in the range -180 to $+180$ for x and -90 to $+90$ for y.

Projected data have had a map projection applied to them, converting the lat-lon values to feet or meters in a planar coordinate system. Typically the x-y values in a projected coordinate system are very large, near a million or more. The layer's coordinate system definition then includes the original GCS and datum, plus a description of the applied projection and its parameters, such as the longitude of the center of the projection. A projected coordinate system definition might read as Continental_US_Albers_Equal_Area_NAD1983. This system would have slightly different coordinates from a Continental_US_Albers_Equal_Area_NAD1927 system, and the two would be slightly offset from each other. Thus, even projected data should share a common datum for best results.

The **extent** of the data layer is the range of minimum to maximum x-y coordinates of the features in the layer. You can view the extent of a data layer in ArcCatalog; and this information can help you decide if the layer is projected or unprojected. Values between $+180$ and -180 would indicate an unprojected coordinate system with map units in degrees, whereas values in the hundred thousand to million range indicate a projected coordinate system with units of feet or meters. You can also view the extent in ArcMap.

Every data set should have a **coordinate system definition,** which records the characteristics of the coordinate system and any projection that has been applied to the data. This definition serves two purposes: it documents this information about the coordinate system for the users, and it also helps ArcMap and ArcCatalog to properly display and manage the data. Making sure that all data sets have correct coordinate system definitions is a critical step in building a GIS database.

ArcGIS helps manage projections by storing the coordinate system information along with the feature data itself. Older versions of shapefiles could not do this. Instead, people tried to record that information in documentation distributed with the data. Commonly, though, this information might be lost or incorrect. It is not an uncommon task, then, to have to determine an unknown coordinate system and assign it to the data, or to correct an incorrectly labeled data set.

On-the-fly reprojection

ArcMap is able to convert coordinate systems while it is displaying data. A data frame may be assigned any desired coordinate system, and then ArcMap will reproject all data in the frame to that coordinate system. For example, your map of the United States stored in decimal degrees of GCS could be displayed in a Lambert Conformal Conic projection by setting the data frame coordinate system. All other maps in the frame would also be converted to Lambert.

In order for this to happen, ArcMap must know the coordinate system of the data as it is stored on the disk. If it does not have this information (i.e., if the data set coordinate system definition is Unknown), then ArcMap will issue a warning message, and it will simply display the data in whatever units it has. Sometimes it will fortuitously match the other data in the frame, but other times it will appear totally out of place. If it matches, then you will know that the Unknown layer has the same, or at least a very similar, coordinate system to the data frame, and you can then set its coordinate system definition.

Converting different datums is more complex. A datum transformation cannot be performed using a set of equations like projections can, but instead relies on a variety of fitting techniques, which are adapted based on the two datums being converted. When ArcMap encounters a datum that does not match the datum of the data frame, it converts it to match the data frame using a default conversion method. However, the default method may not be the best one for the two datums in question, so ArcMap issues a warning when it converts datums. Most of the time you'll be able to simply proceed with the default conversion, but if high positional accuracy is required in a project, it is better to convert all the layers to the same datum using the best transformation method.

By default, the coordinate system of the data frame is set to the coordinate system of the first data set that is added to the frame. The data frame coordinate system can be modified at any time by the user.

Managing coordinate systems

ArcGIS offers a variety of tools for displaying data and managing coordinate systems.

> ➢ You can examine a layer's extent and coordinate system information.

> You can define the coordinate system for a layer using its Properties in ArcCatalog, or by using the Define Projection tool in ArcToolbox.

> You can display a layer in a different projection from the one it is stored in, on a temporary basis in ArcMap, by setting the coordinate system of the data frame.

> You can permanently convert a layer from one coordinate system to another using the Project tool in ArcToolbox.

Projecting data

Sometimes it is inconvenient to work with data sets in many different projections. For one thing, reprojecting the data on the fly takes time, and it slows the drawing speed of the map. Also, when calculating attribute fields containing the lengths of arcs or the areas of polygons, the results are based on the data set map units. If these are degrees, the lengths will be in degrees rather than the more useful meters or feet. Many organizations establish an official projection for their area of interest and use it consistently. Outside data from different coordinate systems are always transformed to the official one. Many circumstances make it best to project data from one coordinate system to another. Fortunately, this task is quite easy. ArcToolbox provides several projection tools.

Fig. 3.11. The Projection tools are found in ArcToolbox.

The projection tools are found under the Data Management section of the Toolbox (Fig. 3.11). The Define Projection tool can be used to define projections, as we have already learned to do in the ArcCatalog properties menus. The Project tool allows the user to actually change the map projection of a data set from one coordinate system to another.

The Project tool works with any ArcGIS data format, although projecting coverages and grids requires an ArcEditor or ArcInfo license. A different tool is used to project rasters. In general, projecting rasters is a more complex and computer intensive process than projecting vector features. The Define Projection tool works on both feature and raster data sets.

WARNING: Many people get confused about the functions of the Define Projection tool and the Project tool. This confusion can ruin data sets, so **pay attention** to the next few paragraphs!

The **Define Projection tool** (or changing the coordinate system in ArcCatalog) only changes the coordinate system definition for a layer—that is, it labels a data set with a coordinate system. It does not change the *x-y* coordinates stored in the file. The Define Projection tool is used ONLY when you need to initially set the projection definition of an unknown layer or to change the definition because the existing one is not correct (it does not match the coordinates actually stored in the file). For example, you might have a data layer that appears greatly offset from the rest of the data layers. You look at its definition and units in ArcCatalog and discover that the coordinate system is listed as UTM, but the *actual* coordinates are really in decimal degrees. Therefore the UTM definition cannot be correct and must be changed to GCS using the Define Projection tool.

The **Project tool** acts on the *x-y* coordinates of a layer and converts them to a different coordinate system. Both the coordinates and the projection definition are changed. The Project tool always produces a new layer as its output; the original data set is unchanged. You use the Project tool if you have a layer in one coordinate system and you want to convert it to a different one. The Project tool should be used only on layers with correct coordinate system definitions.

Many beginners make the mistake of defining the coordinate system when they want to convert a layer from one coordinate system to another, instead of using the Project tool. For example, you might have a shapefile of cities with *x-y* coordinates stored in latitude and longitude, and you want to convert it to a UTM projection. Although the software will allow you to use the Define Projection tool to label the cities with a UTM coordinate system, it will NOT run the calculation to convert the lat-long coordinates in the file to UTM. You will simply have a GCS data set wrongly labeled as UTM. When the data are brought into ArcMap, it treats the layer as though it is in UTM, when in fact it is NOT. The layer is thus incorrectly projected on the fly, and it will appear far away from the rest of the data in the frame. If such a data layer with an incorrect definition is projected using the Project tool, the error is compounded. It may be impossible to go back later and find out what happened.

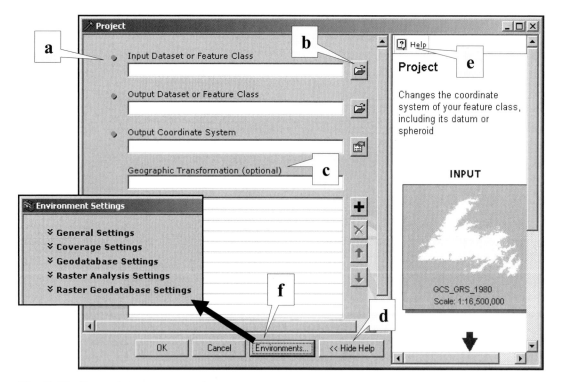

Fig. 3.12. Preparing to run a tool

Using ArcToolbox

In this chapter you will use ArcToolbox (Fig. 3.12) for the first time. This section contains some general information about using tools.

ArcToolbox may be opened and used either in ArcCatalog or ArcMap. When a tool is opened, required inputs are marked with a green dot (Fig. 3.12a). To enter data sets as inputs or outputs, use the Browse button next to the appropriate box (b). If using the tool in ArcMap, you can also select layers in the map from a drop-down list as inputs. Optional inputs are marked (c), and can usually be left to their defaults. Some tools have input parameters that extend beyond the visible pane, and the scroll bar must be used to view and enter them.

To get more information about the tool or an input, click the Show/Hide Help button (d) for a brief description of the tool's function. For more detailed help, especially on the input/output data, click the small Help icon at the top of the Help window (e).

TIP: When using a tool for the first time, it is strongly recommended that you read the detailed help information. It may contain information critical to making the tool work properly.

The operation of tools may be affected by the Environment settings for the geoprocessing environment. For example, the user can set a default output directory or a default *x-y* extent for the output. These settings can be accessed by clicking on the Environments button at the bottom of the tool (f). Beginners are advised to leave these settings alone until they learn more about them.

The input pane for a tool posts symbols to help the user use the tool (Fig. 3.13). A green dot next to an input parameter indicates that it is required and must be entered before the tool will run. A yellow exclamation point indicates that the user should proceed with caution. Placing the cursor over the yellow sign for a moment will display a pop-up message describing the problem—in Figure 3.13 the user is attempting to define a coordinate system for a data set that already has one. A red X indicates an error that will cause the tool to fail. Again, holding the cursor over the X will display the error message.

When the tool begins running, a dialog box appears. If a problem or error occurs, it will be reported in the dialog box. By default the progress dialog will remain open after the tool finishes executing; a check box allows the user to direct it to close automatically upon successful completion of the task. If an error occurs, the dialog will remain open regardless so the error message can be read.

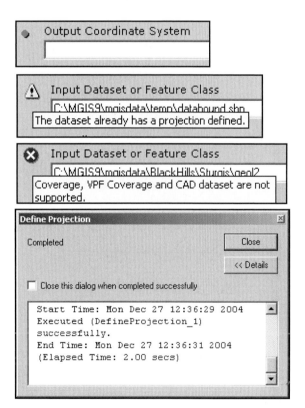

Fig. 3.13. Error handling for tools

Users interested in the full capabilities of ArcToolbox and the geoprocessing environment should consult the extensive sections in the ArcGIS Help (search on the index tab for geoprocessing).

Summary

> ➤ All GIS data has a coordinate system which defines the units and axes used to represent map features as *x-y* coordinates.

> ➤ A geographic coordinate system, or GCS, is based on spherical coordinates of latitude and longitude with units of degrees. A spheroid may be used with the GCS to represent the earth, which is not a perfect sphere.

> ➤ A datum includes a spheroid and its location relative to the earth's surface, and is used to reduce map errors introduced by differences between the spheroid and the earth's actual surface. Common datums include the North American Datums of 1927 and 1983, which are based on different spheroids.

> ➤ Map projections are mathematical equations that convert degrees from a spherical GCS into planar *x-y* coordinates of meters and feet so that the map may be portrayed on a flat sheet of paper. All map projections introduce distortions of area, distance, direction, and shape.

> ➤ UTM and State Plane are two of the most commonly used projection systems in the United States. Each consists of a family of projections designed to minimize distortion over the area covered.

> ➤ Every data set used in ArcGIS should have the correct coordinate system defined and stored with the map features. ArcGIS uses these definitions to help manage and display data correctly.

> ➤ ArcMap can project data on the fly. Each map is converted to the coordinate system of the data frame, which is set by the user. If not set explicitly, it defaults to the coordinate system of the first layer loaded into the frame.

> ➤ Data may be permanently projected from one coordinate system to another using the Project tool in ArcToolbox.

TIP: The ArcGIS online documentation includes a good reference book on map projections and coordinate systems, titled *Understanding Map Projections*. The ArcMap Help section also has some good material about projections.

TIP: When using a tool for the first time, it is strongly recommended that you read the detailed help information. In some cases it contains information critical to making the tool work properly.

TIP: It is considered poor practice to produce maps showing features in geographic coordinate systems, because of the distortion. Use an appropriate projection instead.

TIP: If a layer is open in ArcMap, then a file lock is placed on it to prevent it from being simultaneously modified by another program. Thus, open layers must be removed from ArcMap, or ArcMap must be closed, before trying to change its coordinate system or other properties in ArcCatalog.

TIP: Projecting coverages and grids requires an ArcEditor or ArcInfo license.

Chapter Review Questions

You may need to consult the Skills Reference section to answer some of these questions.

1. If a data set's features have x coordinates between –180 and +180, what is the coordinate system likely to be? What units are the coordinates in?

2. What are the x-y coordinates of a map's origin? _____ What is the x coordinate along the central meridian?_____

3. What is the difference between a spheroid and a geoid? Why don't cartographers use the geoid when creating maps?

4. How can you tell if a map layer is stored using unprojected or projected coordinates?

5. Why is it usually preferable to keep all of a project's data layers in the same datum?

6. What is the difference between a central meridian and the Prime Meridian?

7. What four properties of a map can be distorted by a projection?

8. True or False: A shapefile of the United States with a GCS coordinate system would have an x-y extent that contains entirely positive values. _____ Explain your answer.

9. Explain the difference between defining a coordinate system for a spatial data set, and projecting a spatial data set.

10. True or False: The cylindrical Mercator projection has a central meridian and two standard parallels. _____ Explain your answer.

Mastering the Skills

Teaching Tutorial

The following examples provide step-by-step instructions for doing basic tasks and solving basic problems in ArcGIS. The steps you need to do are highlighted with an arrow ➔; follow them carefully. Click on the video number in the VideoIndex to view a demonstration of the steps.

Examining coordinate systems in ArcCatalog

First, let's learn to examine the coordinate system assigned to a spatial data set.

1➔ Open ArcCatalog and navigate to the mgisdata\Usa directory.

1➔ Right-click the Counties shapefile and choose Properties from the menu.

1➔ Click the Fields tab.

1➔ Select the Shape field and view the Spatial Reference property which gives the name of the coordinate system for the layer.

1➔ Click the ellipses to view the details of the coordinate system.

1➔ Click the X/Y Domain tab and examine its contents carefully.

The X/Y Domain tab shows the extent of the data set, or the minimum and maximum *x-y* coordinate ranges. In this case the *x* units fall in the range –180 to +180, and the *y* units fall in the range –90 to +90, indicating that the *x-y* coordinates stored on the disk are in decimal degrees, and that this data set is thus in a geographic coordinate system, and is unprojected.

Shapefiles actually store projection information in a separate file, with an extension .prj. This information gets displayed as a property of the Shape field in ArcCatalog.

2➔ Close the Spatial Reference Properties and Shapefile Properties windows.

2➔ Open Windows Explorer on your computer and navigate to the mgisdata\World directory.

TIP: If Windows Explorer is currently set to hide file extensions, you will be unable to find the .prj file. To find out, choose Tools > Folder Options > View and make sure the box labeled "Hide extensions of known file types" is NOT checked.

2➔ Locate the country.prj file. It is a simple text file and could be opened in Notepad or Wordpad, but you don't need to open it now. If you did, it would look like this:

GEOGCS["GCS_WGS_1984",DATUM["D_WGS_1984",SPHEROID["WGS_1984",63
78137,298.257223563]],PRIMEM["Greenwich",0],UNIT["Degree",0.0174532925199
433]]

You may not recognize much of this text, but ArcMap and ArcCatalog use this information when managing and displaying shapefiles.

Geodatabases and coverages also store coordinate system information, although they store it in different ways and different locations. Let's look at a coverage projection now.

3➡ Back in ArcCatalog, navigate to the mgisdata\Rapidcity folder.

3➡ Right-click on the rcsoil2 coverage and choose Properties.

3➡ Click the Projection tab and read the projection information.

Coverages store projection information as an INFO format file within the coverage directory. These files are in binary format and cannot be viewed in a text editor.

4➡ Close the Coverage Properties window.

4➡ Find the rcsoil2 folder in Windows Explorer and open it, and find a file called prj.adf.

Geodatabases store projection information inside the single geodatabase file, so it cannot be seen in Windows Explorer. Coordinate systems may be stored for feature datasets or for feature classes, rather than for the geodatabase as a whole. You can examine these coordinate systems in ArcCatalog as well.

5➡ In ArcCatalog, navigate to the mgisdata\Badlands directory.

5➡ Expand the Badlands geodatabase by clicking on the plus.

5➡ Right-click the Transportation feature dataset and choose Properties from the menu to view the coordinate system. Click Cancel to close the window.

5➡ Right-click the Boundary feature class and choose Properties.

5➡ Click the Fields tab and select the Shape field. View the Field Properties to find the coordinate system. Click the ellipses for more information.

5➡ Click the X/Y Domain tab and examine it carefully.

This file has large coordinates and they are certainly not in decimal degrees. In fact, they are in meters. Clearly this data set has a projected coordinate system.

5➡ Click the Coordinate System tab and examine the details carefully. Find where it states the name of the projection, and the projection parameters such as the central meridian and the false easting/northing. Also find the line that tells you the linear units are in meters.

5➡ Close the windows.

Examining coordinate systems in ArcMap

The coordinate system is an important property of every data set. Because GIS relies on x-y coordinates to store and portray data, and because many different coordinate systems exist, one must know which coordinate systems are used and keep track of any transformations between them.

6➡ Open ArcMap and choose to start with a new, empty map.

6➡ Use Save As to save the document with the name of your choice.

6➡ Use the Add Data button to add the country and latlong shapefiles in the mgisdata\World directory.

6➡ Change the data frame name from "Layers" to "World".

6➜ Right-click the country layer and choose Properties, then click the Source tab.

1. What is the coordinate system for this layer? _____

The GCS in the coordinate system name stands for Geographic Coordinate System, which is based on degrees of latitude and longitude.

7➜ Close the Layer Properties window.

7➜ Observe the coordinates shown in the lower part of the ArcMap window and confirm that these are in degrees. The data frame units defaulted to the units of the first data set loaded.

7➜ Use the Zoom In tool to zoom into the tip of Florida in the United States.

7➜ Use the cursor to hover over the most southeastern tip of Florida and observe the *x-y* coordinates (Fig. 3.14).

2. What are the approximate coordinates of the SE tip of Florida?

_____

Fig. 3.14. Hover here

8➜ Right-click the World data frame name and choose Properties.

8➜ Click the General tab.

3. What are the map units of this frame? _____

4. What are the display units of the frame? _____

The map units are dimmed because you can't set them—they are based on the units actually stored on disk for that first shapefile added to the frame. The display units are for information only, and can be set to any units desired by the user. The chosen units appear in the coordinates box and will be reported when using the Measure tool.

9➜ Set the display units to miles.

9➜ Click OK to close the Properties sheet.

5. What are the coordinates for Florida's tip now? _____

The *x*-value tells you the distance Florida lies from the central meridian of the map (in this case the Prime Meridian that runs through Greenwich), and the *y*-value tells you its distance from the reference latitude (the equator).

The coordinate system of a layer indicates the units used to actually store the *x-y* positions of features in the layer. Thus, the country shapefile is stored in latitude-longitude and the units are decimal degrees. Inside the shapefile on the disk, the coordinates making up the county outlines are stored in units of decimal degrees.

10➜ Click the Full Extent button to see the whole world again.

10➜ Right-click the World data frame name and choose Properties.

10➜ Click the Coordinate System tab and notice what it says.

The data frame coordinate system is the same as the layer coordinate system. By default, the frame takes on the coordinate system of the first data layer added to the frame. This default, however, can easily be changed.

10➜ In the folder tree, expand the Predefined folder by clicking on its plus sign.

Notice that two subfolders are present: projected coordinate systems and geographic coordinate systems. Geographic coordinate systems are all based on the latitude-longitude grid units on a sphere. Projected coordinate systems convert spherical units (degrees of latitude and longitude) to Cartesian (planar), typically meters or feet. This conversion is accomplished through a system of mathematical equations that converts one set of units to the other, and many different types of projections are possible.

When a projected coordinate system is defined for the data frame, the data layers in the frame are converted into that system using the appropriate mathematical equation. This process is done on the fly, and while it affects how the layer looks in ArcMap, it does not change the base coordinates in which the layers are stored.   So let's continue choosing a new projection for the data frame to see how this works.

10➜ Navigate through the folders to reach Projected Coordinate Systems > World > Mercator.

10➜ Click on the Mercator (world) projection and click OK.

Notice the dramatic difference in the appearance of the map. The decimal degrees units in the layer data set are being recalculated on the fly to fit into this new coordinate system.

TIP: As you work with the <u>data frame</u> coordinate systems in ArcMap, keep in mind that they are distinct and often different from the coordinate systems of the <u>layers</u> you are working with.

11➜ Zoom into the tip of Florida again.

6. What are the coordinates and units of the SE corner of Florida now?
_____ (Notice that the units, previously miles, were reset when you applied the projection).

11➜ Zoom to the full extent again, and use the cursor and coordinate box to locate the coordinate origin, where both x and y equal zero (approximately—look for where the coordinates change from positive to negative).

Mercator is a cylindrical projection, meaning that it is based on a cylinder of paper wrapped about a sphere. In this example, the $x = 0$ longitude is the Prime Meridian, the longitude that runs through Greenwich, England, and it is in the center of this map. Therefore it is called the central meridian of the map projection. In this projection the central meridian is also the Prime Meridian, but this need not be the case. If you were mapping North America, you would choose a different central meridian that was close to the center of the U.S., for example $x = -100$. The $y = 0$ line is the equator and is called the reference latitude, or the latitude of origin. The equator is often the reference latitude for a projection, but again, it need not be.

Because Florida is west of the central meridian of this projection, its *x*-coordinate is negative.

    7. Which continent would have primarily negative *x* AND negative *y* coordinates in this projection? _____ Which one would have primarily positive *x* and *y* coordinates? _____

        12➔ Use the Insert menu on the main menu bar to add a new data frame to the map. Name it "USA".

        12➔ Add the states shapefile from the mgisdata\Usa folder.

        12➔ Set the data frame coordinate system to Predefined > Projected Coordinate Systems > Continental > North America > USA Contiguous Equidistant Conic.

        12➔ Zoom into the contiguous 48 states.

    8. Which state contains the origin of this map projection? _____

        13➔ Open the Properties for the USA data frame and click the Coordinate System tab. Read the properties of the current coordinate system.

    9. What longitude is the central meridian of this projection? _____ What latitude is the latitude of origin (or reference latitude)? _____

The projection has several other properties as well. False easting and false northing are values applied to translate the coordinates in the *x-y* direction. They are most often used to ensure that all values in the map are positive. This map has no easting or northing applied. The standard parallels are special lines of conic map projections where the paper touches the globe, and the projection is free of distortion along them.

    10. Is the Equidistant Conic projection shown here a tangent or secant conic projection? _____ How can you tell? _____

Changing any of these projection parameters changes the appearance of the map. Let's try an example.

        14➔ Close the USA Data Frame Properties window when finished.

        14➔ Right-click the World frame, and choose Activate from the menu.

        14➔ Turn off the latlong layer by clicking its check box in the Table of Contents.

        14➔ Open the Coordinate System tab in the World data frame Properties sheet.

        14➔ Note the current coordinate system, and then click the Modify button.

        14➔ Change the central meridian from 0 to 60 degrees. Click OK in both menus.

Notice how the map shifts to the left, and now Alaska and British Columbia appear on the right side. The center of the map is now at 60 degrees.

        15➔ Activate the USA data frame and change its projection so that the central meridian is –115° (about the longitude of eastern Nevada). Don't forget the minus sign!!!

Notice that the center of the map changes, and that the states are also rotated up to the right. Examine Figure 3.7 for a clue about why this rotation occurs.

Understanding the datum

Having ArcMap automatically project data can be a great convenience. However, it has some limitations. Let's examine some of these.

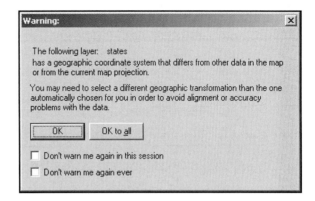

Fig. 3.15. Datum warning

16➔ Activate the World data frame.

16➔ Reset the central meridian of the projection to zero.

16➔ Add the states.shp shapefile from the mgisdata\Usa folder to the map.

16➔ Read the warning carefully (Fig. 3.15).

16➔ Click OK to add the data despite the warning.

In spite of the warning, the data appear in the correct location, matching the world map as they are supposed to.

17➔ Open the Properties sheet for the states layer and click the Source tab.

17➔ Compare the coordinate system for this layer and for the countries.

This states layer has a different coordinate system, GCS_North_American_1983, instead of GCS_WGS_1984. Although both are geographic coordinate systems (GCS), they each are based on a different datum. The states layer is based on the North American datum of 1983 (NAD83), whereas the country layer is based on the datum World Geodetic Survey 1984 (WGS84).

Because the datums are different, the same coordinate location, such as 113.5W, 34.2N, will fall in a slightly different spot, causing a misalignment between the maps. ArcMap supports the automatic reprojection of certain datums. However, methods for doing transformations vary, and the default one chosen by ArcMap may not necessarily be the best one in every case. The warning announces that a datum transformation is being applied. Detailed work requiring high spatial accuracy would require doing transformation using the Toolbox and choosing the best transformation method.

So when this warning sign appears, it does not necessarily mean that you should not go ahead and add the data. It does mean that you should stop and think about whether your application will be affected by possible misalignments due to the different coordinate systems being used. If the answer is yes, then convert the data to matching datums before proceeding. Routines are available to convert from one datum to another. Although exact conversion is not possible in all cases, the accuracy of the data will still be improved.

Understanding map distortion

Think about a sphere for a moment, and imagine its lines of latitude (they go around the middle and include the equator). The circumference of the world at the equator is about 25,000 miles. The circumference at the poles, however, is zero! Latitudes in between have intermediate lengths.

18➔ Set the World frame coordinate system to GCS_WGS_1984 (look in Predefined > Geographic Coordinate Systems > World).

18➔ Set the data frame display units to miles.

18➔ Turn the latlong layer back on.

18➔ Measure the distance across the earth at the equator: _____

18➔ Click the Identify button and click on the topmost latitude line to confirm that it is 90N.

11. According to the map, how long does the 90N latitude line *appear?* _____ What should its *actual* length be?_____

18➔ Zoom into the United States, noting its unusual elongation in the east-west direction. Then return to the previous extent.

When geographic coordinates are displayed on the screen, they are mapped as though they were a simple *x-y* coordinate system going from –180 to +180 in the *x*-direction (longitude) and –90 to +90 in the *y*-direction (latitude). On a sphere, the length of a degree decreases north or south of the equator. The circumference of the earth shrinks to zero at the poles. However, in this planar representation, it is shown having the same length as the equator!

Thus this practice clearly introduces distortions in the shape of the features. Observe the extreme length of Antarctica and the elongated shape of the United States. This east-west lengthening is typical for data sets displayed with geographic coordinates. Because it is undesirable for measuring lengths or computing areas, one frequently uses a map projection to display map data. Even map projections cannot work perfectly, however, and the resulting maps contain some combination of distortions in area, shape, distance, or direction.

TIP: It is considered poor practice to produce maps showing features in geographic coordinate systems, because of the distortion. Use an appropriate projection instead.

19➔ Right-click the states layer in the World data frame and remove it, to avoid getting more warnings when changing projections.

19➔ Change the World frame coordinate system back to Mercator.

19➔ Add the circles shapefile from the mgisdata\World folder.

Examine the circles. They increase in size toward the poles, but as you compare them to the latitude lines it becomes clear that each one has a radius of five degrees. On a globe they would appear the same size; the increase is due to the distortion of this Mercator projection. They do retain their shape as circles, however. Like the circles, Antarctica and Greenland are quite enlarged relative to their true area—Greenland appears larger than South America. Also notice that the latitude lines become further apart toward the poles. In fact, the north and south poles are an infinite distance from the equator in this projection. (Examine the cylindrical projection in

Figure 3.5 to figure out why.) Clearly, area and distance are severely distorted in this projection, and the distortion increases away from the equator. However, at the equator, there is minimal distortion because the sphere actually touches the paper at that point. Mercator does preserve direction and shape, however. The circles and countries retain their proper shapes (as they would appear on a globe), and the longitude lines always point straight up, north.

20➔ Change the coordinate system of the World frame to Sinusoidal (world). (Look in Predefined > Projected > World.)

Sinusoidal is an equal-area projection, meaning that it preserves area, and to some extent, distance. However, it does so at the expense of shape and direction. Notice the squashed look of the United States, and that the longitude lines converge toward the poles, no longer pointing straight up or down. The circles have the same areas now, but have lost their shape.

21➔ Change the coordinate system of the frame to Robinson (world).

Robinson is called a compromise projection because it is designed to minimize distortion in all four properties, but preserves none.

Distortion is unavoidable when using small-scale maps such as world and national maps. For large-scale maps, such as quadrangle maps or city maps, distortion within the map is usually not a noticeable problem if an appropriate projection is chosen and the datums match. There are many excellent projection systems for small areas. Two common systems in the United States are the Universal Transverse Mercator (UTM) system and the State Plane system.

22➔ Activate the USA frame and zoom into the contiguous 48 states.
22➔ Add the shapefile utmzone.shp from the mgisdata\World folder.
22➔ Click OK to the warning.
22➔ Click the utmzone symbol to change it to a hollow shade with a dark orange outline.
22➔ Use the Properties sheet to turn on the labels for the utmzone layer and display the ZONE field.

Notice that Iowa (use Identify if needed) is almost entirely in zone 15, but South Dakota lies in zones 14 and 13. Texas lies in three zones! To map all of South Dakota in UTM one would need to choose either zone 13 or 14. However, distortion increases for areas that lie outside the actual zone. So to map these larger states, it is best to use another projection such as State Plane.

12. What would be the best UTM zone for the state of Nevada? _____

Let's take another look at the projection file we examined earlier. Most of the entries should now make sense.

GEOGCS["GCS_WGS_1984",
DATUM["D_WGS_1984",
SPHEROID["WGS_1984",6378137,298.257223563]],
PRIMEM["Greenwich",0],

UNIT["Degree",0.0174532925199433]]

This data set is stored in a geographic coordinate system with units of degrees, using a datum and spheroid based on the World Geodetic Survey of 1984. The center of the map is at longitude 0, the Prime Meridian.

→ Exit ArcMap. You do not need to save your changes.

Working with map projections in ArcMap

ArcMap makes it very easy to work with projections because it automatically reprojects every layer to the data frame coordinate system. So as long as the map projection is set correctly for the layer, the layers automatically align with each other.

23→ Open ArcMap and begin with a new empty map.

23→ Add the usfsrds2 coverage from the mgisdata\BlackHills\Sturgis folder.

23→ Open the Properties sheet and write the coordinate system and datum of the layer here: _____ Click OK.

23→ Also examine the Extent information, which shows the range of *x-y* coordinates as stored in the file. Close the Properties window.

23→ Add the sd_city shapefile from the mgisdata\Sdakota folder and write its projection and datum here: _____

Even though the two coordinate systems are different, the maps overlay each other correctly because each is projected to the data frame coordinate system.

13. State the data frame coordinate system WITHOUT looking at its properties._____ Why could you know this information without looking?

Take a look at the sd_city coordinate system, GCS_Assumed_Geographic_1. Some data sets, especially shapefiles from the older version of ArcView, do not have projection information stored with the data set. However, if the *x-y* values of the data set fall in the range $x = -180$ to $+180$ and $y = -90$ to $+90$, ArcMap will assume the data are a geographic coordinate system. It has no way of determining the datum, however. So by default it assigns this assumed GCS system based on NAD 1927 as the most likely possibility. However, it may not be correct. If you have more detailed information about the data set, you can explicitly set the coordinate system to its precise value. Imagine that you contacted the data provider for this information and confirmed that the coordinate system is GCS based on NAD 1983. Now we will change it.

TIP: If a layer is open in ArcMap, then a file lock is placed on it to prevent it from being simultaneously modified by another program. Thus, open layers must be removed from ArcMap, or ArcMap must be closed, before trying to change its coordinate system or other properties in ArcCatalog.

24→ Right-click the sd_city layer and choose Remove.

24➔ In ArcCatalog, navigate to the sd_city shapefile in the mgisdata\Sdakota folder, right-click it, and choose Properties.

24➔ Click the Fields tab, and select the Shape field by clicking the grey tab to its right.

24➔ In the Field Properties box, locate the projection name, and click the ellipses button next to it to edit the coordinate system.

24➔ In the Spatial Reference Properties sheet, click the XY Domain tab to view the data extent. Use this tab to confirm that the actual stored coordinates are in fact decimal degrees.

24➔ Next click the Coordinate System tab and choose the Select button. Navigate through the tree to select the Geographic Coordinate Systems > North America > North American Datum 1983.prj. Click Add/OK in all boxes to apply the change.

TIP: Notice that you have just changed only the LABEL for the coordinate system, not applied any mathematical transformation to the actual *x-y* coordinates stored in the data set.

The next day, the data provider calls back and explains that he made a mistake, and that really the coordinate system for the file is based on the NAD27 datum. Before fixing this, you recall that other data layers in the Rapidcity folder, such as rcw24k-rds, are also based on this coordinate system. Although you could repeat the preceding steps to change it back, you decide to import the coordinate system information. Again, you are only changing the coordinate system label, not the *x-y* coordinates themselves.

25➔ In ArcCatalog, open the Spatial Reference Properties sheet for the sd_city layer again.

25➔ Click the Import button this time.

25➔ Navigate to the mgisdata\Rapidcity folder and select the gypsum shapefile. Click Add.

25➔ Examine the updated Coordinate System information in the Spatial Reference Properties window to make sure it is correct. Then click OK in both boxes to apply the change.

Metadata provide another way to quickly examine the coordinate system of data layers in ArcCatalog. First though, make sure that the metadata are automatically updated whenever the viewer accesses it. This way the changes just made in the sd_city layer appear correctly.

26➔ In ArcCatalog, click on Tools > Options in the main menu bar.

26➔ Click the Metadata tab.

26➔ Make sure that the default stylesheet is FGDC ESRI. In the Updating Metadata section, make sure that the box is checked to automatically update metadata when it is viewed. Click OK.

27➔ Click the name of the sd_city in the ArcCatalog tree to select it.

27➔ In the Metadata window, click the Spatial tab. The coordinate system is the first entry. Click on Details to get more information.

27➔ Click on other data sets to examine their metadata.

Projecting data

Next we will learn to project data using the Project tool. We wish to add the sites of interest from the Sdakota folder, in geographic coordinates, to the Rapidcity folder, which are all in UTM coordinates. Projecting always creates a new data set, rather than changing the old one. It changes the coordinate system label and also recalculates the *x-y* coordinates into the new projection. Contrast this function with steps 24 and 25, in which only the coordinate system label was changed. Projecting is done using ArcToolbox.

28➔ In ArcCatalog, switch the display mode from Metadata back to Contents.

28➔ Click the ArcToolbox icon to launch the toolbox.

28➔ Expand the toolbox entries Data Management Tools > Projections and Transformations.

28➔ Double-click the Project tool. Do not select Define Projection by mistake!

29➔ Click the Browse icon for the Input Dataset, navigate to the mgisdata\Sdakota directory, and select the sd_interest shapefile (Fig. 3.16). Click Add.

29➔ Click the Browse icon for the Output Dataset, navigate to the mgisdata\Rapidcity folder, and enter the name of the new shapefile as "interest.shp." Click Save.

29➔ Click the button next to the Output Coordinate System box. The Spatial Reference Properties menu appears. Click Import and select the citybnd coverage in the mgisdata\Rapidcity folder, to ensure that the data are projected to the same coordinate system as the rest of the Rapidcity data. Click Add and then OK.

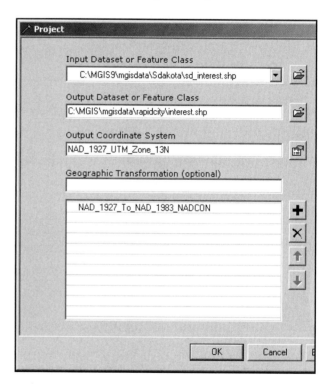

Fig. 3.16. The Project tool

29➔ Notice that a green dot has appeared next to the Geographic Transformation box, indicating that you need to enter information.

29➔ Because the conversion involves projecting from NAD83 to NAD27, you must specify a datum transformation method as well as a projection. Choose NAD_1927_to_NAD_1983_NADCON for the Geographic Transformation.

29➔ Click OK to start projecting. A dialog window appears to inform you of the progress of the tool. Close it if necessary when the tool finishes.

Datum conversions are a whole topic to themselves, beyond the scope of this book. NADCON stands for North American Datum Conversion, and is a reliable all-purpose transformation for the conterminous United States.

30➜ In ArcCatalog, select the sd_interest layer in the Sdakota folder and click the Preview button. Examine the shape of the points.

30➜ Now select the interest layer just created in the Rapidcity folder and compare its shape to the original.

30➜ Also check its metadata to confirm that the coordinate system is UTM Zone 13, NAD 1927.

TIP: If a newly projected shapefile does not appear in the Rapidcity folder, try choosing View > Refresh from the ArcCatalog main menu. If it still does not appear, exit ArcCatalog and start it again. Occasionally ArcCatalog has trouble identifying newly created data sets and must be restarted to show them properly.

Finally, ArcMap has a quick and easy way to project data from a coverage, shapefile, or geodatabase, by exporting the data. When exporting data, you can choose to save the new file in the original coordinate system, or you can save it with the same coordinate system as the data frame. We will convert the U.S. states to a U.S. Equidistant Conic projection.

31➜ Open ArcMap, if necessary, with a new and empty map. Add the states shapefile from the mgisdata\Usa directory.

31➜ Examine the coordinate system of this shapefile in ArcMap.

14. What is the coordinate system of the states shapefile? _____

32➜ Close the Layer Properties window.

32➜ Set the data frame coordinate system to USA Contiguous Equidistant Conic.

33➜ Right-click the states layer and choose Data > Export Data from the menu.

33➜ Choose to Export All Features.

33➜ Choose to use the same coordinate system as the data frame (lower choice).

33➜ Specify the location of the output shapefile as mgisdata\Usa and name it "states_eqc.shp". Click OK. Click Yes to add the new data file as a layer.

33➜ Examine the coordinate system of the new file to confirm that it has been projected.

➜ Close ArcMap and do not save your changes.

TIP: Although you need an ArcEditor or ArcInfo license to use the Project tool on a coverage, this Export function can be used to project a coverage with only an ArcView license!

Working with unknown map projections

In an ideal world, every data set would be perfectly documented with the correct coordinate system and datum, as well as all the other metadata information. Unfortunately, however, we live in a less than perfect world, and must deal with human error and inefficiency. In this world it is not uncommon to obtain a data set and be less than certain about its coordinate system. The author has encountered many scenarios like this, and you will, too:

> Judy from the Planning department sent you the parcels map you requested. They are new at GIS, and the coordinate system is undefined. When you call to ask her what it is, she has no idea. She calls the contractor who digitized the parcels, but that division was sold to another company two months ago and no one seems to know where the files relating to the project are.

> Your field assistants spent two months in the field last summer collecting GPS data. Three months later one of them calls you because he doesn't remember whether the GPS unit was set to collect data in NAD27 or NAD83. He wants to know what he should do.

> In a small county planning department with a GIS staff of one, Joe spent two years compiling roads, buildings, and other data as shapefiles for ArcView 3.2, which does not store information about the coordinate system. He departs suddenly for a new job without leaving any notes about his work. His replacement finds a folder full of data in an unknown coordinate system, and Joe left no forwarding address.

In each of these cases, you would be faced with repeating months of hard work unless you play detective and determine the coordinate systems. This section shows some ways to start.

Imagine that you are a new employee of the Black Hills National Forest, charged with building a GIS for a section of the forest. Your predecessor, Bob, left a directory of data sets with less than ideal documentation, you discover.

> 34➜ Open ArcMap and choose to start with a new, empty map.

> 34➜ Add the raster file L7_Aug20_W84.bil from the mgisdata\BlackHills\Sturgis folder, a Landsat 7 satellite image.

> 34➜ Click Add Data again and also add the coverage usfsrds2 to add both to the map.

> 34➜ Right-click the usfsrds2 symbol and choose a nice contrasting color that shows up well against the image.

> 34➜ Zoom in closer and examine the match between the roads in the coverage and the roads in the image.

Oh, dear! The roads are offset to the south, by about 300 meters. Because the two data sets are this close, you suspect that this is a datum mismatch rather than two different coordinate systems. Now the detective work starts. Begin by investigating the coordinate systems of the two layers.

> 35➜ Open the Properties sheet for the usfsrds2 layer and examine the Source tab. Write the projection here: _____ Close the box.

> 35➜ Open the Properties sheet for the Landsat image, scroll down the information in its Source tab, and locate the spatial reference information as well.

The Landsat image coordinate system has not been defined by the person who created the file. This does not mean that it has no coordinate system—every spatial data set has an intrinsic coordinate system defined by the units used to store its *x-y* coordinates. It only means that ArcMap doesn't know what the coordinate system is because the label is missing. You need to figure it out and create the label. Since ArcMap cannot reproject a data set with an undefined coordinate system, it shows the raw *x-y* values in the file. It is not being projected on the fly to match the other data set in the frame.

The usfsrds2 layer is UTM Zone 13, and pretty close to the Landsat, so it is likely that the Landsat is also in UTM, but has a different datum. Luckily, the Landsat image name offers a major clue: the W84 suggests that this image is based on the WGS84 datum. You'll proceed on this assumption. To check your guess, define the image coordinate system as WGS84 and see if that eliminates the offset.

36➜ Right-click the image name and choose Remove. It is not possible to make changes to a file in ArcCatalog while it is open in ArcMap.

36➜ Click the ArcCatalog button to launch ArcCatalog.

36➜ Navigate to the mgisdata\BlackHills\Sturgis folder and locate the L7_Aug20_W84.bil image.

37➜ Right-click the image and open its Properties.

37➜ Scroll down the information on the General tab, locating the Spatial Reference, currently showing the coordinate system as Undefined.

37➜ Choose Edit, and click the Select button.

37➜ Navigate through the choices to locate Projected Coordinate Systems > UTM > WGS 1984 > WGS 1984 UTM Zone 13N.prj. Select it and click Add.

37➜ Now the new coordinate system shows up in the image properties. Click OK and OK to apply the changes.

37➜ Click and drag the image into the ArcMap window.

Your guess did not seem to work, because the two layers are still offset. However, you have forgotten a small but important detail. The image was the first layer added to the data frame, and so the data frame would have adopted that coordinate system. Because originally it was undefined, the data frame coordinate system is currently also undefined, and ArcMap will not project *any* of the layers.

38➜ Right-click the data frame name and choose Properties > Coordinate System. Notice that it says the coordinate system is unknown, and layers cannot be projected. Click Cancel to close the window.

38➜ Insert a new data frame in the map.

38➜ Add the two layers again, L7_Aug20_W84 and usfsrds2 from the mgisdata\BlackHills\Sturgis folder. (The data frame coordinate system will be set to whichever one is clicked first.)

38➜ Set the road color if necessary, then zoom in and examine the layers again.

The layers match very well now. Once the coordinate systems of both data layers are correct, ArcMap is able to correctly project each layer to the data frame coordinate system. The small offsets still present are probably due to slight errors in digitizing the roads, or errors involved in georeferencing the satellite imagery.

39➜ Right-click the original data frame with the offset layers, and choose Remove.

End of the tutorial.

Exercises

projected NAD – meters

1. Examine the coordinate systems for the files rds_clp.shp and rcwgeology2.shp. For each one answer the following questions: What is the name of the coordinate system? Is it projected or unprojected? What are the map units? What is the *x-y* extent?

2. Open the map document ex_3.mxd in the mgisdata\MapDocuments folder. What is the coordinate system of the data frame in this map?_____ What are the map units of the frame?_____ What are the display units of the frame? _____ What are the map units of the source data? (**Hint**: What is its coordinate system?)_____

3. Measure the distance from Olympia, Washington, to Tallahassee, Florida, in miles. What is it?

4. Explain how you would measure (not calculate) the distance in kilometers instead. Then do it. How far is it in kilometers?

5. Now measure the distance from Olympia to Tallahassee, in miles, with the data frame coordinate system set to North America Lambert Conformal Conic. What is the difference between your answer and the answer for Exercise 3?

6. Set the map projection to Mercator, centered at 103°W, using NAD83. (**Hint**: Modify the Mercator projection listed among the World projections.) What is the distance between Olympia and Tallahassee, in miles, in this projection? How does it differ from the answer in Exercise 3?

7. Create a map of the world and its capitals (natcapitals.shp) in Mollweide projection. Make the continents "Turquoise Dust" color and the capitals small yellow stars. Make the lat-long lines pale grey and put them behind the continents. **Capture** the map to turn in.

8. Examine your map for Exercise 7. Which of the four map properties appear to be fairly well preserved in this projection? Which ones appear poorly preserved? Explain your reasoning for each one.

9. How many different coordinate systems are represented by the shapefiles and rasters in the mgisdata\Rapidcity folder? (Do not include the coverages or the files in the eastpat folder.) List each coordinate system found and the data sets that have it.

10. The shapefile rcw_oops in the mgisdata\Rapidcity folder is supposed to contain roads for the western part of Rapid City, and as such should overlay with the rds_clp shapefile in the same folder. Examine them both, then explain what is wrong with rcw_oops and why it won't overlay with rds_clp. How could you fix the problem? (**Hint**: Compare the coordinate system and *x-y* domain of rcw_oops in ArcCatalog.)

x shape
√ distance
√ area (size)
x direction

Challenge Problem

Using the ArcCatalog Preview function, compare the shape of the citybnd_sp coverage you created with the original citybnd coverage. What is the major noticeable difference? Explain why it occurs.

Skills Reference

Examining the coordinate system for a layer

Every data layer has *x-y* coordinates, which must be in some type of coordinate system. The coordinate system is stored as part of the information about a data set. You can access this information in several ways.

In ArcMap

1. Right-click the layer name and choose Properties.

2. Click the Source tab.

3. Read the coordinate system information (Fig. 3.17).

In ArcCatalog (geodatabases and coverages)

1. Right-click the layer and choose Properties.

2. Click the Projection tab and read the information.

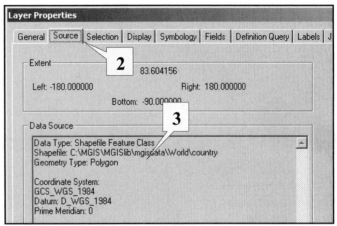

Fig. 3.17. Viewing a coordinate system in ArcMap

In ArcCatalog (shapefiles)

1. Right-click the shapefile and choose Properties.

2. Click the Fields tab (Fig. 3.18).

3. Select the Shape field by clicking on the small grey bar to its left.

4. Read the projection from the bottom of the Field Properties box.

5. Click the ellipses (...) for details or to modify the projection.

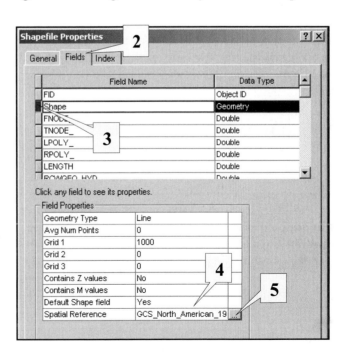

Fig. 3.18. Viewing coordinate systems of shapefiles in ArcCatalog

Setting the map projection of a data frame

1. In ArcMap, right-click the data frame name and choose Properties from the context menu.

2. Click the Coordinate System tab (Fig. 3.19).

3. Use the folder tree provided to select a coordinate system.

4. OR click the Import button to select a spatial data set that has the coordinate system to apply to the data frame.

5. If desired, click Modify to alter the parameters of the existing coordinate system.

Assigning/modifying coordinate systems in ArcCatalog

When a data set has no coordinate system defined,

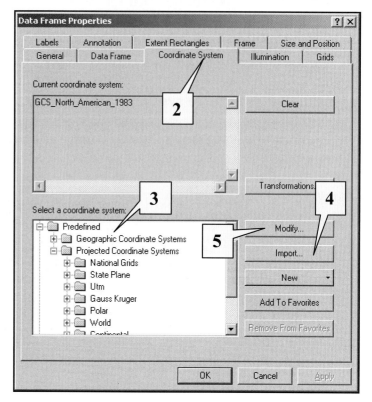

Fig. 3.19. Setting the CS of a data frame

setting the coordinate system allows you to display it with other data layers in ArcMap. It is very important, however, to correctly set the coordinate system to match the actual *x-y* coordinates contained in the file. Be very cautious in changing the map projection if it is already set, unless you know for a fact that it was incorrect to begin with.

TIP: If a layer is open in ArcMap, then a file lock is placed on it to prevent it from being simultaneously modified by another program. Thus open layers must be removed from ArcMap, or ArcMap must be closed, before trying to change its coordinate system or other properties in ArcCatalog.

For shapefiles and geodatabase feature classes

Open the shapefile or feature class coordinate system information tab as described earlier in order to view and change the coordinate system information.

1. Right-click the feature class or shapefile and choose Properties from the menu.

2. Click the Fields tab, select the Shape field, and click the ellipses next to the coordinate system field property.

3. The Spatial Reference Properties menu appears. Select an existing coordinate system from the list provided, import one from another data layer, modify an existing one, or create a new one using the buttons in the menu.

For geodatabase feature datasets

1. Right-click the feature dataset and choose Properties from the context menu.

2. Choose Edit on the General tab.

3. Use the Spatial Reference properties menu to change the coordinate system. Select an existing coordinate system from the list provided, import one from another data layer, modify an existing one, or create a new one using the buttons in the menu.

For any spatial data set

1. Open ArcToolbox if necessary.

2. Choose the Define Projection tool from the section Data Management Tools > Projections and Transformations.

3. Click the Browse button and select the data set to define (Fig. 3.20).

4. Click the icon to open the Spatial Reference Properties window.

5. Use the Spatial Reference Properties window to select a coordinate system from the supported list, import one from another data layer, modify an existing one, or create a new one using the buttons in the menu.

6. Click OK and OK to complete the task.

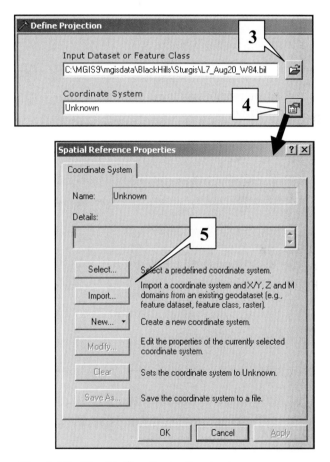

Fig. 3.20. The Define Projection tool

Projecting data from one coordinate system to another

Projecting data changes both the coordinate system definition AND the *x-y* coordinate values of the features, creating a new file in the process and leaving the original unchanged.

1. Start ArcToolbox and choose the Project tool from the Data Management Tools > Projections and Transformations > Feature section (Fig. 3.21).

2. Click the first Browse button to select the data set to be projected.

3. Click the second Browse button to specify a folder, a geodatabase, or a feature dataset to contain the output, and give the output shapefile or feature class a name. Click Save to enter the new name. Click Next.

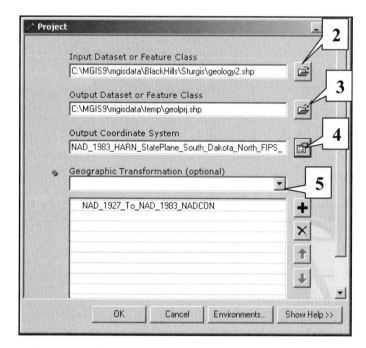

4. Click the button to open the Spatial Reference Properties window to select a coordinate system to project the data to, using Select, Import, or Modify as in Figure 3.20).

5. A green dot will appear by the Geographic Transformation box if one is required to convert from one datum to another. Select a transformation from the drop-down box.

Fig. 3.21. The Project tool

6. Click OK to project the data.

TIP: Projecting coverages and grids requires an ArcEditor or ArcInfo license. If you only have an ArcView license, a red X button will appear by the Input Dataset box.

Chapter 4. Drawing and Symbolizing Features

Mastering the Concepts

Objectives

➢ Drawing features and rasters with different styles and symbols

➢ Using different types of maps to present attribute data

➢ Classifying features based on their attributes using a variety of classification methods

➢ Choosing, modifying, storing, and using different kinds of symbols

➢ Using layers and groups to speed and simplify the display of data

Concepts

ArcMap provides many ways to present and analyze map data, and one of the most powerful techniques is assigning symbols based on one or more attributes. Readers can quickly see spatial patterns not readily apparent from looking at the data. This chapter presents many ways to display features, and it also shows how to edit symbols and save them in groups, as styles.

Types of maps

ArcMap offers several types of maps to choose from. A data layer may simply be portrayed using one symbol, or different features can be assigned different symbols depending on the value of an attribute, such as zoning or population. Such maps are called **choropleth maps.**

When creating maps, it is important to distinguish between **categorical data** and **quantities data** (Fig. 4.1). The type of data controls to some extent the variety and types of maps which can be created.

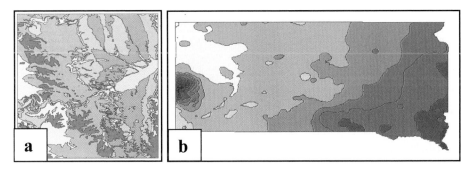

Fig. 4.1. (a) A geology map represents categorical data and is displayed using a unique values type of map. (b) Annual precipitation is numeric data representing quantities and is displayed using a graduated color type of map.

Categorical data (also sometimes called **nominal data**) names or describes objects, such as the name of a state, a land use code, or a type of highway. Categorical data within a map may fall into relatively few groups (such as land use), or they may have nearly as many values as features (such as street names). In the first case, it is often worthwhile to map the data using a different color for each value (a unique values map) in order to emphasize the spatial patterns between the groups. In the second case, there would be so many colors that the map would only be confusing.

Numeric data, referred to as Quantities data in ArcMap, has number values, such as population or rainfall. Numeric data may be represented using several types of maps, including graduated color, graduated symbol, dot density, and chart maps. Numeric data falls into three types.

Ordinal data are simply rankings, such as the top 10 states, and the differences between the rankings may be large or small in magnitude. For example, if we look at rankings of population between states, California (#1) has approximately 13 million more people than Texas (#2), but Texas only has 2 million more people than New York (#3).

Interval data are spread along a regularly spaced scale of measurement units, such as a ruler for measuring distance, or a thermometer for measuring temperature. On an interval scale, equal changes in interval involve equal changes in the object being measured. For example, the amount of energy to heat a thimble of water from 6 to 7 degrees is the same as heating it from 96 to 97 degrees.

Ratio data are also interval data, but they have the additional property that the measurements are related to a meaningful zero point. Precipitation is an example of ratio data: zero precipitation corresponds to a total lack of rain, and two inches of rain is twice as much as one inch. Population is another example of ratio data. Temperature data measured in the familiar Celsius or Fahrenheit scale are NOT ratio data, because in either scale a temperature of zero does not correspond to a complete lack of temperature (heat energy). Any measuring scale that can have negative values, such as temperature and pH, cannot constitute a ratio scale.

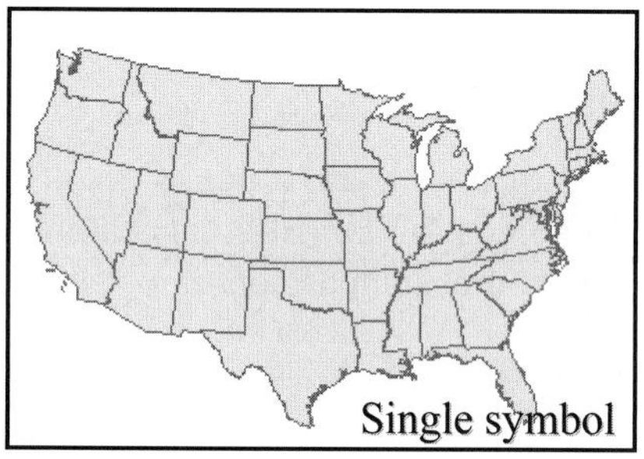

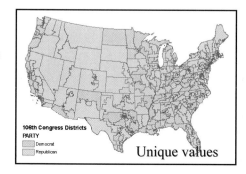

Fig. 4.2. Single symbol map

ArcMap offers a variety of map types to display these different types of data. Some maps are suitable for either categorical or numeric data, whereas others can be used only for one type of data.

Single symbol

In the simplest type of map, all of the features, for example states, are drawn with the same symbol. This type of map is used to show the size, shape, and location of the features (Fig. 4.2).

Categories: unique values

Categories maps assign a different color to every unique value in a field. In the map of congressional districts (Fig. 4.3), the Democratic districts are shown in red and the Republican districts are shown in blue,

Fig. 4.3. A unique values map

based on an attribute field that contains a "D" or "R". Categories maps may be created from string or numeric fields, but they work best when relatively few different values exist in the attribute field. In a unique values map of the United States created from the state name attribute, every state would have its own color because each name is different. A unique values map based on population would have little meaning because each state would have its own population value and get its own randomly assigned color.

In Figure 4.4a, a unique values map was created based on the attribute STATE_NAME. Because every name is unique, every state has a different color, and no patterns are apparent. When the SUBREGION attribute is used, however, one can immediately interpret the pattern to see which state falls into which subregion (Fig. 4.4b). Unique values maps typically show categorical or ordinal data, or numeric data with only a few distinct values, such as land use codes.

Fig. 4.4. (a) A unique values map based on state name yields different colors for each state and shows no patterns. (b) A unique values map based on subregion shows the groupings clearly.

Quantities maps

Symbols for numeric values in quantities maps represent a range of values, such as population. They must use numeric data; they cannot be applied to categorical data. Quantities maps usually show interval and ratio data, although they may also be used to display ordinal data. All of the examples in Figure 4.5 show a single state attribute—population, using different types of quantities maps.

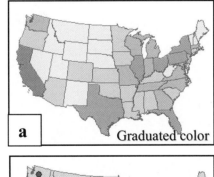

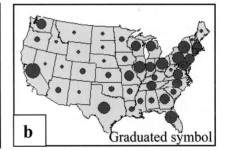

Fig. 4.5. Numeric values may be mapped using a variety of maps based on classifying data into groups and assigning symbols based on color or size to represent the increase in values. Each of these maps shows state population.

Quantities maps rely on a classification scheme: the states are divided into *n* classes based on an attribute, and a different symbol is applied to each class. Each class contains a range of values. For example, data indicating percent whites in a county could be divided into five classes at 20% intervals: 0-20%, 21-40%, 41-60%, 61-80%, and 81-100%. Each class is assigned its own color symbol. A county with 50% whites would be shown using the symbol from the 41-60% class.

Often the symbols are related in some way to make the relationships between classes more obvious, for example using a color ramp that goes from light to dark in the same color, as in Figure 4.5a. Such a map is called a **graduated color map.** Point symbol colors may also be assigned to create graduated color maps for point data. Alternatively, one might create a **graduated symbol map,** using a marker symbol which increases in size (Fig. 4.5b). Graduated color and graduated symbol maps can have any number of classes, although in practice 10 is about the maximum number that can be reliably distinguished by the human eye.

A **proportional symbol** map (Fig. 4.5c) resembles a graduated symbol map, except that the symbol for each state is drawn proportional in size to the population value, instead of dividing the states into classes. Instead of five or six symbol sizes, one has a continuous range of symbol sizes from smallest to largest.

A **dot density** map uses randomly placed dots within a polygon to show the magnitude of a value in the attribute table, such as population. Each dot represents a certain number, for example 1 million people. In that case, California would have about 33 dots in it, and South Dakota would barely have 1! In Figure 4.6, one dot equals 140,000 people. Note that the locations of the dots, which are randomly placed, do not necessarily represent the actual distribution of people in the state.

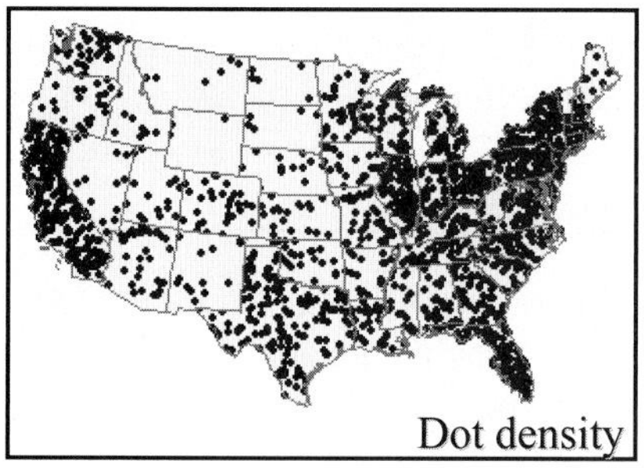

Fig. 4.6. A dot density map showing state population

Chart maps

A chart map expands the number of attributes that can be displayed on a map by replacing a single symbol with a chart representing several attributes. The chart could be a pie chart, bar chart, or stacked bar chart. Figure 4.7 shows a pie chart map with the proportions of whites, blacks, and Hispanics in each state. The pie sizes can be all the same (Fig. 4.7a), or proportional to the sum of the three categories, thus showing the relative number of people in the state as well (Fig. 4.7b). Notice how the pie colors are

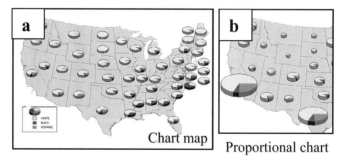

Fig. 4.7. (a) A chart map can represent several attributes at once, such as the proportion of different ethnic groups in a state. (b) The pie size can represent the total population.

chosen to facilitate recognition of the classes: cream for whites, dark brown for blacks, and reddish for Hispanics. Such details help make maps easier for the reader to interpret, and they lessen the need for the reader to rapidly look back and forth from the legend to the map.

Multiple attribute

Multiple attribute maps can show more than one attribute at a time. In Figure 4.8, the SUBREGION field is displayed by unique values, and the circles in each state show the Native American population as a percentage of the total population.

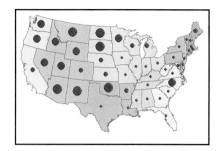

Fig. 4.8. Multiple attribute maps show two attributes.

Normalizing data

Maps created from numeric data using graduated color or graduated symbols may be normalized. Normalization can be done in two ways. The first method divides each value by the total of all the values in the column, giving a percentage. Figure 4.9 uses this method to show the population of each state as a percentage of the total U.S. population. The second method divides the population values by another specified field, such as the area of the state. This approach would produce a map showing the population *density* of each state.

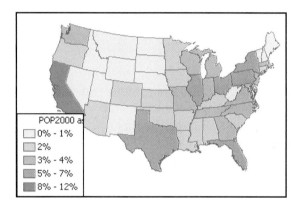

Fig. 4.9. State population normalized by the total U.S. population, showing the percentage of Americans living in each state

Classifying numeric data

Creating graduated color or graduated symbol maps requires classifying a continuous range of data into categories, each of which receives its own symbol. ArcMap offers several methods of classification. The method chosen affects the appearance of the map, so it is important to choose the most appropriate method for the data. All of the maps that follow are based on the same field, Population, yet they present quite different pictures of the data.

Natural breaks (Jenks method)

The Jenks method finds existing groups of values in the data and puts them together, exploiting natural gaps in the data (Fig. 4.10). Note that each class interval is a different size, as shown by

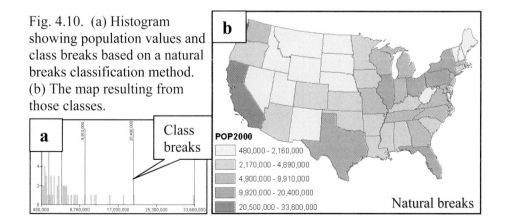

Fig. 4.10. (a) Histogram showing population values and class breaks based on a natural breaks classification method. (b) The map resulting from those classes.

143

the blue lines in Figure 4.10a. The Jenks method works well on unevenly distributed data, such as population. There are many low- and medium-population states, and a few very high-population states. It also works well with almost any data set, making it a natural choice for the default classification scheme in ArcMap.

Equal interval

The equal interval classification takes the range of values in the field and divides them into a specified number of classes of equal size (Fig. 4.11). For example, if the population of the country was 100 million, and five classes were specified, then the classes would range from 0-20 million, 20-40 million, 40-60 million, and so on. This type of classification is very useful for ratio data, such as income or precipitation, because it gives a sense of regularity to the observed increases. However, it is hard to predict how many features will end up in each class. Notice in the population example that nearly all of the states fall into the first class, and that California and Texas are the only ones in their respective classes.

Fig. 4.11. (a) Histogram showing equal interval class breaks for population. (b) Population map based on equal interval classes.

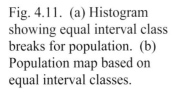
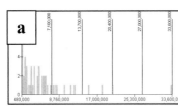
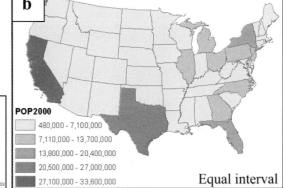

Defined interval

A defined interval classification is similar to an equal interval one, except that the user specifies the size of the class interval, such as 10 million (Fig. 4.12). The software then determines the number of classes needed to encompass the range of values in the field. This method will create simple, rounded values in the classes, which are easy to interpret. It does, however, suffer the same disadvantages as the equal interval classification.

Fig. 4.12. (a) Histogram showing defined interval classes of 10 million people each. (b) Map based on the defined interval classes.

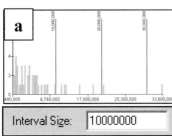
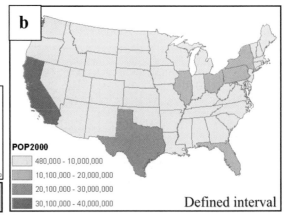

Quantile

A quantile classification puts about the same number of features in each class. This method will create a very balanced map with all classes equally well represented, but some of the classes may be close together in terms of their values. Notice in Figure 4.13a how the first few classes are very close together and low in population, and the succeeding classes encompass larger population ranges.

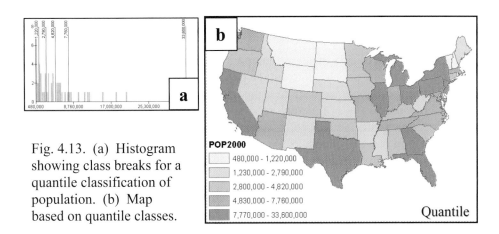

Fig. 4.13. (a) Histogram showing class breaks for a quantile classification of population. (b) Map based on quantile classes.

Standard deviation

A standard deviation classification apportions the values based on the statistics of the field. The user selects the number of standard deviations each class contains, and the software determines the number of classes needed. This method excels at highlighting which values are typical and which are outliers. In Figure 4.14, the yellow states are close to the mean population value, the northern plains states are below average, and Texas, California, Florida, and much of the Northeast are above average in population.

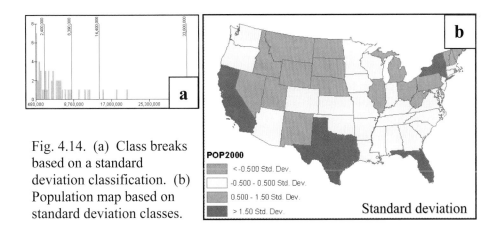

Fig. 4.14. (a) Class breaks based on a standard deviation classification. (b) Population map based on standard deviation classes.

Manual class breaks

Finally, if none of the above options gives the desired map, the user can set class break points manually to any chosen values, by typing them in, or by dragging the blue class break bars to the desired location.

Choosing the classification method

The choice of classification method depends in part on the mapmaker's purposes and on the type of data being presented. Jenks, for example, shows the "nearest neighbors" in a distribution, whereas a defined interval or equal interval map does a better job of portraying relative magnitude. Notice the difference between the population maps in Figures 4.10 and 4.11. The Jenks map suggests that states with large populations are common; only a careful examination of the class breaks in the legend lead the reader to understand the truth—that most states have relatively small populations in comparison to a few with very large populations. Figure 4.11 makes this observation clear at first glance.

Furthermore, certain types of data possess the quality that the relative magnitude of values have an intrinsic meaning to most people. Percentage data, for example, have a physical and psychological meaning—people have an intuitive understanding of the difference between 50% and 100%. Differences in values of median rent have a significance in terms of changes in dollar values that most people grasp immediately. When dealing with such data, it is often wise to use a defined interval map in order to choose classes with logical breakpoints, such as 20%, or 5 inches of rain, or $200, rather than 12.6%, or 1.47 inches of rain, or $187. The reader can then more effectively interpret and assign meaning to the classes.

Using map layer files

Getting a layer to display exactly as desired can take a long time. Layer files (extension .lyr) store information about how a layer is displayed and can be used to quickly give the same symbols to layers in other map documents. A layer file does not store the actual spatial features; it only stores a link to the spatial data file. In Figure 4.15, the layer file SD_Cities.lyr stores the display properties of the sd_cityname.shp data layer. You can preview layer files in ArcCatalog, add them to map documents, and send them to others. As long as the layer files can find the data layers upon which they are based, they can re-create the exact appearance of that layer in any map document. Layers can also quickly set the display characteristics of another data layer if it has the same or similar attributes as the original source layer.

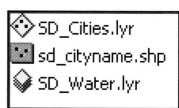

Fig. 4.15. Layer files (shown in yellow)

Layers can be grouped together to form Group layers. The SD_Water.lyr in Figure 4.15 contains two data layers: rivers and lakes. Placing them in groups allows the user to turn all the layers in the group off and on at the same time and to store display properties for the entire group at once.

Editing symbols and using styles

ArcMap comes with many predefined sets of symbols. Additional symbols can be created using the Symbol Property Editor (Fig. 4.16). Symbols are composed of layers, each of which is a character, shape, bitmap, or shade. (These are not the same as data layers.) The symbol in Figure 4.16 consists of a green radioactive marker symbol layer, a yellow circle layer, and a green circle layer. Editing the properties of these layers creates new symbols.

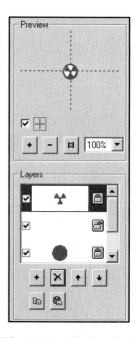

Fig. 4.16. The Symbol Property Editor

Created symbols can be saved for future use by placing them in files called styles. Styles contain groups of symbols that are often used together. For example, a city planner might create a style for making land use maps, so that every employee who makes a map will use the same set of standard symbols, facilitating the interpretation of the maps by others. Standard symbols for streets, traffic lights, sewers, electric lines, and phone lines could also be developed so that every map produced by the planning department appears similar. They could even establish frame borders, text, north arrows, and coordinate systems to ensure that every map produced by the city has the same look and feel.

ArcGIS comes with about 20 standard styles. The default style, ESRI, contains standard circles, squares, and other shapes. Other styles are industry specific, such as the Civic, Geology, and Forestry styles. Figure 4.17 shows some of the Forestry marker symbols. Each one has a name and a category; Deer Viewing and Diving are symbols from the Recreation category.

🔲	Deer Viewing	Recreat...
🔲	Diving	Recreat...
🔲	Dog Area	Recreat...
🔲	Downhill Skiing	Recreat...
🔲	Drinking Water	Facilities
🔲	Environmental Stu...	Warning
🔲	Falling Rocks	Warning
🔲	Fee Area 1	Recreat...
🔲	Fee Area 2	Recreat...

Fig. 4.17. Some symbols from the Forestry style

You are most likely to use Styles to save symbols and use them again later. Modified symbols are stored in the user's personal style, which is saved under the profile directory on the computer. Users can also add symbols to other styles and store them in separate files, providing a means to organize symbols. One style might provide symbols for geologic maps, whereas another style would be used for a planning department project.

Styles are also useful for collecting frequently used symbols into a single style which is easily turned on and used, especially when the symbols come from several different standard styles. By collecting them into a style, you can access them quickly and easily.

Styles can be viewed and managed using the **Style Manager,** which makes it easy to examine the symbols in different styles and to copy, paste, and delete symbols between styles.

Displaying rasters

Raster data falls into two basic categories: image rasters and thematic rasters. Images mainly come from aerial photography and satellites, and the pixels represent different degrees of brightness on a scale of 0-255 DN (digital numbers). A dark shadow would have a low DN and a bright white cement road would have a high DN. Thematic rasters represent measured quantities at a pixel such as land use or rainfall. Thematic rasters also fall into two categories we have discussed before: discrete or continuous. A discrete raster has coded values that define regions, such as geology or land use. A continuous raster has values that change continuously from one location to another, such as elevation or precipitation. Displaying rasters effectively requires an understanding of these basic data categories and the display techniques applicable to them.

Image rasters

Image rasters may be displayed by two different methods. The **stretched** method uses a 255-color ramp (often grey) and assigns each DN a different color. The contrast and brightness of the image can be easily adjusted by applying different stretch schemes. Figure 4.18a shows one band of a satellite image displayed with the stretched method. The other method, **RGB composite,** applies only to images containing more than one band of data. One band is assigned to each of the three color guns on the computer (red, green, and blue) to produce a color image. Each band of the composite may be individually stretched also. Figure 4.18b shows the same image displayed as a true-color composite.

Thematic rasters

Thematic rasters use display options similar to features. A discrete raster is best displayed using a **unique values** classification, where each value of the raster receives its own color, just like a unique values map for vector data. The geology map shown in Figure 4.18c provides a good example. Continuous rasters can be displayed with two different options. The **classified** method divides the range of values in the raster into a specified number of classes, in a way completely analogous to the classification of numeric attributes for vector data as described in this chapter. All of the same classification methods (Jenks, equal interval, etc.) are available. The elevation map in Figure 4.18d was classified into 12 classes for display. Continuous data may also be displayed using the **stretched** option. Since thematic data probably does not fall into the neat 0-255 scale used for image data, however, the values are converted to a 0-255 scale for display. In Figure 4.18e, the elevation is shown with the same colors as in 4.18d, but with the entire range of 256 colors.

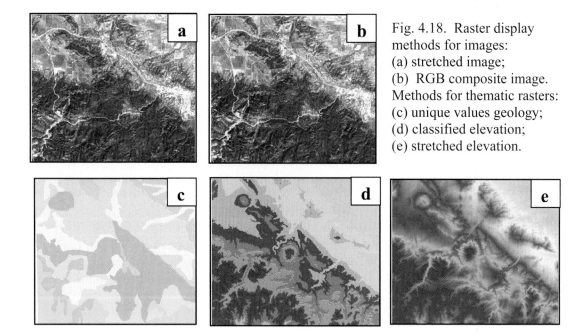

Fig. 4.18. Raster display methods for images:
(a) stretched image;
(b) RGB composite image.
Methods for thematic rasters:
(c) unique values geology;
(d) classified elevation;
(e) stretched elevation.

Summary

➢ Different types of maps offer different views of data: single symbol, unique values, graduated color, graduated symbol, dot density, chart, and multiple attribute maps.

➢ Categorical data (also called nominal data) names features. Numbers may be categorical when they are used as codes, as for land use designations.

➢ Numeric data falls into three categories: ordinal data consisting of rankings, interval data measured on a regular scale, and ratio data which is measured on a regular scale with a meaningful zero point.

➢ Graduated color and graduated symbol maps offer five different classification schemes: Jenks natural breaks, equal interval, defined interval, quantile, standard deviation, and manual.

➢ Map layers store a reference to a data set along with information on how to display the data. Both individual and group layers can be created. Layers can be used to import symbology to new layers as well.

➢ ArcMap has many different symbols for points, lines, polygons, borders, and text. ArcMap comes with many symbol sets, and more can be created with the Symbol Editor.

➢ Point, line, and marker symbols can be edited individually and collected into groups and styles for reuse.

➢ Styles are used to hold and organize related groups of symbols and to facilitate the development of standard sets of symbols for different applications. Users can create, copy, and modify styles to customize them for personal use.

Chapter Review Questions

You may need to consult the Skills Reference section to answer some of these questions.

1. What does it mean to create a map based on an attribute?

2. For each of the following types of data, state whether it is categorical, ordinal, interval, or ratio. Explain your reasoning.

    bushels of wheat per county

    pH measurement of a stream

    soil type

    state rank for average wage

    average maximum daily temperature

    number of voters in a district

    client street address (e.g., 510 Main St.)

    student grade in a class (e.g., A, B+, etc.)

    land parcel ID number (e.g., 1005690)

3. For each of the following attributes, state whether a single symbol, graduated color, or unique values map would be most appropriate. Explain your reasoning.

    precipitation

    rivers

    geologic unit

    land use

    acres of corn planted per county

    county names

4. What is the purpose of a map layer file (.lyr)?

5. List and briefly describe each of the six data classification methods.

Mastering the Skills

Teaching Tutorial

The following examples provide step-by-step instructions for doing basic tasks and solving basic problems in ArcGIS. The steps you need to do are highlighted with an arrow ➔; follow them carefully. Click on the video number in the VideoIndex to view a demonstration of the steps.

Working with symbols and styles

➔ Start ArcMap, if necessary. Navigate to the mgisdata\MapDocuments folder and open the map document ex_4.mxd.

➔ Use Save As to give the map document a new name, and save it frequently as you work.

1. What type of map is currently being used to display the towns? _____

Before we start making attribute maps, let's first learn some ways to edit individual symbols and create new ones. You will also learn how to gather groups of symbols together to create your own styles. First, let's examine some of the styles in more detail.

1➔ Click the Schools symbol to open the Symbol Selector.

1➔ Click the More Symbols button and choose Forestry.

1➔ Click the More Symbols button again and uncheck all the other styles by selecting each in turn, so that ONLY Forestry is being viewed. (Note: your personal style cannot be unchecked, so leave it as is.)

1➔ Make sure that Category is set to All.

1➔ Scroll down the list and examine the different marker symbols available in the Forestry style.

When so many symbols are present in a style, it is handy to group them for easier access. This can be done by assigning each symbol a Category.

2➔ Click on the Category drop-down box to view the categories available for the Forestry style.

2. What categories does the Forestry style include? _____

2➔ Examine each of the categories in turn by selecting it with the drop-down box.

2➔ Then set the Category back to All.

2➔ Uncheck the Forestry style and check the ESRI style again by clicking the More Symbols button and selecting each in turn.

2➔ Click Cancel to close the Symbol Properties window without applying any changes.

Next, we will learn how to create new symbols.

3➔ Click on the Towns symbol to open the Symbol Selector.

3➔ Choose the Circle 2 symbol (near the top).

3➔ Click Properties to open the Symbol Property Editor.

Symbols are composed of one or more layers, each of which contains a basic character. The Circle 2 symbol contains two layers, an outline and a fill layer. The Symbol Property Editor (Fig. 4.19) gives control of each layer individually. Notice that the symbol being currently edited is previewed in the upper left corner of the window.

3➔ Click on the outline layer to select it, and then change its color to dark green.

3➔ Click on the fill layer to select it, and then change its color to a pale yellow.

3➔ Click the plus sign under the Layers box to add a new layer. A black square appears inside the symbol.

3➔ Using the drop-down box at the top of the dialog box, change the character type to Arrow Marker Symbol.

3➔ Experiment with setting the length, width, offset, and angle properties. Watch the changes in the preview area.

4➔ Change the character type to Character Marker Symbol, and then choose the radioactive symbol character (unicode #182— click a symbol to make its unicode index number appear, or type the unicode number in the box).

4➔ Increase the size of the symbol to 22 points, and change its color to dark green.

Fig. 4.19. The Symbol Property Editor

The symbol will now appear as a pie with six slices of alternating colors. Finally, let's add a large halo around it.

4➔ Add another layer. Choose a simple filled circle as the character (unicode #33).

4➔ Click the Down arrow button below the Layers list to move the layer to the bottom of the list, so it will be drawn underneath the other layers.

4➔ Change its color to dark green, and increase its size until a thick border is visible around the yellow/green pie.

4➔ The green outline is no longer needed, so click the layer to select it, and then click the X button to delete it. The symbol should look similar to the one in Figure 4.19.

Once you create a symbol, it is convenient to save it for later use. You can also create new styles, like Civic, which contain many repeatedly used symbols.

> **TIP:** If you are using ArcGIS in a network environment on a computer with multiple users who log in separately, you may experience trouble saving symbols. If this happens, cancel, reopen the Properties box, and try to save the symbol again. If the problem continues, notify your system administrator. You can continue the lesson without saving the symbol.

5➡ Click on OK in the Editor box. The new symbol appears in the upper right corner as the current symbol.

5➡ Click on Save in the Symbol Selector box to save the new symbol permanently. Give it the name "CityStar", and leave the Category box blank.

This symbol is a little too large. However, instead of changing its size by clicking it in the Table of Contents, we can now demonstrate another way to modify symbols: accessing the layer's Symbol Properties.

6➡ Right-click the Towns layer name in the Table of Contents, and choose Properties from the menu. Click the Symbology tab if necessary.

6➡ Notice that the map type is currently set to Features: Single Symbol (in the box labeled Show to the left), and the CityStar symbol is displayed.

6➡ Click on the symbol to open the Symbol Selector.

6➡ Reduce the size of the symbol to 13 points.

6➡ Click OK to close the Properties box.

Now let's learn about using styles to save and access symbols, and to organize frequently used symbols into functional groups. Styles are managed using the Style Manager.

7➡ Click Tools in the main menu bar and choose Styles > Style Manager.

The window on the left contains the default styles, including your own personal style and the ESRI style. It may also contain other styles. Let's examine some of these.

7➡ Click the plus sign next to the ESRI style to expand it.

7➡ Click the folder labeled Marker Symbols.

The ESRI style contains many folders for markers, lines, shades, text, colors, and more. Clicking on the Marker folder shows all of the common marker symbols that appear in the Symbol Selector.

TIP: The folders are color coded. A yellow folder is one for which you have read-write permission, a grey folder has read-only permission, and a blank folder is empty.

7➡ Click the Styles button and view the list of available styles. Notice that the ones currently displayed in the Manager have a check mark.

7➡ Click the Forestry Style so that it is also displayed, and click the plus sign to expand its contents.

Notice that many of these folders are empty—whoever created this style only chose to create a few types of symbols.

3. Which types of symbols does the Forestry style contain? _____

8➜ Click on the Marker Symbols folder in the Forestry style and examine the long list of symbols.

8➜ Click on the small icons to the lower right to change whether the symbols are displayed as large icons, a list, or with details.

8➜ Examine the Color Symbols folder to see specific colors that were created for forestry maps, so that all the maps would look similar if these were used.

8➜ Examine each of the other folders in turn to see their contents.

9➜ Close the Forestry folder by clicking the minus sign.

9➜ Expand the top entry—the one with your name. This is your personal style.

9➜ Click the only folder with symbols, the Marker Symbols folder.

9➜ Notice that the new CityStar symbol is here!

When you saved the CityStar symbol, it was saved to your personal style file. Any other symbols you create are also stored here. You can also use this resource, however, to create a new style.

10➜ Click the Styles button and choose Create New from the bottom of the long list.

10➜ Navigate to the mgisdata\MapDocuments folder, if necessary, and enter the name "Mystyle" for the new style. Click Save.

10➜ Notice that the new style now appears in the list.

10➜ Expand the new style. All of the folders are currently empty.

Use the Style Manager to copy and paste symbols between styles. We will put the new CityStar symbol in Mystyle, along with several symbols from the ESRI and Forestry styles.

11➜ Click your personal Marker Symbols folder to show the symbols in it.

11➜ Right-click the CityStar symbol and choose Copy.

11➜ Click the Marker Symbols folder in Mystyle, and then right-click the window on the right and choose Paste.

You can even edit symbols in the Style Manager to quickly create many new symbols.

12➜ Right-click the blank area again and paste another copy of CityStar into your style.

12➜ Right-click the copy and choose Properties.

12➜ Select the yellow circle layer, and change its color to light green. Click OK.

12➜ Click the Copy of CityStar name and change it to "CityStar2".

Now let's bring some symbols from other styles into Mystyle.

13➜ Locate the School 2 marker symbol in the ESRI style, and copy it into Mystyle.

13➜ Change its color to red.

13➜ Name it "School".

14➜ Locate the Mangrove fill symbol in the Forestry style and copy it to the Fill Symbols folder of Mystyle.

14➔ Change its background color to pale green and the mangrove foreground color to dark green. Change its outline to a thick dark green symbol. Click OK.

14➔ Click Close to close the Style Manager.

To use these new styles, simply load them from among the other standard styles. The standard styles are stored in a common directory in the ArcGIS installation directory—your own styles will be saved wherever you have put them.

15➔ Right-click the Schools layer and choose Visible Scale Range > Clear Scale Range, so the schools are being displayed.

15➔ Click the Schools symbol to open the Symbol Selector.

15➔ Click the More Symbols button to see that the Mystyle style has been added to the list.

15➔ Make sure that Category is set to All.

15➔ Turn off the other Styles, such as ESRI, to make your own symbols easier to find. Click the More Styles button and click each of the checked ones in turn to turn them off.

15➔ Click your new School symbol to select it and click OK.

TIP: If Mystyle is not in the list, choose Add, navigate to the MapDocuments folder, and select it.

Now you can select a red school symbol quickly from your style to display the schools, instead of having to change the color each time. This simple example will hopefully provide some insights into how to employ styles in your own applications.

16➔ Click the symbol of the Schools layer to open the Symbol Selector and turn the ESRI style back on.

16➔ Close the Symbol Selector window.

16➔ Uncheck the Schools layer in the Table of Contents, so that the schools are no longer displayed.

Creating maps based on attributes

Displaying a data layer with a single symbol is only the beginning. Often you can enhance patterns in data and improve the readability of maps by using different symbols, depending on the feature's value in an attribute field.

17➔ Right-click the Towns layer name in the Table of Contents, and choose Open Attribute Table.

17➔ Scroll through the fields, and notice one named CLASS and another called POP100 containing the population. We will use both of these fields to make several kinds of maps.

17➔ Close the table.

Recall that categorical data can be displayed using a unique values map, and that a variety of map types are available for quantities (numeric) data. In this table, CLASS is categorical data and POP100 is quantities data. First we will make a unique values map based on the CLASS field.

17➔ Right-click the Towns layer name in the Table of Contents, and choose Properties from the menu. Click the Symbology tab if necessary.

17➔ Choose Categories: Unique Values as the type of map.

17➔ Choose CLASS as the Value field.

17➔ Click Add All Values to display the categories in the legend box.

17➔ Choose a different color scheme—one with dark, strong colors instead of pastels.

17➔ Click Apply. The markers on the map change, but the Properties dialog box stays open. Move the box aside if necessary to view the map.

TIP: If a class has only a few features, it may be grouped with the other small classes into a class called All Other Values. For example, when mapping land use there might be many residential and industrial parcels, but only a few zoned for parks. To display the parks in the regular legend, click the Add Values button and choose it to be shown in the regular legend. If the All Other Values entry displayed in the legend is not desired, uncheck the box next to it.

Because the markers are small, it is a bit hard to see the differences. Let's make a few changes to these existing symbols to improve the readability of the map.

18➔ Double-click the symbol for City, and the Symbol Selector appears.

18➔ Edit the symbol size from 4 points to 7 points.

18➔ Double-click the other symbols to choose a more contrasting color if needed.

18➔ The CDP class stands for Census Designated Place. Under the Label column, click the CDP text to select it and replace it with Census Desig. Place, which is more descriptive.

18➔ Finally, let's put City at the top of the list. Click on the City symbol to select the row, and click the large Up arrow to move it to the top of the legend.

18➔ Click OK to apply the changes and close the Properties dialog.

The larger cities are now easily distinguished from the smaller towns. Let's go a step further and classify the cities by population.

19➔ Open the Symbology Properties dialog box for the Towns layer and choose Quantities: Graduated Symbols as the map type.

19➔ Choose POP100 as the Value field. Do not choose a Normalize field. Click Apply.

Notice how the population of the towns is immediately apparent now, with a few larger cities and many smaller ones (Fig. 4.20). Again, we can make some adjustments to this default set of symbols to get the effect we want.

20➔ Click on the Template symbol button and choose another color for the symbol from the Symbol Selector when it appears. Click OK, and notice that all the symbols are now the new color.

20➜ Change the maximum symbol size from 18 points to 15 points. Click the Tab key to enter the "15" value—the Enter key causes the dialog box to close.

20➜ Change the number of classes to six.

20➜ Right-click the legend area and choose Flip Symbols from the context menu. Repeat and choose Reverse Sorting. Now the classes are in reverse order.

20➜ Click the Label column heading and choose Format Labels from the context menu.

20➜ Click to ensure that the Category is Numeric, choose Number of Significant Digits and set it to 2, and check the Show Thousands Separators box.

20➜ Click OK to close the format window, and OK again to apply the changes to the map (Fig. 4.20).

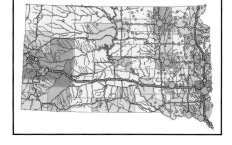

Fig. 4.20. After step 20, your map should look like this.

TIP: Using the Format Labels option to make sure the class labels have evenly rounded values is good cartographic practice. Neglecting this detail presents an unprofessional picture.

4. What type of map is being used to display the Roads layer? _____ Which field is the basis for the map? _____

Next, let's create some maps using a polygon data layer—the counties.

21➜ Open the Symbology Properties for the Counties layer.

21➜ Choose Quantities: Graduated Colors as the map type, and POP2000 as the Value field.

21➜ Pick a different color ramp, if desired.

21➜ Click Apply, and move the dialog window to better see the map (Fig. 4.20).

Now let's experiment with normalizing data before plotting. First we will normalize by the sum of all the values in the population field (i.e., the total population of the state).

22➜ Choose Percent of Total as the Normalization option.

22➜ Click the Label column heading and choose Format Labels. Click the Numeric Options button. Set it to show one decimal place only. Click OK, OK, and Apply.

Now we'll try a different normalization to show the percentage of each county that is Native American (Fig. 4.21).

Fig. 4.21. Your map after step 23

23➔ Choose AMERI_ES as the Value field and POP2000 as the Normalization field.

23➔ Format the labels. Choose Percentage as the category, and click the second option to convert fractions to percentages. Finally, use the numeric options to set it to show one decimal place. Click OK in all boxes.

Classifying data

Now let's try using some of the classification options. Up to now, we have been using the default Natural Breaks classification method with five classes.

24➔ Create a graduated color map of the counties based on POP2000 (be sure to set Normalization back to <None>). Click OK to close the Properties box and view the map.

24➔ Double-click the Counties layer to open the Properties again.

24➔ In the Symbology tab, change the number of classes to 8 and click Apply. Notice how the map changes.

Changing the number of classes is easy. Now let's try changing classification methods. First we'll do an equal interval map.

25➔ Click the Classify button in the Symbology tab.

25➔ The classification method is currently set to the default, Jenks Natural Breaks. Notice how the classes are clustered at the low end of the scale.

25➔ Change the method to Equal Interval, and make sure that 8 classes are indicated. Notice the regularly spaced class breaks in the graph.

5.   What is the size of the class intervals? _____

25➔ Click OK and OK and view the map.

Notice that most of the counties are the lightest color, and the larger classes have only one county in them. Some classes are completely empty. This map gives a very different view of the data, emphasizing that people in South Dakota are highly concentrated in a few areas, and that most counties are relatively unpopulated.

Imagine, however, that you are mainly interested in the demographics of rural South Dakota. You cannot see much in this map because the few very large counties skew the classification. So you decide to increase the number of classes.

26➔ Double-click the Counties layer.

26➔ Increase the number of classes to 12 and click OK.

Hmm, this change does not help very much, because the large counties are still completely dominating the distribution of classes. A better way to solve this problem involves excluding certain counties from consideration. Let's try it.

27➔ Double-click the Counties layer.

27➔ Set the number of classes back to 8.

27➔ Click the Classify button.

27➜ Click the Exclusion button.

27➜ Enter the expression "POP2000" >20000 by double-clicking the POP2000 field, single-clicking >, and typing 20000 in the expression box. Click OK, OK, and OK.

Notice that the excluded counties are not displayed, and the remaining counties now show the variation in rural populations across the state.

Next, imagine that you have been instructed to send a packet of demographic materials to each county based on the number of people in the county. Counties with fewer than 20,000 get one packet, those with up to 50,000 get a different packet, and those with over 50,000 get a third packet. Create a map with these exact classes to determine which counties get which packet.

28➜ Double-click the Counties layer.

28➜ Set the number of classes to 3.

28➜ Click the Classify button.

28➜ Set the method to Manual.

28➜ Click the Exclusion button, and click Clear to clear the exclusion expression. Click OK.

28➜ In the Break Values box to the right, click each value in turn, *beginning with the largest,* and type in "150000" for the highest class, "50000" for the middle, and "20000" for the lowest. (If you start with the smallest it will think the classes are out of order and will reset them after you type in the value.)

28➜Click OK and OK to view the map.

TIP: The highest break value can be determined by reading the maximum data value in the statistics box in the upper right corner. Setting the last class break above the maximum value ensures that all of the data are represented. Notice that no zero class break is needed.

Now imagine that you are investigating the cost of housing in South Dakota. When thinking about dollar amounts, it is easier if the classes are both consistent in size and rounded to even values. Use the defined interval method to get this result.

29➜ Double-click the Counties layer.

29➜ Change the Value field to MEDIAN_VAL.

29➜ Click the Classify button.

29➜ Choose Defined Interval as the method.

29➜ Enter "10000" as the interval size and click OK.

29➜ For an added touch, click the Label heading and choose Format Labels. Select Currency and click OK. Dollar signs and commas are added to the labels.

29➜ Click OK and OK to view the map. Housing prices are more evenly distributed in South Dakota than people.

As a final classification example, we will make a standard deviation map of South Dakota showing the population by county.

30➜ Set up a graduated colors map based on the POP2000 field, with no normalization field.

30➜ Click the Classify field.

30➜ Choose Standard Deviation as the type of map.

30➜ Choose the interval size as half a standard deviation (the number of classes is dimmed because it is determined by the chosen size interval). Click OK.

30➜ If needed, format the labels to have one decimal place. Close the Properties and view the map (Fig. 4.22).

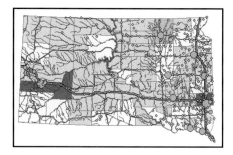

Fig. 4.22. Your map after step 30

Other types of maps

ArcMap offers some other interesting map types. Let's try a couple. The dot density map shows population in a different way than a choropleth map.

31➜ Double-click Counties to open the Symbology tab.

31➜ Choose Quantities: Dot Density as the type of map.

31➜ Select POP2000 as the Value field and click the > button to use it in the map. Notice that more than one field can be selected.

31➜ Right-click the dot symbol next to POP2000 and change its color to black.

31➜ In the Background box, click the button on the right and change the background color to light cream.

31➜ Experiment with the slider bar to set the dot value. For the first map, set the value to about 1 dot per 1000 people.

31➜ Click OK and view the map. Turn off some other layers if needed to get a better view.

For a more detailed map, let's change the dot size and density distribution.

32➜ Double-click the Counties layer.

32➜ Change the dot size to 1 point and the density to 300. Click OK.

As a final example, let's make a pie chart map showing the percentage of whites, blacks, and Native Americans in each county.

33➜ Open the symbology properties for Counties, if necessary.

33➜ Choose Charts: Pie as the map type.

33➜ Locate the WHITE field in the left column, click it to select it, and click the right arrow to move it to the right column. Repeat for the BLACK and AMERI_ES fields.

33➜ Right-click each symbol and choose another color from the pop-up palette—beige for WHITE, dark brown for BLACK, and dark orange for AMERI_ES.

33➜ Choose a background color that harmonizes with the pie colors.

33➜ Click the Properties button, and uncheck the Display in 3D box. Click OK.

33➜ Click the Size button, and decrease the pie size to 20 points so they fit better inside the counties. Click OK.

33➜ Click OK to close the Properties box and effect the changes (Fig. 4.23).

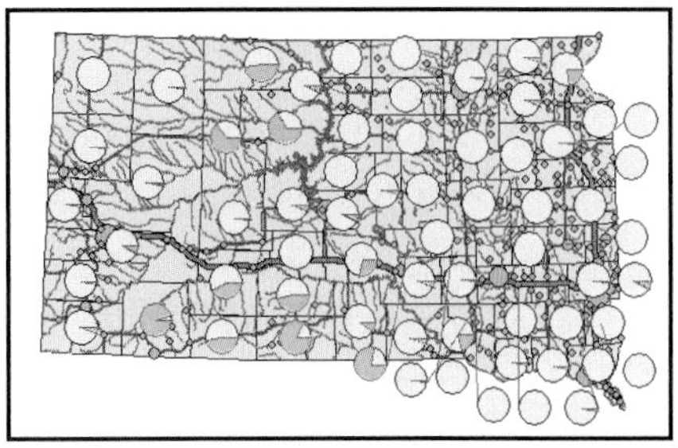

Fig. 4.23. Your map after step 33

Working with layer files

Developing symbol sets can be very time-consuming. Layer files mitigate this problem by storing the display properties of a layer. The layer file can be used whenever that particular symbology is desired, in the same map document or a different one. Let's see how it works by saving our Roads layer with its symbology in a layer file.

34➜ Return the Counties theme to a single symbol map type.

34➜ Right-click the Roads layer and choose Save As Layer File from the context menu.

34➜ Navigate to the mgisdata\Sdakota directory and name it "SD_Roads.lyr". Click Save.

Group layers can store display properties for multiple data sets together. Take a look at the Rivers and Lakes layers. In essence these belong together, but one is necessarily a polygon file and the other is a line file. However, we can group them together so we can turn them on and off together and save them in a group layer file.

35➜ Click on the Rivers name to select it. Hold down the Ctrl key and also click the Lakes layer, so they are both selected.

35➜ Right-click on one of the layers and choose Group from the context menu.

35➜ Click the New Group Layer name and click it again to enter a new name. Call it "SD_Water".

35➜ Click the SD_Water check box to turn off both the lakes and rivers. Click it again to turn them on.

35➜ Right-click the SD_Water group and choose Properties from the context menu.

A group has only two properties tabs: a General tab and a Group tab. The General tab is used to name the group, turn it on/off, and set a scale range, just as for other layers. The Group tab is used to add or remove group members and manage the properties of the individual members.

36➜ Click the Group tab and click the Lakes layer to select it. Use the black arrows to move it up or down in the list to determine the draw order.

36➜ Click the Properties button, and click the Symbology tab.

36➜ Notice the similarity of this tab to the Symbology tab of Layer Properties. However, do not set anything at this time. Simply click Cancel to close the Properties menu. Also click Cancel to close the Group properties menu.

36➜ Right-click the SD_Water group name and choose Save As Layer File.

36➜ Navigate to the mgisdata\Sdakota folder and enter the name "SD_Water.lyr". Click Save.

Now we will start ArcMap again and examine how the layer file works.

      37➜ Open a new map document; choose to save the current one if you wish.

      37➜ Click the Add Data button, navigate to mgisdata\Sdakota, and choose the sd_majrds.shp shapefile.

      37➜ Click the check box to display the roads if necessary.

Notice that by default, the roads are being displayed with a single symbol. The layer file we created had no effect on the roads shapefile itself; it only references the roads shapefile as the source and stores additional information about how to display the data. Now let's try the layer file.

      38➜ Click the Add Data button again, navigate to the mgisdata\Sdakota folder, and choose SD_Roads.lyr.

      38➜ Right-click the SD_Roads layer name and choose Properties from the menu, then click the Source tab. Verify that the source is the sd_majrds.shp file.

6.   What is the coordinate system of the sd_majrds shapefile? _____

      39➜ Close the Properties sheet.

      39➜ Use Add Data one more time to add the SD_Water group layer file. Notice how both lakes and rivers are added together.

      39➜ Right-click the SD_Roads layer and choose Remove. Repeat for the SD_Water layer, so that the Table of Contents is blank again.

Finally, let's use a layer file to specify symbols for a different source layer. We already determined that the road classification is based on a field called FUNC_CLASS. The South Dakota roads came from a much larger file for the entire United States. We can apply the SD_Roads layer file to this larger data set to draw its roads using the same symbols because they both share the FUNC_CLASS field.

      40➜ Use Add Data to add the shapefile majrdnet.shp from the mgisdata\Usa folder.

      40➜ Use the Zoom tools to zoom into a single state to see the roads more clearly.

      40➜ Right-click the majrdnet layer and choose Properties from the menu, then click the Symbology tab.

      40➜ Click the Import button.

      40➜ Fill the button to import from an existing map layer or layer file.

      40➜ Click the Browse button and locate the SD_Roads.lyr file in the mgisdata\Sdakota folder. Choose it and click Add.

      40➜ Click OK in the Import Symbology dialog menu.

      40➜ Confirm that the FUNC_CLASS field will be used to display the layer.

TIP: If a different field is chosen to display the layer, it must contain the same values as the original field, or ArcMap will not be able to match the features with their new symbols. For example, if the US layer had road designations of Primary or Secondary, it would not work with roads classified as Interstate, Highway, and Local.

40➔ Click OK to apply the symbology and close the Properties menu.

➔ Close ArcMap and save changes if desired.

This is the end of the tutorial. However, you may continue and do the optional section on displaying rasters that follows.

Displaying rasters (optional)

Two main types of rasters are available for use in ArcGIS. The first type are called **images** and come in a variety of formats. All images are based on numeric values, and they are frequently stored as binary integers because that is the most efficient format. Images can be displayed, but they cannot be analyzed without an additional software extension called Image Analyst.

Displaying image rasters

➔ Start ArcMap and choose to start with a new empty map.

41➔ Click Tools on the main menu bar and choose Options.

41➔ Click the General tab, and uncheck the box to make newly added layers visible by default. This option prevents new rasters from obscuring old ones, and saves drawing time when you are working with large rasters.

41➔ Click the Add Data button and navigate to the mgisdata\BlackHills\Sturgis directory.

41➔ Add the rasters geolgrid, dem2, hillshd2, and TM_24Sep98_utm.

42➔ Turn on the TM_24Sep_utm raster. It contains a Landsat satellite image showing part of the Black Hills near Sturgis, South Dakota.

42➔ Double-click the raster name to open its Properties sheet.

42➔ Click the General tab and note that a scale range may be set for rasters just as for vector files.

42➔ Click the Source tab, and scroll down and examine the entries.

Notice several things about this raster. The number of **rows** and **columns** are given, as well as the **cell size.** The image contains seven separate bands of information, so in effect this image actually contains seven images in one. The image is stored as **unsigned integer,** meaning it is stored as positive integers (signed integers may contain negative values). The **pixel depth** refers to the number of bits used to store each cell value. This is an 8-bit image, meaning each cell must have a value between 0 and 255. Storing larger numbers would require 16-bit or 24-bit data depth. **Pyramids** are special add-on data structures that permit faster display of the image at different scales. They take a few moments to build, but they save on display time later.

The Statistics section gives information about each band, including the minimum, maximum, and mean cell values, as well as the standard deviation.

43➔ Click the drop-down box to examine the statistics for each band in turn.

7. List the minimum, maximum, and mean cell values for Band 4. _____

Now we will examine some ways to display this image.

44➜ Click the Symbology tab and examine its contents.

Notice the two ways to display the image. Stretched is appropriate for single-band images, or to display one band of a multiband image at a time.

44➜ Click Stretched and select Band 4 as the band to display.
44➜ Notice that the values extend over a smaller range than 0-255.
44➜ For the Stretch Type, choose None, and click OK.

When displaying this 8-bit image, the cell value is assigned a symbol along a scale from 0-255 digital numbers, or DN. Usually the values range from black = 0 to white = 255, with all shades of grey in between. Figure 4.24 shows a histogram of this original image and a stretched image. A histogram graphs the number of pixels occurring for each DN. As the histogram of the original image shows, the values fall primarily in the dark grey regions of DN. In displaying this original image, less than half of the possible symbols would be used and they would be chosen only from the darker colors. The result would be a very dark picture such as the one now on the screen. Therefore, images are typically stretched, to make use of the entire range of colors. The histogram of the stretched image in Figure 4.24 goes from 0 to 255 DN, and the image appears brighter with much better contrast. Stretches may be performed using a variety of methods: several common stretches are available in the Symbology window.

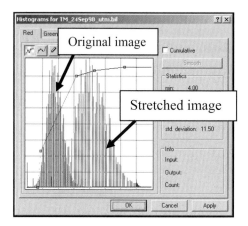

Fig. 4.24. Histogram of original and stretched values in image

45➜ Open the Properties and change the Stretch Type to Standard Deviation and click OK.
45➜ Observe the image, then open the Properties again and click the Histogram button. (Your histogram may not match the one shown in Figure 4.24.)

The stretched image contains a larger range of greys and has much better contrast. The histogram in Figure 4.24 represents the stretch graphically. The black lines indicate the actual cell values in the image, with the height of the line indicating the number of pixels with that value. This image's values are grouped to the left. The colored lines show the display value of these cells—notice that the display values are fully stretched between 0 and 255 to make full use of the range of greys. Essentially the black lines have been stretched apart.

45➜ Click Cancel to close the Histogram box, and OK to apply the changes.
➜ Experiment with the other types of stretches if desired.

RGB composites display three bands at a time. The computer screen has a display gun with three colors: red, green, and blue. In a color composite, one band of the image is assigned to each gun, and the combination creates a full-color picture.

46➔ Open the Symbology properties, if necessary, and choose RGB Composite as the display type.

46➔ Click OK and examine the image.

Currently in this image, the first, second, and third bands are assigned to red, green, and blue, respectively. If the value in Band 1 is high (close to 255) and the other two bands have low values near zero, a bright red color will result. If red and green are both high and blue is low, a mixture of red and green will create a yellow color. All the colors in a computer color image are composed of different proportions of red, green, and blue light, as dictated by the values in the bands.

In a Landsat image, Band 1 records the blue light from the Earth's surface as measured by the satellite sensor, Band 2 records green, and Band 3 records red.

47➔ To see the image as if it were a color photo, open the Properties and use the drop-down boxes to assign Band 3 to the red channel, Band 2 to green, and Band 1 to blue. Click OK.

Band 4 of Landsat measures near-infrared reflectance. Vegetation reflects this wavelength of light strongly. Foresters and others often use a 4-3-1 band combination to assess the presence and health of vegetation.

48➔ Open the Symbology properties again. Assign Band 4 to red, Band 3 to green, and Band 1 to blue. Click OK.

Different band combinations highlight different features in Landsat images. The combination 7-4-1 is often used to contrast vegetation with urban areas and/or dry plains. Try it, and experiment with some other band combinations.

The second type of raster is called a **grid.** A grid is a special raster data format created for use with ArcInfo and ArcView. Whereas images can only be displayed, grids can be both displayed and analyzed. Grid data may be either categorical or numeric. Categorical grids are fundamentally based on numbers, but a text string may be stored as additional identification. For example, a geology grid actually stores numbers indicating each geologic unit, but each number is associated with a text description, such as 1 = Minnelusa Limestone, 2 = Spearfish Fm, etc. Numeric grids may be stored as integers or floating-point values.

49➔ Turn off the Landsat image and turn on the raster dem2. The file is an elevation map of the same area in the Black Hills. Each cell represents an elevation value, in meters, above sea level.

49➔ Double-click the dem2 raster name to open the Properties window, and click the Symbology tab.

Several differences between images and grids are immediately apparent in the Properties window. First, two additional tabs are present, a Fields tab and a Joins & Relates tab. Whereas images are simply arrays of binary data, grids also have an attribute table and can contain attribute information. Second, more display options are available for grids.

50➔ Close the Properties window.

50➔ Right-click the dem2 raster name and choose Open Attribute Table from the
context menu.

Three fields are present. The first is the familiar ObjectID. The second field is called Value and
contains the actual values in the grid. The third field, Count, indicates the number of cells
containing that value. The table is sorted by value. This table indicates that the lowest elevation
in this map is 872 meters and that one cell has that value.

8. How many cells have the value of 1000 meters? _____

The second difference between images and grids concerns the types of display that are possible.
Grids only have one band, so no RGB Composite option is available. Instead, you may use a
classified, stretched, or unique values display type.

The classified and stretched options apply to continuous data. Recall that continuous data can
have a large number of possible values, which usually change smoothly from one to another
across the surface. This dem2 grid contains continuous data, a smoothly varying plethora of
elevation values. The Stretched option is similar to the Stretched option we have already used to
display images such as the Landsat image described previously. However, instead of varying on
a 0-255 scale, the values occupy a wider range. Nevertheless, the stretch concept is very similar.

9. What are the minimum and maximum elevations in this grid? _____

51➔ Close the dem2 attribute table.

51➔ Open the Properties for the dem2 raster and click the Symbology tab.

51➔ Use the drop-down box to change the color ramp used to display the image, and
click Apply. Find one that looks good and keep it.

51➔ At the bottom of the Symbology window, select a contrasting color to display
NoData by clicking the button provided. Notice that the upper left corner of the
image turns that color.

Images and grids often contain areas where the values are unknown. Usually these are stored as a
special data value called NoData. The upper left corner of this DEM has no data because that
area is outside the quadrangle boundaries. However, because a grid must be perfectly
rectangular, the cells must be present and have some value. Setting them to NoData allows the
program to ignore the cells when doing calculations or displaying the grid. Images can also have
unknown values, but they usually do not have the special NoData value. Instead, another value,
typically 0, is used. This 0 value can also have its own display color, set with the check box
provided.

Grids can be shown with color classes, similar to the classification of a field in an attribute table.

52➔ Change the display type from Stretch to Classified.

Notice that the Symbology tab looks very similar to a graduated color classification for vector
features. The Value field is automatically set to Value. The user chooses the color ramp and the
classification options.

52➔ Choose a rainbow color ramp, and set the number of classes to 15. Click Apply and view the map. Zoom in a bit for a better view of the classes.

53➔ Next, click the Classify button.

53➔ Set the classification method to defined interval, and set the interval to 40 meters.

53➔ Zoom into the upper right area and examine the classification. The colors change every 40 meters, rather like having a color map with a 40-meter contour interval.

➔ Experiment with a few other classification schemes for this grid.

The third display option is unique values, which is useful for portraying discrete grids. A discrete grid contains only a few values, and those values usually occupy large, continuous areas and change abruptly from one value to another. Dem2 is not a discrete grid, so we will experiment with the unique values option using geolgrid instead.

54➔ Close the Properties for dem2 if necessary.

54➔ Click the Full Extent button.

54➔ Turn off the dem2 grid and turn on the geolgrid raster.

54➔ Open the geolgrid Properties and click the Symbology tab.

Notice that the grid's default display uses unique values—ArcMap recognized that only a few discrete values were present and automatically applied the correct type of display. The user can modify the individual colors or the color scheme, as for a unique values map for vector features.

54➔ Set the color scheme to pale pastels.

54➔ Click OK to apply the changes and close the Properties.

As a final example, we will take a look at the hillshd2 grid.

55➔ Turn off the geolgrid raster and turn on the hillshade raster.

10. Is this raster an image or a grid? _____ List two clues you could use to answer this question._____

A hillshade map is created from an elevation map by calculating the brightness and shadows on the surface that would be present from an illuminating source. The land surface appears as it would to an observer flying over in a plane on a sunny day. A special display feature called Transparency can be combined with a hillshade to produce very informative backgrounds for maps.

55➔ If necessary, click and drag the geolgrid layer until it is above the hillshd2 layer in the Table of Contents. Turn it on.

55➔ Open the geolgrid Properties and click the Display tab.

55➔ Set the Transparent value to 40 percent.

55➔ Click OK.

55➔ Examine the map. Zoom in and examine an area in more detail.

This display technique shows the land surface shape AND the geology at the same time. Transparency can be applied to either raster or vector layers.

This is the end of the tutorial.

→ Exit ArcMap.

Exercises

Open the ex_4.mxd map document and use it to answer the following questions.

1. Create a map showing hospitals in South Dakota, giving the hospitals the symbol shown with a red foreground and light blue background. **Capture** your map.

2. Add the watershe.shp file from the mgisdata\Sdakota folder, and use it to create a map of South Dakota showing just the rivers, lakes, and watersheds based on the Subregion field. **Capture** your map.

3. Create a group layer containing the rivers, lakes, and watersheds, and call it "sdhydro2". **Capture** the Table of Contents showing your group.

4. Starting with your map from Exercise 3, add a dot density map showing the density of farms in each county, using black 2 point dots, with the ratio 1 dot = 30 farms. Which watershed subregion appears to have the highest density of farms? **Capture** your map. *Missouri - Big Sioux*

5. Create a map of South Dakota counties showing the median rent (field MEDIANRENT) in each county, using a defined interval of $100. How many classes does it contain? _____ *4* Put a title on your map using the New Text tool. **Capture** your map.

6. Create a graduated color map of the *density* of farms in each county. Use a Natural Breaks classification with five classes, and make sure the legend values have only two decimal places. Put a title on the map, including the units. **Capture** the map AND the legend.

7. Add the sd_fedlandp.shp shapefile from the mgisdata\Sdakota folder, showing federal land in South Dakota. Create a map showing the ownership of the land using the colors and legend labels shown here (i.e., your legend should look like this when you are done). Make BIA orange, BLM lavender, BRec blue; DoD dark purple, FS light green, FWS red, and NPS dark green. Make the counties white, and turn off all the other layers. **Capture** the map AND the legend.

▨	Bureau of Indian Affairs
▨	Bureau of Land Managment
▨	Bureau of Reclamation
▨	Department of Defense
☐	Forest Service
▨	Fish and Wildlife Service
▨	National Park Service

Exercise 7

8. Modify the Forest Service symbol in the map from Exercise 7 so that it contains a random pattern of dark green pine trees on the light green background, as shown here. (Hint: the pine tree shown is from the Civic marker style.) Zoom into the southwestern part of the state with the abundant Forest Service land so you can see the effect most clearly. **Capture** your map.

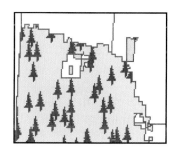

Exercise 8

9. Which geologic unit number (Value) in the geolgrid raster (in the BlackHills/Sturgis folder) has the greatest area?_____ What is its area in square kilometers?_____ What percent of the map area does this unit represent?_____

10. Use the hillshade and DEM grids in the Sturgis folder to create a color shaded relief map of the central area of the dem2 grid, so that the color represents elevation. It should look like the map shown here. **Capture** your map.

Exercise 10. Hillshade image with transparent color overlay representing elevation

Challenge Problem

Using the data in the folder mgisdata\Usa, create a map layout that shows the following:

> The conterminous United States (lower 48 states) showing population by county, with an extent rectangle showing your county of birth. Also show the major roads using the same symbology as in the South Dakota map we used earlier in this chapter.

> An inset map showing a close-up of the county you were born in, the major roads, and the cities displayed with varying size symbols according to population. Label each city. (If there are a large number of towns in your birth county, only label the largest ones, but try to include at least five city labels. If there are not enough towns, expand your view a bit to include at least five towns.) Label the county, too, in larger type.

> Set the map projection of both frames to USA Contiguous Equidistant Conic (follow the path Predefined > Projected Coordinate Systems > Continental > North America).

> Place a callout label showing your approximate birthplace on both maps.

> Put titles on your maps, being sure to include the name of your birth county.

> Be sure to arrange the data frames, titles, and colors for an aesthetically pleasing map.

> Print the map by choosing File > Print when in Layout mode.

Where Maribeth Price was born.

Challenge exercise example

TIP: To make the Chesapeake Bay blue in the sample map, I changed the background color of the data frame to light blue using the Frame tab of the Data Frame Properties sheet.

Skills Reference

Using the Symbol Properties Editor

Symbols are created from one or more layers of symbol objects. For example, the green star in Figure 4.25 is composed of a black hollow outline and a green fill layer. Users can create new layers, put predefined symbols in them, and modify their colors and other properties, to create new symbols. Symbols can also be created from imported bitmap images.

1. Click the Properties button in the Symbol Selector to access the Symbol Property Editor (Fig. 4.25).

2. Add new layers using the + button in the Layer area of the window.

3. Select the type of symbol character to put in the layer.

4. Select the desired character.

5. Modify the size, color, thickness, and other properties of the character.

6. Remove layers or change the order if needed.

7. Click OK when finished creating the symbol.

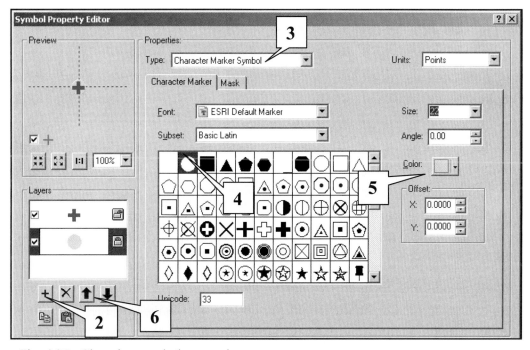

Fig. 4.25. Changing symbol properties

Creating a map based on an attribute

1. Right-click the layer name and choose Properties from the context menu.

2. Click the Symbology tab (Fig. 4.26).

3. Choose the type of map desired. OR, use the Import button to bring in the symbology of an existing layer file and apply it to this data set.

4. Select the field to base the map on.

5. Choose a Color Scheme.

6. If no values are displayed in the window yet, click Add All Values to view the categories.

7. For Graduate Color or Symbol maps, choose a normalization method and field if desired.

8. Double-click a symbol in the list to change its properties using the Symbol Selector.

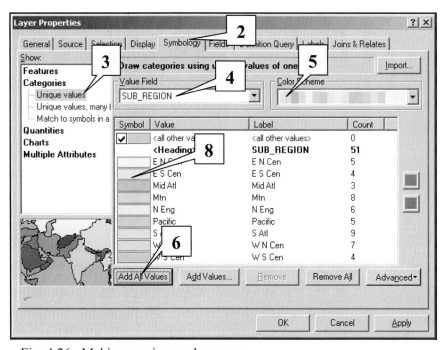

Fig. 4.26. Making a unique values map

TIP: If a class has only a few features, it may be grouped with the other small classes into a class called All Other Values. For example, when mapping land use there might be many residential and industrial parcels, but only a few zoned for parks. To display the parks in the regular legend, click the Add Values button and choose it to be shown in the regular legend. If the All Other Values entry displayed in the legend is not desired, uncheck the box next to it.

Other map types, such as dot density, have different settings in the Symbology tab. Experiment with setting these properties, or look up the recommended settings in the Help files.

Classifying attributes for a map

For Quantities map types, ArcMap applies a default classification strategy of Jenks Natural Breaks with five classes. Use the following procedure to modify the classification.

1. Open the Symbology tab in the Layer properties. Make sure one of the quantities maps is selected for the map type and that a field has been chosen for the map.

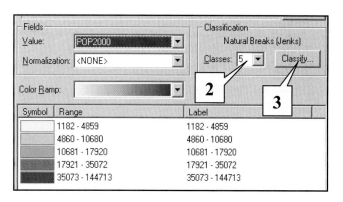

2. Use the drop-down box to change the number of classes, if desired (Fig. 4.27).

3. For other changes, click the Classify button to open the Classification dialog box.

Fig. 4.27. Classification box in the Symbology tab

4. Change the classification method and number of classes, if desired (Fig. 4.28).

5. To exclude certain records, such as those containing zeros or NoData values, click the Exclusion button, and enter an expression defining the records to be excluded.

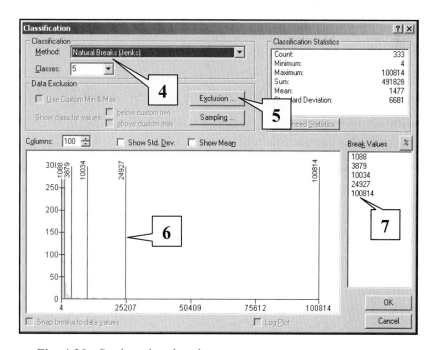

Fig. 4.28. Setting class breaks

6. The Sampling button is used to reduce the number of features used to calculate statistics, and it is useful when data sets are very large. The default sample is 10,000 records.

7. To manually set class breaks, either click and drag on the blue lines in the graph, or type the break values directly into the box on the right. When either of these actions is performed, the classification method automatically changes to Manual. The Statistics shown in the upper left corner can be helpful when assigning classes manually.

8. Click OK when done setting the classes.

Modifying the appearance of the legend

Right-clicking the legend area provides a context menu with several options for modifying the legend (Fig. 4.29).

1. Use **Flip Symbols** to reverse the order of the symbols only.

2. Select a symbol by clicking on it (or multiple symbols using Ctrl-click), and choose **Properties for Selected Symbols** to change their appearance with the Symbol Selector. Use **Properties for All Symbols** to launch the Symbol Selector and modify the appearance of all of the symbols at once, such as changing them all from red to green.

3. Use **Reverse Sorting** to sort the classes and symbols in the reverse direction.

4. Select one or more classes and delete them with **Remove Class(es).**

5. Select two or more classes and merge them with **Combine Classes.**

6. Click **Format Labels** to apply numeric formatting options to the labels, such as changing the number of decimal places or including thousands separators.

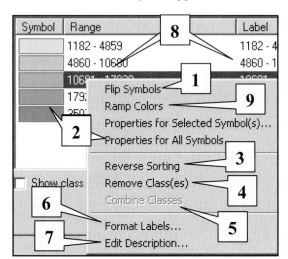

Fig. 4.29. Modifying the legend

7. Click **Edit Description** to enter or change a descriptive headline that appears at the top of the legend.

8. Click on any Range (except the lowest value of the first range) or Label in the legend to type in new values.

9. To create a new color ramp, set the first and last symbols to the desired end colors, then choose the **Ramp Colors** option.

10. When making graduated symbol maps, change the template symbol (the same symbol that increases in size with each class) by clicking the button marked Template.

Creating a map layer

A map layer is a set of stored characteristics for displaying a spatial data set.

1. Set the symbology, labeling, and other display properties of the data layer using the techniques learned in this chapter.

2. Right-click the data layer name in the Table of Contents, and choose Save As Layer File from the context menu (Fig. 4.30).

3. Specify the name and location of the file to be stored.

4. To use the map layer, open it using the Add Data button just as for any other layer.

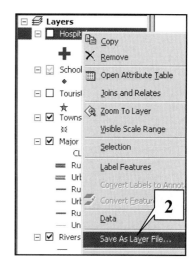

Fig. 4.30. Creating a layer

Creating a map group layer

1. Set the symbology, labeling, and other display characteristics of all the data layers to be included in the group.

2. Click the first group member in the Table of Contents to highlight it. Use Ctrl-click to also highlight the other layers in the group (Fig. 4.31).

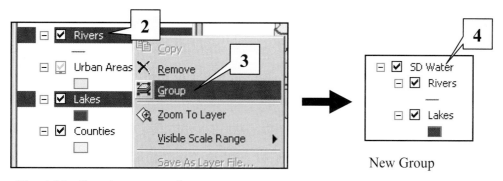

Fig. 4.31. Creating a group layer

3. Right-click one of the group members and choose Group from the context menu.

4. Click on the group name to highlight it, and then click it again to type in a name for it.

5. Right-click the group name and choose Save As Layer File from the context menu.

6. Enter the location and filename of the group layer and click OK.

Chapter 5. Working with Tables

Mastering the Concepts

Objectives

- Knowing types and structure of tables in ArcGIS
- Learning to create, open, close, and save tables
- Modifying appearance of tables and fields
- Editing fields and calculating new values in tables
- Learning to query, calculate statistics, and create summary statistics for table fields
- Understanding cardinality in table relationships
- Creating joins and relationships between tables

Concepts

Overview of tables in ArcGIS

What is a table?

A table is a data structure for storing multiple attributes about a location or object. ArcGIS manages these data tables in an object it refers to as a Table, which is a window that displays information from the tabular data structure and allows the user to work with the information in the file (Fig. 5.1). The data may come from several types of data files, but the Table itself always looks the same and has the same functions, so that users don't need to learn different commands for working with different file types.

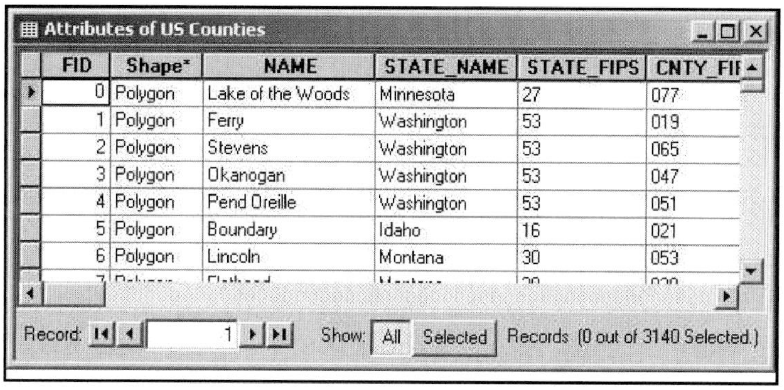
Attributes of US Counties

FID	Shape*	NAME	STATE_NAME	STATE_FIPS	CNTY_FI
0	Polygon	Lake of the Woods	Minnesota	27	077
1	Polygon	Ferry	Washington	53	019
2	Polygon	Stevens	Washington	53	065
3	Polygon	Okanogan	Washington	53	047
4	Polygon	Pend Oreille	Washington	53	051
5	Polygon	Boundary	Idaho	16	021
6	Polygon	Lincoln	Montana	30	053

Record: 1 Show: All Selected Records (0 out of 3140 Selected.)

Fig. 5.1. A table with information about U.S. counties

Tabular data files fall into two main categories, **attribute tables** and **standalone tables.** An attribute table, such as the one shown in Figure 5.1, contains information about features in a geographic data set. A shapefile of counties has this attribute table, which contains information about the state, population, area, and so on for each county in the shapefile. In an attribute table there is always one and only one row of information for each feature in the geographic data set, and the row is linked to the feature itself using a unique ID number, the feature ID or FID (Fig.

5.1). In contrast, a standalone table simply contains information about one or more objects in tabular format, instead of having information about map features. A standalone table might come from a text file, an Excel® spreadsheet, a global positioning system data file, or a database. These tables exist independently of a geographic data set, and they may be only incidentally related to map features. Instead of a FID, a standalone table contains an Object ID (OID) analogous to the row number in a spreadsheet.

In ArcGIS, an attribute table will usually be titled "Attributes of…" and the first few columns of information will be required information associated with the geographic features, such as the type of feature it is and its unique identification number. A data table will usually not contain this specialized information.

The information in a table may come from a number of sources, including dBase (.dbf) files from a shapefile or a database, an INFO file, a comma-delimited text file, or an SQL query from a database server. No matter where the data comes from, within the table it is viewed and edited and queried in exactly the same way.

Parts of a table

Figure 5.2 shows some terminology associated with tables. A table consists of rows and columns. A row is called a record, and it contains information about a single object or feature. A column is called a field, and it stores one type of information. Each field has a name shown in the top row. Field names must contain 13 or fewer characters and should contain only letters, numbers, and underscores. An alternative name called an **alias** can be used to temporarily give a field a more descriptive name that does not have to follow the naming rules. For example, the somewhat perplexing field MEDREN could be given a more understandable alias such as MedianRent. Aliases are displayed with the table in a map document, but they are not stored on the disk.

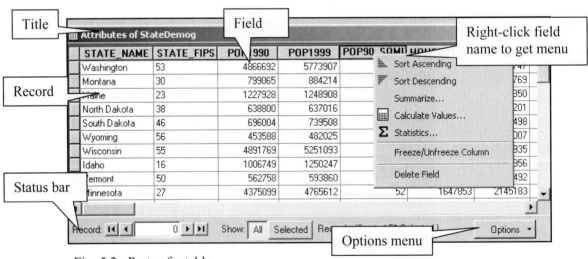

Fig. 5.2. Parts of a table

Before a field may be used, it must be defined. A **field definition** describes the type of data that may be stored in that field. A field can contain only numeric information or only string information, but never both. (Although numbers may be typed in a string field, they are treated as text only, meaning that they cannot be used for calculations.) There are also specialized types of fields that can contain dates, true-false information, or even complex objects such as images. The types of definitions available will vary depending on the source of the table: an Oracle® database supports more field types than a .dbf file.

The top bar of the table contains its title. The bottom row has buttons to navigate through records, options to display all the records or only a selected few, and a report of the number of records in the table. An Options button provides access to table-related commands. In addition, right-clicking on a field name will open a menu of options and actions that can be performed on the field.

In an attribute table, some fields are required and are created and updated by the program which creates the geographic data set. For example, a shapefile has a feature ID (FID) field and a Shape field. **These fields should never be altered by the user.** An ArcInfo coverage will have the cover# and cover-id fields, plus additional fields depending on the types of features in the attribute files. Again, users must not change these fields in any way, or the integrity of the data set may be compromised. Likewise, users must make sure that they never delete records in an attribute table unless the accompanying features in the spatial data set are also deleted. Editing features only within an editing session in ArcMap will meet this criterion.

Table formats

ArcGIS reads several types of table formats. The **dBase format** is used for shapefile attribute tables, but standalone files from database programs may also be opened. **INFO format** tables come from Workstation ArcInfo and an old database program called INFO. INFO formats and file-naming conventions are complex and require the help of ArcCatalog to be managed correctly. **Geodatabase tables** are stored inside geodatabases only, and they use the table format specified by the host database: Microsoft Access in the case of personal geodatabases. ArcGIS can also read several different text formats, including comma-delimited text (fields separated by commas) and tab-delimited text (fields separated by tabs).

Microsoft Excel spreadsheets are a common source of data for GIS systems. Although ArcGIS cannot read spreadsheets directly, converting them to one of the readable text formats is not difficult. Alternatively, spreadsheets can be saved directly in dBase format. Care must be taken, however, in preparing a spreadsheet for conversion to a text or dBase file. The first row should contain field headings, each one less than 13 characters, composed only of letters and numbers, and beginning with a letter. Each field (column) should contain only text or only numeric data, not a combination of both. It is simplest to save the file as a CSV file, creating a comma-delimited text file, and these generally work best. However, saving to dBase is more convenient (when it works), because this is a native ArcGIS format and the table is immediately ready for full use. The Skills Reference section contains detailed instructions for saving a spreadsheet for use in GIS.

TIP: Feature classes with spatial data can be created from tables if the table contains *x-y* coordinates. Lists of objects like climate stations or wells commonly include this information, making it easy to produce point shapefiles of these features. Global positioning system (GPS) units provide an additional common source of spatial coordinates. Chapter 10 discusses methods for creating shapefiles from tabular *x-y* data.

Field types

Unlike a spreadsheet, in which any cell can contain any type of data, a database file used in GIS must contain the same type of information in each field. A new field must be **defined,** and the type of its contents established, before any data are entered. Furthermore, once a field definition is set, it cannot be changed.

The most common field types are numbers, strings, dates, and Boolean fields. The specific ways in which these fields are defined may differ among data files; however, some aspects are the same. The type of data stored in the field, and the space allotted for each number, are two aspects that are defined for every field.

For example, the field *Street* would be a string field because street names typically contain letters. The **field length** defines how many letters can be stored in the name. If the *Street* field is given a length of 10, then any name longer than 10 letters will be truncated to the first 10 letters. Thus "Elm Street" fits perfectly (including the space), but "Maple Street" would be truncated to "Maple Stre" to fit in the field. It is very important to ensure enough space in fields to hold the longest possible case.

When defining a number field, the user must also designate a storage width **(precision)** and the number of decimal places **(scale).** For example, a field with a precision of 5 and a scale of 0 could store a number between –9999 and 99999 (the minus sign takes one space). A field with a precision of 5 and a scale of 2 could store a number between –9.99 and 99.99 (the decimal place also takes one space). When designing a database, try to use the smallest field widths that will store every possible value, because extra spaces increase the size of the database.

Numbers can be stored in a variety of ways. Most databases offer several formats for storing numeric values, and these formats can differ from database to database. The next section describes the common options in general terms.

ASCII versus binary

Computers can store numbers and letters several different ways, and savvy GIS users should know about them. All text is stored as sequences of characters. Each character uses a single **byte** of data. A byte is the basic unit of storage space for a computer—it is composed of a string of eight values which may be zeros or ones. These zeros and ones represent a number in base 2; such numbers are called binary numbers. A single byte can store a binary value from 0 (00000000) to 255 (11111111). Computers use a special code called **ASCII** in which every number, every letter, and every symbol (such as $) is assigned a binary value between 0 and 255. Then to store the word "cat", the computer stores the code for "c", the code for "a", and the code for "t". Thus it takes three bytes of data to store "cat". The word "horse" requires five bytes of data. Numbers can also be stored using ASCII, by storing the ASCII 1-byte code for each numeral. Thus it requires three bytes of data to store "147" and five bytes to store "147.6", or one byte per character. This scheme is one simple and standard way to store information in a database. Text files with a .txt extension and HTML files, among others, are stored in ASCII.

Another scheme involves storing **binary** data. In this case a number is stored in base 2 directly, rather than being assigned one byte per character. The number "3" would be stored in binary as "00000011" in a single byte of information. Recall that a byte, formed of eight bits or single values of 0 and 1, can store a number up to 255, or 2^8-1. Two bytes can store values up to 2^{16}-1, or 65,535. Note that storing 65,535 would require five bytes if it were stored using ASCII codes. Thus binary is a more compact form of storing numbers. It also is faster to compute with binary values, because the computer by design does all its calculations in base 2. If the number is already stored in base 2, the computer does not need to convert it before calculating. Thus it is often advantageous to use binary storage when it is appropriate. Many types of files use a binary encoding scheme, including spreadsheets, word processing documents, and shapefiles.

Precision

Very large numbers such as 1,000,000,000,000 require many bytes to store; so do very small numbers such as 0.0000000000001. People often use scientific notation when dealing with very large or very small numbers; these two would be written as 1.0×10^{12} and 1.0×10^{-13}. Computers can also use scientific notation, usually called exponential or floating-point data. This notation stores values composed of a mantissa (the decimal part of the number) and an exponent. For example, the number 123456789 stored as an exponent would consist of a mantissa of 1.23456789 and an exponent of 8, yielding 1.23456789×10^8 (often written as 1.23456780e08). The computer usually truncates the mantissa at a certain level of precision—when storing values in the trillions, differences of tenths or hundreds are often of little interest. The number might become 1.2345e08. In ArcGIS, a **single-precision** floating-point field stores up to eight significant digits of information in the mantissa. A **double-precision** field stores up to 16 significant digits. Floating-point data types are much more flexible than numeric or binary types because they can store either very large or very small values using the same field.

Different types for different data formats

Every attribute field in a shapefile, geodatabase, or coverage must be defined before use; that is, the field type must be specified (text or numeric) and the field properties set. Once a field is defined, the definition cannot be changed. If a mistake is made defining the field, one must usually delete the field and redefine it. Shapefiles and geodatabases use the same data types, as shown in Table 5.1. (ArcView version 3.x shapefiles and coverages have different allowable data types; see the ArcGIS Help for more information.) When converting from one data format to another, the fields are automatically redefined in the conversion process.

Table 5.1. Field data types for shapefiles and geodatabases		
Field type	**Explanation**	**Examples**
Short	Integers stored as 4-byte binary numbers	255 12001
Long	Integers stored as 10-byte binary numbers	156000 457890
Float	Floating-point values with eight significant digits in the mantissa	1.2893851e12 1.5647894e-02
Double	Double-precision floating-point values with 16 significant digits in the mantissa	1.121141181191416e13
Text	Alphanumeric strings	'Maple St' 'John H. Smith'
Date	Date format	07/12/92 10/17/63
BLOB	Binary large object; any complex binary data including images, documents, etc.	

Queries on tables

Often one wants information about a subset of records in a table, for example, knowing the number of parcels in the city that are designated as commercial. To determine this information a **query** can be performed on the table. In a query, a **logical expression** is used to specify certain criteria (i.e., zoning = commercial), and then the software searches the table and finds the records (parcels) that match the criteria. Those records are returned as a **selected set.** Selected records can be viewed separately, exported to another table, or used to calculate statistics.

In this example, if the field containing the zoning was named ZONE, and the code for commercial property was 492, the logical expression might look something like this:

[ZONE] = 492

Many types of queries are possible, including queries with multiple criteria, such as:

[ZONE] = 492 AND [VALUE] > 300000

This query would find all commercial parcels that have a value greater than $300,000. This chapter presents methods for performing simple queries on tables. Chapter 6 treats the general topic of queries in more detail.

Joining and relating tables

Sometimes it is advantageous to store tabular information separately from geographic features, that is, as a standalone table. For example, a point data set representing climate stations might have the actual climate records stored in a standalone table. This separation has two benefits. First, the stations themselves can be mapped and queried without having to manage many fields of climate information. Second, since each station represents many measurements, the climate data is difficult to store in the station table. Another example occurs when an agency publishes a report such as the grain yields by county. An analyst can incorporate this data into the GIS as a standalone table and create a map, without adding the yields permanently to the database. In both examples, the user needs to integrate the information from the standalone table with information from the spatial data file in order to use the data in the GIS.

Putting information from two tables together is called a **join.** In this operation, the tables are combined using a common field called a **key.** The field name does not need to match, but the data type must match; it is impossible to join numbers to strings. When a join is performed, the two separate tables become one containing the information from both tables (Fig. 5.3). The join is a temporary relationship and may be removed when it is no longer needed.

Joins have a direction: the information from one table is appended to the information from the other table. The table containing the information to be appended is called the **source table.** The table which receives the appended information is called the **destination table.** In Figure 5.3, notice that the destination table on the left, Attributes of US States, is a feature attribute table of a polygon shapefile; notice the FID field. The source table on the right containing demographic data is a standalone table. When the two tables are joined, the demographic data is appended to the US States attribute table.

Join direction matters. Once the demographics are joined to the states shapefile table, then the information could be used to make a map of the population data. If, however, the join had been performed in the opposite direction, with the demographics as the destination and the states as the source, then the resulting table would be a standalone table. In this case a map showing the demographics could not be made because the demographics data is part of the shapefile.

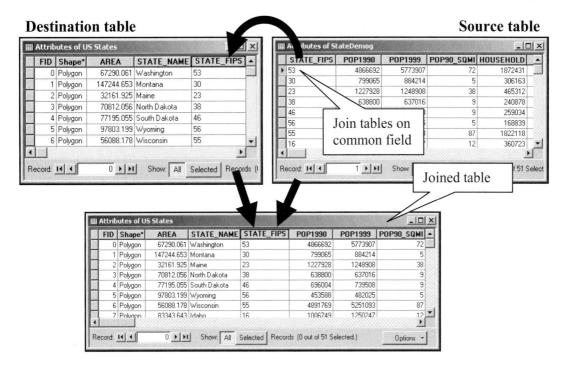

Fig. 5.3. Joining on common field. These two tables are joined into one using the common field STATE_FIPS. The join field appears only once in the joined table. The data from the source table are placed into the destination table.

Cardinality

In joining, the **cardinality** of the relationship between the tables must be considered. The simplest kind of relationship is a **one-to-one** relationship, in which each record in the destination table matches exactly one record in the source table. In Figure 5.3, each state has one corresponding record of demographic data. In a **one-to-many** relationship, each record in the destination table could match more than one record in the source table. For example, a store location could have many employees. In a **many-to-one** relationship, many records in the destination table would match a single record in the source table, such as many cities falling within one state. Finally, a **many-to-many** relationship indicates that multiple records can appear in both tables. For example, a student may take more than one class, and most classes have more than one student.

The direction of the join must be taken into account when ascertaining the cardinality of a relationship. The destination table is the point of reference and comes first; that is, the relationship cardinality is reported as {destination} to {source}. Imagine two tables containing *states* and *counties*. If one performs a join with *states* as the destination table and *counties* as the source, the cardinality is one-to-many, because each state contains many counties. If the join is reversed and *counties* is the destination table, then the cardinality becomes many-to-one, because there are many counties in one state.

TIP: Putting the destination table first when stating cardinality is only a convention, but one adopted by ESRI in its publications and Help documents. Although to some readers it may initially seem backwards, sticking to this convention will cause the least confusion in the long run. It may help to always imagine the destination table on the left, as in Fig. 5.3.

The cardinality of a relationship dictates the ways in which the table can be joined and used. The purpose of a join is to combine the two tables, so for every record in the original table there is one record in the output table. We can call this the Rule of Joining, that there must be one and only one record in the output table for each record in the input table. Consider the four tables shown in Figure 5.4.

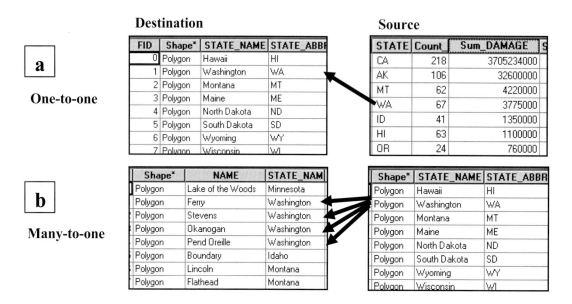

Fig. 5.4. A cardinality of one-to-one or many-to-one permits tables to be combined without violating the Rule of Joining. (a) A one-to-one cardinality. (b) Many-to-one cardinality. Source records may be repeated in the destination if needed.

One-to-one cardinality. In Figure 5.4a, a table containing the number of earthquakes and total damage in each state (source) is being joined to a states attribute table (destination), using the common field provided by the state abbreviation. Each state occurs once in each table, so there is no ambiguity about matching the source records to the destination records.

Many-to-one cardinality. In Figure 5.4b, a table containing state information (source) is being joined to a table of counties (destination) using the common field provided by the state name. Although there are many counties in each state, there is only one record for each state in the source table, and thus no ambiguity in linking the source records to the destination records. The state record does get used more than once, but it still does not violate the Rule of Joining.

One-to-many cardinality. Figure 5.5 shows the reverse of the join in Fig. 5.4b. Now states are the destination table and counties are the source table. Many county records in the source table match each state record in the destination table. The Rule of Joining is thus violated, and it becomes ambiguous which county record should be matched to the state. For this reason, one cannot perform a join if a one-to-many relationship is present. Instead, we perform a different operation called a **relate.**

Destination   **Source**

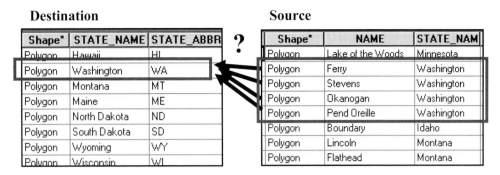

Fig. 5.5. A one-to-many relationship violates the Rule of Joining, because more than one record in the source table matches a record in the destination table.

In a relate, the two tables are still associated by a common field, but the records are not joined together. The two tables remain separate. However, if one or more records are selected in one table, then the associated records can be selected in the other table. For example, selecting the state of Washington in the states table allows the selection of all the counties in Washington in the related table, as shown by the red boxes in Figure 5.5. Although relates are not as powerful as joins, they still have useful functions.

To summarize, a join may be used to combine tables whenever a one-to-one or many-to-one relationship exists between the destination table and the source table. If the relationship is one-to-many or many-to-many, then a relate must be used.

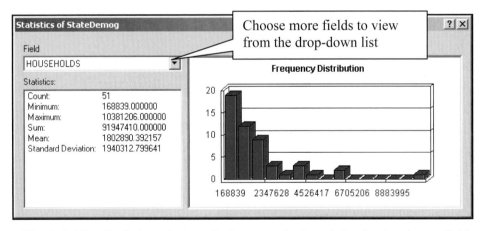

Fig. 5.6. The Statistics window displays standard statistics for the chosen field.

Getting statistics on tables

ArcGIS makes it very easy to explore data using the Statistics command. Right-clicking on a field name and choosing Statistics from the menu yields basic statistical measures, including count, sum, minimum, mean, maximum, standard deviation, and the frequency distribution of the field (Fig. 5.6). If records from a query are currently selected, then the statistics are calculated for the selected set only; if no records are currently selected, then it shows the statistics for all the features in the data set.

Summarizing tables

The ability to summarize values in a table is an important tool in the GIS arsenal. The Summarize command resembles the Statistics command, but it is far more powerful. Statistics merely calculates statistical values for the selected values in a field. Summarize first combines the records into groups based on a specified attribute field, and then calculates statistics for each group. Moreover, the user can choose which statistics to calculate and can choose from one or more statistics for as many fields as desired.

Because many statistics for many fields can be returned from a Summarize, this command produces a table file rather than a simple display on the screen as the Statistics command does. This table can in turn be joined to another table. This combination of Summarize followed by a join works well for taming those unwieldy one-to-many relationships.

We might use Summarize to gather information about the damage and deaths caused by earthquakes. Figure 5.7 contains data on major historical earthquakes in or near the United States. Each earthquake is listed with information including the state in

STATE	DEPTH	DEATHS	DAMAGE	MAG	MMI	LOCATION
MO	0	7	0	7.88	12	New Madrid, Missouri
SN	0	51	0	7.36	12	Northern Sonora, Mexico
AK	0	0	0	8.15	11	Yakutat Bay, Alaska
AK	0	0	0	8.26	11	Southeast Alaska
AR	0	7	0	7.68	11	Northeast Arkansas
CA	20	3000	52400000	7.80	11	Near San Francisco, California
CA	16	12	6000000	7.48	11	South of Bakersfield, California
CA	8	65	50500000	6.62	11	North of San Fernando, California

Fig. 5.7. Earthquake table

which it occurred, the number of deaths it caused, the total damage caused, and the Richter scale (MAG) and Modified Mercalli Intensity (MMI) measuring the energy of the quake. We would like to know which states have suffered the most from historical earthquakes. In particular, we would like to know the total deaths, total damage, and average Richter and MMI scales for each state.

To start the Summarize, we right-click the field we want to use to form the groups. In this case we want to group the earthquakes according to which state they are in, and calculate the statistics for each group. Thus the STATE field is our target. We right-click the STATE field and choose Summarize, which launches the Summarize dialog box shown in Figure 5.8. Next, we choose the statistics we want to calculate (sum of deaths, sum of damage, and average magnitude and MMI). Finally, we specify an output file to contain the results.

The quakesum.dbf output table shown in Figure 5.9 has one record for each state. It also has a Count field indicating the number of earthquakes in each state. A Count field is always generated automatically during a Summarize, whether any statistics are

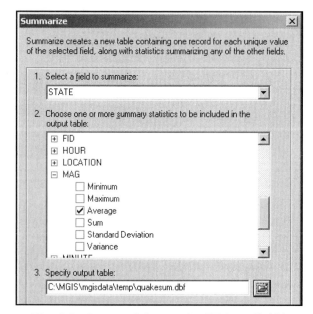

Fig. 5.8. Summarizing on the STATE field in the earthquake table

chosen or not. The output table also includes one field for each of the statistics we requested. To better interpret the data, we sorted the table by the damage field in descending order, to find the most grievously affected states. Our next step might be to join the quakesum.dbf table to the US States layer and create a map showing the total damage or total deaths for each state (Fig. 5.10).

⊞ Attributes of quakesum							
OID	STATE	Count_	Sum_DAMAGE	Sum_DEATHS	Average_MAG	Average_MMI	
6	CA	218	3705234000	3777	5.2575	7.422	
0	AK	106	32600000	125	6.5042	3.7358	
26	MT	62	4220000	32	2.9737	5.9677	
54	WA	67	3775000	15	3.5894	5.8955	

Fig. 5.9. Summarize produces a file which contains the summary field (STATE), the number of earthquakes in each state, and the requested statistics.

Editing and calculating fields

Editing fields

ArcGIS offers two ways to change the values in a table: typing the information directly into the fields, or calculating the value of a field.

Typing information into fields must be done during an edit session in ArcMap. During an edit session, ArcMap keeps track of the changes made and allows the user to undo mistakes. You can type information directly into the fields of an open table, or you can use the Attribute Editor. The Skills Reference in this chapter explains the first method; we will cover the Attribute Editor in Chapter 11 on editing.

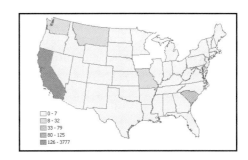

Fig. 5.10. Earthquake deaths by state

Calculating fields

Many values can be entered at once by calculating a field using the Field Calculator. For example, two fields containing the total state population and the number of Hispanics in the state could be used to populate a new field called Hisp_perc with the percentage of Hispanics in each state, using an appropriate expression for the formula.

The first window in the Field Calculator (Fig. 5.11) contains the fields in the table that can be used to create the expression. The functions box contains different functions that can be used in expressions. The functions displayed will depend on

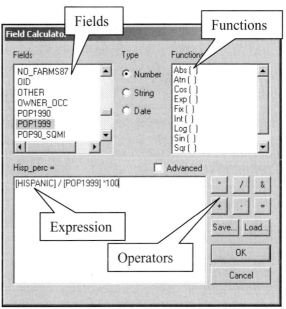

Fig. 5.11. The Field Calculator

whether Number, String, or Date is selected. The expression used to calculate is entered in the large box at the bottom. The operators (+, *, -, /, &, =) appear to the lower right.

The Advanced button allows more complex Visual Basic expressions to be entered. (Visual Basic is a programming language that comes with ArcGIS.) If an equation is complicated, it can be saved and loaded for use another time.

The Field Calculator can be used in conjunction with the selection tools to enter values for large subsets of data at the same time. For example, imagine a table with a field describing soil infiltration as high, medium, or low. You need to replace these terms with an actual numeric value. A new field called Infil_rate must first be created to hold the values. You would then select all of the soils with a *high* rating using a query. Opening the Field Calculator and typing the numeric infiltration value into the expression box would place that number in every record with a *High* rating. By repeating the selection and calculation for each soil type, you could enter all the infiltration values with only three simple calculations.

Chapter 11 introduces methods for editing tables during an edit session. When calculations are done inside an edit session, a mistake can be corrected using Undo. Sometimes doing calculations outside an edit session is more convenient. (The program will warn the user that the calculation cannot be undone.) Often a calculation mistake can be corrected by simply repeating the calculation with the right formula instead of the wrong one, thereby replacing the erroneous values with the correct ones.

Summary

➤ Tables are interfaces designed to display and manipulate tabular data from a variety of sources, including dBase files (.dbf), INFO files, geodatabases, SQL queries, or comma-delimited text files.

➤ Tables associated with spatial data sets are called attribute tables and contain records, one for each feature in the data set. Standalone tables are not associated with map features.

➤ Tables consist of rows and columns of information. A row is associated with one feature in the real world and includes columns of information called fields.

➤ Fields must be defined before they can receive data. Different data formats have different ways of defining fields, and they are converted when the data formats are converted.

➤ Table queries allow the user to select certain records based on one or more criteria. Once selected, these records may be viewed, calculated, exported, or analyzed.

➤ Tables may be joined or related based on a common field, in order to access information in one table from another. Joins may be performed on tables with one-to-one or many-to-one cardinality. Relates must be used on tables with a one-to-many or many-to-many cardinality.

➤ The Statistics function calculates standard statistical values for a given field. If records are selected, only the selected records are included in the calculation.

➤ The Summarize command generates statistics about groups of features within a field.

➤ New fields may be added to tables by defining names and field types. New tables may have data entered by typing in values or by using the Calculate function.

TIP: Data from Excel spreadsheets and other data sources may be imported into ArcGIS from text or dBase formats. See the Skills Reference section for important details.

TIP: Feature classes with spatial data can be created from tables if the table contains *x-y* coordinates. Chapter 10 discusses methods for creating shapefiles from tabular *x-y* data.

TIP: Field names must have 13 or fewer letters and should not contain spaces or odd characters such as @, #, !, $, or %. Letters, numbers, and the underscore character _ are best. Fields must also start with a letter, not a number.

TIP: Exporting a joined table will put both the destination and source table information in the new table.

VERY IMPORTANT TIP: Do not use programs other than ArcGIS to delete or edit attribute tables that are associated with shapefiles. You may corrupt a shapefile and make it unusable.

Chapter Review Questions

You may need to consult the Skills Reference section to answer some of these questions.

1. Describe the difference between an attribute table and a standalone table.

2. List the types of data sources from which tables may display data.

3. Describe how storing the number 255 in ASCII would differ from storing it in a binary representation.

4. Choose the best field type for each of the following types of data in a geodatabase:

   population of countries in the world
   precipitation in inches
   number of counties in a state
   highway name
   distances between U.S. cities, in meters
   birthdays

5. What is the cardinality of each of the following relationships?

   students to college classes
   states to governors
   students to grades
   counties to states

6. Describe the differences between a join and a relate.

7. Describe the difference between using Statistics and Summarize on a field.

8. True or False: A change in the width and order of fields in a table is written to disk and will persist whenever the table is opened in a new map document.

9. True or False: Adding fields to a table in ArcCatalog requires that the table is not currently open in an ArcMap session. Explain why or why not.

10. For each of the following problems, using data sets for the U.S.A., state whether a query, statistics, or summarize would be the best approach to solving it.

   _____ Find all towns with more than 20,000 people.
   _____ Find the total number of volcanoes in each state.
   _____ Determine the total amount of damage caused by earthquakes in the U.S.
   _____ Find the states in which Hispanics exceed the number of blacks.
   _____ Find out which subregion of the country has the most Hispanics.

Mastering the Skills

Teaching Tutorial

The following examples provide step-by-step instructions for doing basic tasks and solving basic problems in ArcGIS. The steps you need to do are highlighted with an arrow ➔; follow them carefully. Click on the video number in the VideoIndex to view a demonstration of the steps.

➔ Start ArcMap if necessary. Navigate to your MapDocuments folder in the mgisdata directory and open the map document ex_5.mxd.

➔ Use Save As to give the document a new name and save frequently as you work.

1➔ Right-click on the US States layer and choose Open Attribute Table from the context menu.

Changing the appearance of a table

First, let's experiment with changing the appearance of this table. Notice how the STATE_NAME field and the FID field are much wider than the data contained in them.

2➔ Narrow the fields to a more suitable width by clicking and dragging the right border of the field.

The FID field information is not very useful, so hide it from sight.

3➔ Right-click the US States layer and choose Properties from the context menu.

3➔ Click the Fields tab.

3➔ Select the FID field and uncheck the box next to its name. Click OK.

Notice that the STATE_ABBR field is also too wide for its data, but that narrowing it will cut off the field name.

3➔ Click on STATE_ABBREV in the Alias column and type in the alias ABBREV; and also give the STATE_NAME field the alias NAME.

3➔ Close the Layer properties window.

Scroll over to the right of the table and look at all the different fields of population data from the U.S. Census. It is difficult to match the values to the right state, however, because the state names quickly scroll out of sight.

4➔ Freeze the NAME column by right-clicking on the NAME field and choosing Freeze/Unfreeze Column.

4➔ Now scroll again and see how much easier it is to interpret the data.

> **TIP:** An unfrozen column will not go back to the place where it originally was in the table. To move it back, remove the table from the map document and add it again.

Now let's use the Sort function to obtain information about the largest and smallest states, using the POP2000 field.

5➔ Right-click on the POP2000 field name and choose Sort Descending from the context menu.

1. Which state has the greatest population? _____

2. Which state has the lowest population? _____ (**Hint:** Either Sort Ascending, or scroll to the bottom of the table.)

Notice that the AREA field has decimal values in it, which make it hard to compare the sizes of different states. Format the field so that no decimals are displayed.

6➔ Right-click on the US States layer and choose Properties from the context menu. Click the Fields tab.

6➔ Click on the AREA field to select it and click the ellipses in the Number Format column.

6➔ Fill the button that says Number of Decimal Places, and choose 0. Click OK and OK.

Using queries on tables

Now let's use the Select By Attributes function to select the states that have more than 5 million people in the year 2000.

7➔ Click the Options button and choose Select By Attributes.

7➔ Enter the query "POP2000" > 5000000. (Field names will be quoted.)

7➔ When done, click on the Show Selected button to view only the selected records.

3. How many states were selected? _____

Notice that when you select the states in the table, the associated features on the map are also selected and highlighted.

4. Which of these populous states are west of the Mississippi River? (**Hint:** Use the Identify button if needed to determine a state's name.) _____

7➔ Click the Show All button. (You cannot clear a selection when only the selected records are being displayed.)

7➔ Clear the selected set by clicking the Options button on the table and choosing Clear Selection from the context menu.

Calculating statistics for fields

Now it is time to calculate and view some population statistics for the states.

8➔ Right-click on the POP2000 field name and choose Statistics from the context menu.

8➔ Notice the statistics that are calculated, and spend some time examining the frequency distribution.

5. What is the population of the largest state? _____

6. The smallest state? _____

7. What is the average number of people living in a state? _____

8. How many people live in all of the states? _____

9. Do most states have fewer than 15 million people? _____

10. Why are there 51 "states" listed in the statistics? _____

> 8➔ Use the drop-down list in the Statistics box to select the POP1990 field.

11. Did the average number of people in a state increase, decrease, or stay the same from 1990 to 2000? _____

> 8➔ Close the Statistics box.

TIP: The frequency diagram cannot be placed directly into a map layout as a graph. However, you *can* capture the window on the screen by holding down the Alt key and pressing the PrintScrn key on the keyboard. This will place the graph on the clipboard so it can be pasted into the Windows Paint program and saved as a JPEG file. The JPEG can then be placed in a layout.

The Statistics command calculates using all records, unless a subset of records has been selected, in which case it uses only the selected subset to calculate the statistics.

> 9➔ Use the Select By Attributes option to select all states with a year 2000 population greater than 5 million. Close the Select By Attributes window.

> 9➔ View the Statistics for the POP2000 field again. Look at the Count statistic and observe that fewer states are included this time—statistics are calculated only for the selected records. Close the Statistics window.

> 9➔ Clear the selected records by clicking on the Options button in the table and choosing Clear Selection.

> 9➔ Close the US States table.

Now let's experiment with calculating some fields in the 106th Congress Districts layer.

> 10➔ Open the table for 106th Congress Districts by right-clicking on the layer name and choosing Open Attribute Table.

> 10➔ Examine the fields.

Notice that the congressperson's party affiliation is listed after his or her name in the NAME field. You would like to make a map showing the distribution of the two parties across the nation, but you cannot do it with the affiliation mixed up with the name this way. Instead, create a new field that contains D for Democrats and R for Republicans. First add the field.

> 10➔ With the table open, click on the Options menu and choose Add Field from the menu.

10➜ Type in the field name "PARTY", choose Text as the field type, and enter a Length of "1" (one). Click OK.

TIP: Field names must have 13 or fewer letters and should not contain spaces or odd characters such as @, #, !, $, or %. Letters, numbers, and the underscore character _ are best. Fields must also start with a letter, not a number.

Next we will select the records with Democratic representatives and use Calculate to enter a "D" in the PARTY field.

11➜ Click on the Options button and choose Select By Attributes from the menu.

11➜ Enter the expression "NAME" LIKE '%(D)%' . Click Apply and close the Select By Attributes window.

11➜ Scroll through the table and notice that all the Democrats are now selected.

LIKE is an operator that searches for matching subsets of characters in a field. The % sign is a wildcard meaning any one or more characters. This expression will find all records that contain (D) regardless of what else is in the field. Since the parentheses are part of the expression, it will ignore other occurrences of D, such as in Danforth. Also notice the double quotes around the field name and single quotes around text strings—these are required.

12➜ Now right-click on the PARTY field and choose Calculate Values from the context menu.

12➜ Click Yes in the warning message to continue.

12➜ In the Field Calculator, enter the expression "D" in the expression box. Make sure you enclose it in *double* quotes.

TIP: Yes, it's confusing. You must put single quotes around strings when *selecting* them, but double quotes around strings when *calculating* them.

12➜ Click on the Options button again and choose Switch Selection to remove the Democrats and select the Republicans.

12➜ Use Calculate Field again to place an "R" in the Republican records.

12➜ Click Options and choose Clear Selection.

Terrific! The fields are calculated and can be used to create a map showing which districts are Democratic and which are Republican.

13➜ Close the 106th Congress Districts table.

13➜ Right-click on the 106th Congress Districts layer, select Properties from the menu, and click the Symbology tab.

Analyzing tabular data

As a final exercise, let's explore the distribution of Hispanics in the United States. Although the HISPANIC field in the US States table could be used to create a graduated color map showing the number of Hispanics in each state (Fig. 5.15), the map is difficult to interpret because the population of Hispanics is typically larger in states with larger populations. The percentage of Hispanics in each state provides a more useful statistic. Creating a new field and calculating the percentage of Hispanics solves the problem.

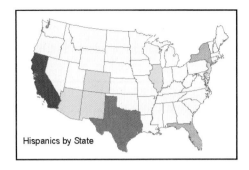

Fig. 5.15. Hispanics by state

25➔ Open the US States attribute table.

25➔ Click on the Options button and choose Add Field.

25➔ Name the field "HISP_PERC" and select Float as its type. Click OK.

> **TIP:** When setting numeric fields, you may take the precision and scale defaults, or you may specify values. The precision refers to the number of digits a field can hold, including negative signs and decimal points. The scale indicates the number of decimal places.

Now calculate the percentage of Hispanics.

26➔ Right-click on the new HISP_PERC field (located to the far right of the table) and choose Calculate Values from the context menu.

26➔ Enter the expression [HISPANIC] / [POP2000] * 100 and Click OK.

> **TIP:** Do all of the records equal zero? You might have accidentally selected a single record by clicking on the table at some point. If records are selected, then they are the only ones that get calculated. Check the bottom status bar to see if any records are selected. If so, choose Options > Clear Selection from the table, and try the calculation again.

27➔ Now create a map showing the percentage of Hispanics in each state (Fig. 5.16). Looks different, doesn't it?

Finally, we will gather some information about Hispanics in the different subregions of the United States, in particular the total number in each subregion and the average percent in each subregion. This is another job for Summarize!

28➔ In the US States table, right-click on the SUB_REGION field and choose Summarize.

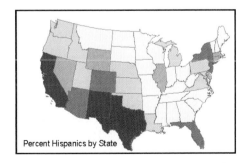

Fig. 5.16. Percent Hispanics by state

28➜ For statistics, click on the plus sign next to the HISPANIC field and choose Sum for the statistic.

28➜ Click on the plus sign next to the HISP_PERC field and choose Average for the statistic.

28➜ Enter the name "SubRHisp.dbf" for the table to be produced, and make sure it is being saved in the Usa folder.

28➜ Click Yes to add the table to the map document.

28➜ Close the US States table.

28➜ Click the Source tab if needed, locate the new table, and open it.

20. Which subregion has the greatest number of Hispanics? _____

This is the end of the tutorial.

➜ Close ArcMap. You do not need to save your changes.

Exercises

Open the ex_5.mxd map document in the MapDocuments folder of the mgisdata directory and answer the following questions.

1. How many counties in the United States have the name Washington? What is their total 2000 population? Which one has the largest area?

2. Calculate the percentage of the population in each state that is black. (Use the POP2000 field for the total population in each state.) Which state has the largest *number* of blacks? Which state has the largest *percentage* of blacks?

3. Which subregion of the country has the greatest number of blacks? Which subregion has the highest average percentage of blacks in a state?

4. Add the table popestmt.dbf from the mgisdata\Usa folder to the map document. This table contains population estimates by county between 1990 and 1998. Examine the table. Which fields could you use to join this table to the US Counties table?

   In the US Counties table, examine the three fields STATE_FIPS, CNTY_FIPS, and FIPS. How are these fields related? Join the tables.

5. Use the POP1990 and POP1995 fields from the popestmt table (now joined to the US Counties table) to select the counties that *lost* population between 1990 and 1995. How many are there?

   In what area of the country are these counties predominantly located? Make a map showing the "losing" counties. **Capture** the map. (**Hint:** Right-click the US Counties layer and choose Selection > Create Layer from Selected Features. Name it "Losing Counties".)

6. Use Summarize to determine the number of counties in each state which lost population. Which three states had the most losing counties, and how many losing counties did each have? (**Hint:** If the summarize doesn't work, try removing the join first.)

7. Imagine that you are the PR director of a group seeking aid for Native American artists, and you wish to contact the *Democratic* Congressional representatives who are from states in which the number of Native Americans and Eskimos (AMERI_ES field in US States) exceeds 100,000. Which states are included in the list? How many representatives will you need to contact?

8. Using the Major Cities layer, determine how many people in the United States live in state capitals. (**Hint:** Look for the designation "State Capital" within the FEATURE field.) What is the smallest, largest, and average population of a state capital?

9. Which state has the highest percentage of people who are married? Which one has the highest percentage of people who are divorced? (**Hint:** You will need at least one decimal place in the percentages for this one.)

10. Which subregion of the United States has the largest number of counties? How many total people did that subregion have in the year 2000? Create a map of the counties showing to which subregion each belongs. **Capture** the map.

Challenge Problem

Imagine that you are planning to open a dating service in Texas. Your preliminary research indicates that such a service would expect to gross $3 for every divorced person, $2 for every single (never married) person, and $1 for every widowed person in a typical county. Calculate the projected gross income for each county in Texas, and create a graduated color map showing the projected income for each county. (Make sure your map shows only Texas counties.) What is the total projected gross income for the entire state? Is there a large cluster of high-income counties that might be a good place to put your office?

Skills Reference

Adding or removing a table

Adding a table means putting it into a map document in ArcMap so that it may be used. Attribute tables are added by adding the spatial data set with which it is associated. Adding a standalone table is done using the Add Data button, just as for spatial data sets.

1. In ArcMap, click the Add Data button or choose File > Add Data from the menu bar.

2. If needed, use the Add Connection button in the Add Data dialog box to add a folder connection to the disk area containing your data.

3. Browse to find the tabular data file. Click to select it and click the Add button. (Shortcut: Double-click the desired file.)

4. To **remove** a table, right-click the table name in the Table of Contents, and choose Remove.

Opening a table

1. To open an attribute table, right-click on the spatial data set *name* (not the icon!) and choose Open Attribute Table from the context menu (Fig. 5.17).

2. To open a standalone table, first make sure that the table is visible by clicking on the Source tab at the bottom of the Table of Contents.

3. Right-click on the table name and choose Open from the context menu.

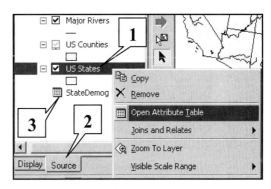

Fig. 5.17. Opening a table

Modifying a table's appearance

You can change certain ways in which the table displays data. Note that none of these operations modifies the actual data on the disk, only the way in which the data are presented.

Adjusting field width

1. Use the cursor to hover over the right edge of the field to resize, until it changes to a bar with a double arrow.

2. Click and drag the edge to the desired width, then release the mouse button.

Sorting on a field

Right-click on the field name to sort by and choose either Sort Ascending or Sort Descending.

Displaying Selected or All Records

A table can show all records with the selected ones highlighted, or it can show only the selected records.

Click on the Selected button at the bottom of the table (Fig. 5.18). Notice that it also reports how many records are currently selected.

Fig. 5.18. Showing selected records only

Hiding fields or creating aliases

Fields can be made invisible to hide them from display. Aliases can be used to give fields more descriptive names that need not follow the restricted naming conventions for fields.

1. If the table is an attribute table, right-click on the data set name and choose Properties from the context menu. If it is a standalone table, right-click on the table name and choose Properties.

2. Click the Fields tab (Fig. 5.19).

3. To create an alias, type a new name in the Alias column.

4. To hide a field, uncheck the box.

5. To format a numeric field, click the ellipses (see the next section).

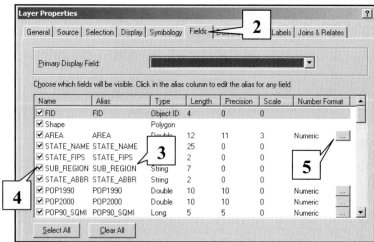

Fig. 5.19. The Fields tab of the Layer Properties window

Formatting field values

You can also change the alignment, decimal places, and other formatting characteristics which determine how a numeric field is displayed.

1. Open the Field Properties tab as described in the previous section.

2. Click on the ellipses in the Number Format column.

3. Choose the numeric category, such as currency or percentage (Fig. 5.20).

4. Specify the number of decimals or significant digits, the alignment, and other options.

5. Click OK.

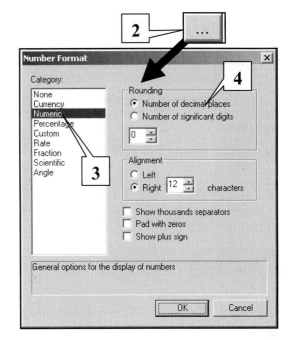

Fig. 5.20. Formatting field values

Freezing or unfreezing columns

Freezing places a column at the left edge of the table and keeps it always in sight as the user scrolls through the table. More than one column may be frozen.

1. With the table open, right-click on the field name of the field to freeze and choose Freeze/Unfreeze Column from the context menu.

2. To unfreeze it, right-click on a frozen field heading again and choose Freeze/Unfreeze Column again.

TIP: An unfrozen column will not go back to the place where it originally was in the table. To move it back, remove the table from the map document and add it again.

Changing table colors and fonts

The user can change how a table looks by changing its font and the colors that are used to highlight selected records.

1. With the table open, click the Options button in the lower area of the table and choose Appearance.

2. Click the color squares to choose a new color from the drop-down palette for the selection and highlight colors.

3. Choose a font, font size, and color.

Selecting records using attributes

A query selects records that meet certain criteria, such as cities having a population greater than five million.

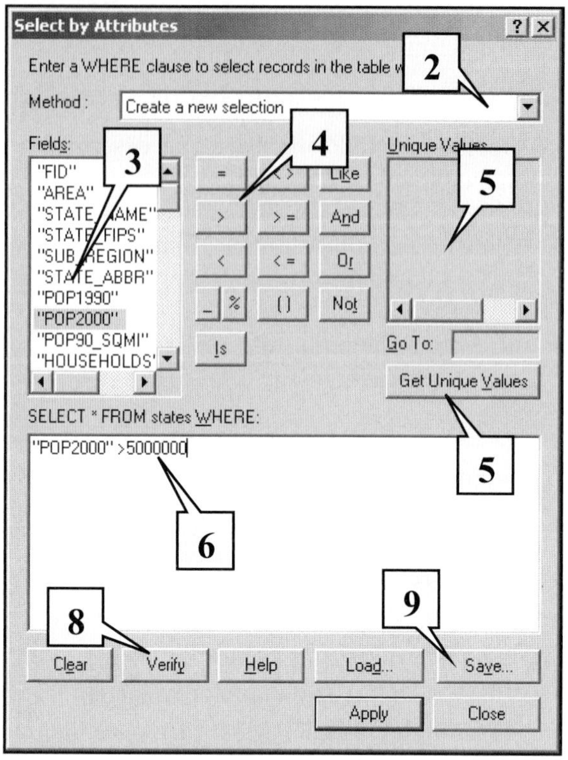

1. With the table open, click the Options button and choose Select By Attributes. A dialog box will appear (Fig. 5.21).

2. Choose the selection method. For a discussion of the different methods, see Chapter 6.

3. Enter a field name from the list into the large query box by double-clicking on the field name.

4. Enter an operator by clicking on it once.

5. To choose a value from the table, click Get Unique Values and then choose the desired value from the list.

6. Alternatively, enter a value by typing it into the query box. If it is a text string, put single quotes around it.

Fig. 5.21. Selecting by attributes

7. Multiple queries use criteria from more than one field, such as finding all cities that are capitals AND have more than five million people. The field name must be repeated for each query (see the examples that follow).

8. Choose to Verify your query if you are uncertain whether it is correctly formulated.

9. Use the Save button to save a query that is long and complex and frequently needed. It will be saved with an .exp extension. Load it again later using the Load button.

10. Click Apply to execute the query. The selected records will be highlighted in the table.

11. To clear the selected records, click on the Options button in the table and choose Clear Selection from the context menu.

Examples of valid queries:
     "POP1990" > 1000000
     "STATE_NAME" = 'Alabama'
     "STATE_NAME" = 'Alabama' OR "STATE_NAME" = 'Texas'
     "POP2000" >= "POP1990"

Getting statistics for a field

Statistics are useful for finding out more about the data in fields. Statistics may be calculated for a selected set of records or for the entire table.

1. If desired, use one of the selection methods described in Chapter 6 to select a subset of records requiring statistical analysis.

2. With the table open, right-click on the field to get statistics for and choose Statistics from the context menu.

3. Examine the statistics and frequency diagram in the Statistics dialog box (Fig. 5.22). If the table has records selected, then statistics will be calculated only for the selected records.

4. Look at statistics for a different field by choosing it from the Field drop-down box.

TIP: The frequency diagram cannot be placed directly into a map layout as a graph. However, you *can* capture the window on the screen by holding down the Alt key and pressing the PrintScrn key on the keyboard. This will place the graph on the clipboard so it can be pasted into the Windows Paint program and saved as a JPEG file. The JPEG can then be placed in a layout.

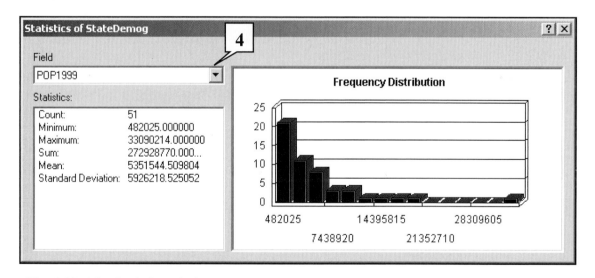

Fig. 5.22. The Statistics window

Summarizing on a field

Summarizing generates statistics for groups of features in a given field, such as finding the average magnitude of earthquakes occurring in each state.

1. With the table open, right-click on the field to summarize by and select Summarize from the context menu.

2. In the Summarize dialog box which appears, verify that the chosen field appears in the first box (Fig. 5.23).

3. In the second box, choose one or more statistics to calculate. Click on the plus sign to expand the list of statistics for a field, and then check the box next to the desired statistic(s).

4. Specify the name of the output table. Click on the Browse button if needed to change the directory where the file will be saved.

5. Click OK.

6. Click Yes to add the new table to the map document.

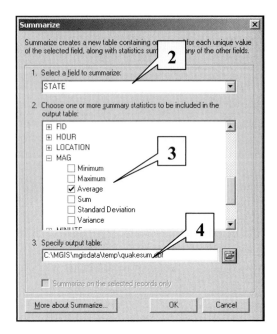

Fig. 5.23. The Summarize window

Exporting a table

Exporting a table means saving it under a new name. You might do this if you don't have write access to a table and want to create your own copy that you can edit. Or you might want to create a new table from only a subset of records in the original table. Or you might have made some edits that you want to save separately from the original table. Or you might be working on an SQL query or INFO file and you want to convert it to a .dbf file for export to another program.

There are two ways to export a table.

1. If the table is open, click the Options button and choose Export from the menu. Or right-click the table name in the Table of Contents and choose Data > Export from the context menu. (Note: The second method works for standalone tables only. If a data layer is right-clicked, then Export will copy all the map features, not just the table.)

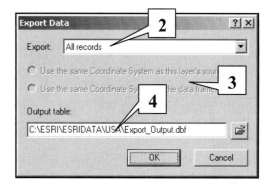

Fig. 5.24. The Export Data window

2. To export the entire table, choose "All records" from the first drop-down box. To export only a previously selected subset, change the drop-down box to say "Selected records" (Fig. 5.24).

3. The two choices regarding the coordinate system are dimmed because in this case you are exporting a table instead of a spatial data set.

4. Choose the directory and name of the output file by typing it in or clicking the Browse button. Click OK.

TIP: Exporting a joined table will put both the destination and source table information in the new table.

Adding or deleting fields

Fields may be added to tables in two ways. The first way uses ArcCatalog to enter and define new fields. This way is most efficient when you are entering many fields at once. You can also add a field during an ArcMap session.

Managing fields of shapefiles or DBF files in ArcCatalog

1. If the table is currently open in ArcMap, save the map document and exit ArcMap. Tables cannot be modified when they are part of an ArcMap session.

2. Navigate to the table in the left side of the ArcCatalog window.

3. Right-click the table and choose Properties from the Context menu.

4. Click the Fields tab (Fig. 5.25).

5. Scroll down the list of fields, if necessary, to the first empty space at the bottom.

6. Click on the empty space and type in the name of the new field.

7. Click on the Data Type box to the right of the new field, and a drop-down list of field types will appear. Choose the appropriate type.

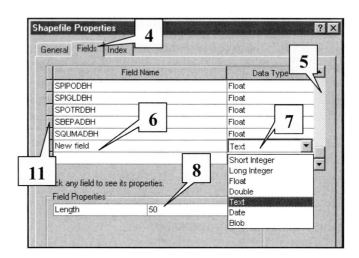

Fig. 5.25. Adding shapefile fields in ArcCatalog

8. View the field property information in the lower part of the dialog box. Beginners are advised to leave these properties alone for the most part. However, the length, precision, and scale properties may be set if needed. Precision determines the number of digits a number can hold, and scale indicates the number of decimal places. (A value with a precision of 6 and a scale of 2 could hold numbers between -99.99 and 999.99.)

9. Repeat steps 5 through 8 until all the fields have been added.

10. Click OK or Apply to enter the changes in the Properties dialog box.

IMPORTANT TIP: Note that field names must have 13 or fewer letters and should not contain spaces or odd characters such as @, #, !, $, or %. Letters, numbers, and the underscore character _ are best. Fields must also start with a letter, not a number.

---

11. To **delete** a field, follow steps 1-4 to bring up the table's Fields tab in the properties menu. Select the field by clicking on the grey tab next to its name, and press the Delete key. Warning: You cannot undo this step!

Managing fields in ArcMap

1. With the table open, click on the Options button and choose Add Field from the menu.

2. Type in the name of the field (Fig. 5.26).

3. Choose the type of field from the drop-down box.

4. View the field property information in the lower part of the dialog box. Beginners are advised to leave these properties alone. However, the length, precision, and scale properties may be set if needed. Precision determines the number of digits a number can hold, and scale indicates the number of decimal places. (A value with a precision of 6 and a scale of 2 could hold numbers between -99.99 and 999.99.)

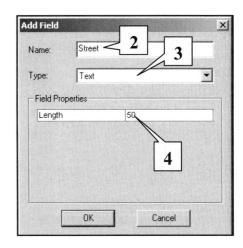

5. Click OK. The field will be added to the end of the table.

Fig. 5.26. Adding a field in ArcMap

6. **To delete a field,** right-click on the field name and choose Delete Field from the context menu. Because this action cannot be undone, you will be prompted to confirm the action. Click Yes to delete the field.

IMPORTANT TIP: Note that field names must have 13 or fewer letters and should not contain spaces or odd characters such as @, #, !, $, or %. Letters, numbers, and the underscore character _ are best. Fields must also start with a letter, not a number.

TIP: You cannot add or delete a field while you are in an editing session; you must stop editing first. If you try to add or delete a field and the option is dimmed out in the menu, then check to see if you are in an editing session.

Calculating fields

Use the Field Calculator to quickly enter information into many records at once, either the entire table or a selected set. The information entered can be as simple as a single number, or it could be a complicated mathematical expression. Calculations may be done outside of an editing session if desired, but then they cannot be undone.

1. With the table open, right-click on the field to be calculated and choose Calculate Values. The Field Calculator will appear (Fig. 5.27).

2. Begin entering the expression in the box to the lower right. Double-click a field name to enter it in the box. Click on operators to include those. Double-click functions to enter them. Numbers may be typed in from the keyboard.

3. For example, the expression in Figure 5.27 was created by double-clicking [Hispanic] in the Fields box, clicking the / operator, double-clicking [Pop1990], clicking the * operator, and typing "100".

4. Click OK.

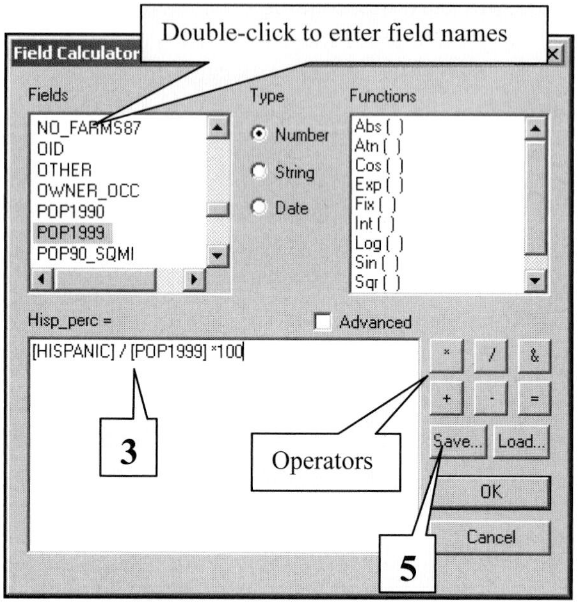

Fig. 5.27. Calculating a field

5. Complex expressions that are used frequently may be saved as *.cal files and loaded again later. Press Save, navigate to the desired directory, and enter a filename. To reload the expression, click Load, navigate to the file location, and select the desired expression file.

Editing fields in a table and saving changes

Fields can only be edited during an editing session. For more information, consult Chapter 11.

1. If not already in an editing session, click the Editor Toolbar button in the main ArcMap button bar to open the Editing Toolbar.

2. Choose Editor > Start Editing from the Editor Toolbar.

3. Open the table if it is not already open, by right-clicking it and choosing Open.

4. Click in the field space of the record to edit and start typing in the changes.

5. Choose Editor > Save Edits from the Editor Toolbar to save during an edit session. When finished typing the changes, choose Editor > Stop Editing and answer Yes when asked whether to save your changes.

Creating joins and relates

Joins and relates are used to link information from two different tables, based on a field that is common to both. The fields may be numeric or string fields, and they may have different names.

In a join, information from the source table is appended to the destination table. The destination table will then have its original fields plus the fields from the source table. The source table field names will be prefixed with the source table name.

1. In the Table of Contents, right-click on the spatial data set or the table that is to become the destination table. Choose Joins and Relates > Joins from the context menu. The Join Data dialog box appears (Fig. 5.28).

2. Make sure that Join Attributes from a Table is the selection action in the first drop-down box.

3. Enter the field in the destination table that will be used as the key to join the records.

4. Choose the source table from the list of tables in the map document, or use the Browse button to locate a file on disk.

5. Choose the field in the source table that will be used as the key.

6. (Optional) The Advanced button is used to control what happens to records in the destination table for which there are no matches in the source table. You may choose to keep unmatched records (in which case the appended fields will be blank) or drop the unmatched records from the table.

7. Click OK.

8. To remove a join, right-click the destination table or layer in the Table of Contents and choose Joins and Relates > Remove Join(s). Then select a specific join to remove, or remove all joins on that table.

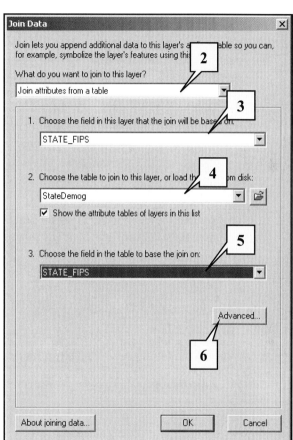

Fig. 5.28. The Join Data window

Creating a new table

New standalone dBase tables must be created in ArcCatalog. Attribute tables are generated automatically when the program creates the associated spatial data set.

1. In ArcCatalog, click on the directory or workspace that is to contain the new file. Make sure you are not inside a coverage directory.

2. Choose File > New > dBase Table from the ArcCatalog menu bar. The new table will appear in the directory with the name New_dBase_Table.

3. Type in the name of the new table. Make sure the table name does not contain spaces or odd characters such as @, #, %, ^, or &. Underscore characters are permitted.

4. To add fields to the table, see the instructions under "Adding or deleting fields" in this chapter.

Deleting a table

Standalone tables should be deleted using ArcCatalog.

1. Make sure that the table is not open in a current ArcMap session; otherwise you will not be able to delete it.

2. In ArcCatalog, navigate to the directory where the table is located.

3. Right-click the table name and choose Delete from the context menu. This action cannot be undone!

VERY IMPORTANT TIP: Do not use Windows™ to delete an attribute table that is associated with a shapefile. You will corrupt the shapefile and make it unusable.

Importing Excel Data into ArcGIS

Preparing the data

Importing Excel data is not difficult, but it requires careful preparation. First, make sure that the spreadsheet meets the following requirements.

- The first row of the worksheet contains the field headings.
- Field headings must start with a letter, must have no more than 13 characters, and must not contain spaces or other shift characters such as %, $, #, etc.
- No blank lines occur between the headings and the data or within the data rows.
- The spreadsheet contains no formulas.
- There are no extraneous data at the bottom of the spreadsheet, such as column totals.
- There are no merged or split cells, and every column contains consistent data types.
- Numeric columns have only numeric values in them, without text characters like "x" or "n/a" to indicate missing values. If present, missing values should be replaced by a numeric NoData marker value such as -99.
- Text fields contain no commas unless all the text is enclosed in quotes.
- Every column has been specifically formatted as text data, or as numeric data with a specified number of decimal places (critical for .dbf files).
- When you are ready to export the data, make sure that no rows or columns are selected. The cursor should be sitting in the upper left cell only.

Once the data are formatted, two different types of files can be created. If one type does not work, try the other. In either case, the method is the same:

- In Excel, choose Save As.
- Set the drop-down file type box to either dBase 4 (.dbf) or comma delimited (.csv) format.
- Type the file name. Make sure it is different from any existing layer. Avoid spaces and other strange characters.
- Save the file.
- Open ArcMap and use the Add Data button to add the new file.

TIP: You must CLOSE the saved .dbf or .csv file in Excel before you will be able to open it in ArcMap.

The .dbf file

The .dbf format is already supported by ArcMap for tables, so this format is the most useful. Once the file is open in ArcMap or ArcCatalog, you can add fields, calculate, and do other table operations normally available for tables. However, this format will cause problems if you have not formatted your Excel columns, so be especially careful.

The .csv file

The .csv file contains rows of data with each column separated by commas. It is quick and easy and works well in most cases. However, a text field with only numbers will be interpreted as numbers in ArcMap even if formatted as text in the spreadsheet. You must also use the Export function in the table Options button in ArcMap to export a .csv file to a .dbf file before you can calculate, add fields, delete fields, and so on.

Chapter 6. Queries

Mastering the Concepts

Objectives

➢ Understanding queries and how they are used

➢ Selecting features interactively with the mouse or graphics

➢ Selecting features using SQL queries on their attributes

➢ Selecting features based on their spatial location with respect to other features

➢ Using definition queries to create subsets of data layers

Concepts

What are queries?

Formally speaking, a query extracts features or records from a data table and isolates them for further use, such as printing them, calculating statistics, editing them, graphing them, creating new files from them, or doing more queries. In the simplest kind of query, you look at a map or table and use the mouse to select the desired record(s). After you select features with a query, the selected features are highlighted on the screen, and the corresponding records in the table are highlighted as well.

Queries fall into two main categories, **attribute queries** and **spatial queries.** An attribute query uses records in the attribute table to test the condition, such as finding all cities with population greater than 1 million people. Attribute queries are sometimes called aspatial queries because they do not require any information about location. A spatial query uses information about how features from two different layers are located with respect to one another. A spatial query might choose cities within a certain county, or rivers that are completely within a state, or hospitals within 20 miles of an airport. Attribute queries can be performed on any kind of table, either attribute tables or standalone tables. A spatial query requires a spatial data layer.

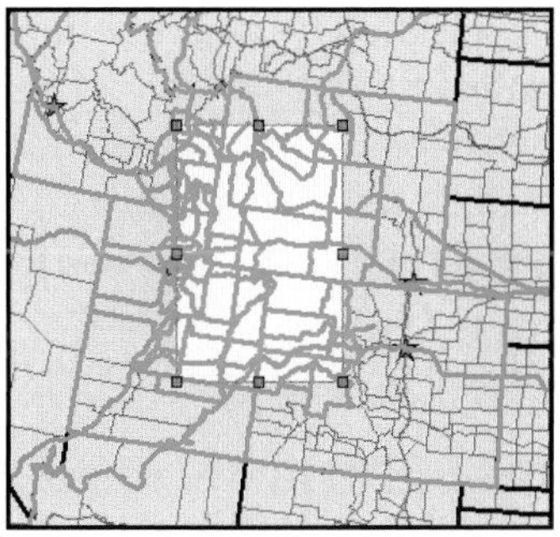

ArcMap has three ways to select features. **Interactive Selection** uses a pointer to select features on the screen. **Select By Attributes** searches for records in the table that match certain criteria based on one or more attribute fields. **Select By Location** searches for features based on their spatial relationship to each other, such as being adjacent to or completely within. In all cases the selected features are highlighted in blue in the map and in the table. The highlight color can be changed.

Fig. 6.1. Selecting states and counties using a rectangle

Interactive selection

In an interactive selection, you use the mouse to click on one or more objects in the map or table, until you have all the desired records. For example, to get all the states beginning with A, you could sort the table on state name, then find and click on each record containing states starting with A. Or, if you wanted the states that have a Pacific Ocean coastline, you could look on the map, hold down the Shift key, and choose California, Oregon, Washington, and Alaska.

You can also create a graphic, such as a box, a shape, or a line, on the screen and use the graphic to select the objects it touches (Fig. 6.1). This technique works well for selecting multiple objects.

ArcMap provides a special tool for selecting map features, called the Select Features tool. By default, the tool selects features from all layers present at the place clicked. In Figure 6.1, notice that all states, counties, rivers, and highways are selected if any parts of them touch the yellow rectangle. Often, however, it is preferable to select only features from a certain layer, such as states. The Selectable Layers option controls which layers can be selected, and it can turn off selection for one or two layers, or for nearly all layers. In Figure 6.2, the same rectangle has been used as before, but the user has set the selectable layers to states only. As a result, only states are now selected.

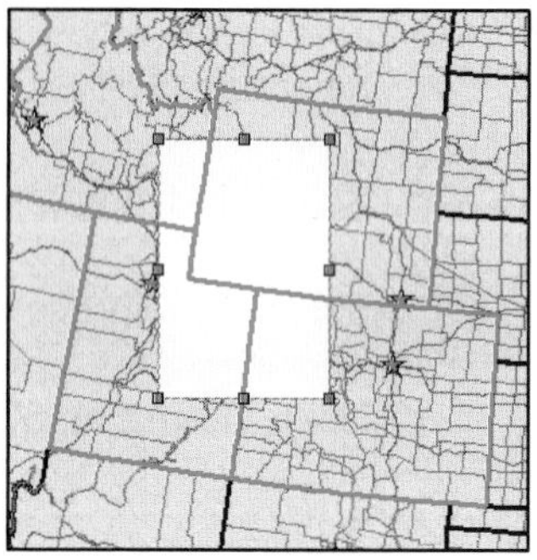

Fig. 6.2. Selecting by a graphic with only states selectable

TIP: The Selectable Layers option only applies to interactive selection with the mouse. It does not affect selecting by attributes or selecting by location as described in later sections because those options always occur on the specified layer(s).

Selecting by attributes

In this method the user specifies a certain condition based on fields in the attribute table, and ArcMap selects the records which meet the criteria, such as choosing all states with a population greater than 5 million. We already introduced selecting by attributes from tables in Chapter 5. In this section we extend selection to features and discuss the more advanced selection options.

A selection by attributes must be based on an **expression.** Expressions are written in a format called Structured Query Language (SQL), which is used by most major databases to perform selections. Although SQL expressions can be very complex and sophisticated, most expressions will be fairly simple ones. ArcMap helps out by providing a query editing box to transform points and clicks into acceptable SQL. Some examples of SQL expressions follow:

SELECT *FROM cities WHERE "POP1990" >= 500000

SELECT *FROM counties WHERE "BEEFCOW_92" < "BEEFCOW_87"

SELECT *FROM parcels WHERE "LU-CODE" = 42 AND "VALUE" > 50000

SELECT *FROM rentals WHERE "RENT" > 700 AND "RENT" < 1500

The SELECT *FROM and WHERE keywords must be present as dictated by the syntax rules of SQL. The lowercase words (e.g., cities) give the name of the table to select from, and the terms in double quotes are the field names. SQL recognizes a variety of comparison operators, including <, >, and =. The conditions AND, OR, and NOT can also be used in building expressions. This same syntax would be used in any SQL-compliant database, so SQL is a very transportable language in which to write queries. When using ArcGIS, it automatically enters the "SELECT *FROM *layer* WHERE" keywords, and the user only needs to enter the actual expression. In future examples we omit the keywords and focus on the expression, but the keywords are implied.

In the final example showing the selection of rental properties within a certain price range, notice that the field RENT appears twice in the expression. SQL will not recognize "RENT" > 700 AND < 1500. The field name must appear in every condition of the expression.

The operators AND and OR must be used carefully to avoid confusion. The AND operator tests to see whether both of the conditions are true. If so, the record is selected; if either condition is false the record is not selected. When using OR, the record will be selected if either of the conditions is true. For example, imagine trying to select parcels with zoning codes 3 and 4. A novice might enter the expression "Zone" = 3 AND "Zone" = 4. This expression yields nothing, because for a single record the zone cannot equal both 3 and 4 at the same time! There is no way that both individual conditions (Zone = 3 and Zone = 4) can be true for the same record. The expression should read "Zone" = 3 OR "Zone" = 4. If either condition is true, then the expression will select the record.

Consider the problem of finding numbers within a certain range, such as towns with population between 1000 and 5000. The expression "Pop" > 1000 AND "Pop" < 5000 is correct in this case, since any number in the range can meet these criteria. If a record with "Pop" = 2000 is being considered, it would test true to both conditions and be selected. However, using "Pop" > 1000 OR "Pop" < 5000 would return ALL of the values rather than the ones inside the range. A record with "Pop" = 500 would test true to the second condition and so be included. A record with "Pop" = 7000 would test true to the first condition and also be included.

SQL does accept three or more criteria. Typically parentheses must be used to enforce the correct order of evaluation. Compare these two expressions for selecting parcels:

("LUCODE" = 42 AND "VALUE" > 50000) OR "SIZE" > 50

"LUCODE" = 42 AND ("VALUE" > 50000 OR "SIZE" > 50)

In the first expression all large parcels > 50 hectares are chosen regardless of their zoning or value, and additional parcels are added if they are code=42 and high in value. In the second expression all of the parcels must have a zoning of 42 and can be either high in value or large in size.

When selecting strings, one must allow for extraneous spaces and capital/lowercase mismatches. For example, the string "Maple St " with a space following St would not match the string "Maple St" or "Maple St." or "Maple st". Often databases have little inconsistencies like this one, because it is very difficult to standardize the formats perfectly, and, of course, people often make typing mistakes. Thus the operator LIKE is often used instead of EQUALS (=) when selecting on string fields. LIKE searches for the specified set of characters within the field and returns any record that contains those characters. These searches are case-sensitive. When using LIKE, a wild card character (%) is used to stand for other letters that may appear. Consider selecting cities based on their NAME field. The query

"NAME" LIKE 'New %'

would select all cities beginning with the word New, such as New London and New Haven, but would not include Newcastle, because there is no space in the latter. The expression

"NAME" LIKE '%new%'

would return only lowercase 'new' in the middle of a word, such as Ken<u>new</u>ick and Pi<u>new</u>ood.

Expressions are constructed using the fields and buttons accessible in the Select By Attributes window (Fig. 6.3). The window menu automatically enters the first part of the expression (SELECT *FROM *layer* WHERE) for you, so in subsequent examples we will drop this part and stick to the part needing to be entered. Notice a few additional options which can be used in this window.

The **Query Wizard** button provides a series of step-by-step menus and complete explanations to help set up a query. This option is good for beginners, although experienced users will often find it faster to simply use the Query window directly.

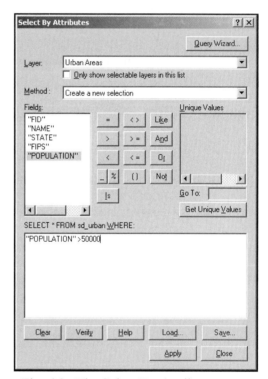

Fig. 6.3. The Select By Attributes window for creating queries

The **Get Unique Values** button adds the values from the table to the window where they may be selected by the user instead of typing them in.

The **Verify** button can be used to check an SQL expression before executing it.

The **Save** and **Load** buttons can be used to store queries for use again later. This option may not be helpful for short queries with only one or two conditions, but remember that SQL queries can be long and complex. An especially complicated or frequently used query might be helpful to keep. An example might be: retrieve all parcels that are low density or medium density residential units with septic systems, which were sold between the dates of January 1 and June 30 of 1999.

The **Clear** button erases old or invalid queries and starts again fresh.

Selecting by location

Selecting by location is a powerful tool unique to GIS systems because it selects features based on their spatial relationship to other features, such as finding wells within five miles of a chemical spill, or finding parcels completely inside a floodplain. A spatial selection uses two layers (such as the parcel layer and floodplain layer) and one spatial condition (such as completely inside). The features of the layer being selected are compared spatially with the features of the second layer to see which ones meet the criteria, and features meeting the criteria are selected. Figure 6.4 shows six different examples of selecting by location, using several of the criteria available. The selected features are shown in light blue in each case.

Fig. 6.4. Examples of selecting one layer based on its spatial relationship to another layer

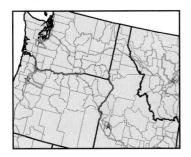

Select counties that contain state capitals.

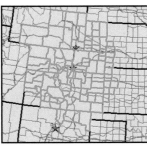

Select counties within 200 miles of Denver.

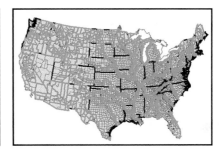

Select counties that intersect rivers.

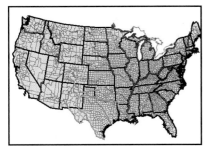

Select rivers that intersect Texas.

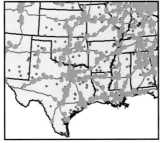

Select cities that are within 20 miles of an interstate.

Select cities within counties named Washington.

Select By Location can test for a variety of spatial relationships, as shown in Figure 6.5. The intersect criterion indicates that features touch in any way. Contains means that one lies completely inside another. A line contains another line if they coincide.

Not all of the test criteria are available in every case, because some criteria make no sense with certain pairs of feature types. For example, selecting points that completely contain polygons is impossible, because a point is a single location and cannot contain a larger area.

```
intersect
intersect
are within a distance of
completely contain
are completely within
have their center in
share a line segment with
touch the boundary of
are identical to
are crossed by the outline of
contain
are contained by
```

Fig. 6.5. Spatial criteria options

The Select By Location window provides the means to set up a spatial query. It reads something like a sentence from top to bottom, such as "I want to select features from the layer Cities that are contained by the features in the layer Rivers" (Fig. 6.5). You may find it helpful to read through queries in this way after setting them up, in order to verify that the selection is accurate.

After you choose the layers and the test criteria, the preview graphics in the bottom of the window explain how the test criterion works in different cases. In Figure 6.6, the user has requested the selection of Cities that are contained by Rivers. The Preview pictures at the bottom of the window show how the criteria "are contained by" apply to points, lines, and polygons: to be selected, points must touch the rivers, and lines must coincide with the rivers, but polygons cannot be selected because a polygon cannot be contained by a river.

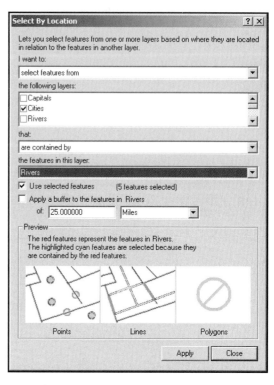

Fig. 6.6. The Select By Location window

Sometimes a problem calls for testing spatial relationships using a *subset* of features rather than the entire layer. For example, a user might want to select the cities within 50 miles of Interstate 80, rather than cities within 50 miles of any interstate. This procedure begins by selecting the desired feature first, I-80, and then performing the spatial query. Be sure to check the box on the Select By Location window to use the selected features in the criteria layer.

Spatial queries may also be performed within a single layer. A spatial query can also test spatial relationships within a single layer, such as finding all the other parcels that are adjacent to city park parcels, or finding all the restaurants that are within two miles of a favorite restaurant.

Choosing the selection method

The selection method applies to all three types of selections (interactive, selecting by attribute, and selecting by location). These methods give greater flexibility in defining selection sets and enable selection using multiple steps. Complex expressions involving many criteria are often less confusing when performed sequentially, rather than entering a long expression full of ANDs and ORs and trying to figure out where to put the parentheses.

ArcMap has four possible methods for selecting features (Fig. 6.7). The methods may be specified in two instances: from the main Selection menu when doing interactive selection, or within the Select By Attributes window.

Create New Selection always selects the specified records regardless of what else is already selected. If other records are selected, they will be replaced with

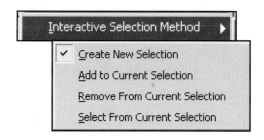

Fig. 6.7. Selection methods

the new selected records. For example, if you have selected green jelly beans, and then choose to select red jelly beans, then you will only have red jelly beans. This method is the default.

Add to Current Selection yields the records that fit the query in addition to any that are currently selected. Thus, if you have green jelly beans already selected, and then select the red ones, then you will have both red and green jelly beans selected. Using this method is equivalent to doing a single selection step with a double expression using the OR operator, as when selecting beans where "color" = red OR "color" = green.

Remove From Current Selection takes away the records that meet the selection criteria. If you already have red, green, and yellow jelly beans selected, and you "choose" green jelly beans, then you will have red and yellow jelly beans left selected. In this option, what you choose gets deleted from your selected set, so you must think slightly backward on this one.

Select From Current Selection chooses the records that fit the criteria, provided that they are already selected. If you had red, green, and yellow jelly beans selected, and choose to select the green ones, you will have the green ones. If you choose to select orange ones, then you will end up with *nothing,* because orange jelly beans were not in the original selection. Performing several selections in a row with this method is similar to performing a series of AND expressions.

A real-life example of applying these selection methods might proceed as follows: Imagine selecting all residential parcels with a VALUE greater than $100,000 that intersect the 100-year floodplain. Because this problem involves selecting both by attribute *and* by location, it occurs in two steps. First, use Select By Location to choose all parcels in the floodplain, using Create New Selection as the method. Then use Select By Attributes to choose LU-CODE = Residential AND VALUE > 100000. However, using Create New Selection as the method would yield ALL residential parcels worth more than $100,000, not just the ones in the floodplain. You need to specify Select From Current Selection in order to obtain the desired records.

One other selection option, **Switch Selection,** takes the selected set and replaces it with the unselected records. In other words, if you have red and green jelly beans available and the red ones are selected, and if you do a switch selection, then you will have green ones selected. Switch Selection is only available in the Options menu *in a table,* because for performance reasons it should be done on a single layer or table at a time.

Selection states

A layer or table can have several selection states, determined by two factors: whether a query has been applied, and how many features are selected as a result of that query. The selection state impacts the behavior of geoprocessing and other commands applied to the layer.

By default a layer has no query applied, that is, it has "no selection." In this state, all of the features are available for processing by a command. If a Statistics command is requested in this state, all of the features would be included in the calculations. In Figure 6.8 the layers shown in normal-face type are in the default state with no selection applied, as they would be shown in the Selection tab in the Table of Contents. Observe that none of these layers is highlighted in the map, and that in tables these layers would appear as having 0 records selected.

Executing a query on a layer pushes it into a new state. The layer has a selection applied, and the number of features or records selected may vary from zero to the full set of features. The layers in boldface in Figure 6.8 have selections. An interactive query on Counties yielded a block of counties selected in Texas, with a total of 37 features selected. A Select All command applied to the Capitals caused all 50 of its features to be selected. A Select By Attribute query for all cities with elevations greater than 10,000 feet resulted in zero features selected, since none of the cities met that criteria. These states are reflected in the map by the selection highlights, and in the table by the number of records selected.

For the three layers with selections, the result of a command, such as Statistics, reflects the number of features selected. The statistics for Counties would be based on 37 records, the statistics for Capitals would reflect all 50 records, and the statistics for Cities would be based on no records. Any other function applied in ArcMap or in ArcToolbox on these layers would also restrict the operation of the function to the selected features or records.

It is important to be aware that the terms "no selection" and "no records selected" have different meanings. In the first case no query has been applied, and all records would be used. In the second case, a query has been applied but no records are selected, so no records would be used.

Also note that the state of no selection and the state of all records selected would have the same result in a processing step—all of the records would be used, even though the number of records shown as selected in the table is zero in one case and all in another.

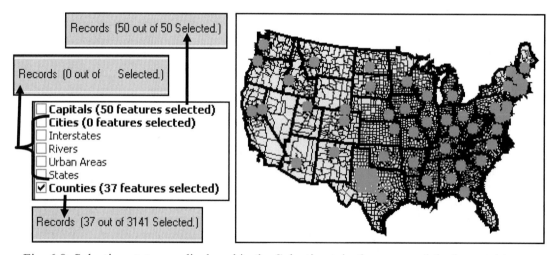

Fig. 6.8. Selection states, as displayed in the Selection tab, the map, and the layer tables.

Definition queries

All of the queries we have talked about so far enable the user to select a given set of features or records and then perform some action on them. In addition, a query can act as a property of a spatial data set, specifying a subset of features which are considered part of the layer. This action makes it *appear* as though the layer contains fewer features than it actually does. For example, a definition query might specify as its criterion "SUBREGION" = 'N Eng', that is, the New England states. As a result of the query, the map would only show the New England states, and the attribute table would contain only the New England records. Although the layer is still using the same source file, and the unchosen records are still stored on disk, they have been temporarily excluded from the view of the data file. A definition query may be removed to once again access all the features in the layer, by deleting the query in the same window in which it was set.

This feature provides a handy way to single out a subset of a data layer without needing to create a new file on the disk. If a geographer was preparing a map of New England he could use a definition query to isolate those states and display them alone without the other states. He would not need to clutter his disk with another shapefile, but could simply use the existing states shapefile and create a definition query to restrict the layer to the features of interest.

In the examples from this chapter, you will practice using different types of queries alone and in sequence.

Using Queries in GIS Analysis

Queries are a fundamental tool in GIS, used in nearly every analysis. They serve a critical function in data exploration, or picking one's way through a data set looking for patterns. They can also serve as the preparatory step to extract certain features prior to performing some other function. Some common applications involving queries might include the following.

- **How many? How much?** Queries can be used to search a table and find which features meet certain criteria. How many houses currently for sale fall into my price range? Combined with the statistics function, this tool becomes powerful. What is the average cost of three-bedroom homes in this town?

- **Where are they? Are there spatial patterns?** Creating a map from selected features and examining their spatial distribution can be illuminating. Where do the wells with high values of contaminants occur? Do they show a pattern relative to known point source pollution emissions? Are they widely spaced or clustered together?

- **Isolate features for more analysis.** Select out the Democrats and calculate an entry in the Party field. Extract the primary roads into a separate layer and create polygons around them at a specified distance (buffering).

- **Explore spatial relationships.** Which parcels lie in the flood plain? Which cities are close to a volcano? Where do roads cross unstable shale units?

- **Raster queries.** Although we have only discussed vector queries, cells of a raster can be queried also. Which cells have a slope greater than 10 degrees? Which cells show an increase in vegetative greenness as measured by satellite images taken in 1998 and 1999? Given two rasters showing land use in 1970 and 2000, where has land use changed?

As you gain experience using GIS, you will use these types of queries and more.

Summary

➢ Queries extract a subset of records or features from a data set based on a set of one or more conditions.

➢ An interactive query uses the Select Features tool and lets the user pick certain features by finding them on the screen and clicking them.

➢ The Selectable Layers option restricts the operation of the Select Features tool to the specified layers. By default, all layers are selected.

➢ Select By Attributes extracts features based on conditions from fields in the attribute table, such as cities with population greater than 1 million.

➢ Select By Location extracts features based on criteria of how two layers are spatially related to each other, such as selecting all cities within 100 miles of interstates.

➢ Four selection methods add greater flexibility to queries. These methods include creating a new selection, adding to the current selection, removing from the current selection, and selecting from the current selection.

➢ The selection state of a layer impacts how many features are used in subsequent commands or tools, such as Statistics. If a layer has no selection, all features are used. If a layer has a selection, then the selected features are used. If a layer has a selection applied that resulted in no (zero) selected features, then no features will be used.

➢ Definition queries temporarily restrict a data layer to a subset of its features. The data layer appears to have fewer features than it really does as long as the definition query is in effect.

Chapter Review Questions

You may need to consult the Skills Reference section to answer some of these questions.

1. What is a query?

2. List three ways to select features interactively.

3. What does it mean to set the selectable layers? What is the default setting?

4. How do you specify that features must be completely within a rectangle or shape to be selected, rather than being partly in it?

5. Write a valid SQL expression to select cities between 1000 and 10,000 people using a field called POP2000.

6. Write a valid SQL expression to select all counties whose names begin with the letter Q.

7. Imagine that you have some trail mix composed of peanuts, raisins, almonds, cashews, dried cranberries, and chocolate candies colored red, green, yellow, and orange. Imagine that you apply the following set of "queries" to the trail mix:

   Create new selection all candies
   Add to selection cashews
   Remove from selection red and green candies
   Select from selection all nuts and candies

   What do you have selected now? _____

8. Imagine doing a spatial query for all cities within 100 miles of a river. What additional steps would you need to take if you wanted only counties within 100 miles of the Missouri River?

9. Imagine that you have a map showing states and cities. Explain how the map would look different under the following two circumstances: (a) you perform an attribute query to select the state capitals; (b) you perform a definition query to select the state capitals.

10. Explain the difference between having no selection and having no selected features. Which one would result in all statistics being zero if the Statistics command were chosen for the table?

Mastering the Skills

Teaching Tutorial

The following examples provide step-by-step instructions for doing basic tasks and solving basic problems in ArcGIS. The steps you need to do are highlighted with an arrow ➔; follow them carefully. Click on the video number in the VideoIndex to view a demonstration of the steps.

➔ Start ArcMap and open the document ex_6.mxd in the MapDocuments folder.

➔ Use Save As to rename the document, and remember to save frequently as you work.

Before we begin, we will add the Clear Selected Features button to the general toolbar. Although you can access this function through the Selection menu, it is faster if you place the button on the toolbar (Fig. 6.9).

Fig. 6.9

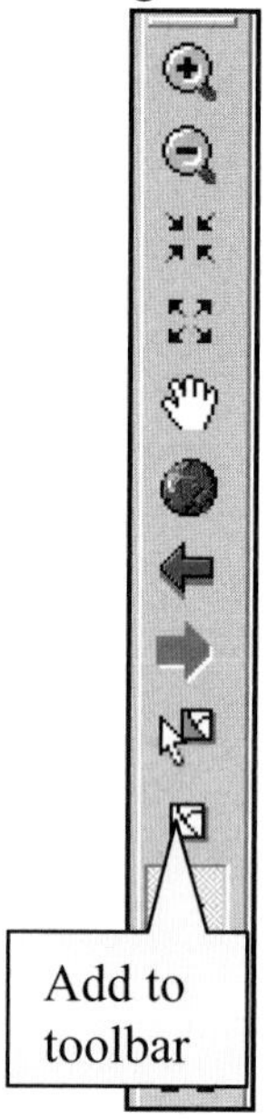

1➔ Right-click on the grey area on or around the toolbars and choose Customize from the context menu.

1➔ Move the window if necessary in order to see the general toolbar (the one with the Zoom/Pan tools on it).

1➔ Click the Commands tab in the Customize window.

1➔ Scroll down the list of Categories on the left to find Selection, and click on it to highlight it.

1➔ Locate the Clear Selected Features command, click on it, and drag it to the general toolbar, just below the Select tool.

1➔ To have this tool appear in all map documents by default, change the Save In file to Normal.mxt (a template containing your map document preferences).

1➔ Close the Customize box.

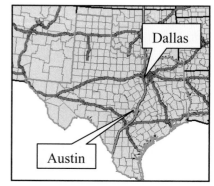

Add to toolbar

Using interactive selection

Now, let's experiment with selecting features with the mouse.

2➔ Choose View > Bookmarks > Texas from the main menu bar.

 2➔ Click on the Select Features tool, and then click on the star that represents Austin, the capital of Texas.

Your results may differ slightly, but you should now have a number of features selected, including the state of Texas, one or more counties near the capital, and an interstate highway running through the capital (Fig. 6.10). Any layer that was near the clicked point was added to the selected set.

2➔ Try clicking several other spots to see what gets selected.

Fig. 6.10. Interactive selection with all layers selectable

2➔ Click and drag a box around the Dallas area, and notice how anything in or that goes through the box is selected.

Setting the Selectable Layers offers control over what gets selected by a click.

3➔ Choose Selection > Set Selectable Layers from the main menu bar.

1. Which of the layers are currently selectable? _____

3➔ Click the button to Clear All, and then check the Interstates box. Click Close.

3➔ Draw a box around Dallas again. This time, only the interstates are selected.

The Table of Contents in ArcMap contains a tab to facilitate the viewing and management of layer queries. Among other functions, the Selection tab in the Table of Contents offers a faster way to change the selectable layers.

3➔ Click the Selection tab at the bottom of the Table of Contents.

3➔ Uncheck the Interstates layer, and check the Counties layer to make it selectable.

3➔ Draw the box again. Now only the counties are selected.

Notice that when a new selection is made, the selection on the other layers disappears. The Selection tab also keeps track of how many features are currently selected for each layer.

Next we will experiment with other ways to interactively select multiple features from one layer, the counties.

4➔ Choose View > Bookmarks > South Dakota from the main menu.

4➔ Click one county in South Dakota.

4➔ Hold down the Shift key and click another county. It is added to the selected set.

4➔ Add three more counties using the Shift-click method.

4➔ Now hold down Shift and click one of the counties that are *already selected*. It will be *remove*d from the selected set.

Imagine that you want to efficiently select all the counties in South Dakota that are west of the Missouri River. We can do this several ways.

5➔ Click the Display tab in the Table of Contents, and turn on the Rivers layer to see the Missouri River.

5➔ Click and drag a box that goes through as many counties as possible in South Dakota, but does not touch any counties east of the river. Do it several times if needed to get a good result. You will not be able to get all the counties, however.

5➔ Hold down the Shift key and draw another box to add more counties to the selected set. Continue, if necessary, until all counties west of the river are selected.

This time, let's make use of the Interactive Selection Method instead.

6➔ Click the Clear Selected Features button.

6➜ Choose Selection > Interactive Selection Method > Add to Current Selection.

6➜ Click on several counties in turn. Each one clicked becomes selected with the previous ones, without needing to hold down the Shift key.

6➜ Use several clicks and boxes to finish selecting all of the west river counties (Fig. 6.11).

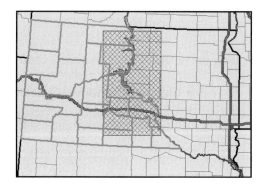

Fig. 6.11. Counties west of the Missouri

7➜ Then add a few east river counties by drawing a box around them.

7➜ Choose Selection > Interactive Selection Method > Remove from Current Selection.

7➜ Click on or draw a box around the east river counties to remove them from the selected set.

At this point you should have all the west river counties selected, and none of the east river ones (Fig. 6.11). Recall that the Statistics command shows information about the selected features.

8➜ Right-click the Counties layer and choose Open Attribute Table.

8➜ Right-click the POP2000 field and choose Statistics.

2. How many counties are selected, and what is the total number of people in them?

_____

8➜ Close the Statistics box and the attribute table.

Now let's get back to testing the different selection methods.

9➜ Choose Selection > Interactive Selection Method > Select from Current Selection.

9➜ Draw a box that encloses the area of the large reservoir in the middle of the state, Lake Oahe. The box should cover approximately the area shown with the hatched symbol in Figure 6.11.

Notice how the west river counties touching the box were selected, but the east river counties were not, because they were not already part of the selected set. Finally, let's use a graphic to select counties.

10➜ Set the Interactive Selection method to Create New Selection.

10➜ Click the tool to Clear Selected features on the general toolbar.

10➜ Click the Down arrow next to the Shape tool menu on the Drawing toolbar, and choose the Circle tool (Fig. 6.12).

10➜ Click on the state capital, Pierre, and drag out the radius of a circle (Fig. 6.12). Watch the circle radius measurement in the lower left corner of the ArcMap window, and try to get it as close as possible to 100,000 meters. The completed circle will be selected, as shown by the blue-handled box around it (Fig. 6.12).

10➜ Choose Selection > Select by Graphics. Any counties touching the circle are selected (Fig. 6.12).

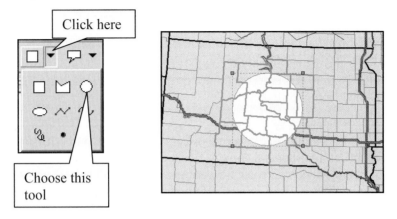

Fig. 6.12. Selection using a graphic (circle)

11➜ Try using the rectangle and line tools to make selections. Try to draw a polygon that selects all of the west river counties in one step.

TIP: If Selecting by Graphic yields nothing, double-check the selectable layers. If it still does not work, just go on. The author has had trouble getting this to work.

When finished selecting, delete the graphic(s).

12➜ Click the Select Elements tool on the Drawing toolbar.

12➜ Hold down the Shift key and click each graphic to select it.

12➜ Click the Delete key to get rid of them.

The Selection tab has a context menu that performs common query functions. Use this context menu to clear the current Counties selection.

12➜ Click the Selection tab in the Table of Contents.

12➜ Right-click the Counties layer and choose Clear Selected Features.

Selecting by attributes

Chapter 5 demonstrated one way to select by attributes, by opening the attribute table, clicking the Options button, and choosing Select By Attributes from the menu. The Select By Attributes window, another method, does not require opening the table, and it can be used on any layer.

13➜ Choose View > Bookmarks > USA from the main menu.

13➜ Switch to the Display tab and turn off the Interstates and Rivers layers.

13➜ Choose Selection > Select By Attributes from the main menu bar.

The Select By Attributes window is identical to the one accessed by the table Options menu, except that you must set the layer to select in the first drop-down box.

14➜ Use the drop-down box to choose Counties as the layer to select.

14➜ Confirm that the selection method is Create New Selection.

14➜ Enter the expression "POP2000" > 1000000 (one million) by double-clicking POP2000 until it is entered in the expression box, clicking the > button, and typing in 1000000 from the keyboard.

TIP: SQL does not gracefully accept extra spaces, parentheses, and other minor syntax errors, and they can be difficult to spot and correct. Most users will find it faster to use the buttons and windows of the query window to enter field names, assign field values, and so on, rather than trying to type expressions in from scratch. If you have trouble making an expression work, try clearing the expression box and reentering the expression. The Verify button can be used to test expressions before actually executing them.

14➜ Click Apply and move the Select By Attributes window to show the map and the window at the same time.

3. How many counties in the Unites States had more than 1 million people in 2000? _____

4. Which county has the most people? _____

One advantage of using this window is being able to query another layer from the same window.

15➜ Click the Select By Attributes box again to activate it.

15➜ In the Select By Attributes box, change the layer being selected to States. Keep the Create New Selection method.

15➜ Change the expression to "POP2000" > 5000000 (5 million). Click Apply.

Notice that both counties and states are now selected, even though Create New Selection was the selection method. This happened because the states and counties are separate layers and are treated separately in this query box. Also notice that the states were selected even though they are not currently a selectable layer. The selectable layer applies only to interactive selection, and it is ignored by the Select By Attributes and Select By Location functions.

Now let's try a selection with double criteria: all counties that lost population between 1990 and 2000, and have more males than females.

16➜ Clear all the selected features using the Clear Features tool.

16➜ Click the Select By Attributes box to activate it, if needed.

16➜ Set the layer to Counties.

16➜ Enter the expression "POP2000" < "POP1990" AND "MALES" > "FEMALES" in the query box. Click Apply.

5. How many counties with more males than females lost population between 1990 and 2000? _____ Where are they mainly located? _____

Now from these, select those counties which have fewer than 10,000 people. But be careful to ensure that the new set of counties is pulled only from the currently selected ones.

17➜ In the Select By Attributes box, change the selection method to *Select from current selection*.

17➜ Click the Clear button to erase the old expression.

17➜ Enter the expression POP2000 < 10000. Click Apply.

Watch as a few of the counties disappear from the selected set.

6. How many counties remain selected? _____

17➜ Close the Select By Attributes window.

17➜ Clear the selected features.

TIP: The Clear Selected Features tool on the toolbar and in the Selection menu clears the selection from ALL layers. To clear the selection for a single layer, right-click the layer in the Selection tab and use the Clear Selected Features command in the context menu.

Selecting by location

Next we will try several examples of selecting by location. First we will select all the counties that are adjacent to rivers.

18➜ Turn on the Rivers layer.

18➜ Choose Selection > Select By Location and adjust the window to show both the Selection window and the map.

18➜ In the first drop-down box, choose the selection method *I want to select features from* (Fig. 6.13).

18➜ Choose Counties as the layer to select from.

18➜ Choose the selection criterion *intersect*.

18➜ Choose Rivers as the layer the selection is based upon.

18➜ Click Apply and view the map.

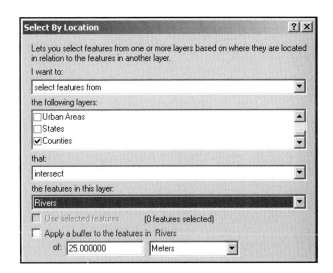

Fig. 6.13. The Select By Location options

7. What percentage of counties in the United States are intersected by rivers? _____

Read the window just as though it were a sentence in order to check the logic. In this example, the sentence says "I want to select features from Counties that intersect the features in Rivers." Next, let's select the counties which contain state capitals.

19➔ Change the window to read "I want to select features from Counties that *completely contain* the features in Capitals."

19➔ Click Apply.

8. Which of these counties containing capitals has the smallest population? _____
What is the capital and which state is it in? _____

Now let's do an example with points and lines, by selecting the cities which are close to interstate highways.

20➔ The cities are not visible because the scale range is set. Right-click Cities and choose Visible Scale Range > Clear Scale Range from the context menu.

20➔ Turn on the Interstates layer to see the results more clearly.

20➔ Change the Select By Location window to read "I want to select Cities that are within a distance of Interstates."

TIP: The layer from the previous query, Counties, has remained checked in this list. Uncheck it before doing the query, or both counties and cities will be selected. The previous layers always remained checked in this window, so get in the habit of checking the list before each query.

20➔ Click the button to apply a buffer to Interstates, and set the buffer distance to 20 miles.

20➔ Click Apply and view the results, closing the selection window.

9. What percentage of U.S. cities are within 20 miles of an interstate highway? _____

In the previous example, we compared *all* the features in one layer to *all* the features in the other. You can also select features based on a subset of the comparison features. For example, let's select all the rivers that pass through Texas. We do this by selecting Texas first, using either interactive selection or selection by attributes, and then performing a Select By Location.

21➔ Clear all selected features.

21➔ Turn off the Cities, Capitals, and Interstates, and turn on Rivers.

21➔ Set the Selectable Layers to States only.

21➔ Click the Select Features tool and select the state of Texas.

22➔ Open the Select By Location window.

22➔ Change the window sentence to read "I want to select features from Rivers that intersect the features in States."

22➔ Check the box, if necessary, to use only the selected features in States.

22➔ Click Apply, and close the Selection window to better view the results.

10. Which rivers intersect Texas? _____

Spatial queries may also be performed within a single layer, using it both as the target and the criterion layer. In this instance the features comprising the criteria must be selected first, before the Select By Location within the layer is performed. For example, select all cities within 400 miles of Kansas City, MO.

23➔ Clear all selected features.

23➔ Turn off the Rivers, and turn on the Cities.

23➔ Use Select By Attributes to select Kansas City. Remember that text queries must have the text in single quotes, or use the Get Unique Values button and select Kansas City from the list. (Check the selection method if you have trouble.)

23➔ Close the Select By Attributes box.

24➔ Open the Select By Location window.

24➔ Change the window sentence to read "I want to select features from Cities that are within a distance of 400 miles from Cities."

Notice that Cities is used twice in the window, and that the Use Selected Features box is automatically checked when the selection and criterion layers are the same.

24➔ Click Apply to make the selection, and close the selection window.

24➔ View the map.

Attribute and location queries gain power when combined sequentially. Let's select all the rivers that intersect states in the Mountain subregion. We do this by first selecting the mountain states using an attribute query, and then selecting the rivers that intersect these states.

25➔ Clear all selected features. Turn off Cities and turn on Rivers.

25➔ Choose Selection > Select By Attributes.

25➔ Set the layers to States, and enter the expression "SUB_REGION" = 'Mtn'. Use the Get Unique Values button to find the 'Mtn' entry.

25➔ Click Apply and close the Attribute Query window.

25➔ Open the Select By Location window, if necessary.

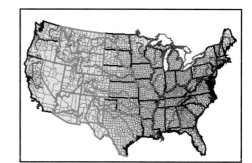

Fig. 6.14. Rivers that cross the Mountain subregion

25➔ Set the parameters to select features from Rivers that intersect the selected features of States.

25➔ Click Apply and view the results (Fig. 6.14).

Notice that many of the rivers extend far to the east to join the Mississippi. Because they do pass through the mountain states, though, they are included. You could try selecting only the rivers that are completely within the mountain states instead.

26➔ Change the selection criterion to *are completely within* and click Apply.

11. Now which rivers are selected? _____

Now hardly any rivers are selected. Ideally, you would like to select the *portions* of rivers that fall inside the mountain states. However, Select By Location can only work on entire features. If a river feature crosses the state boundary without splitting into two features, then the entire river is selected, both inside and outside the state. Unless the data creator specifically forced the features to break at state boundaries, a spatial query will find portions outside the state. This weakness of spatial queries can be troublesome. Chapter 8 presents techniques for splitting features when they cross other features.

Combining attribute and location queries can create quite sophisticated selections. Imagine that a meat packing company is considering adding the capacity to slaughter bison, and wants to survey potential customers between the packing plants in Pierre and Denver to see if the demand for bison meat justifies the added expense of handling these more difficult animals. The company decides to target the survey to counties that are within 300 miles of both Denver and Pierre and have more than 10,000 people.

27➔ Clear all the selected features.

27➔ Turn on the Capitals layer. To make it easier to find the capitals, right-click the Capitals layer and choose Properties. Click the Display tab and turn on Map Tips.

27➔ Use the Selection tab to make Capitals the only selectable layer. Click on the Select Features tool and click on Denver in Colorado to select it.

27➔ Open the Select By Location window, if necessary, and set it to select all Counties within 300 miles of the selected features of Capitals (Denver). Click Apply.

28➔ Right-click the Capitals layer in the Selection tab and choose Clear Selected Features from the context menu. By clearing from the context menu, only the features in Capitals are cleared; the counties stay selected.

28➔ Choose the Select Features tool, hold down the Shift key, and click on Pierre in South Dakota to select it. If you forget to Shift, the county selection will be cleared.

28➔ In the Select By Location window, change the selection method to *Select from the currently selected features*. Make sure the Use Selected Features box is checked, and click Apply.

28➔ Close the Select By Location window.

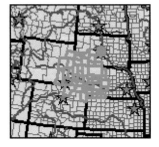

At this point a patch of counties between Denver and Pierre should show as selected (Fig. 6.15). Now we select from these counties the ones with more than 10,000 people.

Fig. 6.15. Counties within 300 miles of Pierre and Denver

29➔ Open the Select By Attributes window.

29➔ Set the layer to select to Counties.

29➔ Set the selection method to *Select from current selection.*

29➔ Enter the expression "POP2000" >10000 and click Apply.

29➔ Close the Select By Attributes window.

29➔ Right-click the Capitals layer in the Selection tab and clear the selected features. The map should now look similar to Figure 6.16.

29➔ Right-click the Counties layer in the Selection tab and choose Create Layer from Selected Features.

29➔ Switch to the Display tab and name the new layer "Packing Survey". Change the symbol if necessary to show up well against the background.

Fig. 6.16. Target counties for the packing survey

12. How many counties will be in the survey? _____ What is the total number of people who live in these counties?

_____

Exporting data

One common operation performed after selecting data involves creating a new shapefile from the selected set. Imagine that you are doing a report for the state on western South Dakota, and want to create maps showing only that part of the state.

30➔ Set the selectable layers back to Counties, use the bookmark to zoom into South Dakota, and select the west river counties again.

31➔ Right-click the Counties layer and choose Data > Export Data from the context menu.

31➔ Choose to export the selected features (Fig. 6.17).

31➔ Fill the button to use the same coordinate system as the source data.

31➔ Click the Browse button and save the file as "westriver.shp" in the mgisdata\Sdakota folder. Click OK.

31➔ Click Yes to add the new shapefile to the map.

31➔ Clear all the selected features.

31➔ Turn on the new westriver layer.

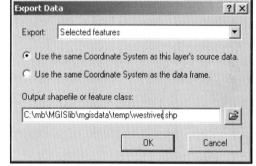

Fig. 6.17. Exporting a map layer

TIP: Use Export to create a full copy of a data set without ArcCatalog, by choosing to export all the features. You can also use it to project from one coordinate system to another by setting the data frame coordinate system to the desired projection, and choosing to use the data frame coordinate system.

Using definition queries

Definition queries make layers look as though they have fewer features than they really do. The features are still within the source data file on the disk, but only a specified subset actually appears on the map and in the attribute table. Definition queries are based on attributes. Let's create a layer showing only Virginia.

32➔ Use the USA bookmark to zoom to the lower 48 states.

32➔ Right-click the States layer in the Display tab and choose Properties from the context menu.

32➔ Click the Definition Query tab.

32➔ Click the Query Builder button.

32➔ Enter the expression in the box "STATE_NAME" = 'Virginia' and click OK.

32➔ Click OK again to close the Properties window, and view the map.

Now only Virginia is being displayed (look for the black outline). We can repeat this step for the Cities and Counties to create a map of Virginia alone.

33➔ Create a definition query to show only Virginia counties.

34➔ Turn on the Cities layer, and create a definition query to show only Virginia cities.

35➔ Turn off all the layers except the Virginia ones.

35➔ Zoom into Virginia and create a graduated color map of the county population and a graduated symbol map of the cities by population.

The map looks a little odd because Virginia has a slight tilt, which is due to the fact that it is off center from the conic map projection being used. We can modify the map projection to fix this problem for a nicer map. First, identify the datum of the source data so that the data frame projection can be chosen to match it.

36➔ Open the Properties for the States layer and click the Source tab. Note that the datum is NAD 1983.

36➔ Set the data frame coordinate system to Virginia State Plane North NAD 1983.

36➔ Admire the map (Fig. 6.18).

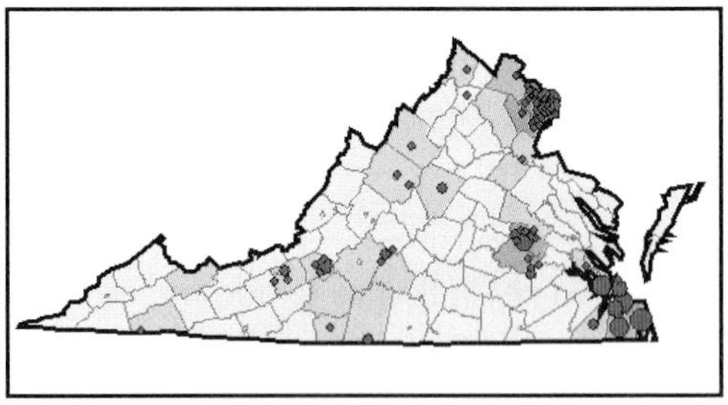

Fig. 6.18. Single-state map of Virginia made using definition queries

Using definition queries gives a similar result to exporting data, but it does not require that a new shapefile be created. This is an advantage if you make a lot of single-state maps and don't want dozens of shapefiles cluttering up the disk. On the other hand, if you use one state over and over again, it is probably worthwhile to use Export to create a separate shapefile of it.

Finally, let us try one query with multiple conditions to make sure we can do those as well as we can single queries. We will add West Virginia to the map along with Virginia.

37➔ Right-click the Counties layer and choose Properties from the context menu.

37➔ Click the Definition Query tab.

37➔ Click the Query Builder button.

37➔ Enter the expression in the box "STATE_NAME" = 'Virginia' OR "STATE_NAME" = 'West Virginia' and click OK.

37➔ Click OK again to close the Properties window, and view the map.

To clear a definition query and show the entire layer again, simply delete the query.

38➔ Open the Counties Properties and choose the Definition Query tab.

38➔ Select the query in the box and press Delete to clear it. Click OK.

You have learned three ways to create a subset of a map, such as a single state. **Exporting** creates a new data file, and the subset can be created using interactive, attribute, or location queries. **Definition queries** use the original file but hide the unwanted states; however they must be based on attribute queries. **Creating a layer file** from a selection is very similar to using a definition query, and it has the advantage that you can use interactive, attribute, and spatial queries to generate the subset. It creates a new layer in the map, but not a new data set on the disk.

This is the end of the tutorial.

➔ Exit ArcMap and save your changes.

Exercises

Use the files in ex_6.mxd to answer the following questions.

1. How many states have counties named for Thomas Jefferson (i.e., how many Jefferson counties are there)? Which state has the Jefferson county with the most people in the year 2000?

2. How many counties in the United States have more men than women? What percentage of the counties do they represent?

3. How many cities in the United States have more Hispanics than blacks, median rents greater than $500, and a population between 250,000 and 500,000? List them (there are less than 10). (**Hint:** Do your queries in more than one step.)

4. What percentage of counties in the United States have a river in or next to them? (Use Intersect.) What percentage of the U.S. population lives in these counties? (Use the year 2000 population.)

5. How many counties have more than 800,000 people and also contain a state capital? List the states these counties are in.

6. How many cities are within 50 miles of a volcano? (Add the volcano.shp data theme from the Usa folder). What is the total number of people living in them?

7. How many other volcanoes are there within 300 miles of Crater Lake, a volcano in Oregon? How many of these volcanoes are also within 50 miles of an interstate?

8. How many cities in the West South Central subregion of the United States are less than 200 miles from Oklahoma City? **Capture** a map showing your selected cities.

9. How many urban areas are *more* than 50 miles from an interstate highway? Which state has the most such areas, and how many does it have?

10. Create a volcano hazards map of Washington and Oregon, showing the counties symbolized by population, the volcanoes, and the cities shown with size based on population. Show ONLY the two states with nothing else appearing outside them. **Capture** your map.

Challenge Problem

Imagine that you are looking for a nice place to live. You like being in a small town in a fairly rural area, but within easy driving distance of an urban area for fun. You would prefer a city with between 20,000 and 40,000 people, no more than 50 miles from an urban area, in a county with a population density of less than 25 persons per square mile.

List the names and states of the cities in the United States that meet all these criteria. There are less than 10!

Hint: If you find yourself making a lot of mistakes, try storing your intermediate steps by creating a layer from the selected features.

Skills Reference

Setting the selectable layers

1. Choose Selection > Set Selectable Layers from the main menu bar.

2. Add or remove layers from the list of selectable ones by clicking inside the check box to the left of the layer (Fig. 6.19).

3. Click Select All to be able to select all the layers.

4. Click Clear All to remove all the checks in order to easily check one or two layers without having to turn all the others off individually.

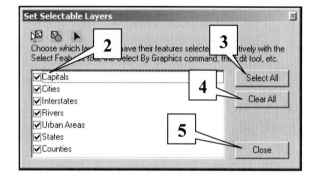

Fig. 6.19. Setting the selectable layers

5. Click Close when finished. The selectable layers remain in effect until changed.

NOTE: You can also set the selectable layers in the Selection tab. See Using the Selection Tab.

TIP: The Selectable Layers setting does not apply to selections based on attributes or location.

Selecting features interactively

1. Set the Selectable Layers, if necessary.

2. If necessary, change the selection method by choosing Selection > Interactive Selection Method and picking one of the methods (Fig. 6.20).

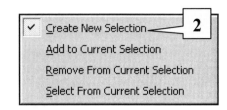

3. Click the Select Features tool. Click on a feature in the map to select it. It will be highlighted in the current selection color (blue by default).

Fig. 6.20. Choosing a selection method

4. To add a feature to the selection, hold down the Shift key and click the feature to be added. To remove features from the selection, hold down the Shift key and click the feature to remove.

5. To add a group of features using a rectangle, click on one corner of the rectangle, drag the mouse to the desired size, and release the mouse button. All features touching the rectangle will be selected.

TIP: Hold down the Shift key while drawing the rectangle to add a group of features to the current selection.

Selecting features using shapes

1. Change the Selectable Layers and/or the Selection Method if necessary.

2. Choose one of the Shape tools from the Drawing toolbar by clicking on the Down arrow and selecting one of the tools (Fig. 6.21).

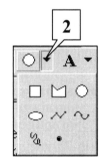

3. Draw the shape on the map using mouse clicks. Double-click to finish the shape.

4. Choose Selection > Select by Graphics to make the selection.

Fig. 6.21. Choosing a shape

Clearing a selection

There are several ways to clear a selection.

To clear features from ALL layers

Choose Selection > Clear Selected Features from the main menu bar.

OR

Click the Clear Selected Features tool, if you have added it to a toolbar.

To clear features from a single layer only

Right-click the layer in the Display or Source tab and choose Selection > Clear Selected Features.

OR

Right-click the layer in the Selection tab and choose Clear Selected Features.

TIP: Clearing a selection is a common task, and the Clear Selected Features button is a helpful addition to the general toolbar (with the Zoom/Pan tools). To add it, right-click the grey menu area and choose Customize from the context menu. Click the Commands tab and then click the Selection section. Drag the Clear Selected Features icon to the toolbar (next to the Select Feature tool is a good spot).

Using the Selection Tab

The Selection tab in the Table of Contents facilitates viewing and managing selections (Fig. 6.22).

1. Check or uncheck the layer boxes to make layers selectable or unselectable.

2. View the selection state of each layer: normal text indicates the default state of no selection applied; bold text indicates that a query is currently applied and that either all features, a subset of features, or zero features are selected.

3. Right-click a layer to open the context menu and choose one of the functions.

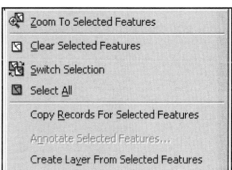

See the discussion in the Concepts section regarding the different possible selection states.

Fig. 6. 22. The Selection tab and its context menu

Selecting by attributes

1. Choose Selection > Select By Attributes from the main menu bar.

2. For a detailed walk-through of the process, click the Query Wizard. Otherwise, fill out the boxes in the window (Fig. 6.23).

3. Choose the layer whose features are to be selected. Check the box to show only the selectable layers in the list, if desired.

4. Choose the selection method.

5. Enter the expression. Double-click a field from the list to enter it into the expression box. Click a condition (such as = or >) to enter it in the box. Type the condition value.

6. To enter a condition value from the table instead of typing it, click the Get Unique Values button to show the values, and click on the desired value to enter it in the box.

7. Click Apply to execute the expression and create the selection.

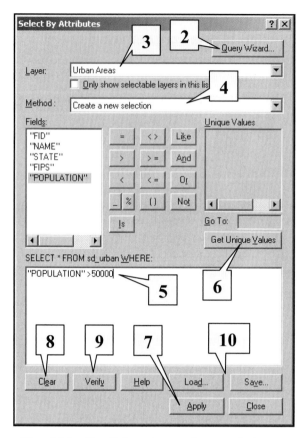

Fig. 6.23. The Select By Attributes window

TIP: Many users find that typing in or editing the expression by hand tends to generate errors. Use the boxes and buttons provided to enter expressions for quickest results.

TIP: When using multiple conditions, the field name must appear in every condition. To find rents between 500 and 1000 dollars, for example, enter "RENT" > 500 AND "RENT" < 1000. You cannot enter "RENT" > 500 AND < 1000.

Other options in this window

8. Click Clear to clear the expression and begin again.

9. Click Verify to check the expression syntax without actually executing it.

10. Use the Load and Save buttons to save expressions and use them again later.

11. Click Close to exit the window without executing the expression.

Selecting by location

1. Choose Selection > Select By Location from the main menu bar.

2. Choose the selection method (Fig. 6.24).

3. Choose the layer(s) from which to select features. Be careful—layers from previous queries remain checked until you uncheck them.

4. Choose the spatial condition.

5. Choose the layer to use as part of the spatial condition (such as within a certain distance of Rivers).

6. To use only the selected features of the condition layer, check the Use selected features box.

7. If using the condition Within Distance Of, or to increase the distance over which the other criteria are applied, enter a buffer amount and distance units, and check the Apply a buffer box if needed.

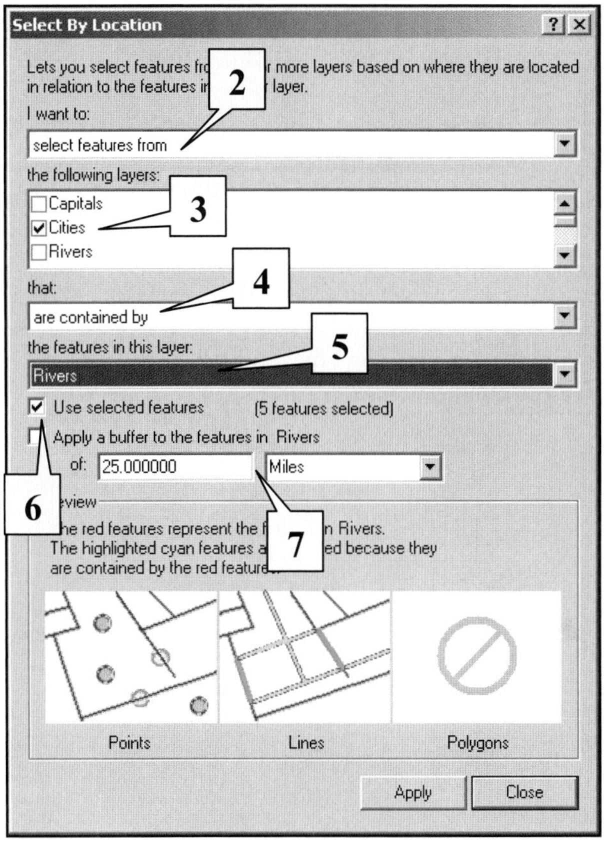

8. Check the pictures to verify that the selection criteria are valid.

9. Click Apply to make the selection.

Fig. 6.24. The Select by Location window

TIP: To use the same layer as the selection layer and the condition layer (for example, to select all restaurants within two miles of the restaurant you own) you must first use an attribute or interactive query to select the feature of interest—your restaurant. Then do the Select By Location, and specify the same layer for step 3 and step 5.

Creating a definition query

1. Double-click the layer to be defined in the Table of Contents, or right-click it to open its properties. Click the Definition Query tab.

2. Click the Query Builder button. The Query Builder is the same menu used for doing a Select By Attributes (Fig. 6.25).

3. Enter the expression. Double-click a field from the list to enter it into the expression box. Click a condition (such as = or >) to enter it in the box. Double-click a value from the values list to enter it, or type in a value from the keyboard. Strings must be enclosed in single quotes.

4. To choose a value instead of typing it in, click the Get Unique Values button to list the values in the field.

5. Click OK in the Query Builder and the Layer Properties windows to execute the expression and create the definition query.

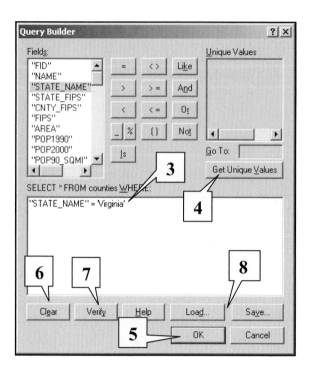

Fig. 6.25. The Query Builder window

TIP: Many users find that typing in or editing the expression by hand tends to generate errors. Use the boxes and buttons provided to enter expressions for quickest results.

TIP: When using multiple conditions, the field name must appear in every condition. To find rents between 500 and 1000 dollars, for example, enter "RENT" > 500 AND "RENT" < 1000. You cannot enter "RENT" > 500 AND < 1000.

Other options in this window

6. Click Clear to clear the expression and begin again.

7. Click Verify to check the expression syntax without actually executing it.

8. Use the Save button to save an expression and use the Load button to recall it again later.

Changing the selection color and other options

1. Choose Selection > Options from the main menu bar.

2. The first box controls how shapes select features when using the Select Feature tool or the Select by Shape tool (Fig. 6.26). By default, any feature partially or completely in the shape is selected. You may also choose to select features completely within the shape, or to select features that the shape is completely within.

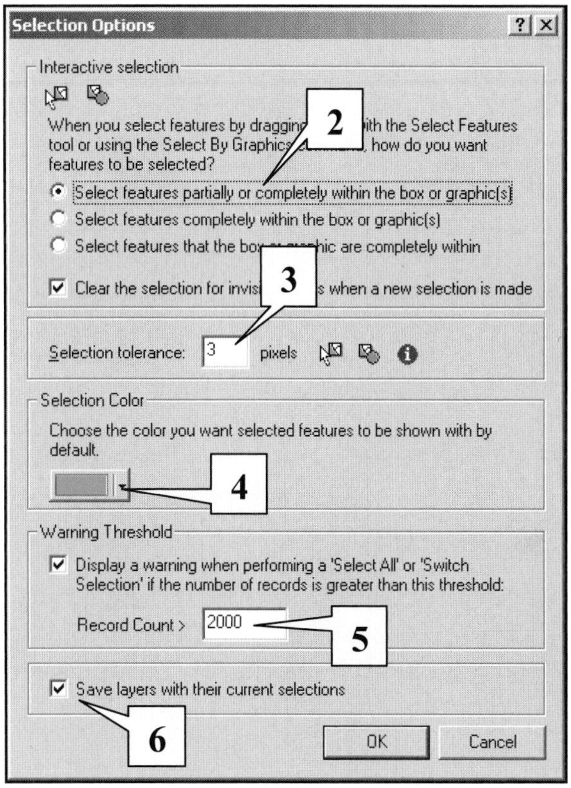

3. The second box sets how close you need to click to a feature before it is selected. If you often have trouble selecting points, increase this distance.

4. This box changes the selection highlight color.

5. In the fifth box, you may alter the threshold which determines how many records will be modified by a Switch Selection command. If the number exceeds the threshold, a warning prompt will be issued. This option prevents you from mistakenly doing a switch that will take a very long time to complete.

Fig. 6.26. Setting the selection options

6. Finally, you can choose whether to save layers with the current selections (the default) or clear the selections of all layers when saving the map document.

Chapter 7. Spatial Joins

Mastering the Concepts

Objectives

> ➢ Learning the purpose and capabilities of spatial joins

> ➢ Performing and interpreting spatial joins

> ➢ Learning to solve problems with spatial joins

Concepts

What is a spatial join?

Chapter 5 presented attribute joins, which are performed on tables—for example joining information in the livestock.dbf table to a map of counties to create a graduated color map of cattle populations in South Dakota. This join was based on a common field, the FIPS code, and had a cardinality of one-to-one. The join resulted in combining the two tables as if they were one table.

A spatial join is similar to an attribute join, except that instead of using a common field such as the FIPS code to decide which rows in the table match up, the location of the spatial feature is used. For example, a point layer containing wells and a polygon layer of geology could be joined to determine the geologic unit each well lies inside. Imagine doing a join between the two tables, with the geology as the source table, and the wells as the destination table. Thus information from the geology table is appended to the wells table. However, instead of using a common field, we use the *location* of the well to match the records. Each well gets the attribute information from the polygon it lies inside. So a well falling inside the Madison Limestone unit gets the attributes of that Madison Limestone geology polygon added to it.

A spatial join uses a location criterion to link the table records together. The previous example uses whether one feature lies inside the other. An alternate criterion is distance—joining the records that lie closest to each other, such as tagging each hotel with its closest restaurant.

Imagine that a community has a porous and fractured geologic unit which serves as a groundwater aquifer, and the city water supply comes from several wells which tap this unit. Outside the city limits, extensive development has occurred in the past 20 years, and many houses dot the rolling hills above the city. The city has become concerned that contamination from faulty septic systems may enter city water supplies through this very porous unit. Assessing the threat requires identifying the number of septic systems which occur in the outcrop area of the aquifer unit.

A spatial join solves this problem perfectly. In the join, each septic system is determined to fall within a certain geologic polygon, and the attributes from that polygon are appended to the septic system's record in the table. As shown in Figure 7.1, septic system 345 falls inside the Minnelusa Formation polygon in the geology layer, so the geology polygon info is appended to its row in the table. Septic system 348 falls inside a gravel deposit unit, and it gets the attributes from the gravel polygon. Once the two tables are combined, several new options are available, such as making a list of septic systems in sensitive areas and sending a letter to the owners, or summing the total number of septic systems in each geologic unit.

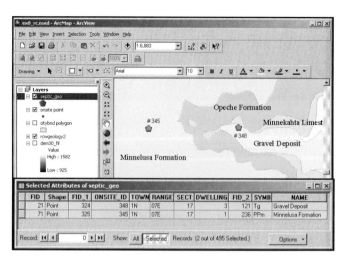

Fig. 7.1. A spatial join gives these septic systems the attributes of the geology polygon that they lie within.

Spatial joins may be performed on any two spatial data layers. Spatial joins always create a new, permanent data layer, rather than being temporary like attribute joins. A user can join points to points, polygons to polygons, lines to points, and nearly any combination of the three types of data. The output layer will always have the same feature type as the destination layer. In the preceding example, polygons (geology) were joined to the destination layer of points (wells), and the output layer contains well point features, with attributes added for the geology.

Types of spatial joins

Spatial joins fall into four main types according to the cardinality of the relationship between the joined layers (simple or summarized) and the choice of spatial criteria (inside or distance). Figure 7.2 shows a matrix of the four possible combinations. A simple join may be used whenever the cardinality is one-to-one or many-to-one, so that the Rule of Joining is maintained. In a one-to-many relationship, a summarized join must be employed.

Consider the simple join types first. In Figure 7.2a, schools and counties are being joined, with schools as the destination layer. The output layer thus contains schools, and each school will have the attribute information of the county it falls inside. With the output layer, one could answer the question, "Which county is a particular school in?" This example may be termed a **simple inside join.** The previous example of the septic systems is also a simple inside join.

If, however, the destination layer is reversed, the cardinality of counties to schools is one-to-many, and a simple join is not possible. A summarized join may be employed in this case. Like the Summarize function in Chapter 5, a summarized join groups the records together, based this time on their spatial location (which county they are in) rather than on an attribute field, and generates of single record of statistics for the group. This single summarized school record can then be appended to the matching county in the destination table, yielding a polygon layer of counties. A Count field is always generated, containing the total number of schools in each county. Combined with the regular attributes of the counties, this Count field might be used to find out how many schools each county has, or what the number of schools is per capita in the county. Figure 7.2b shows this as a **summarized inside join.** Like Summarize, other statistics for the groups may be requested in addition to the Count field.

Distance joins also come in simple or summarized varieties. Consider evaluating the desirability of several hotels based on their distance to local tourist attractions, using a layer of hotels and a layer of attractions with the yearly number of visitors in each one. What is the cardinality of this relationship? This question seems confusing, because there are several hotel sites and many attractions. In a distance join, it is the type of join itself that dictates the cardinality. If we ask the question, "Which attraction is closest to the hotel?" in order to find out whether that attraction has a large visitor pool, we have specified a one-to-one criterion, because only one attraction can be *closest* to each hotel. A **simple distance join** suffices, and each hotel appears in the output table along with the attributes of the attraction closest to it, and the distance between them. In Figure 7.2c, the circles connect each hotel site with its closest attraction.

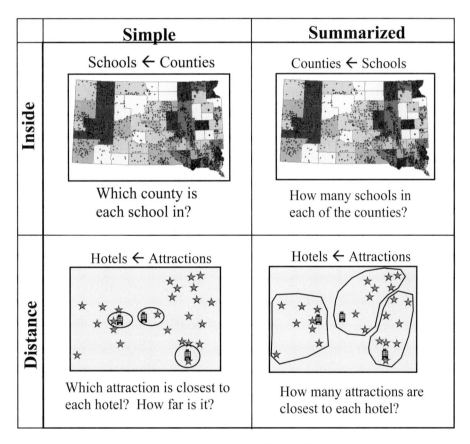

Fig. 7.2. Matrix of spatial joins resulting from different choices of destination table, spatial condition (inside or distance), and cardinality.

A different question may be asked, "How many tourist attractions are closer to this hotel site than they are to any other?" In this case we are interested in evaluating the richness of choice of attractions for each hotel, or perhaps the total combined visitor pool from all the closest attractions. In this case each hotel is connected with many attractions, as shown in Figure 7.2d, and summary statistics for the group, such as the number of attractions and the sum of the visitor pools, may be added to the record for each site. This would be a **summarized distance join.**

Choosing the join type

As in an attribute join, the process begins by deciding which layer is the destination. The user right-clicks the destination layer and chooses to join, specifies that location should be used in the join menu (Fig. 7.3), and enters the desired source layer.

The join menu always offers two ways to decide how to match the fields, taken from the four types shown in Figure 7.2. The Join Data window in Figure 7.3 illustrates the process of joining schools to counties, with counties as the destination layer as described in Figure 7.2a. The Counties layer is the destination, and each county can have more than one school, so it is a one-to-many join. Since this cardinality violates the Rule of Joining, the user must specify one of two ways to combine the tables. Reading the options and imagining the outcome usually provides the user with sufficient clues to make the choice.

Option 1 offers an **inside summarized join**: each county feature receives a statistical summary of the attributes of all the schools inside it, such as the total number of schools, or the sum of the students. Option 2 specifies that a **simple distance join** should be performed, and the county should receive the attributes of the school it is closest to. In this case the second option is nonsensical. All of the schools are inside the county and would be assigned a distance of zero, so it avails little to find the "closest" one. In this example, the summarized join is the appropriate choice.

Figure 7.4 shows the result of the join. The resulting layer is a map of counties with the original county attributes, plus a field called Count_ containing the number of schools in the county. From this field, the graduated color map was created, showing the number of schools in each county. When summarizing, the user can choose from several statistics, such as minimum, maximum, average, and so on. All numeric attributes in the source table are summarized using the chosen statistics and placed in the output table. String fields cannot be summed or averaged, so they are not included in the output table.

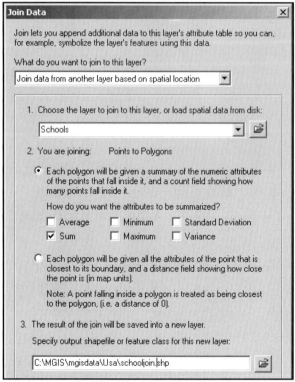

Fig. 7.3. A one-to-many cardinality is handled by either a summarized join or a distance join.

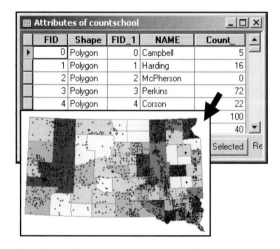

Fig. 7.4. Using a spatial join of schools to counties, one can create a graduated color map showing the number of schools in each county.

The two choices available for a join depend on the geometry of the feature classes being combined (Table 7.1) Two choices suffice, because not all combinations make sense spatially. When combining two point layers, for example, an inside join does not apply except trivially, for one point cannot be inside another unless their locations are identical. In this case the two choices consist of a simple distance join or a summarized distance join.

Table 7.1. Join types available for each combination of feature geometries in a spatial join. The second feature type is the destination layer in each case.

Geometry Type	Join Type	Example
Points to Points	Simple distance	Find the hospital closest to each town.
	Summarized distance	Find all the towns closer to one hospital than to any other hospital.
Lines to Points	Simple distance	Find the water main closest to the proposed building site.
	Summarized inside	Find the total voltage of all electric lines meeting at a substation.
Polygons to Points	Simple inside	Find the soil type that underlies each gas station.
	Simple distance	Find the lake that is closest to each campground.
Points to Lines	Simple distance	Find the elementary school that is closest to each residential street.
	Summarized distance	Find the total number of septic systems closer to a particular stream than to any other stream.
Lines to Lines	Summarized inside	Find the number of roads which cross each river.
	Simple inside	Give a section of hiking trail the attributes of the road it follows for a short distance.
Polygons to Lines	Summarized inside	Give a stream the average erosion index of the soil types it crosses.
	Simple distance	Find the lake closest to a hiking trail, or the national park that a road lies within.
Points to Polygons	Summarized inside	Find the total number of schools and students in a county.
	Simple distance	Find the town that is closest to a lake. A point inside a polygon is given distance of zero.
Lines to Polygons	Summarized inside	Find the total number of rivers crossing a state.
	Simple distance	Find the carrying capacity of the closest power lines to an industrial site.
Polygons to Polygons	Summarized inside	Find the total population of all counties that intersect part of a watershed.
	Simple inside	Find the county that a lake falls completely within.

Examples of join types

We have already examined the options available in joining schools and counties using both simple and summarized inside joins. Next we present an example in which the distance join is the most appropriate choice. Most counties in South Dakota do not have a hospital. The state emergency planning office wishes to know the hospital closest to each county, and how far away it is. The hospitals (points) have been joined to the counties (polygons) with the counties as the destination layer. During the join, option 2 for a simple

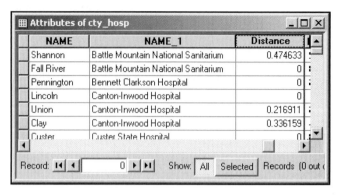

Fig. 7.5. The output table from a distance join of hospitals to counties

distance join is chosen. The output shapefile always has the same feature type as the destination table, polygons in this case. After the join, each county has a field indicating the name of the closest hospital (Fig. 7.5), as well as the distance from the county to the hospital. This example represents a **simple distance join.**

In Figure 7.6, a unique values map was created based on the hospital name. In the northeast part of the state, it is easy to see that the purple counties are closest to the hospital in the purple area, the green counties to the green hospital, and so on. The long, hatchet-shaped county in western South Dakota (Pennington County) has several hospitals. Since all three lie inside Pennington, they technically have a distance of zero, and Pennington's closest hospital was arbitrarily assigned based on which hospital was found in the table first. The lighter pink counties in the central part of the state were assigned to a different hospital in Pennington County and thus appear in a different color.

When a simple distance join is performed, a distance field is automatically appended to the output data set, indicating the distance between the joined features. The distance is reported in the map units of the destination feature class.

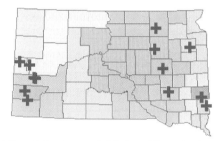

Fig. 7.6. A spatial join of hospitals to counties using the closest option. The colored areas show which hospital the county is closest to.

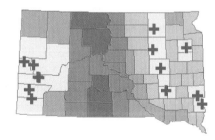

Fig. 7.7. Counties colored according to distance from the closest hospital

The map in Figure 7.7 shows each county displayed according to its distance to the closest hospital—notice that the counties with hospitals inside them have the smallest distance, zero, and the central counties of South Dakota are farthest from any hospital.

Distance joins can also be used between points and lines, as shown in this example of estimating the susceptibility of streams to contamination from septic systems. In Figure 7.8, the point locations represent 1 mile by 1 mile sections, and the size of the symbol indicates the number of septic systems in the section. A join performed between the streams and the septic points can summarize the total number of septic systems closest to each stream—a **summarized distance join.** The thicker the line symbol of the stream, the more septic systems are closest to that stream, and the higher the susceptibility of the stream to contamination.

Coordinate systems and distance joins

Distance joins should always be performed on layers with projected coordinate systems. If a join is performed on a layer with a Geographic Coordinate System (GCS) and units of decimal degrees, two problems arise. First, the distances reported in the table will be in decimal degrees, as shown in the table in Figure 7.5. Degrees cannot be easily converted to miles or kilometers because the conversion factor changes with latitude. Second, and more critical, the result could be invalid.

The distance algorithm relies on the relative *x-y* coordinates of the features to calculate distances and assumes a Cartesian coordinate system to do so. Degrees of latitude and longitude are spherical coordinates, not Cartesian, and the relative distances calculated may be incorrect. Figure 7.9 shows two cases of a summarized distance join of tourist attractions to hotels. In Figure 7.9a, the join was done with both layers in a GCS. Notice that the three attractions in the northeast corner were assigned to Hotel C based on their distance from it

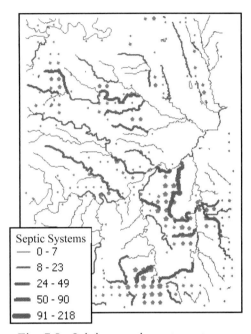

Septic Systems
— 0 - 7
— 8 - 23
— 24 - 49
— 50 - 90
— 91 - 218

Fig. 7.8. Joining septic systems to streams to evaluate stream susceptibility to contamination

in decimal degrees. In the second example, both layers were projected to UTM prior to joining, and the attractions were assigned instead to Hotel B. Since a UTM projection preserves distance within the zone, we know that the second example gives the correct spatial distances between the

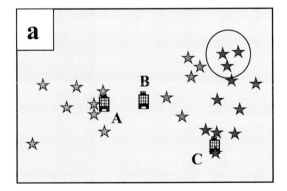
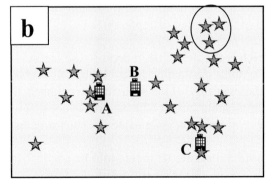

Fig. 7.9. The effect of the coordinate system on a spatial distance join. (a) Three attractions are incorrectly joined to Hotel C (purple stars) when the input layers have a GCS, even though they appear further away. (b) When the layers are projected to UTM before joining, the attractions are correctly assigned to Hotel B (green stars).

attractions and the hotels, and is the valid result. Notice that the spatial distribution of attractions is elongated in Figure 7.9a, as a result of being displayed in a GCS.

It is not sufficient to set the data frame to a projected coordinate system; the layers themselves must be projected. Furthermore, the projected coordinate system should be one that preserves distance over the region of analysis; otherwise the distance analysis will again be incorrect. If you are analyzing the distance of towns to national parks over all conterminous U.S. states, for example, a projection such as USA_Contiguous_Equidistant_Conic should be used, rather than an equal area or a conformal projection.

Setting up a spatial join

Performing the spatial join itself is a simple process; however, determining that a spatial join is called for, and identifying the destination table and the type of join can challenge beginners. This section presents a series of questions to be answered when setting up a spatial join to help produce the correct result. Too many students of GIS simply try random combinations until they hit on the right one—this process is designed to give the right answer the first time.

When faced with a suspected spatial join problem, first try making a simple sketch of the relationships between the layers to be joined. Then answer this series of questions, designed to lead to the correct solution:

> What should the output layer/table look like?
>
> Which is the destination layer?
>
> Should a distance join or an inside join be used?
>
> What is the cardinality of this join?
>
> Should a simple join or a summarized join be used?

Let us demonstrate this process using the problem of estimating stream susceptibility to contamination from septic systems as shown in Figure 7.8. Recall that each point represents a section and is attributed with the total number of septic systems in the section. First sketch the problem, then go on to the questions.

What should the final output layer/table look like? The desired result is a layer of streams. In the attribute table, each stream must be assigned the total number of septic systems that are closer to it than to any other stream. Imagine the fields in the table, with the stream followed by the sum of the number of septic systems.

Which is the destination layer? The features in the output layer are always the same features as those in the destination layer. If streams is chosen as the destination, then the output layer will contain streams. If the septic layer is chosen, then the output layer will contain septic points. Imagining our output table again, we see that what we really want is streams, each of which has septic information assigned to it. Thus streams is the destination layer.

Should a distance or inside join be used? Since we're looking for septic systems closest to each stream, this is clearly a distance join.

What is the cardinality of the join? Since one stream can have multiple septic points closer to it than to any other stream, this is a one-to-many cardinality.

Should a simple or summarized join be used? Since the cardinality is one-to-many, a simple join cannot work. Because the goal is to sum all the septic systems closest to the streams, rather than simply finding the single closest septic to the stream, a summarized join is indicated, with Sum as the statistic.

Now that the questions are answered, the problem setup becomes clear: We need to do a summarized distance join with streams as the destination table and septics as the source table. Even experienced users may find that following this suggested procedure when setting up a join makes it easier to find the right approach. In the next few examples, we will apply this process to set up and solve three different spatial join problems.

Problem 1
Number of earthquake deaths in congressional districts

Imagine that a representative from one of the congressional districts in California is sponsoring legislation to provide earthquake emergency planning funds and is looking for support for the bill. She is planning to throw a big party and invite all of the reps from districts with a significant number of earthquake deaths. She asks one of her staffers, who is a GIS specialist, to draft a list of names for the invitation list. The staffer might approach the problem as follows, using the data in the mgisdata directory.

STOP! Write the answers to the questions, then read on to see if you analyzed the problem correctly.

What should the output layer/table look like? _____

Which is the destination layer? _____

Should a distance join or an inside join be used? _____

What is the cardinality of this join? _____

Should a simple join or a summarized join be used? _____

The earthquake table in quakehis.shp has a state attribute, but none for congressional district. The only way to associate the number of earthquakes with districts is to perform a spatial join.

The goal is a layer of districts with a table containing the number of earthquake deaths for each one. Thus the districts will be the destination layer. The relationship between districts and earthquakes is potentially one-to-many, so a simple join is out of the question. Since the staffer wants to know the total deaths from quakes in each district, the summarize option will be the best, and the proper statistic is Sum. We will therefore do a summarized inside join, with districts as the destination layer.

Figure 7.10 shows the spatial join window filled out for this problem, as well as the resulting table. The table and the map show the selected records from the query—many of the districts are in California, as one might expect.

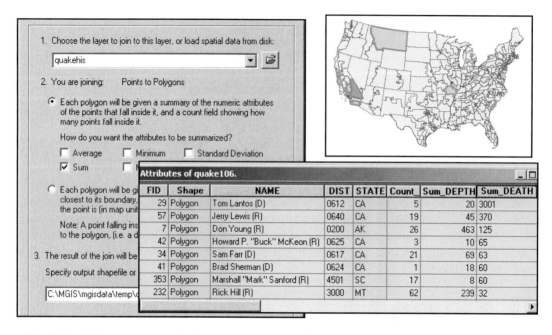

Fig. 7.10. Using a summarized spatial join to find the congressional districts that have had more than 10 earthquake-related deaths

After performing the join, the staffer uses an attribute query to select all of the districts that have had 10 or more earthquake-related deaths. Finding only 12 districts on the list, and knowing that the boss wants a BIG party, the staffer then uses Select By Attributes to find all the districts that have had ANY earthquake deaths. This brings the total up to 29 districts. The guest list should be even bigger, so the staffer decides to use Select By Location to also select any districts that touch one of the districts with earthquake deaths (Fig. 7.11). This brings the guest list to 91, and the staffer can start to make up the invitations.

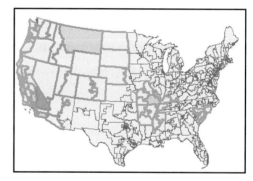

Fig. 7.11. Congressional districts with earthquake deaths or that touch a district with earthquake deaths

Problem 2

Pollution risk of rivers based on county populations

Imagine that Lindsey must do an analysis estimating the risk of pollution to rivers in the United States. She has no detailed information of the sources and types of pollutants, but she can make use of the observation that, generally speaking, large numbers of people are generally correlated with high risks of pollution. Thus she decides to use the population of counties as an estimate of pollution risk. For each river then, she wants to find the total number of people living in counties that contain the river or lie next to it.

STOP! Write the answers to the questions, then read on to see if you analyzed the problem correctly.

What should the output layer/table look like? _____

Which is the destination layer? _____

Should a distance join or an inside join be used? _____

What is the cardinality of this join? _____

Should a simple join or a summarized join be used? _____

The final output layer should contain rivers, with a table listing each river and the total population of counties adjacent to that river. Thus, rivers must be the destination table. Since the counties must actually touch the river, this is an inside join rather than a distance join. The cardinality is one-to-many, since one river can have many counties touching it. Thus an inside summarized join must be used with rivers as the destination table, and the SUM statistic should be requested.

Figure 7.12 shows the spatial join window correctly filled for the join. The table that results shows the river name, the number of counties adjacent to it (COUNT_), and the SUM of the fields, including the POP2000 field which is of primary interest. The figure also shows a graduated symbol map based on the SUM_POP200 field, in which the width of the river represents the total county population living adjacent to it. (The last 0 on the POP2000 field was truncated so that the new field name contained only 13 letters according to field naming conventions.) Notice that ALL of the numeric fields are summed, not just the POP2000 field. One drawback to summarized spatial joins is that the chosen statistic is applied to all the fields, yielding potentially large attribute tables.

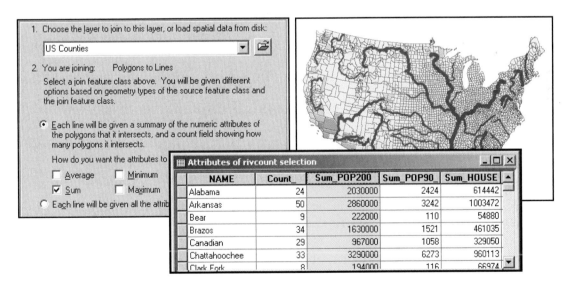

Fig. 7.12. Estimating pollution risk for rivers based on adjacent county populations.

Problem 3

Closest volcano to a city

In this final example, imagine that a professor who specializes in volcanoes has built a website about them. He would like to put a table on his website, so that schoolchildren all over the United States could enter the name of the city they live in and get an information page about the volcano that is closest to them. Since the professor knows nothing about GIS, his graduate student, Cody, gets to make the table.

STOP! Write the answers to the questions, then read on to see if you analyzed the problem correctly.

What should the output layer/table look like? ______________________________
Which is the destination layer? ______________________________
Should a distance join or an inside join be used?______________________________
What is the cardinality of this join? ______________________________
Should a simple join or a summarized join be used? ______________________

The final goal is a cities layer with a table containing cities plus a field indicating the closest volcano and the distance to it. Thus, Cody realizes, Cities is the destination table, and a distance join to the volcanoes field will provide the necessary information. Because the closest volcano is the target, this is a simple distance join.

Figure 7.13 shows the spatial join window settings to produce this analysis, and the result of the spatial join (Cody has made many of the fields invisible to highlight the important ones). The table shows the city name (sorted alphabetically), the state name, the state where the volcano is located, the volcano name and type, and the distance to the volcano.

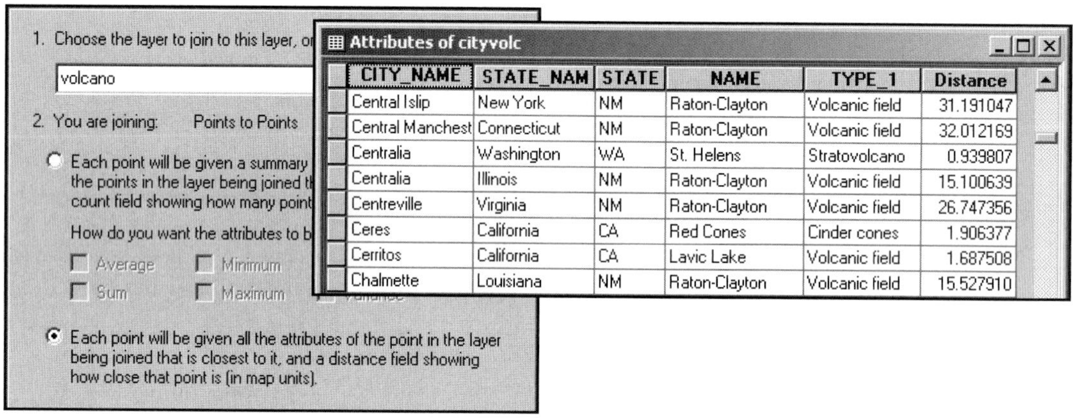

CITY_NAME	STATE_NAM	STATE	NAME	TYPE_1	Distance
Central Islip	New York	NM	Raton-Clayton	Volcanic field	31.191047
Central Manchest	Connecticut	NM	Raton-Clayton	Volcanic field	32.012169
Centralia	Washington	WA	St. Helens	Stratovolcano	0.939807
Centralia	Illinois	NM	Raton-Clayton	Volcanic field	15.100639
Centreville	Virginia	NM	Raton-Clayton	Volcanic field	26.747356
Ceres	California	CA	Red Cones	Cinder cones	1.906377
Cerritos	California	CA	Lavic Lake	Volcanic field	1.687508
Chalmette	Louisiana	NM	Raton-Clayton	Volcanic field	15.527910

Fig. 7.13. A table showing each city in the United States and its closest volcanoes, including the distance to the volcano (in decimal degrees)

Notice the field name TYPE_1 in the output table. It happens that the Cities layer has a TYPE field, and so does the Volcano layer. The output table cannot have two fields with the same name, so during the join the repeated names are automatically converted to unique names. The Cities type field remains TYPE, but the Volcanoes type field becomes TYPE_1. (The city TYPE field was hidden from display to make the figure clearer, but it remains in the table.) Usually the source table field is the one that is renamed.

Also notice that the distance from Champaign, Illinois to the Raton-Clayton volcanic field in New Mexico is approximately 15. These units are clearly not miles or kilometers. In fact, they are degrees, indicating that the coordinate system of the original data is a GCS. Recalling that distance joins should always be done on projected data, Cody realizes that he must repeat this analysis in order to get the correct result. He needs to first project the Cities and Volcanoes layers into a projected coordinate system such as Equidistant Conic *before* performing the spatial join. Then the distance units will be given in meters, and the actual miles can be calculated.

TIP: Always check the distance values after performing a distance join. If the numbers seem very small compared to the number of miles or kilometers you might expect between features, then the units are probably GCS and the layers need to be projected and the join performed again.

Summary

> A spatial join combines the records of two features tables based on the location of the features. A new shapefile or feature class is created by a spatial join.

> An inside join uses the criteria that one feature falls inside another, or in the case of points and lines, on top of each other. A distance join matches the destination layer feature to the record of the closest feature in the source layer. A distance field reporting the distance between the joined features is added to the table.

> A simple join may be used whenever the cardinality of the layers is one-to-one or many-to-one. A summarized join generates summary statistics for all the source features matching the destination features, and is used when the cardinality is one-to-many or many-to-many.

> Four types of spatial joins exist based on the combination of the criterion and the cardinality of the relationship: simple inside joins, simple distance joins, summarized inside joins, and summarized distance joins.

> Distance measurements are reported in map units of the input layers.

> Distance joins should only be performed with layers having projected coordinate systems that do not distort distances. Using layers with a geographic coordinate system may yield incorrect results.

> In the joined table, fields with identical names will be renamed in the output file so that all field names are unique, such as NAME to NAME_1. Usually the source field is renamed.

> Use the following series of questions to help set up a spatial join properly.

What should the output layer/table look like? _____

Which is the destination layer? _____

Should a distance join or an inside join be used? _____

What is the cardinality of this join? _____

Should a simple join or a summarized join be used? _____

Chapter Review Questions

You may need to consult the Skills Reference section to answer some of these questions.

1. What primary characteristic distinguishes a spatial join from an attribute join?

2. Can a spatial join be removed when it is no longer needed? Explain why or why not.

3. What two options may be used to handle one-to-many relationships in a spatial join?

4. If a polygon feature type is joined to a line layer, with the lines as the destination table, what will the feature type of the output layer be?

5. How many output fields will result if a summarized join is specified and a single statistic (e.g., Sum) is selected?

6. Why should distance joins always be performed on layers with a projected coordinate system? What additional consideration should be added?

7. What happens if the two input layers in a join each have a field with the same name?

For the following spatial join problems, answer the series of questions in the text; then state the type of join that should be used: simple inside, simple distance, summarized inside, or summarized distance.

8. Determine the number of parcels within each of the Rapid City watersheds.

9. Find the land use zoning type associated with each septic system in Sioux Falls.

10. Determine the number of counties and the total number of people served by each airport in the United States.

Mastering the Skills

Teaching Tutorial

The following examples provide step-by-step instructions for doing basic tasks and solving basic problems in ArcGIS. The steps you need to do are highlighted with an arrow ➔; follow them carefully. Click on the video number in the VideoIndex to view a demonstration of the steps.

A simple join

➔ Start ArcMap and open the map document ex_7a.mxd.

➔ Use Save As to rename the document, and remember to save frequently as you work.

> **TIP:** In this tutorial when you see the word **STOP**, pause a moment and set up the problem using the questions presented in the Concepts section. Then read on to see if you analyzed the problem correctly. This practice will help you solve the exercises later.

We will begin by practicing and examining the results of some spatial joins. First, we would like to be able to present a list of the attractions sorted according to which county it is in, so that tourists visiting a county know what attractions are available. **STOP** and think it through.

The desired output is a table of attractions with a county field for each one. A simple inside spatial join is indicated, with the tourist attractions as the destination table.

1➔ Right-click the Tourist Spots layer and choose Joins and Relates > Join from the context menu.

1➔ In the top drop-down box, choose to join from another layer based on spatial location (Fig. 7.14).

1➔ Choose Counties as the layer to join.

1➔ Choose to join the point to the polygon it falls inside.

1➔ Click the Browse button, navigate to the Sdakota folder, and enter "tour_county" as the name of the shapefile to be created. Click Save.

1➔ Click OK.

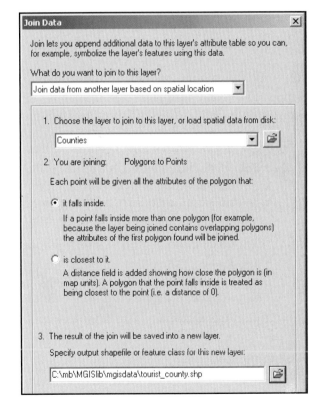

Fig. 7.14. The Join Data window for spatial joins

Notice that the new shapefile appears at the top of the Table of Contents.

      2➔ Turn off the original Tourist Spots layer.

      2➔ Right-click the tour_county layer and choose Open Attribute Table.

      2➔ Scroll to the right to find the fields from the Counties table, including the county name and population information.

The NAME_1 field is so named because both original layers had a NAME field. The field from the Counties layer was given the designation NAME_1 in the output table, so that it would not be confused with the Tourist Spots NAME field. Create an alias for the field to make it easier to remember.

      3➔ Close the attribute table.

      3➔ Double-click the tour_county layer to open its properties.

      3➔ Click the Fields tab, and click in the Alias column of the NAME_1 field. Type in the alias "COUNTY". Click OK.

1. What other two fields received new names in this file? _____

      4➔ Create a unique values map of the tour_county layer based on the COUNTY field, so that the tourist attractions are colored the same when they are in the same county.

      4➔ Click the minus sign next to the tour_county layer to collapse the long list of county names in the Table of Contents.

Suppose we now wish to determine the number of tourist attractions in each county. Now that each attraction has a county, a Summarize will perform this task admirably.

      5➔ Right-click the tour_county layer and choose Open Attribute Table.

      5➔ Right-click the COUNTY field and choose Summarize. Name the output file "tour_sum.dbf" and put it in the Sdakota folder. You do not need to choose any statistics (the Count field will be automatically generated).

      5➔ Choose to add the new table to the map.

      5➔ Close the tour_county table, and open the new tour_sum table.

      5➔ Sort the Count field in descending order.

2. Which three counties have the most tourist attractions? How many does each have?

_____

A summarized join

The previous problem required two steps to determine the number of tourist attractions in each county: first a simple join to assign a county to each attraction, and then a Summarize to sum the attractions for each county. However, we could have found the answer in one step by setting up a slightly different join. **STOP** and think it through.

In this case, the output table should list the counties, each of which has a field indicating the total number of attractions in it. Thus an inside summarized join is indicated, with Counties as the destination table. We will try it this way now.

6➜ Close the tour_sum table.

6➜ Right-click the Counties layer and choose Joins and Relates > Join.

6➜ Set Tourist Spots as the layer to join to.

6➜ Choose the option to summarize the points within the polygons.

6➜ There are no numeric fields in the Tourist Spots table, thus no need to choose a statistic. Leave all the boxes unchecked.

6➜ Name the output layer "tour_county2.shp" and click OK.

7➜ Open the tour_county2 attribute table and scroll to the last field.

The output layer contains the county polygons this time, instead of tourist spots, because Counties was the destination table. One new field was added, Count_, indicating the number of tourist spots found in each county.

7➜ Sort Descending on the Count_ field and verify that the result is the same as before.

7➜ Close the tour_county2 table.

Apparently, spatial joins can be performed more than one way to get the same answer. Usually one way is more efficient than the other, but sometimes only one way suffices.

Distance joins

Point-to-polygon spatial joins are common analysis problems. Let us apply one to the problem of mapping the service areas of hospitals in South Dakota, defined as the counties which are closer to that hospital than to any other hospital. **STOP** and think it through.

To map the service areas, we need a layer of counties with an attribute indicating the closest hospital for each. Then a Unique Values map can be created from the hospital names. Thus Counties is the destination layer, and this problem requires a simple distance join to find the single closest hospital.

8➜ Turn off both tour_county layers and turn on the Hospitals layer.

8➜ Right-click the Counties layer and choose Joins and Relates > Join.

8➜ Set Hospitals as the layer to join to.

8➜ Choose the option to join the closest hospital to the county boundary.

8➜ Name the output file "hospcount.shp" and click OK.

Now we can analyze the results.

9➜ Open the hospcount attribute table and examine all of the fields. Most of these fields came from the Counties layer. Scroll to the end to find the hospital fields added by the spatial join.

9➜ Note that the hospital name field is called Name_1. Close the table.

10➔ Click the Display tab, and move the Hospitals layer over the hospcount layer in the Table of Contents.

10➔ Use the Distance field in the hospcount layer to create a map showing the distance from each county to a hospital, as in Figure 7.7 earlier in the chapter.

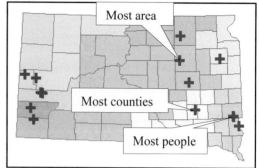

11➔ Use the hospcount layer to create a map based on the hospital name, showing the service area of each hospital. The map should be similar to Figure 7.15.

Fig. 7.15. Hospital service areas

We also wish to know the total number of people in each service area. To find this information, we will summarize the hospcount table using the hospital name field and request the sum of the county population as the statistic. It would also be interesting to know the total area served by each hospital, so we will request the sum of the county areas as well.

12➔ Open the hospcount attribute table.

12➔ Right-click the Name_1 field containing the hospital names and choose Summarize.

12➔ Request the Sum statistic for the POP2000 field and the AREA field.

12➔ Name the output table "hospstat.dbf" and click OK. Choose Yes to add the table to the map. Close the hospcount table.

12➔ Open the hospstat table and examine the fields. Sort the table to answer the following questions.

3. Which hospital serves the most counties? _____ Which hospital serves the most people?_____ Which hospital serves the greatest area? _____

Note that, similar to the tourist attraction problem, the second question regarding the number of people and area could have been solved using a summarized join with a reversed cardinality; that is, joining counties to hospitals with hospitals as the destination layer.

13➔ Before going on, open the hospcount table again and examine the Distance field more carefully. The units are clearly too small to be miles or kilometers.

4. What are the units of this distance field? _____

You may be puzzled, because the map in the data frame certainly looks projected, but the distance units indicate that the layers must be in a GCS.

13➔ Open the Properties of the Hospitals layer and check the coordinate system.

13➔ Check the coordinate system of the data frame.

As we suspected, the layer coordinate system is a GCS. The data frame has been assigned a State Plane coordinate system, but this is not enough to assure a correct distance join. The individual layers should have been projected prior to doing the join. It is very easy to make this mistake without noticing. We will not correct it now; however, we will ensure that we do the next distance join properly.

> 13➔ Close all windows and the table.
>
> 13➔ Remove the hospcount layer, and turn on the Tourist Spots layer once more.

Next, we'd like to analyze the number of visitors at different tourist attractions as a function of the distance from the interstates. As a first step we need to determine the distance of each attraction from the closest interstate. **STOP** and think it through.

We want a table showing each tourist site and its distance from the interstate, so clearly Tourist Spots must be the destination layer. A simple distance join will suffice to solve this problem. First we check the coordinate systems to ensure that we don't make the same mistake as last time. We already determined that the data frame coordinate system is SD State Plane.

> 14➔ Check the coordinate systems of the Tourist Spots and Major Roads layers.

Both layers are in a Geographic Coordinate System with a NAD 1983 datum. We must project them before going on with our join. The Export command is a convenient way to do this when the data frame coordinate system is already projected.

5.   What is the coordinate system of this data frame? _____

> 15➔ Right-click the Tourist Spots layer and choose Data > Export Data.
>
> 15➔ Make sure it is set to export All features.
>
> 15➔ Fill the second button to use the same coordinate system as the data frame. This ensures that the new data file will be projected, since we know the data frame is.
>
> 15➔ Place the output file in the Sdakota folder and name it "sptourist.shp". Click OK.
>
> 15➔ Click Yes to add the new shapefile to the map.
>
> 15➔ Remove the old Tourist Spots layer, and rename the sptourist.shp file Tourist Spots.

Before repeating this process for the Major Roads file, however, notice that it contains both interstates and state highways. Before we can do the join, we need to place the interstates in their own separate layer so that we can join with them alone. Otherwise many of the tourist sites will be assigned attributes of highways rather than interstates. We can do this at the same time that we export the data using the data frame's State Plane projection.

> 16➔ Choose Select By Attributes from the main menu bar.
>
> 16➔ Choose Major Roads as the layer to query, and enter the expression "FUNC_CLASS" LIKE '%Interstate%'. Click Apply. Close the window.
>
> 16➔ Turn on the Major Roads layer to see the selection.
>
> 17➔ Right-click Major Roads and choose Data > Export Data.

17➜ Make sure you are exporting the selected features and using the coordinate system of the data frame.

17➜ Name the output layer "spinterstates.shp", and put it in the Sdakota folder.

17➜ Rename the spinterstates layer Interstates. Give the new interstates a symbol that stands out against the other features in the map.

Now do the spatial join.

18➜ Right-click the destination layer, Tourist Spots, and choose Joins and Relates > Join from the context menu.

18➜ Choose to join by spatial location, and make Interstates the source layer. Choose to join each point to the line that it is closest to.

18➜ Use the Browse button to store the output in the Sdakota folder as tour_dist. Click OK.

19➜ Create a graduated symbol map based on the new Distance field in the new tour_dist layer. Flip the symbols to make the closer tourist spots large and the distant ones smaller (Fig. 7.16).

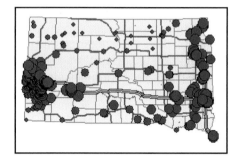

Fig. 7.16. Tourist spots with size indicating closeness to interstate

Examine the distance values in the legend of the map. This time the values are projected units instead of degrees, as we intended. However, we still may not be sure whether the units are meters or feet. We can find out, though.

20➜ Open the tour_dist layer properties and examine the Source tab. Scroll down the coordinate system description to find the linear unit. Close the window when done.

6. What are the map units of this coordinate system? _____

Now that the distance units are known, we could add a new field and calculate the distances in miles, if desired. We'll just go on.

Some polygon joins

Polygon-polygon joins can be done two ways. The first assigns the attributes based on whether the source polygon is closest to the destination polygon, a simple inside join. The second method creates a summary for all source polygons which intersect the destination polygon, a summarized inside join. We will perform examples of both.

Consider the problem of determining the county responsible for the water quality in each South Dakota lake. When the state Fish and Game department receives complaints from citizens about fish kills or water quality in a lake, they can use such a list to quickly identify the county to be contacted. For the most part, each lake resides in one county, constituting a one-to-one join.

Some large lakes may intersect more than one county, so it will be interesting to examine the results in those cases. **STOP** and think it through.

Since we want to have a list of lakes with an attribute indicating the responsible county, in this case Lakes is the destination layer. A simple inside join will solve the problem.

21➔ Turn all layers off except for Counties and Lakes.

21➔ Right-click the Lakes layer and choose to Join.

21➔ Set Counties as the layer to join to.

21➔ Choose the option to join attributes from the polygon that it falls completely inside.

21➔ Name the output layer "lakecount.shp" and click OK.

22➔ Open the lakecount attribute table and examine the results.

7. Which county is responsible for Salt Lake? _____

What happens when a lake does not fall completely within a single county? Lake Oahe, the large dammed reservoir on the Missouri River in the center of the state, provides a good example, because it touches several counties and serves as a boundary between them.

22➔ Find Lake Oahe in the lakecount table and examine its attributes.

8. How was Lake Oahe treated in the spatial join? _____

22➔ Close the lakecount table and clear the selected features.

Another case of polygon-polygon joins involves the possibility of many polygons in the source table falling inside one polygon in the destination table. As an example, consider the number of urban areas falling within each watershed in South Dakota. Urbanized areas often pose special problems for water quality, so knowing the urban population in each watershed provides important clues for designing monitoring programs. We can use a spatial join to determine the total urban population in each watershed. **STOP** and think it through.

The output table will contain a list of watersheds, each with a value for the total urban population. Thus Watersheds is our destination layer. Since more than one urban area can exist in each watershed, the cardinality is one-to-many, and a summarized inside join is required..

23➔ Turn on the Watersheds and Urban Areas layers; turn off all other layers except Counties.

23➔ Right-click the Watersheds layer and choose Joins and Relates > Join.

23➔ Choose to join spatially, set the join layer to Urban Areas, and fill the top button for the summary option. Check the box next to Sum for the statistic.

23➔ Enter the name "shed_urban.shp" for the new layer.

23➔ Click OK.

24➔ Click the Refresh button if needed to display the new layer.

24➔ Open and examine the attribute table of the new shed_urban layer. Close it when finished looking at it.

24➔ Create a graduated color map of the watersheds showing the total urban population.

9. Which watershed UNIT has the most urban areas in it?_____

10. Which watershed UNIT has the most people in urban areas in it?_____

It is even possible to join polygons with very different shapes. Consider trying to find the total population living in each watershed instead of just the urban population. Towns and urban areas fall neatly inside watershed boundaries, but since much of South Dakota is rural, the results would be misleading. The county population data includes all the people, but significant overlaps between counties and watersheds occur. Rarely does a watershed occur completely inside a county, or vice versa.

25➔ Turn off all layers except the original Watersheds and Counties, and examine the relationships between the two.

However, we can still perform a summary spatial join on these two layers. Since we want a total population for each watershed, Watersheds is the destination layer. But which statistic should we request? At first Sum might seem appropriate, because we are trying to find total population. However, consider the watershed labeled in Figure 7.17, containing portions of five different counties. If we sum the populations, we will clearly overestimate the population in the watershed. Instead, an average population for the five counties would provide a better estimate, although not an exact one. An area-weighted average of the county population *densities* would probably give the best estimate, but for now we will use the simpler method of averaging the county populations.

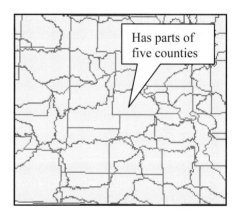

Fig. 7.17. The counties (grey lines) and watersheds (purple lines) have many overlapping areas

26➔ Right-click the Watersheds layer and choose to Join.

26➔ Enter Counties as the layer to join.

26➔ Choose the summarize option for the join method and request the Average statistic.

26➔ Name the output file "watercount.shp" and click OK.

27➔ Open the watercount attribute table and examine the fields. Close the table.

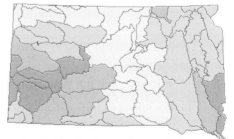

Fig. 7.18. Estimated population of watersheds in South Dakota

28➔ Create a graduated color map from the Avg_POP200 field, as shown in Figure 7.18.

29➔ Create a graduated color map based on the Avg_POP90 (population density) field and compare to Figure 7.18. Is the result similar or substantially different?

Matching issues during joins

Now we will do one more polygon-to-polygon join to demonstrate some issues to be aware of when joining. We would like to have a field in the Urban Areas layer indicating which state each urban area is in. We can do this with a spatial join, but there are a few pitfalls, as we shall see.

30➔ Activate the USA data frame in the same map document by right-clicking the USA frame name and choosing Activate from the context menu.

30➔ Zoom out, if necessary, to see the lower 48 states.

31➔ Join the states to the urban areas, with urban as the destination layer.

31➔ Choose the second option, for polygons completely inside.

31➔ Name the output layer "urban_state" and put it in the Usa folder. Click OK.

Now let's examine an interesting feature of the output layer.

32➔ Use the Find tool to search for Kansas City in the urban_state layer.

32➔ Zoom to the found feature by right-clicking its name in the Find list and choosing Zoom to Feature. Close the Find box.

Kansas City actually occurs within two states, Kansas and Missouri. To which state was it assigned?

33➔ Turn off the original urban layer.

33➔ Use the Identify tool to display the attributes of Kansas City in the urban_state layer.

The state name field is blank, and all the other state fields are either blank or zero. Evidently, NO state was assigned to this urban area because it occupied two states rather than one.

33➔ Close the Identify box.

33➔ Open the urban_state attribute table and sort it in ascending order based on the STATE_NAME field.

Notice that many urban areas failed to join to a state. Some of these are not two-state cities, either. Why didn't these join?

34➔ Close the attribute table.

34➔ Use the Find tool to locate and zoom to Galveston, Texas.

34➔ Close the Find box.

Here we begin to get a clue to the problem. Galveston lies on a barrier island off the coast of mainland Texas. The urban area is slightly more detailed than the state outlines, with the result that the urban area boundaries actually fall outside the state boundaries in some areas (Fig. 7.19). Even this small amount of difference is enough to prevent the join from uniting these two records.

Thus, when joining polygons together, it is important to be aware of possible mismatches that can occur due to the misalignment of polygon boundaries.

This is the end of the tutorial.

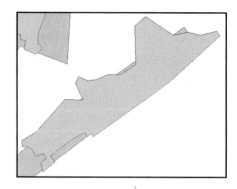

Fig. 7.19. Mismatch of urban areas and state boundaries due to differences in original scale

➔Close ArcMap. You do not need to save your changes.

Exercises

Open document ex_7b in the MapDocuments folder, and use it to answer the follow'

1. Give the name of the watershed in Rapid City which contains the most gas many stations does it have? What is the total number of gas nozzles in the waters...

2. How many gas stations are in a residential land use zone?

3. Which land use category in Rapid City contains the most gas stations? How many gas stations are in the floodway? *summarize w/out choosing anything*

4. Which gas station is farthest from a restaurant? What is the distance (include the distance units in your answer)? What's the name of the closest restaurant?

5. Which restaurant has the most gas stations that are closer to it than to any other restaurant? How many gas stations?

6. Which watershed in Rapid City has the highest number of septic systems, and how many does it have? (**Note:** Each point in the Septic layer represents the center of a 1 mile by 1 mile section, and the DWELLING field contains the number of septic systems in the section. Assume that if the center point is inside the watershed, all the septic systems are inside the watershed.)

7. Using the Streams and West Septic layers, create a map showing the number of septic systems that are closest to each stream. Be sure to use the number of septic *systems,* not the number of septic points, in the map. **Capture** the map.

Use map document ex_7a to answer the remaining questions.

8. Which town in South Dakota is farthest from a hospital? What is the distance, in *kilometers?* *export data first checking -vs a solve as data frame*

9. Assuming that a hospital's service area includes all the towns that are closer to it than any other hospital, determine how many towns each hospital serves. Which hospital serves the most towns? Which serves the most townspeople?

10. Excluding the Missouri River and its dammed lakes, which county in South Dakota has shoreline access to the greatest number of lakes? Which four counties have shoreline access to the greatest total area of lakes? (**Hint:** Find the lake areas first.)

Challenge Problem

Design an analysis to estimate which watershed in South Dakota has the most cattle living in it (in 1992), and do it. The best approach is to determine which counties overlap the watershed and average their cattle populations. Create a map showing your answer, and **capture** it.

Hint: Create a counties shapefile which contains the cattle data as part of it, not simply joined to it, before you begin. You can do this by doing the join and then using Export to create a new shapefile, which will then contain the information from both tables.

Skills Reference

Performing a spatial join

In a spatial join, the attributes of features in the source layer are appended to the features in the destination table. The output layer always contains the same features as the destination table. When choosing the join options in step 4, keep in mind the cardinality of the join to pick the most appropriate one.

1. Right-click the destination layer in the Table of Contents and choose Joins and Relates > Join.

2. Choose to join to another layer based on spatial location (Fig. 7.20).

3. Choose the source layer from the drop-down box, or click the Browse button to choose one on disk.

4. Choose one of the two join options and specify any summary statistics desired, if applicable.

5. Specify the shapefile or feature class file to contain the new output layer.

6. For more information on joins, click the About joining data button.

7. Click OK to execute the join.

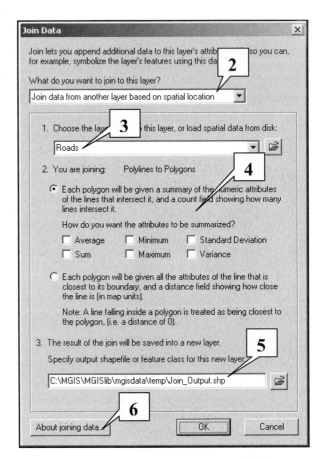

Fig. 7.20. Join Data window for spatial joins

Chapter 8. Map Overlay

Mastering the Concepts

Objectives

➢ Using the Geoprocessing Wizard for advanced spatial analysis including overlay, clipping, and buffering

➢ Using map overlay to locate areas fitting multiple spatial criteria

➢ Understanding differences between spatial joins and overlays

➢ Applying the most efficient strategies when overlaying maps

Concepts

Map overlay

Spatial joins, although powerful, are limited when spatial features do not overlap exactly. Consider the map of roads and land use shown in Figure 8.1. The state government has requested a report of the total miles of road falling into each land use category. At first glance, a spatial join might be considered a quick way to generate this information—by joining the land use polygons to the roads, one might expect to get a field showing the land use type each road crosses.

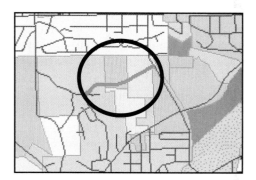

Fig. 8.1. The selected road inside the circle crosses three different zoning types.

However, a problem exists. Note that the single selected road circled in Figure 8.1, shown in blue, crosses three different land use classes. How would a land use type be assigned in this case? If one used the *completely within* join option, no land use class would be assigned for the road at all, because it does not lie entirely inside a polygon. A *summary* would be no good, because a nominal data type such as land use class can't be averaged. The *distance* option would only pick one of the land use classes arbitrarily.

Fig. 8.2. An overlay of zoning and roads splits the roads at the zoning boundaries, so that each road segment lies within only one polygon.

The ideal method would split the road into three sections and assign the land use to each section. Because problems of this sort commonly occur in GIS, analysis routines have been developed to perform this exact function. They are called **map overlay.** The ability to split features that partially overlap is the most important difference between a spatial join and a map overlay. In Figure 8.2, notice how the blue selected roads always end as they cross a land use boundary, and the original single road is now three road sections (two of which are selected). This particular analysis is called an intersection.

Intersection works similarly to spatial join, in that the tables of the two layers are combined into one, based on a common location. Notice that each road segment now has a land use code assigned (Fig. 8.3). However, it differs in that it automatically splits features when they cross each other, thereby enforcing a one-to-one relationship between features, and enabling a perfect correspondence when joining the tables. With the output layer table, it is a simple matter to use the Summarize command to calculate the total length of road in each land use category.

Fig. 8.3. During an overlay, each new road segment retains its original attributes and also receives the land use code and other attributes of the polygon it falls inside.

FID	Shape*	LENGTH	ROADNAME	LANDUSE_ID	LU_CODE
2348	Polyline	0.00964	RANGE RD	72	Office/Commercial
2535	Polyline	0.00964	RANGE RD	70	Medium Density Residential
2564	Polyline	0.00579	CANYON LAKE DR	73	Public
3568	Polyline	0.00574	HILLSVIEW DR	354	Low Density Residential
3569	Polyline	0.00574	HILLSVIEW DR	354	Low Density Residential
3570	Polyline	0.00242		354	Low Density Residential
3653	Polyline	0.00482	CANYON LAKE DR	75	General Commercial
4984	Polyline	0.00733	32 ST	40	Floodway

Selected Attributes of road-lu

Intersection is only one form of map overlay. Other operations are possible, including **union**, **erase**, and **clip**. In an overlay, the first layer is usually called the input layer, and the second is sometimes called the overlay layer. Each operation has certain restrictions as to the geometry type of the input and output layers, as described below.

Figure 8.4 shows different types of overlay performed using two polygon input layers (triangles and circles). A **union** creates all possible polygons from the combination of features in both layers. In a union, both input layers must contain polygons. The **erase** function discards areas of the input layer that fall inside the overlay layer. The input layer may contain points, lines, or polygons, but the overlay layer must contain polygons. A **clip** keeps the areas of the input layer that lie inside the boundary of the overlay layer (rather like a cookie cutter). In a clip, the overlay layer must contain polygons, but the input features may be points, lines, or polygons. The output layer geometry will match the input layer; for example, if the input layer contains points, so must the output layer.

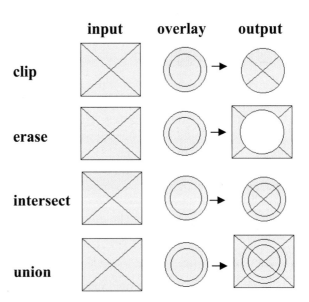

Fig. 8.4. Polygon overlay operations and results. Union and intersect join the attributes of the two input layers. Clip and erase do not.

Unlike intersect and union, which join the attributes of the input and overlay layers together into the output table, erase and clip do not join the tables. The table of the input layer (triangles) emerges unchanged, except that it may have fewer records.

The **intersect** operation is more versatile in the combination of geometries it can handle. Both input and overlay layers may contain points, lines, or polygons (in fact, there is little reason to differentiate between the input and overlay layers). The output geometry may vary, but it cannot exceed the dimensionality of the lowest dimension input. In other words, if the inputs are all polygons, the output may be points, lines, or polygons (Fig. 8.5a). If the inputs are both lines, the output may be lines or points (Fig. 8.5b). If the inputs are polygons and lines, the output may be points or lines (Fig. 8.6). If one input is points, the output must be points.

Overlay functions like intersect and union are distinguished from the others in this chapter because they act like a spatial join; they combine the attributes of the two input layers into a single table. This powerful technique can be used to give the attributes of a polygon layer to lines or points that fall within the polygons, as we saw in the example of land use and roads. In the case of overlaying two polygon layers, we can use it to find areas that fulfill multiple criteria. In the next section we will examine some specific examples of overlaying two or more polygon layers.

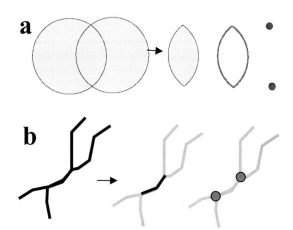

Fig. 8.5. Output geometries of an intersect. (a) Intersecting polygon layers can yield a polygon, lines, or points. (b) Intersecting lines can yield lines or points.

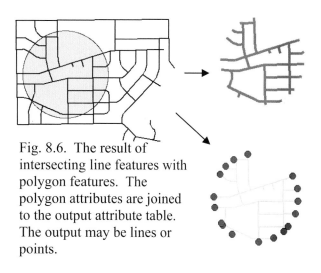

Fig. 8.6. The result of intersecting line features with polygon features. The polygon attributes are joined to the output attribute table. The output may be lines or points.

Examples of using map overlay

The map overlay process is unique to GIS systems, is one of its most powerful tools, and is one of its most commonly used analysis functions. Overlays are often used to apply a set of multiple conditions to an area and determine where they hold, such as finding areas of shale outcrop combined with high slopes in order to create a landslide hazard map. Several additional examples will help demonstrate the process and power of overlay.

Overlay example 1: Hazards testing

Imagine that city engineers have become concerned about possible foundation damage to houses built on certain horizons of the Spearfish Formation which contain large amounts of gypsum, a soft and soluble mineral. They are recommending that homes on this horizon undergo biannual inspections to identify and fix problems early, before the damage becomes severe. They wish to send a mailing to the affected homeowners, explaining the problem.

An overlay analysis can determine which areas of homes need to be notified. Figure 8.7 shows a series of maps in an overlay analysis. Figure 8.7a shows a portion of the residential areas, and Figure 8.7b shows the outcrop area of the gypsum horizon in the Spearfish Formation. These two layers are combined in the overlay, which

Fig. 8.7. Intersecting layers. (a) Residential areas. (b) Areas of the Spearfish Formation. (c) Resulting output layer contains the areas shared by the input layers.

results in a third layer composed of the areas occurring in both input maps (Fig. 8.7c). The output layer contains residential areas that lie on the gypsum horizon.

Let's take a closer look at the intersection process by examining the polygons up close (Fig. 8.8). To perform an overlay, the software takes the polygons from both layers, places them on top of each other, and produces every possible polygon created by combining the lines from both layers. Notice how the orange gypsum polygon with the thick outlines in Figure 8.8 has been truncated at the boundaries of the residential polygon. The areas common to both inputs are preserved in the output file and are assigned the attributes from both the residential and the geology attribute files.

Examine Figure 8.8 again and notice the tiny orange polygon labeled as a sliver. Often when overlaying layers, small errors in digitizing lead to the formation of many tiny sliver polygons. This issue becomes especially troublesome when using layers which share some of their boundaries, such as voting districts and counties. In theory the shared boundaries should match exactly, but in practice few data sets have been corrected to this level of integrity. Depending on the application, users may need to perform additional processing steps to remove these tiny features or merge them with their neighbors. This problem highlights a fundamental rule of GIS—that the output of an analysis can only be as good as the data that are used to generate it.

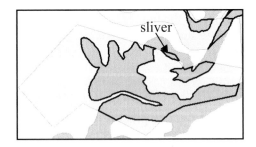

Fig. 8.8. Intersecting makes all possible polygons out of the input layers as they overlay each other.

Overlay example 2: Sensitive habitat analysis

Another analysis might seek to predict sensitive habitat areas for a particular species. Imagine that a rare species of snail inhabits the Black Hills. This snail prefers limey soils in the cool and dense coniferous forest, and is rarely found above an elevation of 1600 meters or below 1200 meters. By combining three maps, a geologic map, a vegetation map, and an elevation map, one can identify the likely areas of snail habitat. Such a map would help biologists to create sampling strategies for counting populations, to analyze whether the habitats are interconnected or widely separated for purposes of genetic mixing, and to help make decisions about forest management to protect the snails.

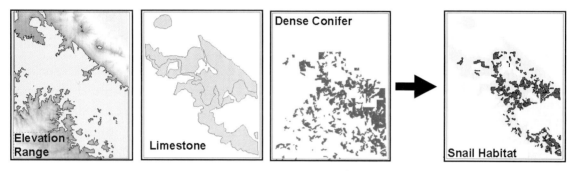

Fig. 8.9. An overlay of elevation, limestone areas, and dense conifer vegetation can help identify areas of potential snail habitat.

The analysis includes a series of steps, using the layers shown in Figure 8.9. First, a digital elevation model (DEM) is queried to create a polygon showing the areas between 1200 and 1600 meters in elevation. Second, a query is applied to a geology map to extract the limestone units as a separate layer. Third, another query separates the dense conifer areas from the other vegetation. Finally, the overlay is performed in two steps, first intersecting the elevation range and limestone layers, and then intersecting the result with the dense conifer map. The final output contains the areas common to all three layers—the snail habitat.

Other spatial analysis functions

Overlay is only one type of spatial analysis available in GIS systems. The remainder of this chapter presents additional commonly used functions. All of these functions are found in ArcToolbox.

Buffering

Buffering is used to identify areas that fall within a certain distance of a set of features. Buffers can be created for points, lines, or polygons (Fig. 8.10). They could be used to find 300-yard drug-free zones around schools, or sensitive protected areas within 100 meters of a stream. Negative buffers can be applied to determine setback limits from the edge of a piece of property. Buffers can be created as simple rings or

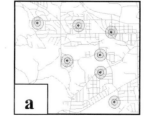
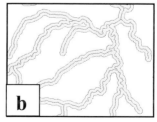

Fig. 8.10. Examples of buffers around (a) wells and (b) roads

as multiple rings. An attribute can even supply different sizes of buffers for different features, for example buffering primary roads by 200 meters and secondary roads by 100 meters. Buffers may be saved as graphics in the map layout or as a shapefile or geodatabase feature class.

Buffering is often combined with overlay to solve a problem. For example, a national forest might be considering adopting a ban on harvesting timber within 200 meters of a road. Creating buffers around the road, and intersecting the buffers with a timber map, could help the managers estimate the total amount of timber rendered not harvestable by such a ban.

Finally, buffering lines and polygons is a time-intensive process, and the time increases quickly as the number of features increases. Any steps to reduce the number of buffers created at one time will facilitate analysis, for example, by performing queries before buffering rather than afterwards.

Clip

A **clip** works like a cookie cutter to truncate the features of one file based upon the outline of another. In Figure 8.11, the county roads file (grey lines) has been clipped by the land use polygon coverage (beige) to produce a shapefile of the roads that fall inside the city planning boundaries (red). The clipping layer must always be a polygon file, but the clipped layers may be points, lines, or polygons. Although a clip appears similar to an intersection, there are two main differences. First, the clip layer's attributes are not appended to the roads file. Second, any polygon edges within the outside boundary of the clip cover are ignored. In other words, only the outside boundary is used for clipping; internal boundaries have no effect on the output layer.

Fig. 8.11. A clip truncates the features of a layer by the boundary of another layer. Here the roads have been clipped to the city boundary.

> **TIP:** To temporarily clip features for display only in ArcMap, use the Data Frame tab of the Data Frame Properties menu.

Append

The **Append** tool is used to combine the features of two or more layers to create a single layer, such as two adjacent quadrangle maps (Fig. 8.12). The appended layers must have the same feature type (i.e., both polygons, both lines, or both points), and the output layer will of course have this same feature type. The two layers must also share the same coordinate system.

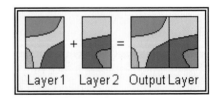

Fig. 8.12. Append combines the features of two adjacent layers.

The treatment of attribute tables during an append requires some consideration. If you wish to combine the attribute information from both layers, the attribute fields must have the same definition and occur in the same order in both tables. If the two tables differ, one can either modify the tables to match (not easy in many cases), or one can combine the features without bringing the attribute information into the output layer.

Although many commonly think of appending as the combination of two adjacent data sets, this stricture is not required. The features may actually occupy the same region, although care should be taken that the features themselves do not actually overlap. For example, one might have six different point shapefiles showing the occurrences of different minerals in a mining area. Appending these files together would yield a single shapefile showing all the mineral occurrences of all types. Of course, care must be taken to ensure that the type of mineral deposit is stored in the attribute table before the merge, so that the type of mineral can still be determined once the points all exist in the same file.

Dissolving

A **dissolve** is used to group features together based on whether they share the same value of an attribute field. For example, the road segments in Figure 8.13 have the same street name (St. Francis), but they are separate road segments. A dissolve based on the street name field would yield a new file in which all streets having that name are one feature. A dissolve can also be used to remove lines between polygon features that share the same value. For example, the map in Figure 8.14 shows ponderosa pine stands of different ages.

Fig. 8.13. Separate road segments may be combined using the dissolve function.

Dissolving on the cov_type field removes the age boundaries and yields polygons based on the cover type (ponderosa) only.

By default, a dissolve will produce multifeatures, in which spatially distinct lines or polygons may constitute a single feature. Each of the orange "polygons" in Figure 8.14 is actual part of a single polygon multifeature.

Before   After

Fig. 8.14. A dissolve removes boundaries between polygons with the same attribute value, in this case, the same tree species.

When features are dissolved, the output layer is an entirely new file. Moreover, this file contains only a single attribute: the field on which the dissolve was based. However, the user can specify additional fields to summarize the information from the original features. For example, in the dissolve shown in Figure 8.14, one might request the average tree age of each output polygon, based on averaging the tree ages of the polygons before the dissolve.

Efficient geoprocessing

Some analysis functions can be time-consuming for the computer to perform. Whenever possible, plan the analysis steps so that the fewest possible number of features are being intersected or buffered. For example, to clip the U.S.A. roads to the boundaries of the state of Texas and buffer them to 50 miles, one should use a Select By Location first to select the roads that intersect Texas, then clip them, and then buffer them. If the steps are performed in this order then the fewest possible features will need to be clipped and buffered. Dissolving extraneous boundaries before overlay can also help streamline analysis. Generally, functions requiring topological associations are more time-consuming than others. Buffer, union, and intersect tend to be the slowest functions.

Coordinate systems and map units

ArcMap can overlay features with different coordinate systems. The output layer will always be in the same coordinate system as the data frame, regardless of the coordinate systems of the two input layers. ArcMap uses its on-the-fly projection capabilities to overlay the data. However, when accuracy is a concern, it is usually better to ensure that all the input layers have the same coordinate system beforehand, and that the data frame has the same coordinate system. This setup will prevent projection errors from creeping into the analysis. In addition, when doing overlays, it is usually advantageous to use a projected coordinate system, rather than a geographic coordinate system, because the shapes and areas of the output are more correctly represented in a projected coordinate system than a GCS.

Another reason for using projected coordinates becomes apparent upon further reflection. Often, one purpose of an overlay analysis includes finding the total area of regions meeting certain criteria, or the total length of lines meeting another. Just as in a spatial join, distances and areas of map features are always reported in the map units of the layer that has been created. Because the distance subtended by one degree depends on the latitude of the feature, degrees are unsatisfactory units for reporting distances and areas. Projected coordinate units such as meters and feet are more readily understood and used. It must be kept in mind, however, that distortions present in the map projection, if significant, will affect distance and area measurements. Thus it remains important to choose a suitable map projection with minimal distortion of area and distance when these measurements are planned.

Geodatabases and coverages automatically create and update fields containing the area and perimeters of polygons, or the lengths of lines. Shapefiles do not maintain this information. However, users may create AREA or LENGTH fields for shapefiles and calculate the values manually. However, employ caution when using information from these fields. If polygons in a shapefile are clipped, dissolved, or intersected, or undergo any other operation that changes their shape, the AREA fields will not automatically be updated. The user must manually update the fields again to ensure they are correct. Never assume that an AREA or PERIMETER or LENGTH field in a shapefile contains the correct values, unless you yourself have created them and kept them up to date. The tutorial demonstrates how to update these fields.

The spatial analysis functions described in this chapter offer powerful tools for doing spatial analysis and solving problems with maps. In the examples section we will experiment with the many ways these functions can be used.

Summary

➢ Spatial joins have one disadvantage, in that features which partly overlap each other cannot be satisfactorily resolved.

➢ Map overlay resembles a spatial join, but it splits features when they partly overlap. This function enforces a one-to-one relationship between features when their attributes are joined in the output table.

➢ Map overlay comes in two basic types. An intersection keeps all the features that are common to both input layers. A union keeps all the features from both layers regardless of whether they appear in both. Attributes from both layers are joined together in the output.

➢ Erase and clip are similar to overlay functions, but they do not join the attributes.

➢ Buffers are polygon constructions that enclose all the area within a certain distance of a set of features. Buffers may be created for points, lines, or polygons, and they may be constructed as single or multiple rings.

➢ Clipping truncates the features of a layer based on the outlines of another layer, such as clipping a road map to the city limits.

➢ Append allows two data sets with the same feature type to be combined as a single data file, such as merging two adjacent quadrangles to make a single data file.

➢ A dissolve combines features within a data layer if they share the same attribute. This function can be used to convert many street segments into a single line feature, or remove boundaries between parcels with the same zoning.

➢ Map overlay and spatial analysis are generally best done using a projected coordinate system, especially if one anticipates determining areas and lengths as part of the analysis.

➢ Shape lengths and areas are stored and updated automatically in coverages or feature classes in a geodatabase. Lengths and areas of features stored in shapefiles must be calculated and updated manually.

TIP: To temporarily clip features for display only in ArcMap, use the Data Frame tab of the Data Frame Properties menu.

Chapter Review Questions

You may need to consult the Skills Reference section to answer some of these questions.

1. What is the most important difference between a spatial join and a map overlay?

2. Why would it be superfluous to use the summarize capability of spatial joins when doing a map overlay?

3. If you have files of rivers and soils and plan to do an overlay, state which layer is the input layer and which is the overlay layer. What feature type will the output layer be?

4. What is a buffer? What two choices are offered for the output format?

5. What function would you use to create a map of a study area, such that all the features in the map stopped at the study area boundary?

6. What attribute fields will be present in a layer resulting from a dissolve?

7. Why is it usually advantageous to use a projected coordinate system when doing map overlay?

8. How can you determine the areas of polygons in a shapefile?

9. What determines the coordinate system of the output when overlay is used?

10. When merging adjacent map layers, such as two geology maps, how does the user ensure that the attributes are correctly transferred to the output layer?

Mastering the Skills

Teaching Tutorial

The following examples provide step-by-step instructions for doing basic tasks and solving basic problems in ArcGIS. The steps you need to do are highlighted with an arrow ➔; follow them carefully. Click on the video number in the VideoIndex to view a demonstration of the steps.

To demonstrate map overlays, we will do the problem described earlier in the chapter concerning the rare Black Hills snail. As described, we will use overlays to define the potential snail habitat. This habitat will be defined by three criteria: on a limestone geology unit, in dense coniferous forest, between the elevations of 1200 and 1600 meters.

➔ Start ArcMap and open the map document ex_8a.mxd.

➔ Use Save As to rename the map document. Remember to save it often as you work.

Preparing to overlay

Our first step is to create layers which contain polygons where each condition holds, and nothing elsewhere. We will begin with the geology layer. The limestone areas include the Madison Formation and the Upper Paleozoic units. We will use an attribute query to place these units in a separate layer.

1➔ Choose Select By Attributes from the main menu bar.

1➔ Set the layer to Geology, and enter the expression "NAME" = 'Upper Paleozoic' OR "NAME" = 'Madison Limestone'. Click Apply and close the query box.

1➔ Right-click the Geology layer and choose Selection > Create Layer from Selected features. Name the layer "Limestone".

Next we need to select the dense conifers and create a layer from them. We will do the select by attributes in two steps so the expression is less complicated. First we will select the ponderosa pine (TPP) and white spruce (TWS). Then we will select the dense areas (class contains C or 5) from the already selected set.

2➔ Turn on the Vegetation layer, and turn off the Limestone layer.

2➔ Open the Select By Attributes box again and set the layer to Vegetation.

2➔ To select just the conifers, enter the expression "COV_TYPE" = 'TPP' OR "COV_TYPE" = 'TWS'. Click Apply.

3➔ Change the selection method to *select from current selection*.

3➔ Clear the expression, and enter another one that says "SSTAGE96" LIKE '%C%' OR "SSTAGE96" = '5'. Click Apply.

3➔ Close the Select By Attributes window.

3➔ Convert the selected features to a new layer and name it "Dense Conifer".

3➔ Turn off the Vegetation layer.

The elevation range has already been prepared. You are almost ready to begin.

4➡ Turn off all layers except the Elevation Range layer, the Limestone layer, and the Dense Conifer layer.

4➡ Zoom into the middle right part of the map and examine the Dense Conifer polygons.

Notice that these polygons have many boundaries inside them because they are divided based on age and density as well as species. The amount of time needed to intersect a layer is proportional to the number of features in the layer, and the process can be very slow when large numbers of features are involved. To streamline the process, we are going to remove the unnecessary boundaries between these polygons, using the Dissolve function in ArcToolbox. (You may wish to reread the section on using ArcToolbox in Chapter 3 before going on.)

5➡Click the ArcToolbox icon on the main menu bar.

5➡ Navigate the ArcToolbox tree through Data Management Tools > Generalization > Dissolve and double-click the Dissolve tool to start it (Fig. 8.15).

When using a tool for the first time, it is wise to Show Help for the tool to find what it does and what the parameters mean.

6➡Click Show Help on the Dissolve tool and read the description.

6➡Click on the Input Features box. The Help message changes to describe the Input Features parameter.

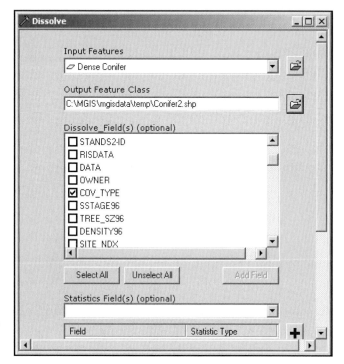

Fig. 8.15 The Dissolve tool

6➡ Click on and read the descriptions for the other input parameters in the window.

6➡ To bring the tool description back, click on the gray menu area.

More detailed information can be found by clicking the Help icon at the top of the tool. For now, proceed with using the tool.

7➡ Click the drop-down arrow in the Input Features box and set it to Dense Conifer.

7➡ Click the Browse button to set the Output Feature class in the BlackHills/Sturgis folder and name it "Conifer2".

7➡ Check the box to dissolve on the COV_TYPE field.

7➡ Notice that other options lie further down in the tool, but we don't need to change any of them.

7➔ Click OK. A processing window will appear to inform you of the status of the analysis. If any errors or warnings are encountered, messages will appear in the window.

7➔ When the tool finishes, click the Close button to close the message window. Check the box if you want it to close automatically every time it finishes.

TIP: When using a tool, be aware that some options may be out of sight, and you must scroll down to see them. Often the defaults are fine, but get in the habit of checking before you finish.

8➔ Examine the output file and note that most of the intervening boundaries have disappeared.

8➔ Right-click the Dense Conifer layer and choose Remove.

8➔ Rename the Conifer2 layer "Dense Conifer".

Intersecting polygons

Now we are ready to do the overlay. We will use Intersect because we want to find the areas common to all three criterion layers. Because the tool can only intersect two layers at a time, we must use it twice.

9➔ Open the ArcToolbox > Analysis Tools > Overlay > Intersect tool (Fig. 8.16).

9➔ Click Show Help and examine the general tool description and the input parameters descriptions.

1. What options are there for the Join Attributes parameter? Which one do you think is best to use?

    _____

    _____

    _____

    _____

    _____

10➔ Click on the drop-down button under Input Features and choose the Dense Conifer layer. It will be added to the list of Features.

10➔ Click the drop-down button again to add the Limestone layer to the list.

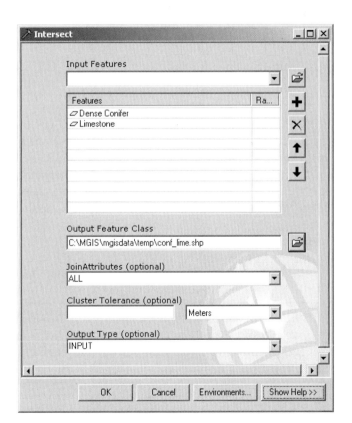

Fig. 8.16. The Intersect tool

If you have an ArcInfo license, you can intersect multiple layers in one step. Add the Elevation Range layer now and name the output "conf_lime_elev.shp". Then skip Step 11. If you have an ArcView or ArcEditor license, continue on to do it in two steps.

10➔ Enter the Output Feature Class, placing it in the Sturgis folder and naming it conf_lime.shp.

10➔ Set the Join Attributes option to NO-FID. This is the best option for most cases, for it combines the attributes from both tables, but leaves out the FIDs which are generally not that useful in the output layer. However, leaving the default set to ALL is also acceptable.

10➔ The remaining options can be left on their defaults. Click OK, and close the message window after it finishes, if necessary.

10➔ Examine the output, noticing how polygons outside the limestone area have disappeared.

11➔ Use the Intersect tool to intersect conf_lime and the Elevation Range layers, naming the output "conf_lime_elev.shp", and putting it in the Sturgis folder.

12➔ Zoom back out to the full extent of the data.

12➔ Rename the conf_lime_elev layer "Snail Habitat".

12➔ Remove the Limestone, Dense Conifer, and conf_lime layers, as they are no longer needed.

12➔ Zoom into the extent of the Snail Habitat, and turn on the Roads layer.

Overlay of lines in polygons

The wildlife biologists have discovered that the snails have a three-week breeding season in early June. During this period they are very active and seek the open, drier areas offered by roads. As a result, many are crushed. The Forest Service wants to consider closing primitive roads that traverse through snail habitat during the breeding season, to lessen the number of snails crushed. Before they can do this, however, they need to assess which roads must be closed. You will make a map highlighting the proposed road closures. This process involves a line-in-polygon intersection.

13➔ Use Select By Attributes to select the Primitive roads "TYPE" = 'PR'.

> **TIP:** The ArcToolbox tools automatically use the selected features from input layers in ArcMap. Thus we can skip the step of creating a separate layer, as we did for geology and vegetation. However, creating the intervening selection layers can be helpful in the long run if for some reason you make a mistake and must repeat portions of the analysis.

14➔ Use the Intersect tool to intersect the Roads layer (with the primitive road selection applied) and the Snail Habitat layer.

14➔ Name the road intersection layer "Proposed Closures".

15➔ Create a map similar to Figure 8.17, with the closure roads highlighted in red and the other roads in black. Turn off the Elevation Range layer and clear the road selection. Set the transparency of the Snail Habitat layer to 50% to help bring out the road patterns.

Clipping layers

Now for the final step. The vegetation layer does not extend as far to the north as the rest of the data, and it may give readers the false impression that the analysis is valid there.

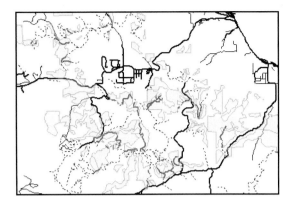

Fig. 8.17. Map showing proposed road closures in snail habitat areas

Clipping the Roads layer at the edge of the vegetation layer will prevent any misunderstanding. In other words, the extent of the vegetation defines the study area.

Now you know that you need the Clip tool. Let's practice finding tools when you know the name but not where to look for it in the Toolbox.

16➔ In ArcToolbox, click the Index tab. Type "clip" in the box.

16➔ Two functions appear, clip (analysis) and clip (management). The words in parentheses refer to the area of the Toolbox.

16➔ Open each tool and click the button to Show Help.

16➔ It becomes clear that the clip (management) tool is for rasters, so clip (analysis) is the one to use. Close both Clip windows for the moment.

16➔ To learn where the Clip tool resides in the Toolbox, click clip (analysis) to select it and then click the Locate button.

17➔ Double-click the clip (analysis) tool to open it.

17➔ Set the input features to Roads. Set Vegetation as the clip features.

17➔ Store the output in the Sturgis folder as "road_clip.shp". Click OK.

You can quickly transfer the symbology of the old Roads layer to the clipped roads.

18➔ Open the properties of the road_clip shapefile and click the Symbology tab.

18➔ Click the Import button, and choose to import symbology from a layer or layer file.

18➔ Click the drop-down box to set the layer to Roads.

18➔ The road classification is based on the TYPE field. The clipped roads use the same field, so leave the TYPE name in the box. Click OK and OK.

19➔ Remove the old Roads layer.

19➔ Rename the road_clip layer "Roads".

19➔ Click and drag the Roads layer below the Snail Habitat layer to more easily see the proposed road closures.

Working with buffers

The Forest Service biologist wants to do one more analysis. To prevent snail death on the major roads, which cannot be closed, they are considering thinning the tree stands within 200 meters of the roads. This thinning would make the stands less ideal snail habitat, which might keep the snails away from the roads. The biologist wants to prepare a map showing the stands to be cleared and the total percentage of snail habitat that the clearing would eliminate.

20➜ Use Select By Attributes to select the primary and secondary roads from the Roads layer, using the expression "TYPE" = 'P' OR "TYPE" = 'S'.

20➜ Create a layer from the selected roads, and name it "Major Roads".

TIP: Buffers are time-consuming to create. When planning to buffer, it is helpful to reduce the number of features as much as possible before buffering. For example, in this case we will use the Elevation Range layer to clip the roads before buffering, since all of the habitat areas are inside this region, anyway.

21➜ Open the Clip tool. Set Major Roads as the Input Features, and Elevation Range as the Clip Features.

21➜ Save the result in the Sturgis folder as "Major_Road_Clip.shp".

21➜ Turn off the other road layers to examine the new file.

Now buffer the clipped roads.

22➜ Locate the Buffer tool in ArcToolbox and open it.

22➜ Set the Input Features to Major_Road_Clip, and the Output Feature Class to "roadbuf.shp" in the Sturgis folder.

22➜ Set the Linear unit to meters, and type "200" in the box.

22➜ Scroll down if necessary and set the Dissolve Type to ALL. Click OK.

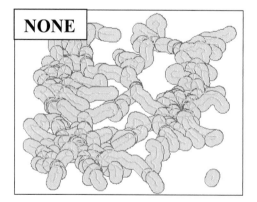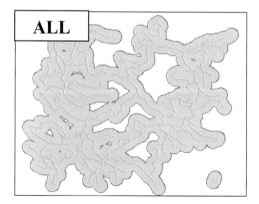

Fig. 8.18. By default the Buffer tool creates undissolved buffers using the NONE option as shown on the left. In this case the ALL option produces the desired result.

By default the Buffer tool does not dissolve boundaries between buffers. In many cases, however, especially when buffering lines, the user would want to dissolve the individual buffers

290

to create merged polygon buffers. To do so, the Dissolve Type in the tool must be set to ALL. Figure 8.18 shows the difference in results obtained using the default NONE option and the ALL option.

Now intersect the road buffers with the snail habitat to determine the potential areas to be cleared.

23➜ Open the Intersect tool.

23➜ Choose Snail Habitat and Roadbuf as the Input features.

23➜ Specify the Output file as "Proposed_Thin" in the Sturgis folder. Click OK.

Determining lengths and areas of features

Now the final step is to determine the area of snail habitat that will be eliminated by the thinning. Geodatabases automatically keep track of the areas, lengths, and perimeters of features, but shapefiles do not.

24➜ Open the Proposed_Thin attribute table and examine the AREA and PERIMETER fields.

Although these fields exist, they originally came from one of the original files intersected to create this final layer: vegetation, geology, etc. Note all the different FID and other fields present in this file, as a result of all the intersections, buffers, and more. Since shapefiles do not automatically update areas and perimeters of polygons, the values in these fields DO NOT record the true area of the new polygons. In a shapefile these fields must be updated manually.

Updating the area, perimeter, and length fields of shapefiles is done in the layer's attribute table and uses the Field Calculator and a small scrap of Visual Basic for Applications (VBA) code. Rather than learning VBA or memorizing the scrap of code, we'll just learn how to access the Help explaining this process. In the future, you need only remember how to access the Help to perform the function again.

25➜ Choose Help > ArcGIS Desktop Help from the main menu bar.

25➜ Click the Index tab, and type in "area".

25➜ Locate the entry "area, calculating for polygons" and double-click it.

25➜ Click on the entry "How to make field calculations" to expand it.

25➜ Click on the entry "Updating area for a shapefile" to expand it, and read it.

> **TIP:** You can also find this entry using the keyword "length" in the index. Notice that this Help also contains similar instructions for updating perimeters and lengths in shapefiles.

26➜ Resize and position the windows so that you can see both the Help window and the attribute table for the Proposed_Thin layer.

26➜ Highlight the scrap of VBA code in the Help window and use Ctrl-C to copy it.

26➜ Right-click the AREA field and choose Calculate.

26➜ Check the Advanced box in the calculator window.

26➜ Paste (Ctrl-V) the code into the large box.

26➜ Type or copy the final variable name, dblArea, into the small box at the bottom.

26➔ As you click the final OK, watch the areas update in the attribute table.

1. What is the total area of the proposed thinning? _____

The units of the AREA field will be in the same coordinates as the map units of the shapefile's coordinate system, in this case, square meters.

27➔ Repeat the procedure to update the polygon areas for the Snail Habitat layer.

2. What is the total area of snail habitat?_____

3. What is the percent reduction in habitat if the proposed thinning is done? _____

This is the end of the tutorial.

➔ Close ArcMap and save your changes.

Exercises

Open the map document ex_8b in the MapDocuments folder, and use it to answer the following questions.

1. Create a map highlighting the residential streets of Rapid City, that is, the streets inside a residential zoning category. Zoom into the main part of town for better legibility. **Capture** your map.

2. The water resources office is concerned about fertilizer runoff contaminating the Minnekata aquifer. Create a map showing agricultural areas in the Minnekata Limestone. Remember to minimize the number of polygons you are intersecting! **Capture** your map.

3. What is the total length of streets, in km, that fall inside the Rapid Creek floodplain (LU-CODE = 'Floodway')?

4. What is the total area of agriculture inside the Minnekata formation, in sq km?

5. What is the total length of streams (include both perennial and intermittent) crossing the Minnelusa Formation?

6. Clip the septic systems and the land use layers with the watersheds, and create a map of Rapid City showing those features. **Capture** your map.

7. Create a map showing the drug-free zones around schools (areas within 300 meters of a school). Show the roads and land use in the map also. **Capture** your map.

Use map document ex_8a to answer the following questions.

8. The Forest Service is concerned about septic systems contaminating the Madison Limestone. Create a map showing private land (OWNER = PVT in the Vegetation layer) on the Madison. **Capture** the map.

9. Create a cover type map which shows only the dominant vegetation species (COV_TYPE) without any subdividing lines. What is the total area (GISACRES) of aspen (TAA)? Use a white symbol for the unknown areas. **Capture** the map.

10. The Forest Service is considering adopting a policy to not cut timber within 200 meters of a primary road (TYPE = P). Assuming that cutting timber includes all "SSTAGE96" LIKE '4%', determine the total area lost to harvesting if this policy were adopted, in sq km. **Capture** a map showing the off-limits areas.

Challenge Problem

Imagine that you have always wanted a vacation home in an aspen stand with a little pond. Create a map showing all aspen stands (COV_TYPE = TAA) that lie on low infiltration geology units (INFIL = LOW) and are within 500 meters of a primary or secondary road (TYPE = P or S). **Capture** your map.

Skills Reference

Using tools in ArcToolbox

1. Click the ArcToolbox icon to open it (Fig. 8.19).

2. If using a tool for the first time, click the Show Help button and read about the tool.

3. With the Help showing, click on a parameter box to get more information about it.

4. Click the Help icon to get the full help information about the tool.

5. Green dots indicate a required parameter.

6. Input/output features may be set by clicking the Browse button and locating the spatial data set, or by making a choice from the drop-down list (ArcMap only). If the drop-down list is used, only the selected features will be used in the tool.

7. Click the Environments button to change the Environment settings (only recommended for advanced users knowledgeable about the geoprocessing environment).

8. If a yellow warning or red error icon appears after entering a parameter, make the cursor hover over it to see the message about the error or potential error.

9. Click OK to start the tool. Keep track of the messages in the progress box. Close the box when finished.

10. To close the dialog box automatically upon completion of the task, check the box.

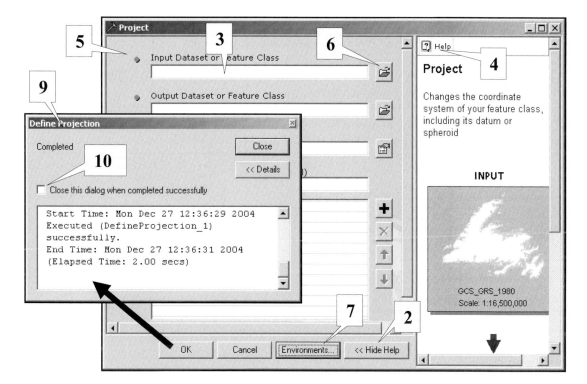

Fig. 8.19. Using tools in ArcToolbox

Performing an intersection

1. Choose the Analysis Tools > Overlay > Intersect tool from ArcToolbox (Fig. 8.20).

2. Enter the Input Features by clicking a choice from the drop-down list or by clicking the Browse button to select a fiie from the disk.

3. Enter a second Input Features layer.

4. Enter the name and location of the Output Feature Class.

5. (optional) Set the JoinAttributes option to NO-FID, the best choice in most cases. ALL joins all attributes from both tables, NO-FID joins all attributes except the FIDs, and FID-ONLY joins only the FIDs.

6. Leave the other optional parameters alone unless you know what you're doing.

7. Click OK.

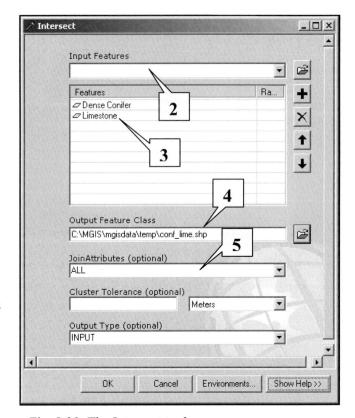

Fig. 8.20. The Intersect tool

TIP: Two input layers must be entered if the user has an ArcView license. With an ArcInfo license, three or more layers may be entered and the intersections will be performed sequentially.

Creating buffers

Note: Buffering is a time-intensive process, especially for lines and polygons.

1. Choose Analysis Tools > Proximity >Buffer from ArcToolbox.

2. Choose the Input Features to buffer (Fig. 8.21).

3. Specify the name and location for the Output Feature Class.

4. Specify the units and value for a fixed buffer distance, OR

5. Choose a field from the input layer attribute table containing the buffer distances.

6. (optional) Change the Side and End type settings.

7. Set the Dissolve Type—use ALL to dissolve boundaries between overlapping buffers, NONE to leave them as is.

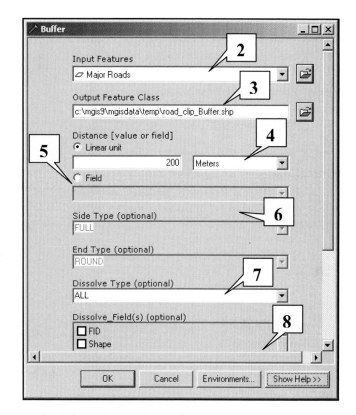

Fig. 8.21. The Buffer tool

8. (optional) Choose one or more Dissolve Field(s). Buffers sharing the same value in the field will be dissolved.

9. Click OK.

Clipping a layer

1. Choose Analysis Tools > Extract > Clip from ArcToolbox (Fig. 8.22).

2. Specify the Input Features layer to be clipped.

3. Specify the Clip Features layer representing the clip boundary.

4. Specify the name and location of the Output Features.

5. Leave the other options at their defaults.

6. Click OK.

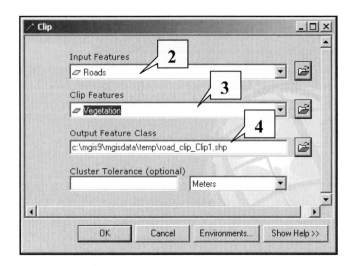

Fig. 8.22. Clipping a layer

Appending layers

For best results, the layers to be appended should have attribute tables with the same fields in the same order.

1. Choose Data Management > General > Append from the ArcToolbox (Fig. 8.23).

2. Specify the layers to be appended together.

3. Specify the *existing* output feature layer to contain the appended features.

4. Choose the Schema type to specify the handling of attributes. Use TEST if you know that the attribute tables match (i.e., they have the same fields in the same order). Use NOTEST if the attribute tables don't match—the features will be appended without the attributes.

5. Click OK.

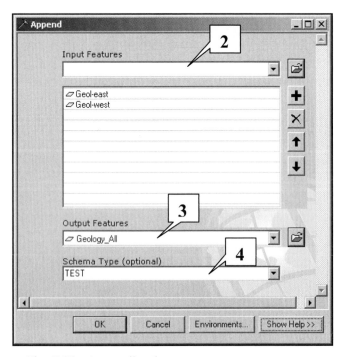

Fig. 8.23. Appending layers

Dissolving

1. Choose Data Management Tools > Generalization > Dissolve from ArcToolbox.

2. Specify the Input Features layer to be dissolved (Fig. 8.24).

3. Specify the name and location of the Output Features to be created.

4. Choose one or more attribute field(s) on which to base the dissolve. This field will appear in the output layer; all other fields will be dropped.

5. To get summary statistics, check the boxes to choose the fields and statistics to place in the output layer. Specify as many fields and statistics as desired, or none at all.

6. (optional) Scroll down and uncheck the Multipart box to not allow multipart polygons in the output, that is, polygons that are physically separate remain separate in the output even if they have the same dissolve attribute value.

7. Click OK to begin dissolving the layer.

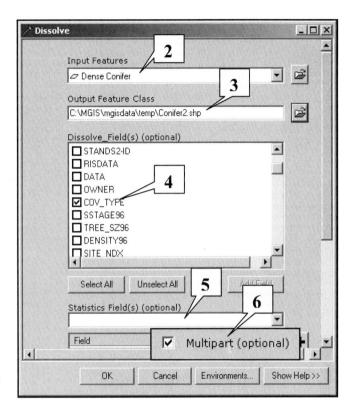

Fig. 8.24. Dissolving a layer

Creating/updating areas or lengths of shapefile features

Shapefiles do not maintain area and perimeter fields for polygons, or length fields for lines. Users must create and calculate these fields manually. Many analysis functions such as Intersect or Dissolve will modify the shape of the features and render the values in the fields incorrect unless they are updated by the user.

1. In ArcMap, open the attribute table of the shapefile.

2. If the area, perimeter, or length field does not already exist, use Options > Add Field to create one. For best results use the Double data type for the field.

3. Open Help > ArcGIS Desktop Help from the main menu bar, and click the Index tab (Fig. 8.25).

4. Search for the text "area" and locate the entry "area, calculating for polygons." Double-click it. (You can also search for length or perimeter to find the Help.)

5. Click on the entry "How to make field calculations" to expand it.

6. Click on the entry "Updating area for a shapefile" to expand it.

7. Resize and position the windows so that you can see both the Help window and the attribute table.

8. Follow the steps in the Help window. (Rather than typing the VBA code, simply select it in the Help window and copy/paste it into the Calculate box.)

9. As you click the final OK, watch the areas update in the attribute table.

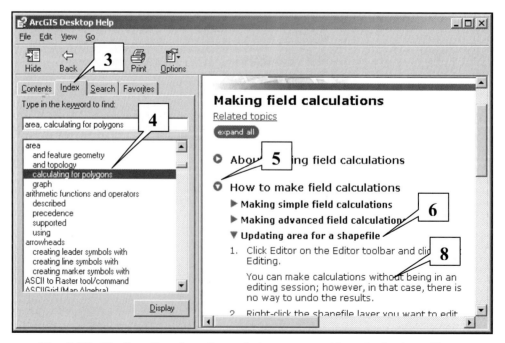

Fig. 8.25. Finding directions for updating areas and lengths in shapefiles

Chapter 9. Presenting Data

Mastering the Concepts

Objectives

> ➤ Learning how to make a good map

> ➤ Reporting information from tables

> ➤ Creating map layouts and printed maps

Concepts

GIS analysis often results in information to be shared with others in the form of maps or reports. Whether you are creating a large poster-style map, a page-sized map, or a report, a few guidelines help in devising a map design which expresses the essence of the data and gets the message across. This chapter introduces some basic ways to communicate ideas to others.

Tips for making maps

A map can be a work of art as well as a vehicle of information, and these two aspects support each other. By designing the map carefully and using some basic guides to style, you can maximize the transfer of information and also present yourself as a competent professional. A poorly designed map gives the opposite impression.

When pen and ink were involved, people had to spend time designing the maps in order to make a good map with the minimum of effort. Now that computers can quickly create and modify maps, the design step is often ignored. People may not even take advantage of the computer's wonderful ability to try out different designs, and many bad maps result. High-quality maps require planning, attention to detail, and aesthetic sense. The basic process follows these steps:

> ➤ **Determine the objectives of the map.** What is being communicated? Who is likely to use the map and under what circumstances? The use of the map will guide some important design decisions, such as how big the text must be. A map being projected on a screen to a large audience needs simplicity, large text, and bold colors. A highway travel map needs more details.

> ➤ **Decide on the data layers to be included.** Computers provide a big advantage here because one can easily add or remove layers as the map takes shape and change a design decision if it does not appear to work in practice.

> ➤ **Plan a layout** that includes all the frames and other map elements (legends, etc.) needed. Following some simple guidelines about placing the elements yields more elegant, pleasing maps.

> ➤ **Choose colors and symbols** that create the right effect and maximize readability. Symbols with psychological relevance can be used to make maps easier to interpret, such as making water features blue. The choice of symbols can enhance a message or hide it.

> **Create the map,** paying attention to the details, colors, layout, and aesthetic qualities.

Determine the objectives of the map

Maps serve a variety of purposes. A one-page map can present information in a report or a paper. A poster-sized map can display important information at a public meeting or an information booth. Some maps are presented to the public and must be meticulously planned and carefully executed. Others are rough copies to take into the field for collecting more data. Some questions to help guide the map plan are suggested in the following paragraphs.

Who will be using the map? Maps for the general public should be simple, aesthetic, and carefully explained using legends and supporting text. These maps usually benefit from orthophotos or shaded relief backgrounds that show land surfaces and help the reader easily identify landmarks and other features of interest. Consideration should be given to possible human disabilities in mapmaking, such as poor eyesight or color blindness. Maps for professional interest may be more complex and make more assumptions about the readers' level of knowledge.

Under what circumstances will the map be used? A page-sized map in a report should be legible to most readers with normal eyesight in good light. A travel map might be used at night or while driving a car, so enhanced readability helps its purpose. A poster map should be tested to ensure that all its text is legible to a person standing several feet away—an 18-point font is considered a minimum size for this type of map. A map used for field work might have many details to help the user find the right locations easily. One issue that is seldom considered is the distribution of the map. Maps in reports or articles are often read as a facsimile—consider designing the map so that its main features are apparent even as a black-and-white copy. Designing a good black-and-white map is even more challenging than using all of those beautiful colors.

What objectives should the map achieve? Maps do not always simply convey information. The design can contribute to the messages intended by the map creator or detract from it. For example, a city map showing point-source pollution sites in big red symbols suggests danger and may unnecessarily alarm nearby residents. Such a map might cause unwanted publicity and action by interest groups out of proportion to the actual risk. On the other hand, the map could alert the city council to a real danger in order to encourage appropriate action.

How sensitive is the map information? Are privacy issues involved? Many GIS mapmakers, carried away with the ability to create dozens of maps, do not stop to consider the possible effect of certain types of information, especially when layers are put together. For example, laying a fault map on top of the city parcels layer might have serious unintended consequences for the property values of parcels near faults. The public may not be aware that most faults are inactive and do not pose a serious threat. Recently the author was advising a student who was creating maps for Jewel Cave National Monument, and they made a decision to omit the usual latitude-longitude markers on maps of the cave passages because it might be deleterious to the cave system if the general public could determine the location of the caves with high precision and seek unauthorized entrance. These issues also arise with maps of fossil sites or archeological digs when it is important that the exact locations of sensitive material remain out of the public domain. One must also be cautious in putting private information on public maps, such as the names of parcel owners, or property values. Although some of this information is public, the ease with which a GIS can be searched, or can print out large volumes of information, justifies a more cautious approach.

Choosing the map layers

The objectives of the map usually dictate the choice of the main layers, but the addition of other layers can often improve a map (or diminish it!). Generally, the more ancillary information on a map, such as roads, towns, landmarks, topography, and vegetation features, the easier it is to find locations on the map and relate it to what we know. A geologic map without topography or roads does little to help a geologist evaluate hazards in a development area. However, too much information can drown the message or make the map difficult to interpret.

A good design rule of thumb is to present the most important layers with the clearest and largest symbols, so that the critical aspects of the layer are emphasized. Minor layers can be presented with less obtrusive symbols and smaller text. One should not have to hunt to be able to see the major patterns of import in the map. Figure 9.1a attempts to show the distribution of septic systems and gas stations with respect to watersheds in Rapid City. The satellite image in the background helps people relate the watersheds to known landmarks and neighborhoods, but it interferes with the interpretation of the map. A redesigned map in Figure 9.1b uses the transparency function to tone down the satellite image and changes the size and color of the marker symbols for enhanced visibility. Now one can easily identify which watersheds are most at risk of contamination, which is the purpose of the map.

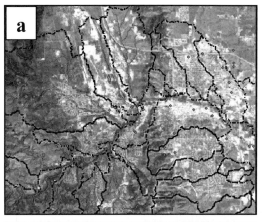

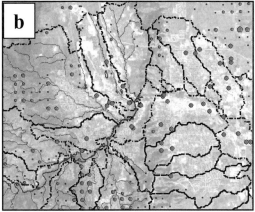

Fig. 9.1. Watersheds, septic systems (orange), and gas stations (green) in Rapid City. (a) The satellite image drowns out the important information. (b) Making the image transparent and changing the symbols makes interpretation much easier.

Planning the layout

The basic elements in EVERY map should include the map itself, a legend, a north arrow, and a scale bar. Optional elements include titles, graphics, pictures, neatlines, and graticules (latitude-longitude markers).

Balance and alignment are very important in designing the map. Balance means that the elements are evenly arranged on the page and are a good size relative to each other. Filled space should be fairly evenly distributed with empty space. Take a look at the map shown in Figure 9.2. All of the basic elements are there, but notice the poor layout. The elements are crowded into the lower left corner. Their edges are uneven. Alaska is cut off in

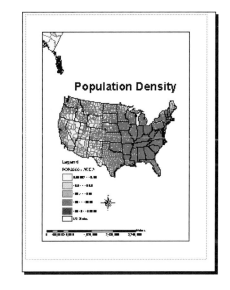

Fig. 9.2. A poorly designed layout

the corner. The legend is very large compared to the size of the map, and the legend boxes are large compared to the legend text. The scale is almost as long as the map. The elements are not aligned, but form arbitrary stair-step lines along the map edges. This is a poorly designed map. It looks like someone made it at 2:00 a.m. the morning before the report was due! However, a few simple changes can dramatically improve its appearance (Fig. 9.3).

> Choose a landscape layout to complement the east-west length of the United States and reduce blank space on the page. This choice also increases the size of the map for better visibility.

> Distribute the elements evenly on the page.

> Reduce the size of the legend and scale bar to emphasize the greater importance of the map.

> Use **neatlines** (boxes around map elements) to enclose and balance the space.

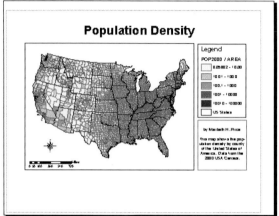

Fig. 9.3. A revision of the map in Figure 9.2 enhances its quality and professionalism.

Choosing symbols

Symbols can highlight or obscure the meaning of a map. Here are some important guidelines when choosing symbols.

> Natural earth tones such as browns, greens, blues, and tans are more pleasing and easy on the eye than strident reds, hot pinks, and wild purples. Such colors should be saved for special emphasis or contrast and should occupy a small area of the map.

> Light pastels are best for large areas and should take up most of the map. Dark or bold colors should be used sparingly to highlight important features.

> Take advantage of the psychological aspects of symbols. Blues and greens are cool and calming, and are easily understood when applied to water and vegetation, respectively. Reds and oranges tend to denote danger, heat, and agitation. Browns are steady and solid, or can suggest dryness when combined with greens for vegetation.

> Use symbols that mimic the phenomenon being mapped. Water should almost always be blue. A reader viewing the map in Figure 9.4 would probably at first glance interpret the Gulf of Mexico as land. NDVI maps showing the amount of vegetation are often displayed with browns and tans in low-vegetation areas, light greens in moderate areas, and dark greens in heavy areas, mimicking the actual appearance of the ground conditions.

Fig. 9.4. Blue makes features look like water.

➢ Maps showing increases in an attribute, such as county population, city size, or road traffic loads, should have symbols that follow an easily recognized pattern. Increases for polygon features are best shown using a single or dichromatic color ramp. Too many colors require that the reader's eyes dart back and forth between the map and legend to interpret the map, but a single color ramp immediately makes the phenomenon obvious. For lines, increasing thicknesses can be used to illustrate increase, and increasing marker size is typically used for points.

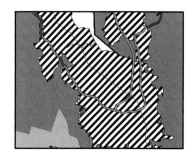

Fig. 9.5. A vibrating moiré pattern

➢ Avoid "vibrating" color patterns such as thick alternating white/black lines (Fig. 9.5), because they are tiring and unpleasant to look at. They are called moiré patterns. However, by thinning the lines, increasing the separation, and making the white transparent, we have a good symbol to display the urban areas without masking the population density of the counties (Fig. 9.6).

➢ Emphasize important information by using dark and bold colors, large sizes, or high contrast with surrounding map elements. De-emphasize less important information using pale colors, turning black to grey, or using transparency.

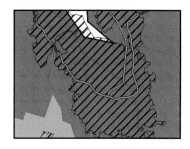

Fig. 9.6. Eliminating the white background provides a useful see-through pattern.

Examples

In Figure 9.7 we see the population density and highways of the San Francisco area. In Figure 9.7a, a rainbow color ramp was used, but the sense of increase is not obvious, and the reader would have to move his eyes back and forth from legend to map to interpret the data. Also, the lower-density counties are being dramatized in dark red, drawing attention to them. Unless they are the focus of the map for some reason, this balance is distracting. The preponderance of mixed bold colors is also tiring to look at. Overall, this map is aesthetically poor and difficult to interpret.

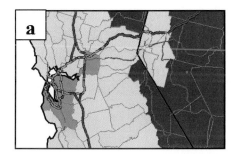 

Fig. 9.7. Map of county populations in the San Francisco area. (a) A map with poorly chosen symbols. (b) A redesign of the map which emphasizes the spatial patterns instead of obscuring them.

In Figure 9.7b, the rainbow color ramp is replaced with a graduated set of browns, making it obvious which counties have high populations. Most of the map is pale, thereby emphasizing the high population areas with their dark brown color. This color balance seems far more restful and pleasant. The highways are shown in a contrasting green for good legibility, and the urban highways, which are shorter and more difficult to spot, have been put in bright green to enhance their visibility. The major interstates are shown as thick lines and the highways as thin lines, delineating at a glance the relationship between these two road classes. These symbols are pleasant to view and enhance the readability of the map. In Figure 9.8, an additional refinement has been added; the county outlines have been made light grey. This change de-emphasizes the relatively unimportant county boundaries, reduces clutter, and brings out the roads more clearly.

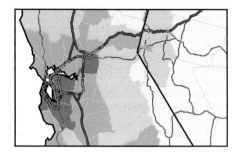

Fig. 9.8. Grey county boundaries highlight the more important roads.

This world population map illustrates some other guidelines (Fig. 9.9). The ocean is blue, but the mapmaker has disregarded the guideline to reserve dark, bold colors for highlighting small, important areas. The red-orange graduated color serves to highlight the population but contrasts noisily with the dark blue. The latitude-longitude lines, which are not the focus of the map, criss-cross the population data and create a busy feel. Let's see how a redesign can improve this map.

In Figure 9.10, a light blue color for the oceans instantly creates a more relaxed, agreeable atmosphere. The latitude-longitude lines have been de-emphasized by making them grey and drawing them behind the continents instead of on top of them. The graduated color shades contrast nicely with the ocean, are brown/orange to suggest earth, and are predominantly light. The darker populous countries are naturally emphasized.

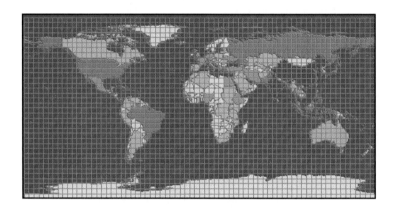

Fig. 9.9. A map of the world. Dark colors and busy lines give this map a heavy and chaotic feel.

Creating the map

ArcMap provides a wonderful environment for experimenting with map designs. It offers detailed control of almost every aspect of the map, down to the formatting of values and spaces in the legend. Several additional considerations will help achieve that professional look.

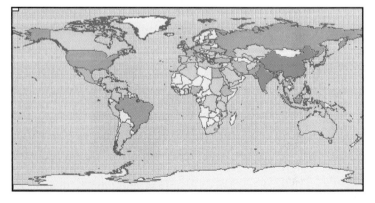

Fig. 9.10. A map of the world. The pastel colors and de-emphasized latitude-longitude lines are far more pleasant.

The layout of the legend often reveals the difference between an amateur and a professional quality map. The legend on the left in Figure 9.11 illustrates the kind of effort to avoid. No self-respecting cartographer would use the population classes shown here, a jumble of random numbers. How could one possibly get a quick understanding of the classes? The layer heading cities.shp would mean little to someone not familiar with ArcGIS file formats. The POP1990 label might not be understood by some people as population data. And look how the outline practically runs over the text, destroying the balance of the legend. This legend demonstrates some classic mistakes made by beginners, or an amateur in a hurry!

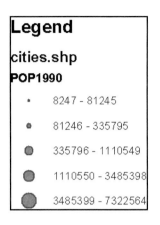
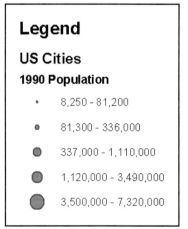

The redesign on the right side of Figure 9.11 has replaced the layer and attribute names with more understandable labels. The class numbers have been formatted to three significant digits and the thousands separator has been included for additional clarity.

Fig. 9.11. Attention to details such as labels and numeric formats enhances professional quality.

Finally, the Legend Properties window was used to increase the gap between the legend neatline and its contents. This legend is clear, balanced, and professional, and it only took one minute more to make the changes.

Finally, ArcMap provides tools to add pictures, charts, and explanatory text to maps. These additions can provide the finishing touches to a high-quality map. However, the same considerations of legibility, balance, and importance still apply.

Maps and reports in ArcGIS

ArcGIS provides two basic tools for creating maps and reports. The layout mode in ArcMap facilitates map design by incorporating data frames and map elements such as legends, north arrows, scale bars, text, and more. Each map document can contain one map layout. Map documents may also be saved as templates, which can be used to re-create a design again and again, or to give an organization's maps a common format and look.

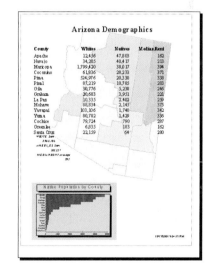

Reports may be designed using either the simple report utility packaged within ArcGIS or the more sophisticated Crystal Reports software that comes with it. This chapter covers the basic reporting tool. Creating a report includes choosing fields to present, selecting simple statistics to report, and setting up the detailed report layout and formatting. Reports can be included in map layouts or printed separately (Fig. 9.12).

Fig. 9.12. A layout with a report and a chart

ArcGIS also supplies a basic charting tool to create a variety of different types of charts, such as bar charts, pie charts, and scatter charts. Like reports, graphs can be created and included in layouts (Fig. 9.12).

Working with map elements

Any proper map layout contains many elements. **Map elements** are items placed in a map layout by the user to provide information and make the map attractive. Examples include the data frame itself, titles, legends, scale bars, pictures, reports, and graphs. Elements are placed on maps during a layout session using the Insert menu on the main menu bar (Fig. 9.13). Choosing to place an element usually launches a series of one or more windows in which the user defines the properties of the element, such as the size and symbol of a north arrow. These properties can also be modified after the element has been placed. Elements can be resized and moved to different locations. Right-clicking the object and choosing Properties allows editing of the element's properties, such as changing the north arrow color or angle. The tutorial includes many examples of placing and modifying map elements in a layout.

Fig. 9.13. Inserting map elements

The Layout toolbar

The Layout toolbar (Fig. 9.14) provides basic zoom and pan tools which allow the user to zoom and pan across the layout page in order to examine areas in greater detail, or to fit the entire page on the screen to check the map balance. The tools look similar to those on the regular tools menu, but they affect the view of the page, rather than the portion of the map shown in the data frame. These Zoom/Pan tools have no effect on the printed map scale or on the view in the data frame. Beginners should be careful to select the appropriate zoom tool depending on whether they wish to resize the data view and change the map extent or scale, or simply change the size of the Layout view.

Fig. 9.14. The Layout toolbar has Zoom/Pan tools and buttons which affect only the size of the layout on the screen, not the printed map scale.

Working with map scales

GIS spatial data acquires a scale as soon as it is drawn on a computer screen or printed on paper. Controlling the scale to achieve the desired effect is an important aspect of mapmaking.

The map scale in a data view is determined by the current zoom setting and the size of the window on the computer screen. Usually the zoom tools are used in the data frame window to control what features appear in the map, and the precise scale is not important.

In Layout view, the size of the data frame window, plus the scale in the Data view, controls the final scale of the printed copy. The data frame Properties window has a Data Frame tab which controls how scaling is performed in the map layout.

Automatic scaling is used most frequently. The user decides on a zoom scale in the data window and then sets the size and position of the data frame on the layout page, using his eye and aesthetic sense to balance the map size and other map elements. The printed scale is allowed to range freely to give the desired effect, and the user does not try to impose a specific scale. As shown in Figure 9.15a, in Automatic mode a data frame can be freely resized using the Zoom/Pan buttons, and the map is automatically scaled such that the portion visible in the Data view fills the frame at the largest possible size in Layout mode.

Fixed scale may be used to specify a particular scale. For example, if the user plans to overlay the map on a USGS topographic sheet, then the scale of the map must be exactly 1:24,000. In this mode, the features appear at the specified scale, and the data frame boundaries determine the extent of the map. Parts of the map may be cropped by this method if the user is not careful to ensure that the size of the data frame is sufficient to display all of the desired area. In Figure 9.15, the user started with the map shown at the top of 9.15a and set the specific scale to 1:32 million. As a result, the map features were reduced in size and portions of Alaska became visible in the frame. When the frame was subsequently resized, the features remained at the same scale and were cropped to fit in the new frame. If a data frame is set to a fixed scale, the zoom buttons are dimmed and the user is unable to zoom either in the Layout or the Data view. The user may pan, however, because panning does not change the map scale.

Fixed extent ensures that all the features within a specified *x-y* coordinate region are displayed in the frame. In Figure 9.15c, the user started with the top map in 9.15a and set the data frame to Fixed Extent. The current *x-y* values of the frame became the set values in the Properties menu. Alternatively the user could type specific *x-y* coordinates into the boxes to establish the extent. All the Zoom and Pan tools are dimmed in Fixed Extent mode. The data frame may be resized using only the data frame size and position tab in the Properties menu. After resizing, the features get smaller in order to fit inside the reduced frame so that the fixed extent is preserved.

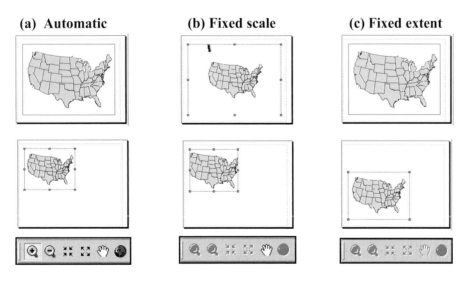

(a) Automatic   **(b) Fixed scale**   **(c) Fixed extent**

Fig. 9.15. Scaling options for data frames, shown in Layout mode. The effects of resizing the data frame differ in each case. The Zoom/Pan tools are only active in Automatic mode.

Setting up scale bars

All maps should contain either a scale bar showing the size of the features or at least a statement of the scale: "This map is shown at a scale of 1:62,500." Since ArcMap keeps track of the scale of the features on the layout page, it automatically constructs a suitable scale when the user requests one, and it even resizes the scale bar as needed if the map scale changes. The user does have several options to set, including the style of the scale bar, the units to be used, and the number and/or size of the divisions into which the scale bar is divided. Setting the options appropriately requires some knowledge of scale bar terminology.

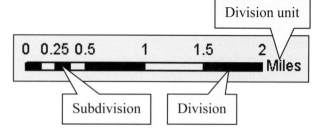

The **divisions** are the sections into which the scale is divided. The scale bar in Figure 9.16 has four divisions. The first division may be subdivided, yielding what are called **subdivisions.** This scale bar also has four subdivisions. The **division units** are the units of measurement used to divide the scale bar, in this case miles.

Fig. 9.16. Terminology of a scale bar

The **division value** is the length of each division, expressed in division units. This scale bar has a division value of 0.5 miles.

The division units, the map scale, the division size, and the length of the scale bar are interrelated, so that changing one of them may cause changes in one or more of the others. In the preceding example, if the number of divisions was increased, then the scale bar would have to be longer. Zooming in within the data frame, that is, changing its scale, would cause the scale bar to increase in size. Three options control how the scale bar will respond to changes in one of the parameters. These options are set using the scale bar's properties.

Adjust Width keeps the division size and number the same, and the scale bar is fixed in size for a particular scale. Thus the scale bar in the example would always have four divisions of 0.5 miles each. The bar would not change if it was resized; it would only change if the map scale changed.

Adjust Division Value changes the width of the divisions when the scale bar is resized and leaves the number of divisions constant. If the bar in Figure 9.16 was enlarged, the division value would increase from 0.5 miles to a bigger value (and not necessarily a nicely rounded one).

Adjust Number of Divisions leaves the division value constant and changes the number of divisions to the size indicated for the bar. If the bar in Figure 9.16 was enlarged, then the number of divisions would increase from four to a larger value (assuming the size increase was large enough to allow another division to fit).

In the rest of this chapter, you will learn to use the layout, report, and graph tools in ArcMap to produce quality output.

Summary

> For best results, making maps requires planning, knowledge of some basic guidelines, and attention to detail.

> The purpose and audience of a map influence the design process. However, nearly all maps should contain at least a title and contact information, scale bar, north arrow, and an appropriate legend.

> Maps should contain sufficient detail for interpretation, without excessive detail which detracts from the purpose.

> Layouts should be designed with attention to symmetry, size, and balance of the elements. Neatlines and shades can be used to help distribute space well. Alignment of objects along page margins is an important detail.

> Careful choice of symbols enhances the map. Guidelines for symbols include using primarily pastels and earth tones except when emphasis is desired, using larger and bolder symbols for the more important features of a map, using colors and symbols that naturally reflect the nature of the features, and avoiding clutter and clash.

> ArcMap provides tools for creating map layouts, reports, and charts.

> Scaling of the map is controlled by the data frame properties. Three methods may be used: automatic scaling, a fixed scale, or a fixed extent.

> Scale bars use divisions, division units, and division widths to control the size and markings of the bar.

TIP: Certain map projections, particularly conic and transverse cylindrical, do not preserve direction. In such projections the north arrow should point at the appropriate angle instead of straight up. ArcMap does not tilt the arrow automatically. When selecting the north arrow, click the Properties tab to set it to the correct angle. If the direction changes significantly across the map, such as for national and world maps, the latitude-longitude lines should be included for reference, rather than using a north arrow.

TIP: Colors shown on the computer screen often look somewhat different than colors printed on paper. Plan to print at least one draft of the map to make sure that the colors look acceptable. Colors can also vary from printer to printer.

TIP: Choose File > Export map to create an image file of a layout, in a variety of common formats including JPEG and PDF. These images may be placed in word processing or slide documents, or shared with other users. See the Skills Reference section for detailed instructions.

TIP: To save a map as a template for use with another map document, choose File > Save As and change the file type to ArcMap Templates (.mxt). See ArcGIS Help for more information on templates.

Chapter Review Questions

You may need to consult the Skills Reference section to answer some of these questions.

1. List four questions about map objectives that would influence the design of a map.

2. What two opposing factors must be considered in choosing which layers to put on a map?

3. What is meant by map balance?

4. What types of colors generally work best for maps?

5. How can the psychology of colors be used to enhance a map's meaning?

6. What are moiré patterns, and why should they be avoided?

7. What three options control how a scale bar changes when it is resized? Explain what happens in each case.

8. List three common pitfalls that amateurs make when creating legends.

9. What is a map template?

10. What is a neatline and what is its purpose?

Mastering the Skills

Teaching Tutorial

The following examples provide step-by-step instructions for doing basic tasks and solving basic problems in ArcGIS. The steps you need to do are highlighted with an arrow ➔; follow them carefully. Click on the video number in the VideoIndex to view a demonstration of the steps.

➔ Start ArcMap and open ex_9a.mxd (Fig. 9.17).

➔ Use Save As to rename the document, and remember to save frequently as you work.

 1➔ Switch to Layout view by clicking the Layout button.

 1➔ Click the Full Page button in the Layout menu, if necessary, to see the entire layout page.

Setting up the map page

First we are going to change the page layout and turn on the grid so we can align the objects more easily.

1➔ Choose File > Page and Print Setup from the main menu bar.

1➔ Examine the options to see what these are.

1➔ Fill the button to change the page to Landscape orientation.

1➔ Uncheck the button to scale the map elements proportionally to changes in page size, to make the frames stay the same size and shape.

1➔ Click OK.

2➔ To set the grid, choose Tools > Options from the main menu, and click the Layout View tab.

2➔ Notice the options for setting rulers: the page units (currently inches), the grid, and snapping.

A grid is a set of points a specified distance apart (now set to 0.25 inches). When snapping is on, objects within the snap tolerance (currently 0.2 inches) will be automatically snapped to the closest grid location. This feature simplifies aligning map elements.

2➔ Check the box to Show the grid.

2➔ Check the box to Snap to the grid, and make sure the tolerance is 0.2 in.

2➔ Click OK to close the Options box.

Now let's see how snapping works.

 3➔ Use the layout Zoom In tool (not the map Zoom In tool), to zoom into the data frame showing South Dakota.

 3➔ Click the Select Elements tool on the Drawing or Standard toolbar. Click the South Dakota frame to select it, and drag it to a new location. Notice that the frame corner snaps to a grid point when it gets close.

3➔ Click the Zoom Whole Page button to view the entire page again.

Next we will set up the basic layout design. Notice the dotted lines around the edge of the page. These guides mark the edges of the printable area. Keep the map inside these guidelines.

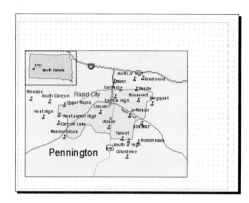

 4➔ Use the Select Elements tool to resize and move the two data frames until they look approximately as shown in Figure 9.17, with the South Dakota map inside the upper left corner of the Rapid City frame.

Fig. 9.17. Arrange the frames like this

Although it is simple to arrange the data frames by clicking, dragging, and resizing, sometimes precise sizes and locations are needed. Set these options using the data frame properties Size and Position tab.

5➔ In the Table of Contents, right-click the Rapid City data frame and choose Properties.

5➔ Click the Size and Position tab.

5➔ Set the X and Y position to 0.75 (inch). The page has coordinates of inches, with 0.0 at the lower left. Hence this step will give the lower left corner of the data frame exactly 0.75 inch margins.

5➔ Set the frame size to 7 inches wide and 6 inches high. Click OK.

 TIP: If the map does not redraw completely, use the Refresh button, located next to the Layout and Data view buttons to redraw it.

5➔ Adjust the South Dakota frame to fit neatly in the upper left corner of the Rapid City frame, with the left and top boundaries matched.

Use the Zoom and Pan tools in the regular toolbar to adjust the data frames even when in Layout view. Notice how much empty space lies below the schools in the map. Move the map down a bit to adjust this.

 6➔ Click on the Rapid City frame to make sure it is the active frame.

6➔ Click the Pan tool on the standard toolbar (NOT the Layout toolbar).

6➔ Click and drag inside the Rapid City frame to move the map down a little bit.

The tools on the standard Tools toolbar operate on the data frame, whereas the Layout tools operate on the layout. It takes a little practice to always use the right one the first time. To see the difference, do the following exercise:

 6➔ Still using the Tools Pan tool, move the map inside the Rapid City data frame around a little, as you did before. (Put it back in the starting spot before you stop.)

6➔ Now switch to the Layout Pan tool, put the cursor inside the Rapid City frame, and move the map again. Do you see the difference?

6➔ Repeat the exercise, using first the Fixed Zoom In and Fixed Zoom Out on the Tools toolbar, and then the Fixed Zoom In and Fixed Zoom Out on the Layout toolbar. (Put everything back in their original scales when finished.)

Adding a legend to the map

Next we will add a legend to our map to illustrate the symbols for the Rapid City schools.

7➔ Choose the Select Elements tool and click the Rapid City frame on the map layout to make sure it is active. The legend is always added to the active frame.

7➔ Choose Insert > Legend from the main menu bar. A Legend Wizard will appear.

7➔ Examine the list of layers and make sure you want them all in the map. Counties seems superfluous, so select it on the right and click the < button to remove it from the list.

7➔ Next establish the order of layers. To move a layer, click on it to select it and click the Up or Down arrow until it is in the desired location. Put Schools first, Major Roads next, then Urban Areas, then Rivers, then Lakes.

7➔ Click Next.

8➔ Leave the Legend Title as is, except click the button to Center it. Click Next.

8➔ Choose the Triple Graded border for the Legend. Click Next.

8➔ To choose a different style patch, click on the Lakes layer and choose the Water Body area patch. Click the Rivers layer and choose the Flowing Water line. Click Next.

8➔ This section gives very detailed control of the spacing between different elements of the legend. It is fine to leave the defaults on this step. Click Finish.

8➔ The legend appears in the middle of the map. Click and drag it so it is next to the map frame and even with its top (Fig. 9.18).

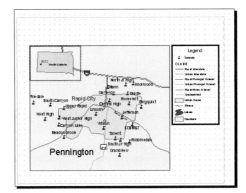

Fig. 9.18. Adding the legend

If you made a mistake creating the legend, don't worry. You can easily go back and change its properties. First let's change the text style of the Legend title.

9➔ Use the Layout Zoom In tool to zoom into the legend area.

9➔ Click the Select Elements tool, and double-click inside the Legend frame.

9➔ Click the Legend tab, which controls the symbols of the Legend title, the default patch and line symbols, and the spacing between elements.

9➔ Click the Symbol tab and set the Legend style to Country 1. Increase its size to 16.

9➔ Click Properties and choose Center for the Horizontal Alignment. Click OK, OK, OK, and view the legend again.

Okay, now let's change the patch style for the Urban Areas layer.

10➜ Double-click the legend again, and choose the Items tab.

10➜ Click the Urban Areas layer to select it, and click the Style button.

10➜ Click the Properties button and click the General tab if necessary. Check the box to override the default patch. Choose the Urbanized Area patch. Click OK, OK, and OK to view the changes.

Next, the CLASS heading over the roads is not very descriptive, and it looks rather ugly with its large capital letters. Change it to Road Class instead to make it clearer. This text is set, not in the legend, but in the Table of Contents.

11➜ Click the CLASS text under the Major Roads layer in the Table of Contents to select it, and then click it again to edit it. Type in "Road Class" and press Enter.

Finally, let's learn how to change the font and size of the labels in the legend.

12➜ Double-click the legend to open the Legend Properties.

12➜ Click the Items tab, and choose the Schools layer.

12➜ Click the Style button, and then click the Properties button.

12➜ Click the Label Symbol button.

12➜ Set the style to Country 1, the size to 12 point, and make it bold.

12➜ Click OK four times to see the changes.

Fonts can be changed for individual layers one at a time. Unfortunately, there is no way to change them all at once. Rather than doing the rest right now, we will go on to create more elements of the map.

Placing a scale bar on the map

A scale bar is a necessary element of a map, and several styles are available. The scale bar is placed in the active frame, so make sure that the correct frame is activated before creating the scale bar!

13➜ Click the Zoom Whole Page button to view the entire page again.

13➜ Click the Rapid City frame to make sure it is active.

13➜ Choose Insert > Scale Bar from the main menu.

13➜ Choose the Alternating Scale Bar 1 style bar.

13➜ Click Properties, click the Scale and Units tab, and make sure that the Division Units are set to miles.

13➜ Click OK and OK. Move the scale bar down to the lower left corner of the Rapid City frame.

13➜ Use the Layout Zoom In tool to zoom into the scale bar to examine it more closely.

> **TIP:** To change the properties of a scale bar after it has been created, double-click the scale bar using the Select Elements tool, or right-click it and choose Properties.

The scale bar might show up rather short, with overlapping labels, or it might be long with uneven units such as 2.6 miles. If it appears unsuitable, try resizing the scale bar to improve it.

14➜ Click the Select Elements tool.

14➜ Click the right boundary of the scale bar (a double horizontal arrow will appear) and drag it to the right to increase or decrease the length of the scale bar.

14➜ Repeat until the scale bar is about 2 miles long.

Unless you were lucky, you probably ended up with some fairly uneven labels, instead of nice, round numbers. We can modify the scale bar to get exactly the properties we want. First we'd better review terminology, however, as shown in Figure 9.19.

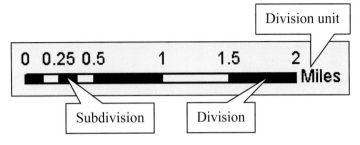

Fig. 9.19. Terminology of a scale bar

The **divisions** are the number of main units into which the scale is divided. The scale bar in Figure 9.19 has four divisions. The first division may be subdivided, yielding what are called **subdivisions.** This scale bar also has four subdivisions. The **division units** are the units of measurement used to divide the scale bar, in this case miles. The **division value** is the length of each division, expressed in division units. This scale bar has a division value of 0.5 miles.

The actual length of the scale bar is determined by the map scale, the number of divisions, and the division units. Thus creating a suitable scale bar requires careful calibration of its length, units, and divisions. Any of the three can be adjusted to get the desired result. As you adjusted the length of the bar, ArcMap kept the number of divisions constant and adjusted the division value. Other options are possible. As we view this sample scale bar, we might decide to create a scale bar with six divisions of 0.5 miles each, and five subdivisions for the 1:100,000 scale map.

1. What would the length of each subdivision be in the scale bar just described? _____

15➜ Double-click the scale bar to open its properties, and click the Scale and Units tab.

Notice that the division value is dimmed out. In order to set it explicitly, we must change the way ArcMap adjusts the bar when resizing.

15➜ Under When Resizing, set the drop-down box to Adjust Width.

15➜ Set the Division value to 0.5 miles.

15➜ Set the number of divisions to 6.

15➜ Set the number of subdivisions to 5. Click OK.

15➜ Examine the new scale bar. Try to resize one end.

Notice that the resize handles don't appear, and if you move an edge the entire scale bar moves, rather than getting longer or shorter. This happens because we specified Adjust Width. Once we dictated the number of divisions and the division unit, the size of the bar was determined. It will only change size if we change the scale of the map.

16➔ Open the scale bar properties again, click the Scale and Units tab, and set the When Resizing option to Adjust Number of Divisions. Click OK.

16➔ Try resizing the scale bar again. This time, the number of divisions changes.

16➔ Adjust it until the scale bar is two miles long. Notice that the division units do not change.

Adding other map elements

Many other elements can be added to the map using the Insert menu. Keep in mind that the element will be inserted into whichever data frame is currently active. If you make a mistake and insert an element into the wrong data frame, simply click the Delete button to get rid of it.

2. How can you tell that map elements are automatically selected immediately after they have been added? _____

3. How would you delete an element if it was not selected? _____

Adding a north arrow

17➔ Click the Zoom Whole Page button to see the entire page again.

17➔ Choose Insert > North Arrow from the main menu bar.

17➔ Click the symbol you like best.

17➔ Click the Properties button and set the color to a dark blue. Click OK, OK.

17➔ Click the Select Elements tool and move the north arrow to the lower right corner of the Rapid City frame.

TIP: Certain map projections, particularly conic and transverse cylindrical, do not preserve direction. In such projections the north arrow should point at the appropriate angle instead of straight up. ArcMap does not tilt the arrow automatically. When selecting the north arrow, click the Properties tab to set it to the correct angle. If the direction changes significantly across the map, such as for national and world maps, the latitude-longitude lines should be included for reference.

Adding a title

18➔ Choose Insert > Title from the main menu bar. A centered text field appears on the map.

18➔ Type in "Rapid City Area Schools" and press Enter.

18➔ Using the Drawing toolbar, set the font size to 48 and make it Bold.

Adding other text

19➔ Choose the Insert > Text choice in the main menu bar.

19➔ Type "by Your Name" and press Enter.

19➔ Move the text to a centered position below the legend.

20➔ To add wrapped text, click the New Rectangle Text tool (one of the label tools) and click and drag to create a box under the legend, aligning the edges with the data frame and the legend.

20➔ Double-click inside the new text box to open its properties, and type the following text into the box, without using any Enter keys: "This map shows the names and locations of schools in the greater Rapid City area. The data were obtained from the ESRI Data and Maps CD-ROM set." Click OK.

20➔ Resize the text box to fit neatly around the text, and move it down to the location shown in Figure 9.20.

Adding a picture

21➔ Choose Insert > Picture from the main menu bar.

21➔ Navigate to the mgisdata\MapDocuments folder and choose the file schoolkids.jpg. Click Open.

21➔ Resize and move the picture to occupy the upper right corner of the Rapid City frame.

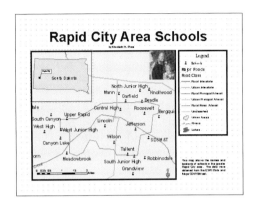

Fig. 9.20. The map layout after step 21

TIP: Information on allowed image formats can be found in the online help index by typing the entry "rasters" and choosing the subheading "formats, supported."

Adding neatlines and shades

22➔ Choose Insert > Neatline from the main menu bar.

22➔ Fill the button to place it around all elements.

22➔ Set the gap to 12 points (the gap controls the distance between the elements and the neatline).

22➔ Choose the Triple Graded border from the drop-down box.

22➔ Choose the Linear Gradient fill from the Background drop-down box (Fig. 9.21).

22➔ To change the gradient color, click the small icon to the upper right of the background symbol box.

22➔ Click Properties, then click Change Symbol.

22➔ Click Properties again, and choose a light green color ramp. Click OK in all the boxes.

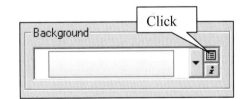

Fig. 9.21. Changing the gradient color

Reviewing and printing the map

Now we're just going to tidy up a few more details before printing the map. First of all, the Pennington text is distracting, so we'll remove it from the map. Then we will add a background color and stronger frame boundary to the South Dakota map for a neater appearance.

23➔ Click the Rapid City frame with the Select Elements tool to activate it.

23➔ Switch to Data view.

23➔ Select the Pennington text and press the Delete key. Use the scroll bars if needed to move down and find the text.

23➔ Switch back to Layout view, and adjust the map location using the Pan tool, if necessary.

24➔ Right-click the South Dakota frame in the Table of Contents, and choose Properties. Click the Frame tab.

24➔ Set the border to Triple Graded to match the others. Set the background color to 10% Grey. Give it a drop shadow of 50% Grey. Click OK

24➔ Open the Rapid City frame Properties and give it a Triple Graded Border.

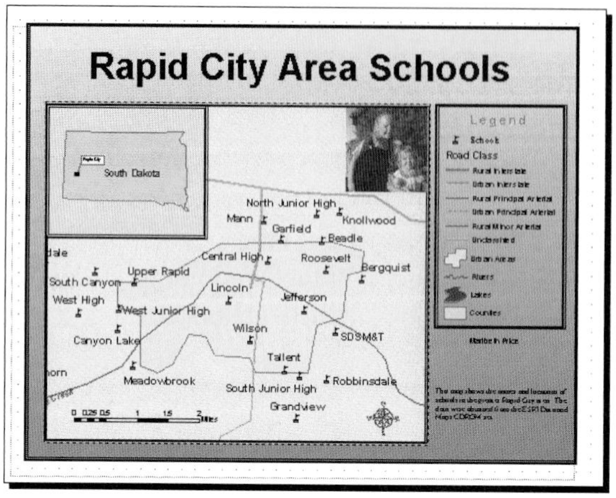

Fig. 9.22. The final layout should look similar to this.

25➔ Examine the map for balance and aesthetics (Fig. 9.22). Then SAVE it before you print, lest a printing glitch cause you to lose your map edits.

25➔ To print, choose File > Print from the main menu bar.

25➔ Click Setup to change the printer, if necessary. Adjust the other settings if needed. When ready, click OK.

Congratulations! You have just created your first map in ArcGIS!

TIP: Colors shown on the computer screen often look somewhat different than colors printed on paper. Plan to print at least one draft of the map to make sure that the colors look acceptable. Colors can also vary from printer to printer.

TIP: Choose File > Export map to create an image file of a layout, in a variety of common formats including JPEG and PDF. These images may be placed in word processing or slide documents, or shared with other users. See the Skills Reference section for detailed instructions.

Creating a map from a template

Map templates are used for two purposes. First, they provide a quick way to make a nice-looking map. Second, they can establish a similar look and format for a group of maps.

➔ Open the ex_2.mxd map document in your MapDocuments folder.

26➔ Click the Change Layout button in the Layout toolbar and click the General tab.

26➔ Examine the layout choices. As you click each name in turn, a thumbnail view shows what the layout looks like.

26➜ Click the thumbnail icon to view the map thumbnails all at once.

26➜ Notice the directory in which these templates are stored by viewing the pathname in the box below the windows. They are stored within the ArcGIS program directory.

26➜ These are general layouts. Click the Industry tab to see maps for certain industries, such as the military. Also view the World and USA templates.

26➜ Return to the General tab.

26➜ Choose the LetterLandscape map template. Click Finish and view the new map (Fig. 9.23).

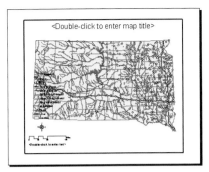

Fig. 9.23. Initial result of using the Letter Landscape template

The map switches to Layout view, and the selected layout has been applied. The user can finish adding titles and adjusting the placement of the elements (such as the legend) until the map is done.

TIP: To save a map as a template for use with another map document, choose File > Save As and change the file type to ArcMap Templates (.mxt). See ArcGIS Help for more information on templates.

This is the end of the tutorial. You may continue on to do the next three optional sections on creating graphs and reports.

➜ If stopping, exit ArcMap. You do not need to save your changes.

Making graphs (optional)

Imagine that you have been commissioned to write a report on the Native American population in Arizona. You need to create a bar graph showing the Native and Caucasian populations of each county. Then you will write a report on several demographic factors, including the median rent.

➜ Open the map document ex_9b.mxd in the MapDocuments folder.

➜ Use Save As to rename the document, and remember to save frequently as you work.

27➜ Choose Tools > Graphs > Create from the main menu bar. The Graph Wizard appears.

27➜ Pick the Graph Type as Bar and the subtype as Stacked Bar (note the names appear under the pictures as you select each subtype). Click Next.

27➜ Choose Counties as the layer containing the data.

27➜ Under the fields to graph, uncheck Area, and check White and AMERI_ES. Use the Up arrow to put AMERI_ES on top. Click Next.

27➜ Title the graph "Native Population by County".

27➜ Check the box to label the *x* axis and choose the field Name (of the county) to label it with. Click Finish. The bar graph appears in a separate window.

Although the graphing process was easy, we need to make a few changes to the graph. First of all, the default colors are garish. We also see that the biggest county is dominating the graph, and we decide we want to remove the whites so we can see the differences in the native population better. Fortunately these changes are easy to make.

28➔ Place the cursor on the window bar of the graph. Right-click it and choose Properties to display the map properties.

28➔ Click the Data tab, and uncheck the box to display the white population. These high values are masking the native populations.

28➔ Click the Type tab, and choose the Pareto Bar subtype, which sorts the bars from largest to smallest. Click OK to see the changes.

29➔ Open the graph properties and click the Appearance tab, then click the Advanced Options button.

29➔ Click the Markers tab. Choose a darker green color.

29➔ Click the Background tab.

29➔ Choose Graph from the Apply To list by clicking its button (the changes to the other options will be applied to the element chosen from the list).

29➔ Choose the Drop Shadow style for the graph.

29➔ Click the Graph Title button, and choose Drop Shadow for it as well.

29➔ Choose a light grey background color for the Graph Window. Click OK.

29➔ Uncheck the box to Show Legend.

4. Why is it rather silly to have the legend showing in this graph? _____

30➔ Take a moment to examine each of the other tabs to see what options are available. For more information about making graphs, consult the software documentation.

30➔ Click OK and OK to view the changes (Fig. 9.24).

30➔ Examine the graph, and then close it using the X button in the upper right corner. You will place it in the layout later.

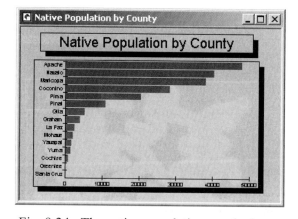

Fig. 9.24. The native population graph chart

Creating a report (optional)

ArcGIS comes with a basic report creator and a special report creator called Seagate™ Crystal Reports™. In this lesson we will learn the basic reporting tool.

Reports are created and saved within the map document. Right now we will create a report on demographics in these Arizona counties.

31➔ Choose Tools > Reports > Create Report from the main menu bar. The Report Properties dialog box appears. This window is the tool for creating reports.

31➔ For the layer or table to report, choose Counties.

31➔ From the list of fields choose NAME and click the single arrow to move it to the list of fields to report.

31➔ Also add WHITE, AMERI_ES, and MEDIANRENT to the list to be reported.

31➔ If needed, use the Up/Down arrows to set the order of the fields to match as they have been listed.

31➔ At the bottom right, click the little arrow next to Show Settings.

The window expands to show a preview of the report being created. Currently the fields overlap, and it is clear we need some changes. Ideally we would like to have all the fields in one row so that the entire report fits on one page. First we will change the field widths. (Because the entire page does not fit inside the video capture window, you will be unable to see the settings in the video clips.)

32➔ Click the Display tab.

The box on the left contains an expandable list of all the elements, objects, and fields in the report. If you click on an element to select it, its settings are displayed in the box to the right. You can then modify the settings. To see the changes, click the Update Settings button at the top of the display area on the right.

32➔ Click the plus to expand the Fields, and choose the NAME field.

32➔ Click in the Width box and set the width to 1.9 inches. (Use Enter after setting each field.)

32➔ Click the WHITE field to select it, and set the width to 1.2 inches.

32➔ Click the Number Format box. A set of ellipses will appear to the right.

32➔ Click the ellipses to open the number formatting menu. Check the box to show thousands separators. Click OK.

33➔ Set the width of AMERI_ES to 1.2 inches and format its values as for WHITE.

33➔ Set the width of MEDIANRENT to 1.2 inches.

➔Click the Update Settings box in the top of the viewer area to see the changes.

Now the values fit across the page, but some of the titles are cut off. We will change the field headings to smaller words in lowercase so they fit better.

34➔ Click the NAME field, click in its Text box, and change its text to County.

34➔ Repeat to change WHITE to Whites, AMERI_ES to Natives, and MEDIANRENT to MedianRent.

➔Click Update Settings to see the changes.

Now we will add a few more elements to this report.

35➔ Expand the Elements tree.

35➔ Click to select Title. Type "Arizona Demographics" in the Text box.

35➔ Click the Font box, then click the ellipses that appear. Set the font style to 24 point bold. Click OK.

36➔ To add some space between the fields and the title, click the Field Names entry, and set the Top value to 0.5 inches. Click Update Settings.

36➔ Click the element Date, and turn on its check box. Click the box next to its Section property, and choose Bottom of Page from the drop-down menu which appears.

Use the Summary option to generate statistics for the fields. Let's put the total population for all counties and the average median rent at the bottom of the report.

37➔ Click the Summary tab.

37➔ Check the boxes for Whites Sum, Natives Sum, and MedianRent Average.

37➔ Make sure that the section is set to End of Report.

37➔ Click Update Settings to see the new addition.

37➔ Click the Display tab again.

37➔ Expand the Summary and Report entries to see the new data. For now, leave the settings for these entries as they are.

For one final touch, sort the counties by the Native population, so it is easier to see any relationship between that field and the MedianRent field.

38➔ Click the Sorting tab.

38➔ Click on the word None in the AMERI_ES row, and change it to Descending when the drop-down box appears.

Save the report format in case it is needed again.

39➔ Click the Save button. Navigate to the MapDocuments directory and name the report "ArizonaReport".

Now it's time to generate the report itself.

39➔ Click the Generate Report button.

Placing graphs and reports in layouts (optional)

The report appears in its own window. At this point you could print it, copy it to the clipboard, export it as a .pdf, .rtf, or text file, or add it to a map. We will do the latter.

39➔ Click Add. Since there is only one page, you do not need to specify a range. Click OK to add the page.

39➔ Close both report windows.

39➔ Set the map to Layout view, if necessary.

TIP: You cannot save report settings for later use; you can only save the report text. Once you close the Report Properties box, you are no longer able to edit that report; you would have to start again from scratch. **Never choose Close unless you are certain you are finished with the report and don't need to edit it again.**

Notice the report has appeared. We will now create a final layout for this project.

40➔ Click the Select Elements tool, and click on the report text to select it. Resize it slightly smaller to fit better on the page.

40➔ Move it so that it is centered in the upper part of the page.

TIP: To remove a report from a layout, click the black arrow Pointer tool on the Drawing toolbar. Click on the report to select it and press the Delete key.

41➔ Right-click the Counties layer and choose Properties. Click the Symbology tab.

41➔ Set up a graduated color map showing the AMERI_ES population, normalized by the total population. The monochrome orange ramp would look nice. Format the values to show as percentages with no decimal places.

41➔ Click the Display tab of the Counties properties, and set the transparency to 60%. Click OK.

42➔ Use the Zoom/Pan tools to place the Counties map centered behind the report. Leave space at the bottom for the graph.

42➔ Right-click the Layers frame and choose Properties. Click the Frame tab and set the Border style to <None>. Click OK.

42➔ Turn off the layers for Voting Districts, Cities, and Arizona, so that only the counties are being displayed.

Finally, add the graph to the bottom of the layout.

43➔ Choose Tools > Graphs > Manage from the main menu bar.

43➔ Click the Native Population by County graph to select it. Click the Show in Layout button.

43➔ Close the Graph Manager, and move the graph to the bottom left of the layout. Also resize it if necessary for better balance.

44➔ Different colors for the graph would harmonize better with the map. Double-click the graph in the layout to access its properties.

44➔ Click the Appearance tab and click the Advanced Options Button.

44➔ Click the Background tab. Set the Palette to 128 Rainbow, and choose a light orange color for the background.

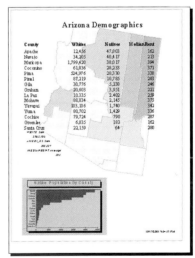

Fig. 9.25. The Arizona map and report

44➜ Click the Markers tab. Notice the palette has changed here, too. Choose a brown color for the bars. Click OK and OK. View the map (Fig. 9.25).

You could continue formatting this layout, adding descriptive information and working on the balance, but for now we've demonstrated the graphs and reports functions, and we can end here.

➜ Use File > Save As to save the map document in the MapDocuments folder under a new name, such as "Arizona.mxd".

This is the end of the optional tutorials.

➜ Exit ArcMap.

Exercises

Imagine that you are planning to open an Internet-based dating service in the Mountain (Mtn) subregion of the United States. Your preliminary research indicates that such a service would expect to gross $3 for every divorced person, $2 for every single (never married) person, and $1 for every widowed person in a typical county. You must do the following:

➢ Calculate the projected yearly gross income for each county in your study area.

➢ Calculate the total projected yearly income for each state.

➢ Create and print a layout containing the following:

- A graduated color map showing the income by county for the entire Mountain subregion. You'll want to include other layers such as state boundaries, and possibly towns, urban areas, or other layers to enhance the map. Take care that it doesn't get too cluttered, however!

- (Optional) A graph showing the total income for each state.

- (Optional) A report detailing the number of people in each group (single, divorced, widowed) for the top 15 earning counties.

In your work, pay close attention to details of layout, colors, formatting, balance, and all the other aspects we have learned about. Make the map as aesthetic and professional as you can. Be sure to include all the required map elements, as well as your name. Be as creative as you want.

Challenge Problem

Create an attractive graduated color map of median rent, showing the entire U.S.A. with all 50 states. Use additional data frames to show Alaska and Hawaii individually at larger scales, each with its own appropriate coordinate system and scale. Be sure to state the coordinate systems used in each data frame on the map.

Consider: Ideally, each data frame should use the same classification breaks in the legend, so that you need only put the legend on the map once. How do you make sure this happens?

Skills Reference

Setting up the map page

1. Choose File > Page and Print Setup from the main menu bar.

2. Check the printer and use Properties to change it if needed (Fig. 9.26).

3. Set the printer paper size, source, or orientation options if needed.

4. Set the map page width, height, and orientation.

5. To use the settings from the current printer for the map page size, check the box. Otherwise, set the page size. See the tip below.

6. If you plan to change the map size and want the map elements to be resized along with the page, check the box.

7. If desired, check the box to show the printer margins in the Layout.

8. Examine the page preview and adjust settings if necessary.

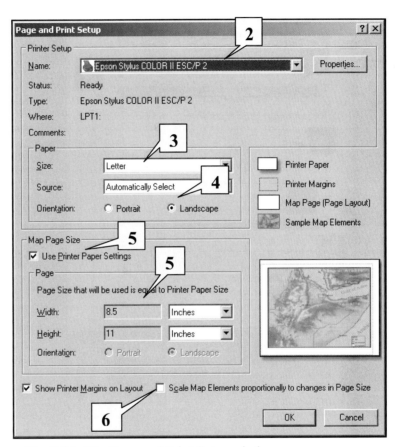

Fig. 9.26. The Page and Print Setup window

TIP: Using the Printer Paper Settings may cause a map document to issue a warning when it is opened on a system where the original printer is not available. Thus this option is best avoided when planning to share map documents with others. If you receive such a warning, simply proceed. Later you can uncheck this box in page settings, or set the printer to one on your system, to avoid the warning in the future.

Setting the scale or extent

The Data Frame properties tab provides three options for setting the map scale of the frame: automatic, fixed scale, or fixed extent. Automatic scaling is the default.

1. Right-click the data frame name to open its properties window. Click the Data Frame tab. Choose the desired scaling method (Fig. 9.27).

Setting a fixed scale

2. Fill the second button and type the desired scale in the box. The map will be displayed at the specified scale both in Layout and in Data view mode.

3. Click OK. The zoom tools will be deactivated in order to preserve the scale.

Setting a fixed extent

4. First, use the Zoom/Pan tools to set the Data view to the desired extent, OR determine the *x-y* coordinates of the desired extent.

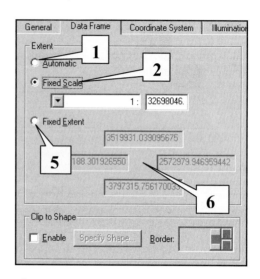

Fig. 9.27. Scaling options in the data frame properties window

5. In the Data Frame tab, fill the third button. The boxes will be filled with the current *x-y* extent of the data frame.

6. Change the *x-y* coordinates if desired.

7. Click OK. The Zoom/Pan tools will be deactivated in order to preserve the extent.

Clipping to a layer

1. Open the Data Frame Properties and click the Data Frame tab.

2. Check the Enable box under Clip to Shape (Fig. 9.28).

3. Click the Specify Shape button.

4. Choose to Outline Features and select the layer defining the outer boundary.

5. OR enter coordinates for a custom rectangle.

6. Select a border to outline the clip feature, if desired.

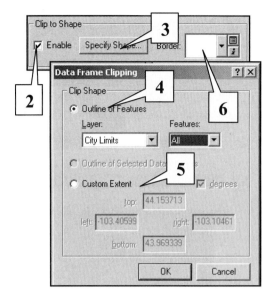

Fig. 9.28. Clipping layers in a data frame

Composing the data frames

The map layout has at least one data frame, which can be dragged and resized to position it on the page.

1. Make sure the Layout view button, at the bottom left of the display window, is clicked (Fig. 9.29).

 2. Choose the Select Elements tool from the Drawing menu or the Zoom/Pan menu.

3. Click on a data frame to activate it. Blue handles and dashed lines will appear to indicate that it is the active frame.

4. Click and drag on the data frame to move it to a new location (Fig. 9.30).

5. Click and drag a side handle to increase or decrease the size in one direction.

6. Click and drag on a corner handle to increase or decrease the size in two directions.

Layout view

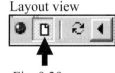

Fig. 9.29

Fig. 9.30. Click and drag the active frame to move it or change its size.

Using the Layout toolbar

The Layout toolbar provides tools for zooming around the layout page (Fig. 9.31). It has no effect on the zoom in the data frame. The function of the buttons, from left to right, is as follows:

1. The Layout Zoom In tool is used to enlarge a portion of the layout. Click the tool and then click and drag a box around the area to zoom to.

Fig. 9.31. The Layout toolbar

2. The Layout Zoom Out tool zooms out a specified distance from the layout centered on the clicked point.

3. The Layout Pan tool moves the layout within the window. Click and drag on the layout to move it.

4. The Fixed Zoom In tool zooms in a specified amount centered on the current display when you click anywhere on the layout.

5. The Fixed Zoom Out tool zooms out a specified amount, centered on the current display when you click anywhere on the layout.

6. The Zoom Whole Page tool zooms so that the entire layout page can be seen.

7. The Zoom to 100% tool shows the layout at the same scale as it will be printed.

8. The Go Back to Extent tool returns to the previous extent.

9. The Go Forward to Extent goes to the next extent. This button is only available if you have clicked the Go Back to Extent tool at least once.

10. This drop-down box sets a particular percent enlargement for the layout.

11. The Change Layout button launches the Template window so you can add or change the layout of the map using a predefined template.

Adding a legend

1. Use the Select Elements tool to click on the data frame containing the layers to appear in the legend. The legend is always created from and placed in the active frame.

2. Choose Insert > Legend from the main menu bar.

3. Choose which layers will be included in the legend. To add a layer, click it in the box on the left and click the > button. To remove a layer from the legend, click it in the box on the right and click the < button. Choose the number of columns in the legend. Click the Preview button to examine a preview of the legend. Click Next.

4. Modify the legend title text and formatting to desired settings. Click Next.

5. Specify a legend border, background, and drop shadow if needed. Click Next.

6. Click a layer to modify its symbol size and patch style. Click Next.

7. Modify the spacing if needed (usually not necessary). Click Finish.

8. The legend appears in the map, selected. Click and drag the legend to the desired location, and resize it if desired. Resizing will change the size of the text and boxes within.

TIP: Converting the legend to simple graphics may help adjust the fine details. Right-click the legend and choose Convert to Graphics from the context menu. The legend is converted to a grouped set of graphics. To ungroup them and work with each element individually, choose Draw > Ungroup from the Drawing toolbar.

Changing the appearance of a legend

Legend properties can be edited and modified after the legend is created.

To open the Legend Properties window, double-click the legend, or right-click it and choose Properties. This window has four tabs: Legend, Items, Frame, and Size and Position.

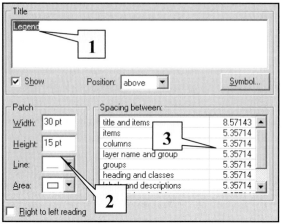

The Legend tab

1. Modify the title of the legend if desired (Fig. 9.32). Choose not to show it by unchecking the box, or change the text of the title to something else, or change the position of the legend, or change the font by clicking the Symbol button.

Fig. 9.32. The Legend tab

2. The Patch refers to the default size and symbol styles for the legend. Change the size by typing in new values in the Width and Height boxes. Change the shape of the default Line or Area patches using the drop-down boxes. The patch changes affect the entire legend. The patch style for individual layers can be changed in the Items tab.

3. Adjust the default spacing between the elements in the legend by typing in the desired values, in points.

The Items tab

This tab gives you individual control of every layer in the legend.

1. Change which layers are displayed by adding or removing them from the Legend Items box (Fig. 9.33).

2. To modify other properties of a single layer, click on it to select it.

3. Change the position of the layer in the list by clicking on the Up or Down arrows.

4. To place the selected layer in a new column, check the box.

5. Set the number of columns for this layer to span in the legend.

6. To change the style of the legend layer, click the Style button. The Legend Item Selector window appears.

7. Choose one of the predefined styles. To further modify the chosen style, click the Properties button. The Legend Item window appears.

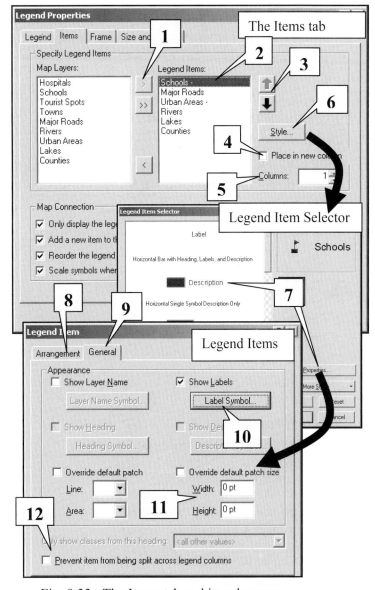

Fig. 9.33. The Items tab and its submenus

8. Click the Legend Item window Arrangement tab to change the relative position of the patch, the layer name, and the description.

9. Click the Legend Item window General tab to set specific properties of the item style.

10. Change the font style of the text by clicking the Label Symbol button.

11. Change the patch shape, or the patch width and height, if desired.

12. You can also prevent the legend item from being split across a column break.

The Frame tab

The Frame tab can set a border around the legend and/or give it background shading or a drop shadow (Fig. 9.34).

1. Choose a border style using the drop-down box and change its color, if desired.

2. To modify the existing border, such as changing its pattern or color or thickness, click the Border Selector button or the Edit Border button.

3. Set the gap distance between the map and border to the desired X and Y values.

4. Choose to round the corners by entering a percentage greater than zero. The higher the percentage, the more rounding.

5. Set the background shade if desired, using the same steps as for the border.

6. Set the Drop Shadow desired, using the same steps as for the border.

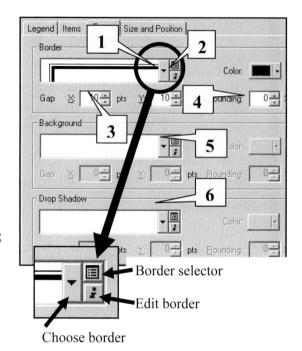

Fig. 9.34. The Frame tab

The Size and Position tab

The Size and Position tab lets you specify an exact position and size for the legend in page units, which are inches by default (Fig. 9.35).

1. Set the position in page units, relative to the lower right corner of the page.

2. The anchor point indicates which part of the legend sits the XY distance from the corner. To place the lower right corner of the legend at 3 inches from the left and 3 inches from the bottom, enter 3,3 and click the lower left anchor point. To center the legend at 3,3, enter 3,3 and click the center anchor point.

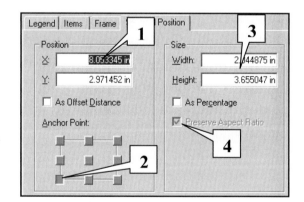

Fig. 9.35. The Size and Position tab

3. Set a specific width and height of the legend if desired, either in page units or as a percentage of the page size.

4. If Preserve Aspect Ratio is checked, the shape of the legend will remain constant if it is resized.

Adding a scale bar

1. Using the Select Elements tool, click to activate the desired frame. The scale bar will be placed in the active frame and will be sized according to the scale of the active frame.

2. Choose Insert > Scale bar from the main menu. The Scale Bar Selector window will appear (Fig. 9.36).

3. Choose the desired scale bar.

4. To modify the scale bar, click the Properties button to open the Scale Bar window (Fig. 9.37).

5. The Numbers and Marks tab controls the spacing of numbers and marks on the bar. These values can usually be left as defaults.

6. The Format tab controls the font of the scale text and the style of the scale bar. The defaults are usually fine.

7. The Scale and Units tab controls the length and divisions of the scale bar (Fig. 9.37).

8. Set the units for the scale bar to miles, kilometers, or some other unit.

9. Choose the When resizing… option. One or more of the input boxes here may be dimmed, depending on which option is chosen.

10. Set the division value, the number of divisions, and the number of subdivisions as applicable.

11. Check the box to place the subdivisions before the zero point, rather than in the first division.

12. The chosen units will be labeled on the scale bar, e.g., Miles. Choose the label position, change the label text and font symbol, and set the gap between the scale bar and the label if desired.

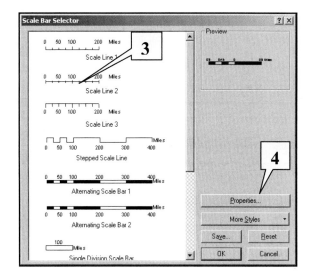

Fig. 9.36. The Scale Bar Selector

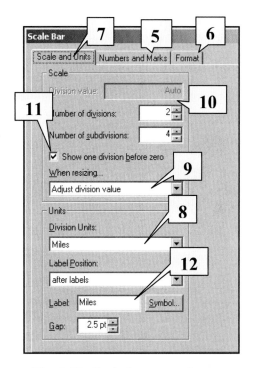

Fig. 9.37. Scale bar properties

TIP: To change the properties of a scale bar after it has been created, double-click the scale bar using the Select Elements tool, or right-click it and choose Properties.

Adding a north arrow

1. Choose Insert > North Arrow from the main menu and choose the desired symbol.

2. Click the Properties button to modify the symbol. Click OK to place it.

3. If necessary, click and drag it to the desired location, or resize it.

TIP: In some map projections, north is not straight up. Set the north arrow marker angle manually to point true north if needed.

Adding titles and text

1. Choose Insert > Title or Insert > Text from the main menu bar. These commands are similar except that the default font size is larger for the Title.

2. Enter the text in the box on the map and press Enter.

3. The text remains selected. If needed, change its size or font using the Drawing toolbar, or click and drag it to a new location.

4. To modify the text after it has been created, double-click it with the Select Elements tool, or right-click it and choose Properties. You can change the text, set the font and size, or specify a particular position for the text.

Adding pictures

1. Choose Insert > Picture from the main menu bar.

2. Navigate to the folder containing the picture and click on it to select it.

3. Click Open to add the picture.

4. Resize and/or move the picture to the desired location using the Select Elements tool.

TIP: Information on allowed image formats can be found in the online help index by typing the entry "rasters" and choosing the subheading "formats, supported."

Adding neatlines, backgrounds, and shadows

Neatlines are lines which enclose one or more map elements. They are used to balance the space and provide alignment between irregular map elements for a more aesthetically pleasing map. Most objects in layouts, including data frames, legends, and scale bars, have a tab in their properties to set up borders, backgrounds, and drop shadows. To access these tabs, open the element's properties, either by double-clicking it or right-clicking it, and choosing Properties.

To create a neatline or shaded box as a separate object on its own accord, do the following:

1. Choose Insert > Neatline from the main menu bar. The Neatline window appears.

2. Choose the desired Placement option (Fig. 9.38).

3. Set the border, background, and drop shadow styles.

4. To further modify the available styles, click the Border Selector or Edit Border buttons as shown.

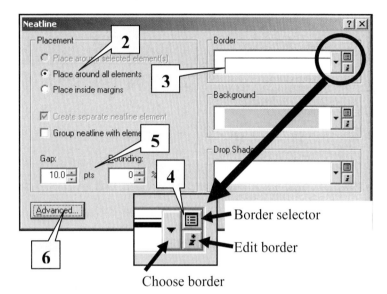

Fig. 9.38. Setting up neatlines and boxes

5. Set the gap between the neatline and the elements, and enter rounding if desired. The higher the percentage, the more rounding occurs.

6. The Advanced button allows customized editing of the symbols, gaps, and rounding for each border, background, and shadow.

Changing map elements

Nearly all map elements can be modified after they are created.

1. Open the properties of the element by double-clicking it, or by right-clicking it and choosing Properties from the context menu.

2. Use the tabs and windows to modify the properties.

Adding graphics to layouts

The Drawing toolbar (Fig. 9.39) provides functions for creating and modifying objects on a layout. These objects may also be created within a data frame itself, in which case they will be scaled if the map changes size. If they are in the layout, they will be unaffected scale changes.

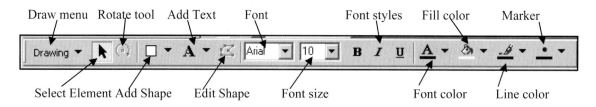

The Drawing toolbar contains common functions found in other programs (Fig. 9.39). A small black triangle on a button indicates that the button contains a menu.

Draw menu Rotate tool Add Text Font Font styles Fill color Marker

Select Element Add Shape Edit Shape Font size Font color Line color

Fig. 9.39. The Drawing toolbar

Printing a map

1. To preview a map to see how it will look on paper, chose File > Print Preview from the main menu bar.

2. To print, Choose File > Print from the main menu. The Print window appears (Fig. 9.40).

3. To change the printer or its properties, click the Setup button.

4. Set the number of copies to print.

5. Choose the desired tiling options if the map is larger than the printer paper.

6. Preview the layout placement on the page, and if all is well, click OK.

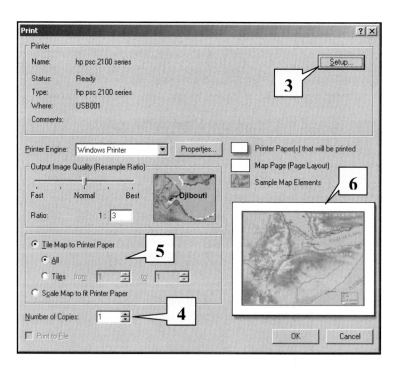

Fig. 9.40. The Print window

TIP: Colors shown on the computer screen may look different than on paper, or they may vary from printer to printer. Plan to print at least one draft copy to forestall surprises.

Exporting a map as a picture file

You can export a map as an image file to put it on a web page or inside another document, or as a PDF to share it with others.

1. Choose File > Export Map from the main menu bar.

2. Navigate to the folder where you will save the picture file (Fig. 9.41).

3. Choose the type of picture to save.

4. Enter a name for the file.

5. Click the gray arrow for more export options, such as the resolution or the output quality. The options will vary for different image formats.

6. To avoid having a white border around the picture, check the box to Clip Output to Graphics Extent.

7. Click Save.

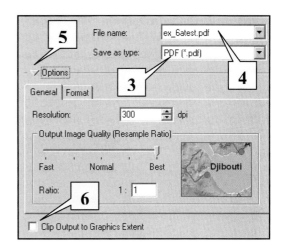

Fig. 9.41. Exporting a map as a picture

TIP: If you plan to enlarge the map before printing, it is usually desirable to increase the picture resolution from the default. Note that this option can substantially increase the file size, however.

Creating a graph

Creating a new graph

1. Choose Tools > Graphs > Create from the main menu bar. The Graph Wizard will appear to step through the process of creating a graph (Fig. 9.42).

2. Choose the graph type in the box on the left, and the graph subtype in the box on the right. Note that clicking each graph subtype causes a brief description to appear below the icons. Click Next when done.

3. Choose the layer or table to be used. If desired, check the box to use only selected records for the graph.

4. Choose the fields to be included in the graph. To change the order of fields, click on a field name to select it, and use the Up/Down arrows to move it.

5. Choose whether to put records or fields as the data series (on the *x* axis). Click Next.

6. Add a title and subtitle for the graph.

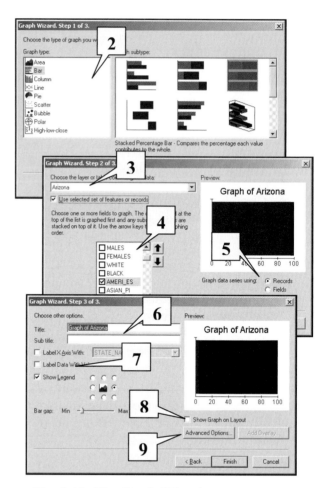

Fig. 9.42. The Graph Wizard

7. Choose whether to show labels for *x* and *y* axes, and the location of the legend.

8. To show the graph on the layout, check the box.

9. Click the Advanced Options button to change graph properties such as colors, fonts, backgrounds, titles, scale ranges, and more. See the next section for more details.

10. Click Finish to create the graph.

11. Click the X in the upper right corner of the window to close a graph. To open it again, choose Tools > Graph from the main menu and select the title of the graph to open.

12. To place a graph on the layout, open the graph. Right-click the blue bar at the top of the graph and choose Show on Layout.

Modifying the properties of a graph

1. To open the chart properties, right-click the blue bar at the top of the graph and choose Properties from the context menu. A window with three tabs appears. The tabs are the same windows used to create the graph and are called Type, Data, and Appearance.

2. To change the type, layout, or data series used in the graph, click the appropriate tab and make the changes as described in the previous section.

3. To change fonts, colors, scales, backgrounds, and other details, go to the Appearance tab and click the Advanced Options button. A new window appears with seven tabs: Titles, Axis, Fonts, Markers, Trends, Error Bars, and Background.

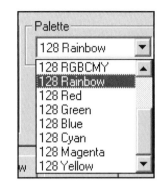

Fig. 9.43. Setting the color palette

There are many options for changing the graph, which you can experiment with to become familiar with them. Here are directions for a few common tasks.

4. **To get help** on one of the options in the graph window, click the Help button.

5. **To change the color of a symbol** or bar, choose the Markers window. On the map in the left corner, click the symbol to change. Then set the color, pattern, or other options to the desired setting.

TIP: When setting colors anywhere in the Advanced Options menus, the choice of colors is determined by the palette, which is chosen in the Background tab on the lower right. Different palettes offer different sets of colors (Fig. 9.43). The first 16 colors are always a default set, followed by the chosen palette colors.

---

6. **To change the fonts,** click the Fonts tab. Choose one of the graph text objects by filling in the appropriate button. Set the typeface and size. Click another text object and change the font for it also. Smart Scale automatically adjusts the font size to fit in the space designated for that text object, so the longer the text, the smaller the font size.

7. **To change the graph background,** click the Background tab. Choose one of the objects to apply the changes to by filling in the button. Set the desired changes. Choose another object, if desired, and change it also. Change the graph space background and outline using the section titled Graph Window.

Creating a report

A report lists certain fields in a table, along with formatting and optional summary statistics, to provide a neat and organized view of tabular data. Creating a report has several steps. A simple report can be created quickly. More time spent adding titles and formatting values can improve the report's appearance.

1. To begin creating a report, choose Tools > Reports > Create from the main menu bar. The Report Properties sheet appears with five tabs: Fields, Grouping, Sorting, Summary, and Display (Fig. 9.44).

2. The Fields tab is the only one available at first, because the fields to report must be chosen before other options are set. Start by choosing a table from the drop-down list.

3. Select the fields to include in the table. Click the left/right arrow keys to add or remove fields from the list.

4. To change the order of the fields, click one to select it and then click the Up/Down arrows to move it up or down in the list.

5. To use only a selected set in the report, check the box.

6. To see a preview of the report, click the Show Settings arrow. Click it again to hide the preview.

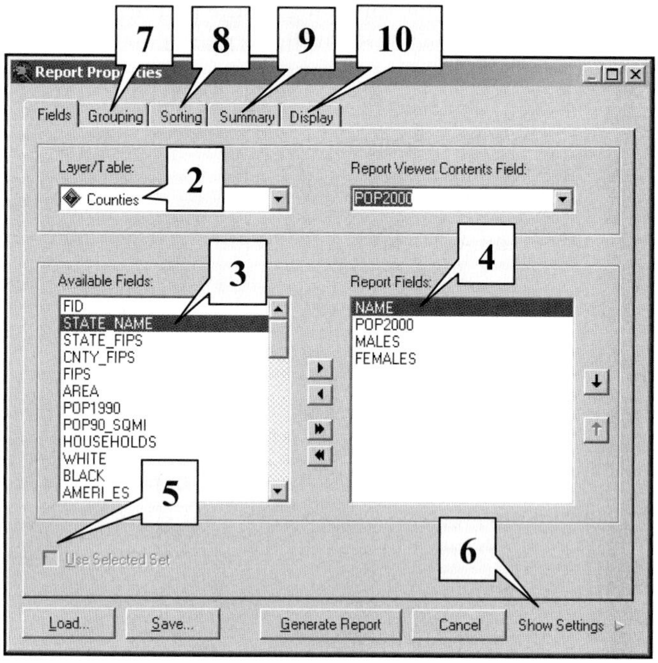

7. Click the Grouping tab, if desired, to group values together based on a common value in another field. For example, you could summarize the value of parcels according to their zoning type.

Fig. 9.44. The Fields tab in the report creator

TIP: Some fields in the Report Creator have values inside them, such as sorting by ascending or descending order. To change the value, click on the field to bring up a drop-down list of available values, and choose the one desired.

---

8. Click the Sorting tab to sort entries, if desired. You may sort by up to three columns at once. Choose the first sort field, click the field that says None, and change it to Ascending or Descending. The sort order will appear in the last column. Add another field to sort by in the same way. To change the sort order, reset columns to None and do them again in the correct order.

9. Click the Summary tab to calculate summary statistics for the report. Choose the desired statistics for each field, and set their location with the drop-down list.

10. Click the Display tab to modify column widths, fonts, spacing, and other details.

The Display tab contains two boxes (Fig. 9.45). The box on the left contains an expandable list of report elements, such as the title, the column headings, the list of fields, and so on. The box on the right contains one or more properties of the element currently selected from the list.

11. Click one of the report elements in the box on the left. Use the +/- signs to expand or collapse the list as needed.

12. If an empty box appears next to a map element, the element is not currently included in the report. Click on the empty box to put a check mark in it and add the element. To remove an element from the report, uncheck its box.

13. Change any of the desired settings in the box on the right by clicking on the entry and either typing in the information or choosing a value from a drop-down list.

14. To view the effect of changes, click the Show Settings arrow to see a preview of the report.

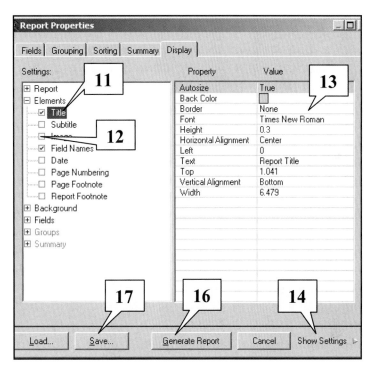

Fig. 9.45. The Display tab in the report creator

15. Click Update Settings at the top of the Preview window to see changes reflected in the report. No changes show in the Preview window until you click Update Settings.

16. When you are finished setting up the report, click Generate Report to create it. This report is a temporary creation. If you don't like its appearance, close the report and continue changing the settings. Continue until the report looks acceptable.

17. When satisfied with the report, click Save to save it on disk as an ArcGIS report format. Or click Generate Report to open the Report Viewer and add it to the layout.

> **TIP:** You cannot save report settings for later use; you can only save the report text. Once you close the Report Properties box, you are no longer able to edit that report; you would have to start again from scratch. **Never choose Close unless you are certain you are finished with the report and don't need to edit it again.**

Using a report

Once the report is generated or loaded from the disk, it appears in the Report Viewer. At this point you have several options for using it (Fig. 9.46).

Fig. 9.46. The Report Viewer toolbar

1. Click the Print button to print it.

2. Click the Export button to save it as text, rich text, or a .pdf file.

3. Click the Add button to add the report to the layout. The pages will be added to the layout as separate EMF images.

4. Click the Copy button to copy the report to the clipboard.

5. Use the zoom tools and drop-down box to examine parts of the report in more detail. The Page Up and Page Down buttons move from page to page in the report.

6. The Stop button stops the progress of adding a report, in case you accidentally added a very long one to the layout.

TIP: To remove a report from a layout, click the Select Elements tool on the Drawing toolbar. Click on the report to select it and press the Delete key.

TIP: The Report function can be used to export certain records and columns into one of three formats: a portable document format file (.pdf), a rich text file (.rtf), or a text file (.txt).

Creating a map from a template

A map template is a set of data frames, titles, styles, and other map elements which are already formatted and ready to receive the data in the data frame(s). Use a map template to quickly create a map using a standard format. You can also save your own map as a template to create a similar map again.

1. Click the Change Layout button in the Layout toolbar.

2. Click one of the tabs to see a choice of templates (Fig. 9.47). Templates you have created will be stored in the My Templates tab.

3. Click on a template to see a preview of it.

4. Use the Browse button to navigate to another directory containing more templates saved elsewhere.

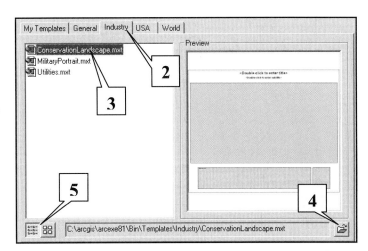

Fig. 9.47. Choosing a map template

5. Click the Thumbnail button to see icons of all the templates; click the List button to go back to the original list view.

6. Click the desired template and choose Next or Finish.

TIP: If the template has more than one data frame, it will prompt the user to assign the data frames in the map document to the data frames in the template.

---

7. If the data frames need assigning, click each of the frames in the list on the left (Fig. 9.48).

8. Use the Move Up and Move Down buttons to put them in the same order as the numbered frames in the new layout.

9. Click Finish.

10. Finally, change any titles or other map elements in the template which need to be customized.

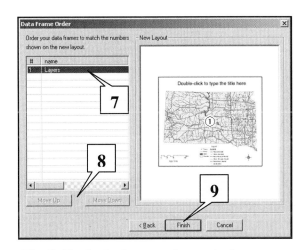

Fig. 9.48. Assigning map frames to the template frames

347

Chapter 10. Geocoding

Mastering the Concepts

Objectives

 - Understanding the geocoding process

 - Knowing the different styles of geocoding

 - Setting up an address locator service

 - Understanding geocoding options

 - Converting addresses to locations

 - Converting *x-y* coordinates in a table into point locations

Concepts

What is geocoding?

As everyone knows, a street address tells a postal worker where to deliver mail. It contains a type of spatial information. However, additional knowledge on the part of the post office must be added to the address in order to deliver mail. Unlike a latitude-longitude point, which uniquely locates a house on the globe, a street address requires information about the city and state, the location of the street, and the sequence of house numbers. A newly hired postal worker delivering a piece of mail in a strange city would need a good map in addition to the address.

Likewise, geocoding combines map information with street addresses in order to locate a point uniquely. It enables someone to convert a list of addresses into points on a map (Fig. 10.1). Geocoding has many uses.

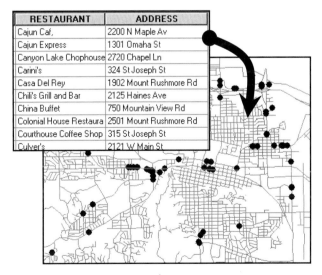

Fig. 10.1. Geocoding can convert restaurant addresses in a table to points on a map.

911 Addressing

One critical use being developed by many government agencies concerns finding street addresses given over the phone during an emergency call, in order to quickly route response vehicles to the correct location. Routing of vehicles is accomplished by providing a map, or even a GPS location in vehicles equipped with GPS units. Because lives are at stake, this application requires highly accurate address information, and it is one of the most challenging and expensive applications to implement.

Driving directions

Many people these days have gone to one of the popular websites such as MapQuest to find directions from a starting address to a destination. The results include both step-by-step driving directions and a map showing the route. Other people may have tried the new systems installed in cars, such as OnStar or Magellan, which guide people to their destination from their current position as identified by GPS. Both of these applications are built upon geocoding technology.

Crime Analysis

Police records of crime incidents include the address, and these locations can be converted to points on a map through geocoding. Analysts could use these data to help identify "hot spots" in order to help plan patrol routes for officers or to target efforts to develop neighborhood watch programs. Police departments can glean information about spatial patterns for different types of crime, or target the time of day particular crimes tend to occur. Such information can be critical in developing strategies to reduce crime and protect residents and businesses.

Marketing

Many companies collect address information about their customers as a regular part of conducting business. Converting addresses to locations and analyzing their distribution might help a company provide better service by opening new stores in underserved areas. Mail-order businesses can locate areas with low sales for targeted marketing efforts. Sales and service companies can better plan where to place agents to serve their clients. Recently a study helped a university in South Dakota to understand the national distribution of its students, and it is helping them to plan marketing strategies to increase national enrollment (Fig. 10.2). A whole host of business applications are available once information about customer locations is known, and many companies now routinely use GIS in their operations.

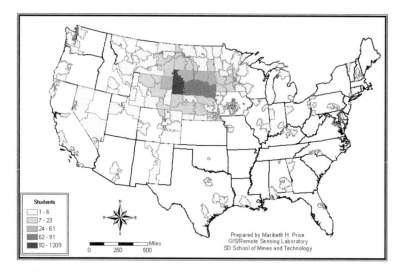

Fig. 10.2. National distribution of students attending the South Dakota School of Mines and Technology in 2004

How does geocoding work?

Just as a stranger requires a map to find an address in a new city, the computer also requires specific map data in order to turn an address into a location. This spatial data is called the **reference layer.** Geocoding typically occurs using a street layer composed of polylines with specific attributes indicating each street's name, and the address ranges present on that street. However, other types of geocoding can use point or polygon reference data, such as matching store franchises to points representing cities, or matching a list of owner addresses to parcel polygons.

The computer requires specialized software to interpret addresses. Consider that humans automatically recognize a variety of address styles. An address might be written 123 Maple Street, or 123 Maple St., or 123 N. Maple St., or even 123 Maple, and the postmaster would have no difficulty finding the house. When compared by computers, however, each of these addresses is different. Thus, geocoding programs must interpret different styles of addresses, and even manage typographical and other errors, such as someone entering Maple Rd instead of Maple St. They accomplish this task by dividing the address into different components and then scoring how well they match the reference components. A high score (>80) indicates a good match, a medium score (60-80) indicates a tentative match, and a low score (<60) suggests a poor match.

In order to interpret an address, ArcMap begins by parsing the address into individual fragments of information. The address "235 W. Main Str." could be divided into four pieces, including the address number (235), the direction prefix (W.), the street name (Main), and the street type designation (Str.). After separating the components, each fragment is standardized so that it has the best chance to match the information in the reference layer. All letters may be capitalized (MAIN), periods removed (W), and common abbreviations reduced to a single standard (ST). This process is process called **address standardization.** This step is necessary because a computer, confronted with the problem of comparing (Str.) in the address table to (ST) in the reference layer, will conclude that the two do not match, when in fact they do. A human would hardly notice the difference, but computers can be tiresomely literal in their comparisons. An address may include any of the following components:

Prefix direction. A direction indicator that comes before the street name, such as *E* Maple St., or *North* Main St. Some towns and cities use prefixes for directions, others use suffix directions (e.g., Main St. North). This part may have any of the standard directions, N, E, S, W, NE, SW, SE, and NW. The complete spellings North, East, South, and West are also recognized.

Prefix type. An indicator of the road type which comes before the street name, such as *Highway* 85. In this case 85 is the street "name" and Highway, or Hwy, is the prefix type.

Number. This is the address number, such as *123* Maple St.

Street name. The name of the street, such as 123 *Maple* St., or *Mount Rushmore* Road.

Street type. This part consists of a type indicator that comes after the street name, such as Main *St.* or Maple *Ave.* The software recognizes a variety of street types (Street, Lane, Place, Road, etc.) and understands most common abbreviations, such as St, St., Rd, Av, and Ave.

Suffix direction. A direction indicator that comes after the street name, such as Florman St. *E* or 87th St. *North*. This part may have the same possible values as the prefix direction.

Zip code. This part contains a standard U.S. zip code, such as 57702, or optionally the Zip+4 codes used by some cities.

Table 10.1 shows some examples of addresses and their standardizations. Note that in addition to separating the address into components, the text is converted to capital letters to facilitate comparisons in case-sensitive databases. The complete words in the directions and types are also converted to standard abbreviations, so Road becomes RD, and the periods are omitted.

Table 10.1. Examples of address standardization

Address	Prefix Dir	Prefix Type	Number	Street	Type	Suffix Dir
123 Maple Road			123	MAPLE	RD	
15 Center St. East			15	CENTER	ST	E
314 Hwy 85 N		HWY	314	85		N
234 E. St. Patrick St.	E		234	SAINT PATRICK	ST	

Once the address is standardized, the components are compared to attributes in a spatial data layer, such as a line shapefile containing the city streets. The shapefile must contain certain attribute fields to compare with the standardized components. The score for each component match is averaged to give an overall match score for the address, in terms of percentage. A score of 100 indicates a perfect match. A zero indicates that none of the components match. Usually a score of 80 or better is considered a good match, and a point will be created for the address. When the score is lower than 60, the geocoding software looks for other possible candidates. Typically candidates have scores above 30. The user can examine the candidates individually to find out if any of the candidates might provide a reasonable match.

	FID	Shape*	PREFIX	STREET	SUFFIX	SUFFIX2	FRADD	TOADD
	3634	Polyline	E	VAN BUREN	ST		400	499
	3635	Polyline		RACINE	ST		900	999
	3636	Polyline		EGLIN	ST		1700	2699
	3637	Polyline		MEADOWLARK	RD		700	799
	3638	Polyline	N	SPRUCE	ST		800	899
	3639	Polyline	E	VAN BUREN	ST		500	599
	3640	Polyline	N	LACROSSE	ST		850	899
	3641	Polyline		MEADOWLARK	RD		800	899
	3642	Polyline					0	0
	3643	Polyline		MEADOWLARK	RD		800	899

Fig. 10.3. The attribute table of a shapefile that could be used as the basis for geocoding

Figure 10.3 shows an example of a street shapefile that could be used for geocoding. Notice that most of the fields described previously are present, although the names are not exactly the same. Also notice that instead of a single address, the street has two fields (FRADD and TOADD) which describe the range of addresses that fall along the street. The FRADD is the From address and contains the lower range value; the TOADD is the To address and contains the higher range value.

To actually create a point feature based on an address and street shapefile, the following logic is used. The number of the address is tested to see if it falls within the range of addresses on that street section. If it does, the point locating the address is placed along the street in proportion to its value between the range endpoints. In Figure 10.4, for example, Meadowlark St. has an address range of 700 to 799. If matching the address 750 Meadowlark St., the point feature would be placed approximately halfway between the street ends, because 750 is about halfway between 700 and 799. Likewise, 725 Meadowlark St. would be placed a quarter of the distance along the street, and so on. This method of placing points is called the **One Range** method because a single address range is used for each street.

For more advanced geocoding, a **Two Range** method can be used. In this method each street has four attributes defining its address ranges, a Left From and Left To address range for the left side of the street, and a Right From and Right To address range for the right side. This method can provide more accurate results in many cases; however, the effort involved in setting up the streets shapefile attributes is more time-consuming.

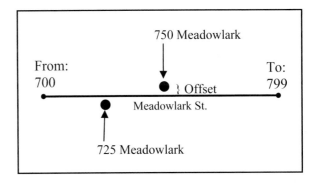

Fig. 10.4. Placing a located point along the street based on its proportional distance from the ends

When using the Two Range method, the addresses will be located on the left or right side of the street, as dictated by the address ranges. In addition, the user can specify an **offset** distance, such that the actual point is placed slightly to the side of the street instead of directly on top.

The assumption that a house or other structure is located along a street segment at a distance proportional to its address number does not always work as well as it should. Parcel sizes may vary along the street, or some address numbers may be skipped entirely. This problem occurs most severely in rural areas where the spacing of addresses may be very irregular. If Meadowlark St. in Figure 10.4 represents a mile-long section line, and all the houses are clumped together at one end, then the actual location and geocoded location may be quite different. Geocoded locations should always be considered approximations.

One Range and Two Range are examples of different geocoding styles. The style chosen will depend on the way in which the streets layer has been set up for geocoding. Other styles are also available as described in the next section. The user must ensure that the attributes in the reference layer support the desired geocoding style.

Although geocoding is most often associated with placing addresses, in fact it applies to many types of data. Points can be located by city and state, by zip code alone, or even by township/range information. These different styles of geocoding have two common themes: they convert attributes to point locations, and they allow the user to link data that are similar but are not necessarily an exact match.

Available geocoding styles

Geocoding takes a variety of approaches, from a specific house location on one side of the street, to simply finding a city and country on a world map. The available options depend on the attributes of the reference layer, the setup in the address table, and the type of matching to be done. ArcMap offers several predefined styles for geocoding. Each style requires a different set of attributes in the reference layer. To choose an appropriate style, the user must be familiar with the organization of the reference data.

Single Field

A single field locator uses one field to match to the reference layer, which typically contains points or polygons, although lines could also be used if desired. Any single field, such as the zip code, a state name, or a place name, can be used as the key to match the locations. This method expands geocoding from addresses to *any* type of information. For example, a table might contain information about petroleum wells, keyed to specific well IDs like "TenRun Hole No. 6." Such IDs may change slightly from database to database, depending on who typed them in (Ten Run Hole #6, or Ten Run Num 6). Performing an attribute join to link the table with a point layer of well locations would yield many missing values because the matches are not exact for each ID. Geocoding would work better because it can link close matches, not just exact matches.

US Alphanumeric Ranges

Many people are familiar with using grid zones to locate streets on a map by consulting a list of streets, finding its zone (such as H7), locating the zone by means of letters and numbers on the map margin, and scanning the streets in the zone to find the one in question. This geocoding style uses this approach to reduce the search time while geocoding. Each address is attributed with its zone identification, and the geocoding algorithm need only search inside that zone. The required fields are the grid zone, the street name, and the From and To ranges for both sides of the street.

US Cities with State

This style accommodates national scale matching of location based on the city and state name only. Cities may be matched to points or polygon features. This style could create a map showing the locations of customers across the United States. The required fields are the city name and a state name or state abbreviation.

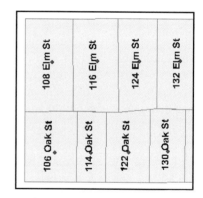

US One Address

This style uses points or polygon reference data, set up so that each address corresponds to one point or polygon (Fig. 10.5). For example, it could be used with points representing buildings, or polygons representing parcels. The required fields include street number and street name. The user may optionally include prefix and suffix directions and other address components.

Fig. 10.5. The US One Address style codes to point or polygon layers such as these parcels.

US One Range

This style uses a single address range along a street, rather than having ranges for both the left and right sides of the street. The required fields are the street name, the From address, and the To address. Prefixes, suffixes, and street types may also be used if desired.

US Streets

This style represents the Two Range option discussed previously, and relies on a reference layer with two address ranges, one for each side of the street. The required fields include the street name, a From and To address for the left side of the street, and a From and To address for the right side. Prefixes, suffixes, and street types may also be used if desired.

World Cities with Country

This style allows matching of a city and country on a world map. The reference layer may contain cities as points or as polygons. The required fields include the city name and country name or country abbreviation.

ZIP 5-Digit and ZIP+4

The ZIP style matches zip codes to points or polygons in a zip code reference layer. The only field required is the 5-digit zip code. The ZIP+4 style works the same way, but has an additional field containing the 4-digit extension. Note that the ZIP code and the 4-digit extension must be in separate fields.

With Zone

Many of the above styles offer a zone option. A zone is an additional piece of information in the reference layer, such as state, city, or zip code, which is added to the other required fields for that style. The addition of the zone allows matching over larger areas. For example, the US Streets style can only find addresses within a single city because only the street address is included; there is no city name to distinguish between Main St. in Rapid City and Main St. in Sioux Falls. To match street addresses across all of South Dakota would require a city name or zip code as a zone, provided that such a field exists in the reference layer and the address table.

The geocoding process

Before starting, the user must first verify that a geocoding layer is available and that an address locator (called a geocoding service in previous versions) has been set up for this layer. Setting up an address locator service is covered in the next section. The user must also have a table of addresses to be matched.

To begin geocoding, start an ArcMap session and load the address table and the reference layer. An address locator is loaded from ArcCatalog, and then geocoding can begin.

The first geocoding attempt is usually done in batch mode, that is, the locator tries to match all of the addresses it can automatically. When finished, it issues a report of how many addresses were matched, and how many remain unmatched. You then have the option of examining the unmatched addresses interactively to find more matches.

Interactive mode examines each unmatched address and suggests one or more match candidates. A candidate has a lower score than a match but scores high enough to provide a possible match. If no candidates appear, you can try lowering the candidate cutoff score or relaxing some of the geocoding options, such as the spelling sensitivity.

Once you have matched all the addresses that can be matched, the software will create the output shapefile of point locations. The attribute table will also contain information, such as the standardized address, the match score, and the status of the record (matched or unmatched).

Occasionally the address table or the geocoding layer contains errors or inconsistencies. In this case the user may decide to halt geocoding, fix the problems, re-create the address locator if necessary, and then try again to match the addresses. For example, the original reference layer prepared for this textbook had many streets named for saints, such as Saint Patrick St. In the reference layer as well as the address table, the street names were abbreviated to St. Patrick, St. Anne, etc. However, during geocoding the software automatically converted *St* to *Saint* during the address standardization. As a result, dozens of addresses remained unmatched. To streamline matching, the author edited the streets layer to replace all the St. abbreviations with Saint.

In another example, some students were analyzing crime patterns using street addresses and the StreetMap USA product. The police reports used the standard address prefixes for Rapid City (*West* Main St.), but StreetMap uses suffixes (Main St. *West*). In this case the easiest solution involved editing the spreadsheet containing the addresses so as to place them in the address file with suffixes instead of prefixes.

These examples demonstrate that geocoding is not a simple push-button operation, but often requires some investigation and ingenuity on the part of the user. A thorough understanding of the geocoding process, address formats, and reference data being used helps in detecting and solving barriers to matching.

Setting up an address locator

Before any geocoding can be performed, the user must create an address locator. An address locator specifies a streets layer, plus some definitions of how the locator will work. To create one you must decide which geocoding style will be used, assign the correct fields in the shapefile to the required matching fields, and set the options which control how the locator performs. Once an address locator is defined, it can be used over and over again by many users.

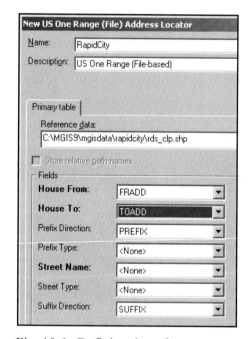

Fig. 10.6. Defining the reference layer for a new address locator

Address locators are created using ArcCatalog. The first step includes choosing the style of the locator and specifying whether the geocoding layer is a shapefile or a geodatabase layer. The second step includes defining the parameters of the locator. Figure 10.6 shows the first part of the menu for defining a locator based on the US One Range for shapefiles style. You must enter the name and location of the shapefile which contains the streets and attributes. Next, you must assign the correct attribute from the streets table to the fields needed for geocoding. In this case, for example, the FRADD field in the shapefile is used as the House From field. The fields shown in boldface are required; the others are optional.

The fields shown in the setup will vary depending on the style of locator requested. A zip code locator, for example, would need fewer fields, as would a city-state style of geocoding.

The next part of the menu (Fig. 10.7) contains various options for how the locator will work. You can accept these default options, modify them immediately, or change them during geocoding.

Input address fields

You can specify the names of fields in the address table which are assumed to contain the addresses to be matched. It saves time if the locator automatically recognizes the address field in the address table being matched.

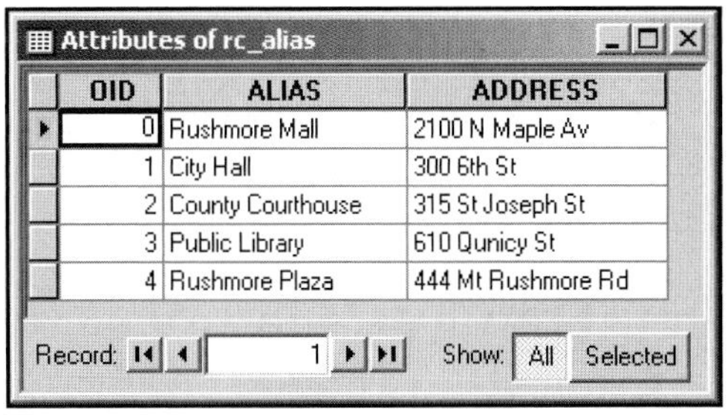

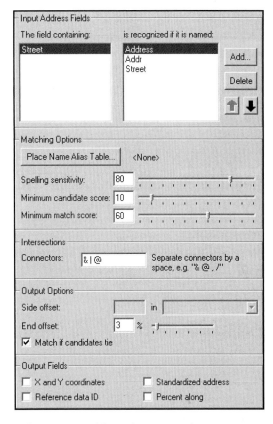

Fig. 10.7. Address locator options

Alias table

You can specify an alias table, which allows certain names to substitute for addresses (Fig. 10.8). For example, the address table might only list Rushmore Mall as the "address" for a series of stores, or the place name "City Hall" might often appear instead of the actual address. An alias table links these common name listings with the actual address at which they occur, that is, Rushmore Mall sits at 2100 N. Maple Av. and City Hall sits at 300 Sixth St. Then if the address file contains "Rushmore Mall" as the location of a store, the location can be found.

Attributes of rc_alias

OID	ALIAS	ADDRESS
0	Rushmore Mall	2100 N Maple Av
1	City Hall	300 6th St
2	County Courthouse	315 St Joseph St
3	Public Library	610 Qunicy St
4	Rushmore Plaza	444 Mt Rushmore Rd

Record: 1 Show: All Selected

Fig. 10.8. An alias table contains common place names with their addresses.

Matching options

These options control the default behavior of the locator, although they can be adjusted during actual geocoding, if desired. The **spelling sensitivity** specifies how closely a street name must match before it is considered a valid candidate. The value basically reports the percentage of correct characters. With this option, small misspellings or typing mistakes do not prevent a match from being made, a definite plus considering many people's typing skills. Thus, Pierce St would probably still be recognized even if it was spelled Peirce St. by mistake.

The **minimum candidate score** is used to find possible candidates when a good match can't be found. During interactive matching, the user examines each unmatched address individually. During this process, the computer suggests possible candidates with lower scores than a good match. If no candidates are displayed, lowering the minimum score sometimes helps. The minimum score can also be increased if too many candidates are routinely showing up.

The **minimum match score** is the lowest score that will constitute a match. In some applications it is acceptable to lower the default in order to match as many records as possible, and the cost of a mismatch is not important. To a company analyzing customer locations, a few wrong addresses are not likely to significantly affect the analysis. In other cases, the minimum score might be raised. An emergency management team could find a poorly matched address to be a matter of life and death. In this case, poor matches should be identified and individually considered by the dispatcher to avoid sending the team out to a bad location.

Intersections

Geocoding can also match intersections, such as the corner of 5th St. and St. Joseph St. This option specifies the symbols that are considered to be connectors, so that the address locator can properly interpret the address as an intersection rather than a regular address. The default values are **&**, |, and @ characters, but additional characters may be used. However, the separation character should not be one that might ordinarily appear in an address.

Output options

This section controls parameters related to the output file. First, for a Two Range style, the user can set an offset distance to place the locations slightly off the street to the left or right. A setback from the street endpoints may also be specified, so that no point ends up exactly at the end of a street. Finally, in the situation when an address matches more than one street with the same score, marking the check box dictates that the address will be matched to the first street found.

Output fields

The user can request additional fields in the output file for more information about the locations. These fields include the x-y coordinates of the point, the reference data ID (the feature ID of the street it was matched to), the standardized address, and the percent distance along the street where the point was placed.

The reference data

The reference layer provides the key link that allows the interpretation of an address to a more universal coordinate system—the Cartesian x-y coordinates of a map projection. It does this by linking the specific attributes of a spatial feature stored in a table, such as the street name or address range, to the x-y coordinates stored for the feature. As such, the validity of the attributes is of prime importance in making this translation.

The requirements for reference data vary widely with the type of address locator being used. A reference layer might be as simple as a point layer of cities with states or zip codes in the attribute table. Street address geocoding probably has the most stringent requirements because it needs address ranges for one or both sides of a street. Such information is tedious to gather, and requires significant preparation and maintenance time to ensure that it remains up to date and has minimal errors. U.S. Census data, or TIGER data, is the most common starting point for assembling reference layers, although it requires a fair amount of conversion, assembly, and error checking before it is ready to use. Reference data are also available from a number of commercial vendors, with considerable variability in completeness, quality, and price.

Adding x-y coordinates

Another useful GIS function involves taking *x-y* coordinates from a table and converting them to locations on a map. The *x-y* coordinates must be given in some real-world coordinate system, such as latitude-longitude or UTM meters. The coordinate system details must also be known, including the geographic coordinate system and datum, and the projection. Global positioning systems (GPS) frequently provide this type of data. A GPS is a unit which receives signals from four or more special satellites and uses the information to calculate its position. Other examples of *x-y* data sources might include benchmarks, surveyed points, or locations measured on a map. In any event, the procedure for adding the points is identical.

The layer added from *x-y* points is called an **event layer.** This terminology comes from considering each entry in the table to be an event, such as an earthquake or a traffic accident. Not all tables contain events, of course, but the terminology remains. An event layer does not constitute an actual shapefile or feature class; it is merely the combination of the table and the located points on the map.

Event layers appear similar to point feature classes, but they do not constitute a feature class in the truest sense, for they remain a table of objects represented by spatial points. Some geoprocessing functions will not accept event layers as input. In many cases users will want to export the event layer to a shapefile or geodatabase features class for use in a tool, or for permanent storage. When exported, all the attribute fields in the original table become attributes of the feature class.

Summary

➢ Geocoding converts addresses in a table to points on a map, using reference map layers which contain information about street names and address ranges.

➢ The process breaks down addresses into standard components, compares these to the attributes in the reference layer, and scores how closely the components match possible candidates in the reference layer.

➢ Once a match is found, the address location is placed along the street at a distance proportional to the address value between the street address ranges.

➢ Geocoding styles vary on a number of factors, such as whether zip codes are used, and whether the address ranges are given for the street as a whole (One Range) or for both the left and right sides of the street (Two Range).

➢ Before geocoding, you must create an address locator by choosing a reference layer and setting the desired coding options.

➢ The process proceeds first by matching as many addresses as possible in batch mode. Then the user may interactively match single addresses or change the geocoding options and try for another batch match. The process may be repeated until as many addresses as possible have been matched.

➢ Event themes are sets of point locations created from tables containing x-y coordinates. They may be exported to shapefiles or geodatabases for permanent storage.

TIP: Under certain network configurations, users may not be able to create address locators due to a problem in the ArcGIS software. For a workaround, consult the section in the Skills Reference of this chapter titled *Solving Network Problems*. The workaround must be implemented by the system administrator.

TIP: To edit a reference layer, three steps are required to implement the changes. First, use ArcCatalog to delete any address locators based on that layer. Second, use Windows Explorer to delete the *.aig file in the same directory as the reference layer (e.g., if the reference layer is streets.shp, then delete the streets.aig file). Finally, rebuild the address locator. The *.aig file is a geocoding index created when the locator was built. After the reference layer is edited, the index is out of date. Deleting it ensures that a new one will be made when the address locator is created again.

TIP: Geocoding occasionally has some odd effects with editing routines. If editing problems start to occur after a geocoding session, save the work, exit ArcMap, and start the program again.

Chapter Review Questions

You may need to consult the Skills Reference to answer some of these questions.

1. What two data objects are required for geocoding?

2. Describe the difference between One Range and Two Range geocoding.

3. If the 1200 block of Maple Street is 1000 ft long, how many feet from the 1200 corner would the address 1275 Maple be located during geocoding?

4. What is a street type? Give some examples.

5. Which components are present in this address: 1298 Bear Valley Road NW?

6. How does batch mode matching differ from interactive matching?

7. What is an address locator, and what does it include?

8. What is an alias table and what is its purpose?

9. True or False? Once a geocoding result is completed, you can no longer change the locations or rematch additional addresses.

10. If the address table coded road intersections as Maple St – Elm St, what step would you need to take to correctly locate intersection addresses?

Mastering the Skills

Teaching Tutorial

The following examples provide step-by-step instructions for doing basic tasks and solving basic problems in ArcGIS. The steps you need to do are highlighted with an arrow ➔; follow them carefully. Click on the video number in the VideoIndex to view a demonstration of the steps.

➔ Start ArcMap and open ex_10.mxd in the MapDocuments folder.

➔ Use Save As to rename the document, and remember to save frequently as you work.

Setting up an address locator

For our first example, we will geocode some restaurants in Rapid City. The addresses were obtained from the telephone book and typed into a .dbf file for geocoding. However, before we can start, we need to decide what style of geocoding we will use and create an address locator.

1➔ Open the Streets attribute table and examine the fields.

1. For each of the following address components, write down the name of the field in the table which contains that information. If there is no field with that information, write None.

   Prefix direction   _____

   Prefix type   _____

   Street numbers   _____ and _____

   Street name   _____

   Street type   _____

   Street direction   _____

2. Based on the fields just listed, which geocoding style is best to use? _____

   1➔ Close the Streets attribute table.

   1➔ Open the restaurants table and examine the fields and records.

3. Which field contains the restaurant addresses? _____

4. Are there any intersections for addresses? _____ What connector symbol is used? _____

   1➔ Close the restaurants table.

   1➔ Open the Properties of the Streets layer and examine the Source tab contents.

5. What is the source layer for Streets and where is it located? _____

6. Is the reference layer a coverage, shapefile, or geodatabase feature class? _____

Now we have all the information we need to set up an address locator using the Streets as the reference layer.

362

→ Start ArcCatalog.

2→ Locate the Address Locator entry in the catalog tree and click the plus to expand it.

2→ Double-click on the entry to Create New Address Locator.

2→ For the geocoding style, choose US One Range (File) and click OK.

2→ Name the address locator "RapidCity".

2→ Click the Browse button and navigate to the mgisdata\Rapidcity folder. Select the rds_clp.shp shapefile and click Add.

Next, we specify the fields to be used for the address components. Note that the required fields are in boldface type.

3→ Choose the FRADD field for the House From component, and TOADD for the House To component.

3→ For the Prefix Direction, choose PREFIX. For the Prefix Type, choose <None>.

3→ For the Street Name choose STREET. For the Street Type choose SUFFIX. For the Suffix Direction choose SUFFIX2.

3→ Examine the geocoding options on the right side of the window.

The default geocoding options are fine for our purposes, so we are finished setting up the locator.

3→ Click OK to create the locator.

→ Close ArcCatalog and return to the ex_10.mxd document in ArcMap.

TIP: Under certain network configurations, users may not be able to create address locators due to a problem in the ArcGIS software. For a workaround, consult the section in the Skills Reference of this chapter titled *Solving Network Problems*. The workaround must be implemented by your system administrator.

Geocoding

It is time to geocode those restaurants!

4→ Choose Tools > Geocoding > Geocode Addresses from the main menu.

4→ Click the Add button because the locator you created is not listed yet.

4→ Click on the drop-down box to locate and click the Address Locators folder.

4→ Click the RapidCity address locator (note that its name is prefixed by your user name).

4→ Click Add to finish adding the locator.

4→ Click the newly added locator to highlight it, and click OK.

The geocoding window appears.

5→ Choose restaurants as the address table.

5→ Select ADDRESS as the address field.

5➡ Specify the name and location of the output shapefile. Put it in the Rapidcity folder and name it "rceats.shp".

5➡ Click OK to begin geocoding.

5➡ Examine the results in the Review/Rematch window.

7. How many restaurants were matched with a score of 80 or better? _____

8. How many restaurants are unmatched? _____

At this point you have several ways to proceed. You could change the geocoding options and try to rematch. However, it is usually instructive to examine the unmatched addresses in interactive mode to find out why they won't match.

6➡ Under the Rematch Criteria, fill in the button to Unmatched Addresses.

6➡ Click the Match Interactively button.

6➡ Examine the six unmatched records (Fig. 10.9).

Notice that two of these entries contain no addresses, only the place name "Rushmore Mall." An alias table can help match place names.

Click here to select

FID	pe	Status	Score	Side	
	Point	U	0		I-90 & Haines Ave
3	Point	U	0		RUSHMORE MALL
9	Point	U	0		720 LaCrossta St
33	Point	U	0		RUSHMORE MALL
45	Point	U	0		5th & St Joseph
63	Point	U	0		312 East Blvd N

Fig. 10.9. The unmatched restaurants

7➡ Click the Geocoding Options button at the bottom left of the window.

7➡ Click the Place Name Alias Table button. The Alias Table window appears.

7➡ Click the Browse button and navigate to the mgisdata\Rapidcity folder. Select the file named rc_alias.dbf and add it.

7➡ Choose ALIAS as the Alias field, and ADDRESS as the Address field.

7➡ Click OK in the Alias Table window, and again in the Geocoding Options window.

7➡ Click in the cell to the left of one of the Rushmore Mall entries.

Notice that several match candidates appear in the window now. The highest score is listed first, N Maple Ave, with a score of 100. It would be easy to match these two mall addresses interactively, but in many cases the addition of the alias table could have fixed dozens of addresses. So we will proceed with a method for the general case, and perform another batch match with the alias table present.

8➡ Click Close to close the Interactive Match window. The Review/Rematch window appears again.

8➡ Click the Match Automatically button. Watch the unmatched records go from 6 to 4.

8➡ Click Match Interactively again to continue matching the other addresses.

9➡ Click the address that reads 720 LaCrossta St.

No candidates appear for this address. However, we can try relaxing the spelling sensitivity to see if we can find any candidates.

9➔ Click the Geocoding Options button in the lower left corner.

9➔ Use the slider bar to set the spelling sensitivity to 50. Click OK.

Now a candidate appears, the 700 block of N LaCrosse St. With a little consideration, you decide that LaCrossta is probably a misspelling of LaCrosse, and that these two addresses match. Of course, familiarity with the city can help a great deal when deciding about matches.

9➔ Click the LaCrosse St candidate to select it (Fig. 10.9).

9➔ Click the Match button.

Notice that after clicking Match, the U (unmatched) in the address listing changes to an M (matched), and the match score is also displayed. If you decide later that this match is not a good one, click the Unmatch button to unmatch the location. We will leave the spelling sensitivity at 50 for the rest of this matching session to facilitate finding candidates.

Let's try matching another one.

10➔ Click on the address 312 East Blvd N. No candidates appear.

10➔ Examine the standardized address.

10➔ Click the Modify button to examine the standardized address in more detail.

An examination of the standardized address clarifies the problem. ArcMap misinterpreted the street name, EAST, as the prefix direction, BLVD as the prefix type (which we are not using), and N as the street name. To fix the problem, edit the standardization.

10➔ In the Edit Standardization window, delete the E in the prefix direction box and also delete the BLVD from the prefix type box.

10➔ Type "EAST" in the box for the street name and "BLVD" in the box for street type.

10➔ Type "N" in the box for the suffix direction.

10➔ Press the Enter key.

Now two candidates appear. The first one is a perfect match with a score of 100.

10➔ Click the first candidate with the 100 score, and click the Match button. Close the Edit Standardization box.

Next we will try to match the two intersections.

11➔ Click the intersection address 5ᵗʰ and St Joseph to select it.

11➔ Click the Modify button to see the standardization. Note that this time there are entries for two streets, not one.

11➔ Edit the standardization until it looks like Figure 10.10.

11➔ Choose the candidate with the highest score and match it. Close the Edit Standardization box.

TIP: Occasionally the standardization goes blank in the place next to the Modify button. When this happens, no candidates can be found. To fix it, close the Edit Standardization box. Try clicking on another address, then click the original address again, until the standardization appears. Then click Modify again and proceed normally.

As you edited the boxes in the standardization, the candidates started to appear and rearranged themselves as more data were entered. A perfect match appeared after all of the boxes were filled. Now we do the last unmatched address.

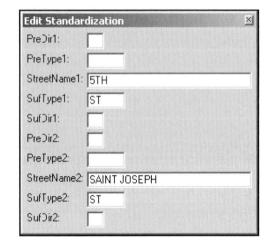

12➔ Click the intersection address I-90 & Haines Ave. No candidates appear.

12➔ Click Modify and edit the standardization.

Haines Ave appears correctly, but the I ended up in the wrong box.

12➔ Put "I-90" in the single box for StreetName1.

Fig. 10.10. Editing an intersection

Still no candidates. Sometimes a little ingenuity is required in thinking of alternatives.

12➔ Change "I-90" to "INTERSTATE 90". Still no candidates.

Let's try *really* lowering the spelling sensitivity in a last-ditch effort to find a match.

12➔ Click the Geocoding Options button and change the spelling sensitivity to 10.

Aha! A candidate finally appears. The creator of this reference layer used INTSTE as an abbreviation for INTERSTATE, a rather odd choice. Changing the reference layer attribute to INTERSTATE would ensure easier geocoding in the future.

TIP: To edit a reference layer, three steps are required to implement the changes. First, use ArcCatalog to delete any address locators based on that layer. Second, use Windows Explorer to delete the *.aig file in the same directory as the reference layer (e.g., if the reference layer is streets.shp, then delete the streets.aig file). Finally, rebuild the address locator. The *.aig file is a geocoding index created when the locator was built. After the reference layer is edited, the index is out of date. Deleting it ensures that a new one will be made when the address locator is created again.

12➜ Match the final record with the candidate and close the Edit Standardization box.

12➜ Click Close to finish interactive matching.

➜ Examine the Review/Rematch window once again.

Reviewing tentative matches

Now the window reads 85 good matches, and 11 matches with scores < 80. Just for information, let's look at the matched addresses with the lower scores, to make sure that the matches look reasonable, and to see what kinds of issues lower the match score.

13➜ Change the Rematch Criteria button to Addresses with score < 80.

13➜ Click the Match Interactively button.

13➜ Click the first address, 535 Mountain View Dr, to examine its candidates.

Note that the address has a T for tentative, instead of an M for matched. As you view the candidates, you see that the reference layer contains Mountain View Rd instead of Mountain View Dr. This is a reasonable guess, as people often mix up Rd and Dr in addresses. However, be careful about assuming this in every case. Harney Blvd and Harney Ct probably don't match, because a boulevard is generally larger and busier than a quiet residential court, and a user is less likely to have typed in this mistake. In such a case you might want to unmatch the record.

The second address was previously matched by hand, the LACROSSTA street typing error. The third address, 1803 LaCrosse St., has the prefix direction missing.

13➜ Examine the remaining addresses and candidates one by one.

9. What is wrong with each of the following addresses?

    1301 Omaha St   _____

    710 Meridian Ln   _____

    2250 Haines   _____

None of the tentative matches seem to have any major reason to indicate that they are not valid, so finish matching.

13➜ Click Close and examine the Review/Rematch window again.

13➜ Click Done.

The Review/Rematch window closes and reveals the shapefile of Rapid City restaurants. Notice the concentration of restaurants along certain streets and in certain parts of town, as one might expect.

To better investigate the results, give the layer a spatial index and turn on Map Tips.

➜ Save your map document after making sure you have given it a different name.

➜ Close ArcMap and start ArcCatalog.

14➜ Locate rceats.shp in the mgisdata\Rapidcity folder where you put it, and open its Properties.

14➜ Click the Indexes tab, and click Add to add a spatial index. Click OK.

14➜ Close ArcCatalog and open your geocoding map document again.

15➜ Open the Properties for rceats.shp.

15➜ Click the Display tab, and turn on Map Tips.

15➜ Click the Fields tab, and set the Primary Display Field to RESTAURANT. Click OK to close the Properties window.

15➜ Use the cursor to hover over several restaurants and see their names.

➜ If desired, zoom in closer to a part of the map and examine the locations in more detail. Return to the original extent when finished.

➜ Save the changes to the document.

Geocoding U.S. cities

Next we will practice our geocoding skills by matching a table of precipitation data for U.S. cities to our shapefile of U.S. cities. When finished, we'll be able to make a map of monthly normal precipitation totals for the United States. As before, we must create the address locator as our first step.

10. Which style of geocoding will we use this time? _____

16➜ Start ArcCatalog and navigate to the mgisdata\Usa folder.

16➜ Use the Preview tab to examine the cities.shp attribute table.

11. Which field will we use for the city name? _____ Which field do we use for the state? _____ (**Hint:** look at the far right of the table.)

17➜ Locate the Address Locators entry in the ArcCatalog tree.

17➜ Double-click the entry to Create New Address Locator.

17➜ Choose the US Cities with State (File) style. Click OK.

17➜ Name the locator "US_City_State".

17➜ Click the Browse button and specify the cities.shp file in the mgisdata\Usa folder as the reference layer.

17➜ Specify the CITY_NAME field for the Cities entry, and the ST_ABBRV field for the State.

17➜ Click OK to create the locator.

➜ Close ArcCatalog.

12. Why is it better to use the state abbreviation than the state name? _____

Now let's prepare to do our geocoding and examine the table we'll be using for the "addresses."

18➜ Return to ArcMap and your geocoding map document with the restaurants.

18➜ Insert a new data frame and name it "US Weather".

18➜ Add the cities and states shapefiles from the mgisdata\Usa folder to the new data frame. Also add the table normal_precip.dbf.

19➜ Change the data frame coordinate system to USA Contiguous Albers Equal Area Conic.

19➜ Zoom into the conterminous United States.

19➜ Open the normal_precip table and examine the fields. The precipitation values are given in inches for each month, and a yearly total is also given.

19➜ Close the normal_precip table.

All right, let's give it a try.

20➜ Right-click the normal_precip table and choose Geocode Addresses.

20➜ In the Choose an address locator window, click the Add button and add the US_City_State locator just created. Click Add and then OK.

20➜ In the Geocode Addresses window, specify CITY for the City field and STATE for the State abbreviation field.

20➜ Click the Browse button and specify the location and name of the output shapefile. Put it in the mgisdata\Usa folder and call it "cityprecip.shp".

20➜ Click OK to start geocoding.

20➜ Examine the Review/Rematch window.

13. How many matched records are there? _____ How many unmatched? _____

That is not a bad effort. Let's take a look at the unmatched records to see if we can get more cities matched.

21➜ Fill the button for Unmatched Addresses and click Match Interactively.

21➜ Choose the first city, ANNETTE, AK and see if any candidates appear. No!

21➜ Click the Geocoding Options button and reduce the spelling sensitivity to 10. Click OK. Anchorage, AK appears, but that is clearly not a good match.

Perhaps some of these cities do not appear in the reference layer. Write down the names of the first five cities: _____

22➜ Click Close to stop interactive matching, and Done to close the Review/Rematch window.

22➜ Use the Find button to search the CITY_NAME attribute in the Cities layer for the first five cities you wrote down. Close the Find window when finished.

Apparently our guess was correct, that these cities are missing from the reference layer. However, this is not true of all the cities. Let us go back and examine the geocoding result again, making use of our ability to review and rematch a finished geocoding result.

23➔ Choose Tools > Geocoding > Review/Rematch Addresses > Geocoding Result: cityprecip from the main menu bar.

23➔ Click Yes to open an editing session.

23➔ Click Match Interactively.

23➔ Scroll down the list to number 35, Eureka, CA. Click to select it. A candidate appears and it looks like a good one. Match it.

14. Why didn't this city get matched? (Hint: Look carefully at the end of the state abbreviation. Click Modify if needed, to get a better look.) _____

Even that small typographical mistake was enough to prevent a match! Let us examine each of the unmatched addresses in turn to see if any more can be matched. However, none of the cities from 1-34 are matchable, so simply proceed on from Eureka. First, however, relax the spelling sensitivity to facilitate finding candidates.

It is helpful to know that the cities with the precipitation data come from airports which sometimes are named for multiple cities. For example, the next matchable address is Lexington, KY, which can be matched to Lexington-Fayette. Some have multiple entries; for example, you can match three different stations to New York City (two airports and Central Park).

24➔ Set the spelling sensitivity to 20.

24➔ Click to select the next city, Mt. Shasta. No candidates.

24➔ Keep clicking each address in turn until you find some candidates. If any of them look like good matches, match them.

24➔ When you are finished, close the Interactive Match window.

24➔ Examine the results in the Review/Rematch window.

When finished, you should have about 208 good matches, 15 tentative matches, and 50 unmatched addresses. Once matched, the data can be used to analyze the distribution of rainfall across the United States.

25➔ Click Done to close the Review/Rematch window.

25➔ Turn off the Cities layer to see the precipitation locations better.

25➔ Create a graduated symbol map showing the annual precipitation (field ANN) in symbols that increase in size with precipitation (Fig. 10.11).

15. Can you think of two reasons why the geocoding method works better to accomplish this task, as opposed to joining the precipitation data to the Cities layer based on a common CITY_NAME attribute?

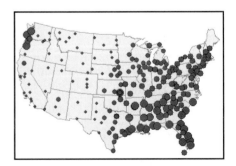

Fig. 10.11. Annual precipitation of the matched cities

Recall that during the interactive matching, you found three entries for New York City, based on the three different airports. How were the points created for these three cities?

26➜ Zoom into the southernmost tip of New York state and find the point on the western side of Long Island (Fig. 10.12).

26➜ Use the Identify button to click on the point.

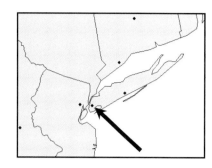

Notice that three entries come up in the Identify window. Click on each entry in turn, and note that each of the three airports in the precipitation table was geocoded as its own point. The three points lie directly on top of each other, however, and appear as one point.

Fig. 10.12. Click the geocoded point for New York City.

26➜ Close the Identify window and zoom back to the previous extent of the conterminous United States.

We have accomplished our task, but let us briefly see some applications of the data we have created. If you have Spatial Analyst, do the following steps. If not, skip down to step 30.

27➜ Choose Tools > Extensions from the main menu bar.

27➜ Turn on the Spatial Analyst extension and open the Spatial Analyst toolbar.

28➜ Choose Spatial Analyst > Options.

28➜ Under the General tab, set the Mask layer to the States shapefile, and choose to save in the same coordinate system as the data frame.

28➜ Under the Extent tab, choose Same As Display.

28➜ Under the Cell Size tab, choose As Specified Below, and enter "10000" (meters) in the cell size box. Press the tab key to enter the value and see if this gives a reasonably sized grid. If the number of rows or columns is greater than 1000, increase the cell size.

28➜ Click OK to close the Options window and keep these settings.

29➜ Choose Spatial Analyst > Interpolate to Raster > Inverse Distance Weighted from the Spatial Analyst menu.

29➜ Specify the Geocoding result as the input points and ANN as the Z value field. Leave the defaults on everything else. Click OK.

29➜ Turn off the States layer to view the raster result.

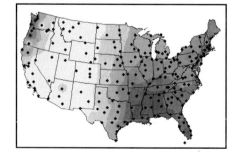

30➜ If you don't have Spatial Analyst, load the grid ann_precip_in from the folder mgisdata\MapDocuments\Results.

Fig. 10.13. Rainfall grid interpolated from city precipitation data

30➜ Create a graduated color map of the grid (Fig. 10.13).

With this grid, you could use a zonal statistics function in Spatial Analyst to find the average precipitation for each state, or for each county.

TIP: Geocoding occasionally has some odd effects with editing routines. If you have been geocoding and you start to experience problems with editing, save your work, exit ArcMap, and start the program again. We will do this now to ensure that the rest of the tutorial session works well.

→ Save the map document.

→ Exit ArcMap. If you are asked whether to save your edits, do so.

Adding x-y data

As our final example, we will use the Add XY Data function to create points from latitude and longitude values in a table. In the Sdakota folder we have a table containing climate stations for the state. These stations were downloaded from a website of the National Climatic Data Center. They come from a cooperative network of stations, and so are called the COOP stations. Each climate station has a latitude and longitude. The coordinates are degrees latitude/longitude, in datum NAD 1983, like the rest of our South Dakota data.

→ Start ArcMap again and open the ex_10mine.mxd document.

31→ Insert a new data frame and name it "South Dakota".

31→ Add the sd_counties and the table coop_sta.dbf from the mgisdata\Sdakota folder.

31→ Open the coop_sta table and examine its fields.

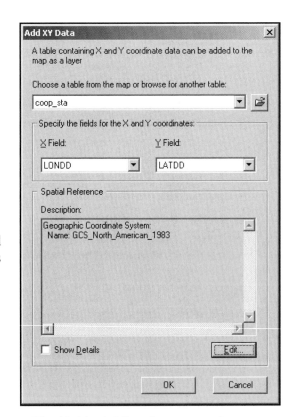

Notice that there are four fields with lat-lon values. The first two fields, LAT and LON, come from the original downloaded text file, and are given as single numbers representing degrees and minutes; for example 4526 represents 45 degrees, 26 minutes. ArcMap cannot interpret these values correctly as geographic map coordinates, so these values have been recalculated to decimal degrees as shown in the LATDD and LONDD fields. The calculations were done in several steps using Microsoft™ Excel®, but only the final columns were retained. The spreadsheet was then saved as a .dbf file so it could be opened in ArcMap. We are ready to create a point map of the stations.

Fig. 10.14. Adding the coop stations from *x-y* data

32→ Close the coop_sta table.

32→ Choose Tools > Add XY Data from the main menu bar.

32→ Fill out the fields as shown in Figure 10.14. Set the X field to LONDD, the Y field to LATDD, and the coordinate system to GCS NAD 1983.

32→ Click OK to add the points.

When *x-y* data are added to a map, they take the form of an event layer. Each pair of *x-y* coordinates is referred to as an event. An event layer only exists inside the map document in which it was created. To save it permanently and make it available to other map documents, export the event layer as a shapefile or geodatabase feature class.

33➜ Right-click the coop_sta event layer and choose Data > Export Data.

33➜ Specify the name and location of the output shapefile (or geodatabase feature class). Call it "coopstat.shp" and put it in the Sdakota folder.

33➜ Click OK.

33➜ Choose Yes to add the new shapefile to the document.

33➜ Right-click the original event layer and Remove it from the map document.

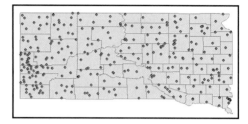

Now the coop stations shapefile is ready for use to analyze South Dakota's climate (Fig. 10.15).

Fig. 10.15. South Dakota Coop climate stations

This is the end of the tutorial.

➜ Save your map document.

➜ Exit ArcMap.

Exercises

1. In the mgisdata\Rapidcity folder, locate the file gas_loc.dbf. Geocode these gas stations using the rds_clp shapefile in the same folder. **Capture** a map showing the gas stations.

2. The gas_loc table also contains latitude-longitude coordinates in NAD27. Create another layer of gas stations based on these coordinates. **Capture** this map also. Are the two layers the same?

3. Convert the lat-lon stations to a shapefile in UTM 13, and determine how many of the lat-lon gas stations are more than 100 meters from the geocoded positions. (You can assume that the closest station is the same one, although this might not in fact be the case.) What is the minimum, maximum, and average distance between points in the two data sets?

4. In the mgisdata\Usa folder, you will find a file called customers.dbf. These data come from the test marketing of a catalog for natural landscaping to minimize watering of plants (xeriscaping). Create a map showing the location of the customers. **Capture** your map.

Challenge Problem

Search the Internet for a data set which interests you, is geocodable, and contains at least 50 data points. Then geocode it. With your answer, include the source of the data, the type of locator used, and the reference layer used.

Skills Reference

Creating an address locator

1. Examine the reference layer to ensure that all the fields required for the geocoding style are present, and identify the names of the fields.

2. Start ArcCatalog, if necessary.

3. In the folder tree, navigate to the bottom of the list and click the plus next to Address Locators to expand it.

4. Double-click on Create New Address Locator.

5. Choose the style of locator (Fig. 10.16). Choose a (File) locator if the reference layer is a shapefile or a (GDB) locator if it is a geodatabase. Click OK.

6. Name the address locator.

7. Specify the name and location of the reference layer.

8. Set the required fields (bold) and optional fields to the appropriate fields in the attribute file of the reference layer.

9. If the field name containing the addresses in the address table is different than the ones shown, add it to the list.

10. Choose a place name alias table if one is available.

11. Change the matching options if desired.

12. Enter a different set of connectors if needed.

13. Set the side offset, if using Two Range, and the end offset.

14. Check the optional fields to add to the output file.

15. Click OK to create the locator. It will be saved with its name appended to your user name, e.g., John Smith.AddressLocator, and will appear in ArcCatalog as an address locator.

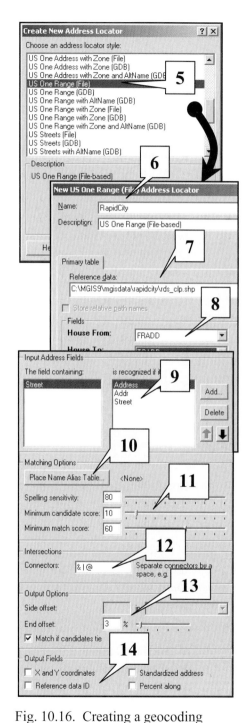

Fig. 10.16. Creating a geocoding service

Geocoding addresses

1. Start ArcMap, if necessary, and add the address table and the reference layer (optional) to the map.

2. Choose Tools > Geocoding > Geocode Addresses from the main menu bar.

3. If the desired locator does not appear in the menu, click Add (Fig. 10.17).

4. Click the Look in drop-down box to find the Address Locators folder. Click it, and choose the locator. Click Add to close the menu and add the locator to the list.

5. Click the desired locator to highlight it, and click OK.

6. Specify the table containing the addresses (Fig. 10.18).

7. Specify the field containing the addresses to be matched.

8. Specify the name and location of the output file.

9. Click the Geocoding Options button to change the spelling sensitivity, minimum candidate score, or minimum match score.

10. Click OK to begin matching.

11. When the matching is complete, the Review/Rematch menu will appear with the results of the batch match. See the next section titled "Reviewing and Rematching Addresses" to continue.

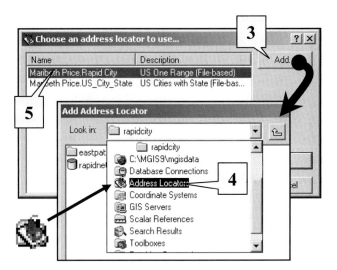

Fig. 10.17. Adding an address locator

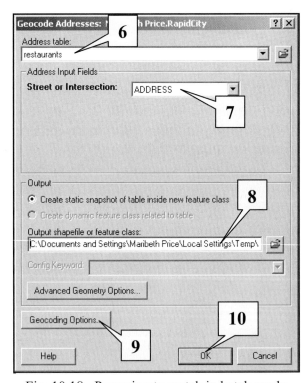

Fig. 10.18. Preparing to match in batch mode

Reviewing and rematching addresses

1. Examine the Review/Rematch window for the match statistics, and decide whether to match addresses interactively or to change the geocoding options and try another batch match.

2. Choose which addresses should be rematched (Fig. 10.19). You can rematch all addresses, or only unmatched addresses, or addresses below a certain score.

3. Click the Geocoding Options button to change the spelling sensitivity, minimum candidate score, or minimum match score.

4. To rematch automatically in batch mode, click Match Automatically.

5. OR to review addresses individually and match them interactively, choose Match Interactively. Continue with the instructions in the following section on "Interactive Matching."

6. OR click Done when satisfied with the matching results.

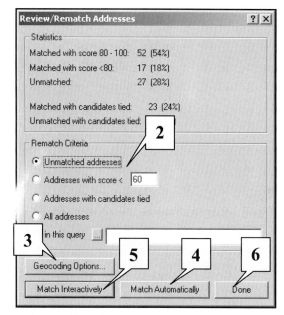

Fig. 10.19. The Review/Rematch window

Rematching a geocoding result

You can reopen a completed matching session to rematch more addresses.

1. It is best to terminate any editing before rematching addresses. Choose Editing > Stop Editing from the Editor menu and save any changes if needed.

2. Choose Tools > Geocoding > Review/Rematch Address > and choose the geocoding result to rematch.

3. Click Yes to open the workspace for editing.

4. The Review/Rematch window will appear. Proceed as described in the section titled "Reviewing and Rematching Addresses."

Interactive matching

1. Choose an address to rematch from the list by clicking the box to its left (Fig. 10.20).

2. If a list of candidates appears in the bottom box, click the closest match to choose it, and click the Match button. (The score for each candidate is shown to help identify the closest match.) The address will be matched, and the U in the upper list will change to an M. The match score will also be shown.

3. If no candidates appear, try clicking the Geocoding Options button to modify the spelling sensitivity or the minimum candidate score.

4. If still no candidates appear, click the Modify button next to the standardized address to see if the address is being interpreted correctly.

5. Change the address standardization if needed. Press Enter after typing each change to make it take effect and allow new candidates to appear.

6. If still no candidates appear, the address is probably unmatchable. It may be incorrect, or the reference layer may be incomplete or incorrect. Click another address and begin again.

7. If you are trying to decide between several candidates, click the Zoom to Candidates to view them on the map. Click Original Extent to go back to the previous extent.

8. When you are finished matching, click Close to return to the Review/Rematch window.

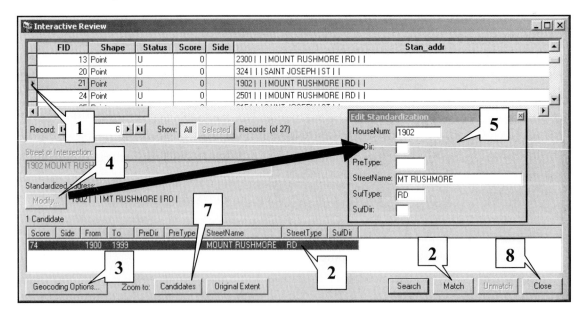

Fig. 10.20. Matching addresses interactively

Adding x-y data

1. Make sure the table has fields for x and y coordinates, and that you know the coordinate system of the coordinates.

2. In ArcMap, choose Tools > Add XY Data from the main menu bar.

3. Select the table containing the x-y coordinates (Fig. 10.21).

4. Specify which fields contain the x-y coordinates. Recall that longitude is the x coordinate and latitude is the y coordinate!

5. Click the Edit button to specify the coordinate system of the points in the table.

6. Click OK to create the points.

To save the event layer permanently as a shapefile or feature class in a geodatabase, do the following steps:

7. Right-click the new event layer and choose Data > Export Data from the context menu.

8. Choose to export All features or Selected features (Fig. 10.22).

9. Choose to use the original coordinate system or the coordinate system of the data frame.

10. Specify the name and location of the output shapefile or feature class.

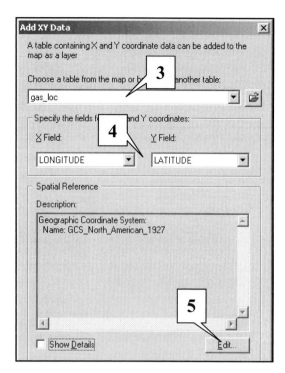

Fig. 10.21. Adding x-y data to a map

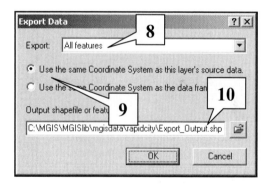

Fig. 10.22. Exporting the event layer as a shapefile

Solving network problems

The author has found that under certain network configurations in ArcGIS, the users cannot create address locators. The error message states that the user does not have appropriate permissions, because the software is attempting to save the locator to the profile of the user who originally installed the software. This section describes a workaround for this problem by modifying a registry key to write the address locator to a different location. ESRI has been notified of this problem.

The *system administrator* should perform this fix.

1. Log into the workstation as the person who installed the software.

2. Create a directory to store the address locators, such as C:\geocodinglocator. If it is a local drive, all users should have write access to it. Note that the address locators of ALL users will be added to this directory. Alternatively, the network home folder for each user may be specified using a system variable.

3. Run Regedit and navigate to the location

HKEY_CURRENT_USER\Software\ESRI\CoreRuntime\Locator\Settings

4. There is a key called LocatorDirectory; modify it to point to the address locators directory you chose.

5. Copy the user profile from the user who installed the software to the Default User profile.

6. Delete the user profiles of all users who need access to geocoding. When they log in again, the default user profile will be copied to their personal profile, incorporating the changes to the registry. **Note that all profile customizations for individual users will be lost.**

Chapter 11. Basic Editing in ArcMap

Mastering the Concepts

Objectives

➢ Understanding the basic editing process

➢ Using snapping to ensure topological integrity of features

➢ Adding features to map layers

➢ Using the sketch tools and context menus to precisely position features

➢ Entering and editing attribute data

Concepts

Editing in ArcMap provides the ability to modify and update existing layers of data, or to create entirely new layers. For example, if a housing subdivision is added to a city, then the new roads must be added to the city's roads layer. Likewise, new parcels, sewer lines, and other infrastructure need to be added to the city database to ensure that it is up to date. Or imagine that the city council decides to create a new layer showing the garbage collection zones where none had existed before.

Editing overview

ArcMap has a special Editor toolbar that collects the editing functions in a convenient package. The editing process begins by opening this toolbar. Next you must turn on the editing functions and identify which folders and layers will be available for editing. Setting several options, such as snapping, makes editing easier. Then you are ready to actually start adding or changing features.

The Editor toolbar provides a variety of tools for creating features, such as adding a simple polygon, finding the intersection of two lines, or adding one line parallel to another. Existing features may be modified several ways, such as relocating their boundaries, splitting them in half, or moving them from place to place. Attributes may be entered as new features are created. Edits may be saved during an editing session and again when it is finished.

The Editor toolbar

The Editor toolbar controls the process of editing (Fig. 11.1). The first button on the left shows the editing menu, used for turning editing on and off, setting options, and performing certain types of edits. The Select button selects one or more features for editing. The Sketch tool enters vertices when creating or modifying features. The Task bar controls how the Sketch tool works and performs a variety of editing functions, such as creating new features or modifying existing ones. The target window specifies which layer the edits will affect. The Split and Rotate buttons provide easy access to common editing commands. The Topology Edit tool is used to edit features with shared boundaries. And finally, the Attributes button launches an editing menu for the feature attributes. In the next few sections, we will examine in more detail how these editing menu elements work.

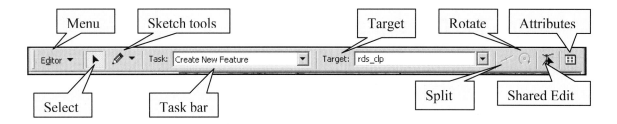

Fig. 11.1. The Editor toolbar contains many editing options and functions in one neat package.

General information about editing

What can you edit?

ArcGIS offers different levels of editing capability, according to whether you've purchased ArcView, ArcEditor, or ArcInfo. ArcView can edit shapefiles and personal geodatabases only. ArcEditor or ArcInfo can also edit SDE databases, coverages, geometric networks, and feature topology. All of these levels are accessed through the same ArcMap interface.

> **TIP:** Coverages cannot be edited in ArcMap. They have an older topological data model and cannot utilize the topology editing tools. Use the ArcEdit program in Workstation ArcInfo instead, or convert the coverages to shapefiles or geodatabases for editing.

ArcMap can edit several layers at once, as long as they are all in the same folder or geodatabase. This capability makes it easier to view and edit related layers simultaneously or to copy features from one layer to another.

Editing and coordinate systems

ArcMap can display and edit layers containing different coordinate systems. As long as the coordinate systems are correctly defined for each layer, the edits will automatically be converted to the coordinate system of the layer being edited. Consider editing a shapefile of roads stored in a geographic coordinate system (GCS) while displaying the roads with a digital orthophotoquad (DOQ) in Universal Transverse Mercator (UTM) projection. As the roads are edited, the coordinates will automatically be converted into decimal degrees before they are saved into the roads shapefile.

Performance and reliability

Although ArcMap has many useful capabilities, such as editing across coordinate systems, and editing multiple files simultaneously, one must be somewhat cautious in taking advantage of these benefits. Editing coordinate features can be a complicated process, and software does not always work perfectly under arduous conditions. If an editing scenario is demanding extensive system resources by having many files open, or by doing many projection transformations during an edit session, the system performance and reliability may suffer adversely. Moreover, because spatial data files are complex constructions of multiple files, the price of an editing glitch may be the integrity of the data. An ill-timed system glitch might not only lose recent changes, but might also corrupt the entire data set. Thus, you are encouraged to use the following procedures while editing:

➢ **Always have a backup copy of the file being edited,** stored elsewhere on the disk or on a different medium. This precaution is especially important when working on data sets that would be expensive or impossible to replace should something go wrong.

➢ Shapefiles are the simplest and most robust type of data file to edit and the hardest to corrupt. When feasible, create and enter data as shapefiles first, and then, if necessary, convert them to geodatabases or coverages later. However, geodatabases offer special capabilities for editing; if these are being used then using shapefiles is not an option.

➢ Keep the number of open files in an ArcMap editing session to a minimum. A prolonged editing session should be conducted in its own map document and include only the layers that are actually needed.

➢ Avoid editing across different coordinate systems whenever possible. On-the-fly projection takes additional system resources and can degrade editing performance. Projecting vectors is preferable to projecting rasters, however, because raster projection is more time-consuming.

➢ Save edits frequently in case of a system glitch. Every 20 to 30 minutes usually works well. ArcMap does NOT automatically save edits at regular time intervals, like some programs do.

Snapping features

When creating line features that connect to each other, such as roads, one must take care that the features connect properly. In other words, the features must have topological integrity. Two tools can be used to ensure topological integrity: snapping and integration. Snapping makes sure that features are topologically correct as they are added. Integration can correct errors after the fact, but requires an ArcEditor or ArcInfo license.

Lines which fail to connect are called **dangles.** Although you may not be able to see the gap between the lines when displaying the map, the gap will exist unless the endpoints of each line (nodes) have exactly the same coordinate values (Fig. 11.2). Even though the map may look correct, certain functions, such as tracing networks or locating intersections, will not work properly. It is impossible to intersect lines properly by simple digitizing.

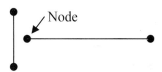

A dangle—the two lines fail to connect.

Correct topology—the horizontal line intersects the vertical one, creating three lines.

Fig. 11.2. Topological relationships between lines

The snapping option ensures that the nodes of lines and the vertices of polygons match. When snapping is turned on, it affects features which are being added or modified. If you place the cursor within a specified distance of an existing node or vertex, then the new feature is automatically snapped to the existing one—that is, the coordinates are matched at that one point (Fig. 11.3). When you have snapping turned on, you will notice the cursor jumps to an existing vertex whenever you move the mouse within the snap distance, and remains there until you move away.

The snap distance can be changed to suit your preference. However, caution must be used to specify a realistic distance. If the snap distance is too small, then you will have difficulty making features snap. If it is too large, then you may find yourself constantly snapping to objects when it isn't needed.

Three types of snapping can be used and set independently for each layer (Fig. 11.4). When snapping is turned on for a layer, it affects the features being added to every other layer, not just itself.

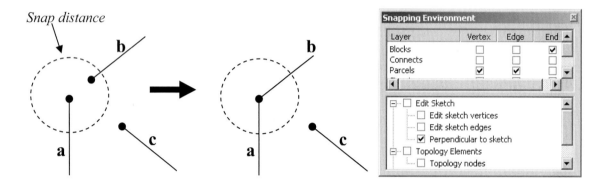

Fig. 11.3. Snapping. Line "b" snaps to "a" because it falls inside the snap distance. Line "c" remains unconnected.

Fig. 11.4. Setting the snapping environment

End snapping is the most restrictive: it only allows the new line to be snapped to the endpoints of an existing line. End snapping can ensure that new streams connect to the ends of existing streams.

Vertex snapping allows the endpoints of the new line to be snapped to any vertex in an existing line. It can ensure that adjacent polygons connect only at existing corners.

Edge snapping constrains the feature being added to meet the edge of an existing feature. In this case, the vertex being added could be placed anywhere along an existing feature. Edge snapping can ensure that a street ends exactly on another street, even if the new street falls in between two vertices of the existing one.

Figure 11.5 illustrates how the snapping effect changes depending on which type of snap has been set. The vertical line already exists in the layer, and the horizontal line is being added. The dashed circle shows the snap distance. If end snapping is turned on, then the new line will snap to the endpoint of the existing line (Fig. 11.5a). If the edge snap

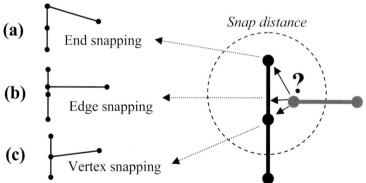

Fig. 11.5. Where will it snap? The snap type dictates where the new horizontal line will end.

is turned on, then the line can be snapped anywhere between the end and the vertex (Fig. 11.5b). If vertex snapping is set, then the new line will connect to the closest vertex (Fig. 11.5c), although if the line had been closer to the end, it would have snapped to the end.

If more than one kind of snapping is turned on, then the most inclusive one takes precedence. To ensure snapping only to ends, therefore, both edge and vertex snapping must be turned off.

Creating adjacent polygons

Two adjacent polygons should always share the same boundary. In shapefiles and geodatabases, the boundary gets stored twice, once for each of the polygons. However, the vertices should exactly match in both features. If this condition holds, the polygons are said to share a coincident boundary (Fig. 11.6). If they fail to match, gaps or overlaps are generated, which constitute topological errors and may cause problems during display or analysis.

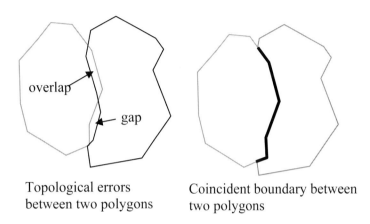

Topological errors between two polygons

Coincident boundary between two polygons

Fig. 11.6. Topological relationships between polygons

Although snapping can help create coincident boundaries between polygons, it grows annoying to digitize every boundary twice. The Editor toolbar has a special method for adding adjacent polygons. The Create New Feature task is used to enter a complete polygon with no neighbors. To add an adjacent polygon, you would set the task to AutoComplete Polygon, and digitize only the new part of the polygon. The editor then ensures that the polygon is finished and shares the same boundary as the first polygon (Fig. 11.7).

Draw first polygon with Create New Task.

Add new polygon with AutoComplete.

Two polygons with coincident boundary.

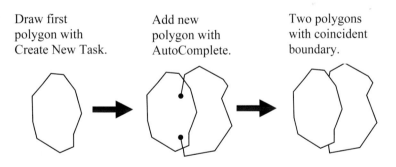

Fig. 11.7. Creating a coincident boundary between two adjacent polygons with the AutoComplete task

Editing features

Editing must be performed within an existing data layer or table. Feature classes must be first created in ArcCatalog before any features can be added to them. New empty layers cannot be created within ArcMap.

An editing session must be initiated before any changes to a file can be made. This requirement protects the user from accidentally making changes to a file without realizing it. Also, because ArcMap can only edit within one directory or one geodatabase at a time, opening the session establishes the folder or geodatabase being edited.

The Editor toolbar controls which layer is being edited. On the toolbar, a drop-down box specifies the **target layer.** Any feature created will be placed in this layer. The type of target layer also dictates the types of edits which can be made. You can create a new feature in any kind of layer (point, line, or polygon), but the AutoComplete Polygon task would be dimmed if the target layer contains points or lines.

Basic editing uses three components of the Editor toolbar: the Select tool, the Pencil tool, and the Task bar (Fig. 11.8). The Task bar specifies which editing action is taking place, such as creating a new feature, or reshaping an existing feature. The Pencil tool provides the input to the task, such as sketching the vertices of the new feature. The Select tool can specify which features will be affected by the task.

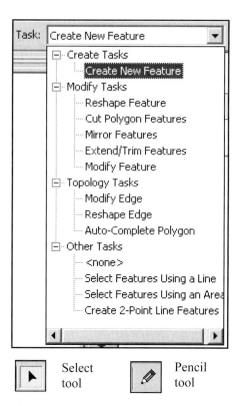

Fig. 11.8. The Select tool, Sketch tool, and Task bar work together.

For example, to create a new feature you first select the Create New Feature task, and then enter the shape of the feature using the Pencil tool. To modify a feature (change the shape defined by its vertices), you must first specify the Modify Feature task, then select the feature using the Select tool, and then move the vertices with the Pencil tool.

At any point during a sketch, you can bring up context menus to help you locate the next vertex with specific characteristics (Fig. 11.9), such as making the next segment have a specific length, or a specific bearing, or making it perpendicular or parallel to an existing line, or deleting part or all of a sketch. The context menus provide a variety of ways to enter features.

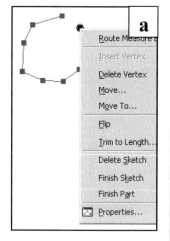

Fig. 11.9. A sketch and its context menus.
(a) The Vertex menu.
(b) The Sketch menu.

Two context menus are available. The **Vertex menu** appears when you right-click a vertex or segment of the sketch (Fig. 11.9a). This menu can add, delete, or move vertices. It can also flip and trim the sketch. The **Sketch menu** appears when you right-click any location off the sketch (Fig. 11.9b). This long menu provides detailed functions for specifying exact angles, lengths, distances, and more. Both menus allow the user to delete or finish the sketch.

About the sketch tools

The sketch tools create a preliminary shape of a feature, analogous to an artist sketching a figure lightly in pencil prior to inking in the final lines of the picture. This sketch is not actually added to the target layer until it is "finished." In Figure 11.9a, the Pencil tool has been used with the Create New Feature task to enter the vertices of the polygon. When it is complete, the user right-clicks the sketch to show the Sketch menu and chooses Finish Sketch. The sketch then becomes a regular polygon.

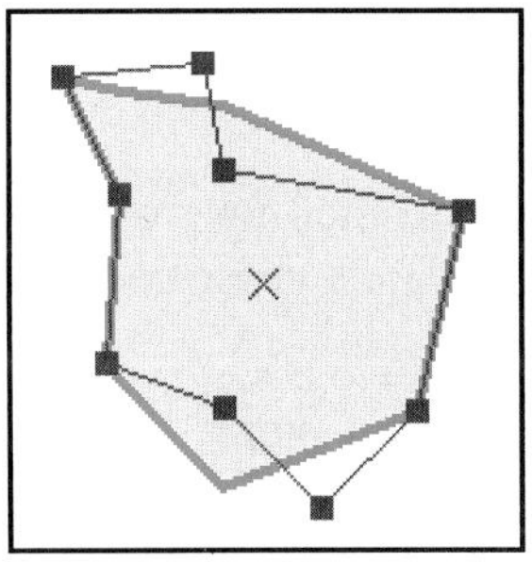

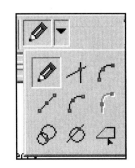

Fig. 11.10. Modifying a feature

Sketches also appear when reshaping or modifying an existing feature. Figure 11.10 shows an original feature as a shaded polygon, and the lines and square vertices show a sketch which will become the new shape. Reshaping permits the user to change vertex locations, insert new vertices, or delete old ones. The sketch converts to a feature when the user finishes making changes.

The Pencil tool is one of five different sketch tools, which can be chosen by clicking the black drop-down arrow next to the current sketch tool (Fig. 11.11). Each sketch tool has its own way to enter vertices for a sketch. The functions of the other tools are explained in Chapter 12.

Fig. 11.11. Nine sketch tools

Editing attributes

Editing features often includes editing their attributes. The user can modify attributes in two ways. In the first way edits are typed directly into an attribute table, as discussed in Chapter 5. The second method uses the Attribute Editor to display and edit the values of many features at once.

The Attribute Editor is accessed through a button on the Editor toolbar (Fig. 11.12). The left side of the editor shows the selected records; if no features or records are selected, then both areas will be blank. The right side of the editor shows the attributes of the feature highlighted on the left. In Figure 11.12, for example, the attributes are shown for the first road, St Charles St. Clicking another feature displays its attributes instead.

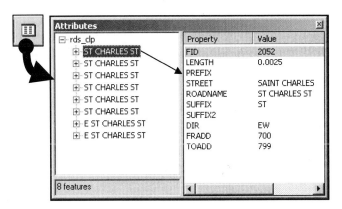

Fig. 11.12. The Attribute Editor lets you view and edit many records at once.

To edit the attributes of a single feature, you simply highlight it on the left and type the changes into the fields on the right. You can also change an attribute for ALL of the selected features, such as changing the street suffix from ST to RD for all of these streets. To edit multiple records, click the rds_clp entry at the top of the list to highlight it. The attribute fields go blank, and edits typed in the fields would be applied to every record in the list.

The field used to display the records on the left is called the primary display field, and it can be set in the Fields tab of the layer's properties. To have the streets listed by the STREET field instead of ROADNAME, simply change the primary display field. (This is the same field used for Map Tips.)

Like other types of edits performed during an edit session, attribute changes can be recalled using the Undo command. In addition, the edits must be saved before they become final.

Saving work

During an editing session, ArcMap records all of the changes as temporary files. This approach allows the user to undo edits. ArcMap offers two ways to save work permanently (Fig. 11.13).

Choosing Save Edits from the Editor toolbar during an editing session will store changes and continue with editing. Note, however, that once edits are saved, the Undo command can no longer reverse them. Users can Undo every step sequentially up until the last save. Saving edits also clears any selected features and closes the Snapping and Attribute Editor windows.

Fig. 11.13. Options for saving edits

When finished editing, choose Stop Editing from the Editor toolbar menu. You will be prompted to save any remaining edits. If you say Yes, all the recent edits are saved and the editing session is closed. If you say No, the edit session is closed and the changes are discarded.

It is important to realize that saving edits and saving a map document are two different operations. Saving edits inscribes changes to the layer files only. Changes to the map document, such as the addition of layers or changes in symbols, are NOT saved.

Conversely, saving the map document does not save edits. However, if you are closing the map document with unsaved changes to one or more layers, you will be prompted to save the edits.

Summary

➢ Editing can create or modify map features and their associated attributes, from simple changes to a few features to creating an entirely new layer.

➢ Editing sessions are initiated and controlled using the Editor toolbar.

➢ You may edit multiple layers of data, but only one geodatabase or folder may be edited at a time.

➢ Edits are stored in the same coordinate system as the layer, regardless of the coordinate system of the data frame. However, editing is streamlined when the layer and frame coordinate systems match so that no reprojection need take place.

➢ Basic editing uses the Select tool, one or more of the Sketch tools, and the Task bar. The Task bar indicates the type of edit being performed, the Select tool selects features to edit, and the Sketch tool provides the input points and segments.

➢ The Pencil tool creates preliminary figures, allowing them to be shaped and modified until the desired shape is reached. Finishing the sketch creates the new feature.

➢ Different sketch tools provide a variation of functions for placing vertices. Context menus associated with the sketch tools add even greater functionality.

➢ Snapping ensures that features connect properly with each other by automatically matching the coordinates of features within a certain distance of each other (the snap distance). Features may be snapped to endpoints, vertices, or edges.

➢ When adding adjacent polygons, you must ensure that they have coincident boundaries by using snapping or the AutoComplete task.

➢ The Attribute Editor facilitates editing the attributes of selected features.

➢ Edits may be saved during or at the end of an editing session. Saving edits clears the Undo list and the selected features.

TIP: Coverages cannot be edited in ArcMap. They have an older topological data model and cannot utilize the topology editing tools. Use the ArcEdit program in Workstation ArcInfo instead, or convert the coverages to shapefiles or geodatabases for editing.

Chapter Review Questions

You may need to consult the Skills Reference to answer some of these questions.

1. How are editing functions and options accessed in ArcMap?

2. If a data layer is in UTM coordinates, and the data frame is set to State Plane coordinates, in which coordinate system will the edits be stored?

3. True or False: Because you can Undo edits, it is not necessary to make a backup copy of any data set before editing. Explain your answer.

4. When you draw a new feature, how do you specify which layer it belongs to?

5. What is a dangle, and how are they prevented?

6. What is the purpose of the AutoComplete task? What errors does it prevent?

7. How many context menus can you access while sketching? How do you control which one you get?

8. How do you change the attributes of many records at once in the Attribute Editor?

9. What is the easiest way to create polygons with all 90-degree angles?

10. What is the difference between the Deflection and the Deflect Segment entries in the Sketch menu?

Mastering the Skills

Teaching Tutorial

The following examples provide step-by-step instructions for doing basic tasks and solving basic problems in ArcGIS. The steps you need to do are highlighted with an arrow ➔; follow them carefully. Click on the video number in the VideoIndex to view a demonstration of the steps.

> ➔ Start ArcMap and open the map document ex_11.mxd in the MapDocuments folder.
>
> ➔ Use Save As to rename the document, and remember to save frequently as you work.

This map document contains a number of layers for a neighborhood in Rapid City. These layers represent typical files that might be created for a city GIS database; they include street edges, parcels, buildings, and water connectors. You will also create a shapefile of water mains and laterals. We will practice editing features using these files. First, however, create some layers to practice in, so that mistakes won't mess up the "real" layers.

These feature classes are all stored in the eastpat folder, and you will create the new practice shapefiles in this folder as well. When creating the new files, you'll assign them the same coordinate system as the streets shapefile in the eastpat folder.

1➔ Click the icon on the main menu to open the ArcToolbox window.

1➔ In ArcToolbox, navigate to the Data Management Tools > Feature Class > Create Feature Class tool.

2➔ Click the Browse button for the Output Location. Navigate to the mgisdata\Rapidcity folder and select the eastpat folder. Click Add.

2➔ Type the Output Feature Class name "practicept".

2➔ Set the Geometry Type to POINT.

2➔ Leave the Template Feature Class blank. This field could be used to assign the field definitions and coordinate system of another layer to this new feature class.

2➔ Scroll down if necessary to locate the Spatial Reference box. Click the Browse button, and then click the Import button to import the spatial reference from an existing file.

2➔ Navigate to the mgisdata\Rapidcity\eastpat folder and select the streets.shp file. Click Add. Click OK to close the Spatial Reference Properties window.

2➔ Click OK to finish creating the practicept shapefile.

The new shapefile is automatically added to the map document. Now create two additional practice shapefiles, one for lines named practiceln, and one for polygons named practicepo. Be sure to set the correct geometry type (line or polygon) and the coordinate system for each.

1. What coordinate system is the new shapefile in? _____

> 3➔ Repeat the preceding steps to create a practice line shapefile named "practiceln". Make sure the feature geometry is set to POLYLINE.

4➔ Also create a practice polygon shapefile named "practicepo". Make sure the feature geometry is set to POLYGON.

Some simple editing functions

First we need to set up our editing session.

5➔ Close ArcToolbox.

5➔ Turn off the Orthophoto layer for now.

5➔ If the Editor toolbar is not visible, click the Editor button.

5➔ From the Editor toolbar, choose Editor > Start Editing.

The files are all from the same folder, so you will not be prompted to choose an editing space.

Points are the least complex features to edit, so we will begin with those. We don't need any snapping for these points, so we will check to make sure all snapping is off first.

6➔ Choose Editor > Snapping from the Editor toolbar.

6➔ Make sure all of the boxes are unchecked.

6➔ Close the Snapping Environment window.

7➔ Check the target layer and make sure it is set to practicept.

7➔ Check the Task bar and make sure it says Create New Feature.

7➔ Click the Pencil tool.

7➔ Click near the middle of one of the blocks defined by the Streets layer to add a point.

The point appears as soon as you click the mouse. Since points have only one "vertex," you do not need to finish the sketch or anything else. Notice that the point is selected.

8➔ Add a point to the rest of the blocks by clicking near the center of each in turn.

As always, the last point added is selected. As you add each new point, the previous selection gets unselected. Now you should have seven unselected points and one selected point (Fig. 11.14).

9➔ Open the attribute table for the practicept layer.

Notice that the final point is selected both in the map and in the table. Our next step is to add a field to this table and edit its attributes. However, we cannot add a field while a layer is being edited, so we must first stop editing and save our edits so far.

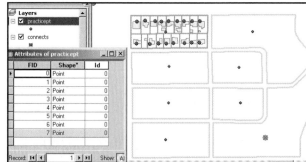

Fig. 11.14. Adding points to each block. The final point added is selected in both the map and the table.

9➜ Choose Editor > Stop Editing from the Editor toolbar.

9➜ Say Yes when prompted to save the edits.

Notice that the selection disappears when you stop editing. Now add the new field to the table.

10➜ In the table, click Options > Add Field.

10➜ Name the field "Name", set the field type to Text, and set its Length to 10. Click OK.

10➜ Resize the table so that only the fields are visible. It will be easier to see the map with the smaller table.

10➜ Move the table so that you can better see the map.

Now turn editing back on and enter the block names.

11➜ Choose Editor > Start Editing from the Editor toolbar.

11➜ Set the Selectable Layers to just the practicept layer.

11➜ Click the Select tool on the Editor toolbar.

11➜ Click on the point in the upper left block to select it.

11➜ Click in the highlighted record of the ID field in the table, and enter the number "1".

11➜ Click in the highlighted record of the Name field, and enter "BLOCK 1".

It would be nice to see our handiwork, so let's turn the labels on for practicept.

12➜ Open the properties for practicept and click the Labels tab.

12➜ Check the box to turn the labels on, and set the Label Field to Name.

12➜ Change the symbol to Arial 12 point bold.

12➜ Click Placement Properties, and fill the button to place the label on top of the point. Click OK.

12➜ Click OK again to close the Properties window.

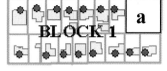

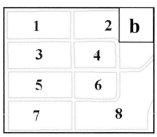

The block name should appear in the middle of the first block (Fig. 11.15a). Now add the other block numbers and names to the table.

13➜ Click to the left of the second row in the table to highlight it. The corresponding point will be highlighted in the map.

13➜ Enter the number and name of the block again, using the diagram in Figure 11.15b as a guide.

Fig. 11.15. Labeling the blocks by number

13➜ Repeat until every point is correctly labeled. You may need to click the Refresh button to make the labels appear correctly.

14➜ When finished adding the block names, close the table.

14➜ Click the practicept layer twice, and rename it "Blocks".

Now, save the edits made so far, but don't close the editing session.

14➜ Choose Editor > Save Edits from the Editor toolbar.

All of the edits added to practicept are now permanently stored in the file. Just for good measure, save the map document also.

14➜ Choose File > Save from the main menu bar.

Next we switch the target layer and start creating lines.

15➜ Change the target layer on the Editor toolbar to practiceln.
15➜ Zoom into Block 3.
15➜ Turn off the Blocks layer for now.

Whenever you start a new editing task, it is good practice to check the target layer, task, and Sketch tool to make sure they are correct for the next edits.

16➜ Make sure that the target layer is practiceln, the Task bar reads Create New Feature, and the Pencil tool is clicked.
16➜ Click somewhere on the screen to add the first vertex of a line.
16➜ Continue clicking until you have added five or six vertices.
16➜ Double-click to end the sketch and create the first line.
16➜ Add several more lines until you are comfortable with this task. (For now, do not let them cross the street edges.)

Well, that's not so hard, is it? Now let's try using some simple techniques to move, delete, rotate, and split the lines.

TIP: Many of the features you create in these exercises are simply arbitrary shapes done for fun and practice. When editing actual features, pay more attention to precise and careful entry of vertices, and try to achieve the greatest possible accuracy.

17➜ Set the Selectable Layers back to All layers.
17➜ Click on the Select tool on the Editor toolbar.
17➜ Click one of the new lines.
17➜ Press the Delete key.

TIP: Sometimes features do not disappear when you press Delete. If this happens, click the Select tool on the Editor toolbar and press Delete again. The Delete key does not work with some of the cursor tools, but it always works if the Select tool is active.

18➜ Select another of the new lines.
18➜ Place the cursor on top of the selected line in order to make small crosshairs appear.
18➜ Click and drag the line to a new location.

18➜ Move some of the other lines to get the hang of this task.

Ok, now let's try rotating the lines.

19➜ Select one of the lines.

19➜ Click the Rotate tool on the Editor toolbar.

19➜ Place the cursor near the selected line. Click and drag with a circular motion to rotate the line about the X in its middle.

19➜ Rotate it several times to get the feel of it.

20➜ Place the cursor on top of the X in the selected line until it changes to a crosshairs symbol. Then click and drag the X to a new location away from the line. Release the mouse button.

20➜ Click and drag with a circular motion again to rotate the line about its new center.

20➜ Continue playing with rotation until it feels natural.

Next we will experiment with the Split tool, which divides a line into two separate features. The Split tool can only be used on lines; it will be dimmed if the target layer contains points or polygons.

21➜ Select one of the lines (Fig. 11.16a).

21➜ Click the Split tool on the Editor toolbar.

21➜ Click near the center of the selected line to split it. The two pieces will briefly flash, and both will still be selected.

21➜ Click on one side of the split line to show that it is in fact split (Fig. 11.16b).

21➜ Move the selected line away to emphasize the split (Fig. 11.16c).

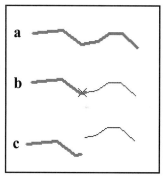

Fig. 11.16. Splitting a line

It is also easy to make copies of features.

22➜ Select one of the lines.

22➜ Choose Edit > Copy from the main menu bar, or press Ctrl-C.

22➜ Choose Edit > Paste from the main menu bar, or press Ctrl-V. The copy will be placed exactly on top of the existing feature.

22➜ Move the selected feature to verify that there are two identical lines now.

You can copy features from one layer to another. Both layers must have the same feature type. Always set the target layer to the target, not the source, before beginning a copy.

23➜ Change the target layer to Streets.

23➜ Select one of the lines from practiceln.

23➜ Press Ctrl-C, and then Ctrl-V.

23➜ Click and drag the new feature to a different location.

23➜ Click anywhere off all the lines to clear the selected features.

Notice that the original line has the practiceln symbol, but the new line has the thick pink symbol of Streets. However, we don't want this line in Streets, so get rid of it.

24➔ Select the pink line in Streets, and press the Delete key.

24➔ Change the target layer back to practiceln.

24➔ Draw a box around the remaining lines with the Select tool, and delete them also.

Working with snapping

Snapping makes editing much easier by ensuring that all the features that are supposed to touch actually do. It also aids analysis by preventing topological errors between features. In ArcMap, snapping is quite flexible and can be administered precisely while you are in the middle of sketching features. Let's see how it works. First, we check the snap tolerance and turn snapping on for the desired layers.

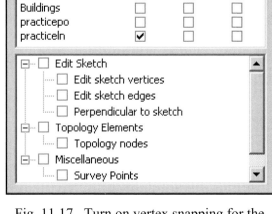

Fig. 11.17. Turn on vertex snapping for the practiceln layer

25➔ Choose Editor > Options and click the General tab.

25➔ Make sure that the snap tolerance is set to 15 pixels. This distance is larger than one would usually use, but it will make clearer what is going on. Click OK.

25➔ Choose Editor > Snapping from the Editor toolbar.

25➔ Check the Vertex box for the practiceln layer (Fig. 11.17).

25➔ Leave the Snapping window open for now, but position it to show both the map and the window at the same time.

TIP: You can dock/undock the Snapping Environment window just like other windows.

Now we are ready to use snapping.

26➔ Click on the Pencil tool.

26➔ Use the Pencil tool to draw a line with distinct angles and obvious vertices. Don't connect the ends. Double-click to finish the line.

26➔ Move the cursor toward one of the vertices, and watch it jump to the nearest vertex when it gets close.

The cursor stays at the vertex until you move at least 15 pixels away. If you click to add a vertex while the cursor is "frozen" to the vertex, then the resulting new vertex will exactly match the existing one. Try it.

26➔ Click on one of the vertices to begin a new line.

26➔ Create another vertex that is off the existing line.

26➔ Come back and snap to another vertex of the first line.

26➔ Double-click to finish the sketch.

Now let's see how the different types of snapping operate. You have used vertex snapping so far.

27➔ In the Snapping window, uncheck the vertex box, and turn on end snapping for practiceln.

27➔ Move the cursor over the different vertices. Now snapping only occurs at the endpoints of the lines, not at the vertices.

27➔ Turn off end snapping, and turn on edge snapping for practiceln.

27➔ Move the cursor over the lines, and notice that snapping occurs anywhere along the line. Make the cursor follow along the edge.

27➔ Make a sketch that is snapped to one of the previous lines at several locations. Finish the sketch.

We will make extensive use of snapping during the rest of the chapter.

28➔ Uncheck all the boxes in the Snapping window and close it.

28➔ Choose Editor > Options from the Editor toolbar.

28➔ Click the General tab and set the snap tolerance back to 7 pixels. This value works well for most editing.

Using the sketch context menus

Now that you can perform most of the simple editing functions, let's take a look at some ways in which the Sketch menu enhances editing capabilities, such as editing simple mistakes during a sketch.

29➔ Make sure that the target layer is practiceln, the edit task is Create New Feature, and the Pencil tool is selected.

29➔ Enter some vertices of a new line. Notice that the last vertex is always red and other vertices are green.

29➔ Right-click on top of the last red vertex to open the vertex context menu (Fig. 11.18).

29➔ Examine the menu entries for a moment. This context menu is called the Vertex menu.

29➔ Choose Finish Sketch.

Choosing Finish Sketch from the context menu does the same thing as double-clicking at the end of the sketch. Obviously double-clicking is faster. But the Vertex menu offers some other useful functions.

Fig. 11.18. The Vertex menu

30➔ Click on the map to start a new sketch, and add some vertices.

30➔ Right-click the last red vertex to open the Vertex menu.

30➔ Choose Delete Vertex from the Vertex menu.

Notice that the last vertex is deleted, and the second-to-last one is now red. This menu can delete any of the previous vertices.

30➔ Right-click on any of the green vertices, and choose Delete Vertex.

30➔ Right-click another vertex and examine the menu again.

Now, you might want to get rid of the menu without actually selecting one of its entries.

31➔ Try clicking on a location on the map but off the menu.

The menu went away, but you inadvertently added a vertex at the place clicked.

31➔ Right-click the extra vertex and choose Delete Vertex.

31➔ Right-click another green vertex to open the menu again.

31➔ This time, click on a plain grey area at the top of the ArcMap window. This time, the menu goes away without adding a vertex. That is the best way to get rid of the menus.

Notice that all this time we are still in the middle of creating this sketch. You can carry out quite a few actions while sketching, with no negative effects on the sketch. Next, let's learn to add a new vertex to the sketch.

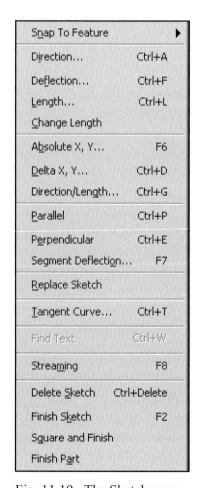

32➔ Right-click on a segment of the sketch, *between* two vertices.

32➔ Choose Insert Vertex from the context menu.

32➔ Watch the new vertex appear where you had right-clicked.

32➔ Right-click another segment of the sketch between two vertices.

The Vertex menu appears again. This time, notice how Add Vertex is available and Delete Vertex is dimmed. Before, when you clicked ON a vertex, Delete Vertex was available and Add Vertex was dimmed. Except for this difference, the menu is the same. Clicking ON the sketch gives the Vertex menu. To get the Sketch menu, click OFF the sketch.

32➔ Click the grey area again to make the menu go away.

32➔ This time, right-click OFF the sketch, somewhere on the map. The Sketch menu appears (Fig. 11.19).

32➔ Examine the entries in this longer menu.

Fig. 11.19. The Sketch menu

Some of the entries, such as Finish Sketch, are found in both menus. This menu contains many more functions. With so many options, it can be easy to make some major mistakes as we experiment with these choices. You can completely delete a sketch and start over at any time.

32➔ Choose Delete Sketch from the menu.

Okay, time to experiment!

33➔ Start a new sketch with several vertices.

33➔ Right-click off the sketch to open the Sketch menu.

33➔ Choose the second entry, Direction.

33➔ Type "90" in the Direction box and press Enter.

As the mouse moves, notice that the cursor is constrained along a vertical line. You have forced your next segment to follow the angle you have specified. The angles are measured counter-clockwise from the east (Fig. 11.20).

33➔ Click along the line to enter a vertex. The new segment is exactly vertical.

33➔ Right-click off the sketch and choose Direction again from the menu.

33➔ Type "-45" in the box and press Enter.

33➔ Click along the line to enter a vertex.

33➔ Double-click to finish the sketch.

Fig. 11.20. Angles measure in degrees from horizontal (east).

2. What angle would you enter to create a line going S25°W? _____

Next, let's create a segment of a specific length.

34➔ Start a sketch and add several vertices.

34➔ Right-click off the sketch and choose Length from the Sketch menu.

34➔ Type "30" into the box and press Enter.

3. In what units are these distance values?_____

34➔ Move the cursor around, noticing how the last segment always remains the same length.

34➔ Click somewhere to enter the vertex.

TIP: If the last segment goes off the page, or is too small to see, try entering a different length. Simply right-click off the sketch again and type in a different length value.

The Sketch menu also combines the length and direction functions to enter a line with a particular bearing and direction. We will use this to enter the next segment of this same sketch.

35➜ Right-click off the sketch and choose Direction/Length.

35➜ Type "45" for the direction, and "30" for the length (you can use different values if needed to keep the sketch on the screen). Use the Tab key or mouse to switch between the boxes.

Notice that you don't even have to click to enter the vertex. Once you specify an angle and length, the position of the next vertex is fixed, and it is placed automatically.

35➜ Create several more segments using the Direction/Length option.

35➜ Double-click the sketch to finish it.

What if you don't know the absolute bearing of a segment, but only its angle relative to another segment? We have tools for this situation also. First, let us take the situation of specifying that our new segment falls at an angle from the previous segment.

36➜ Start a sketch and enter several vertices.

36➜ Right-click off the sketch and choose Deflection from the menu.

36➜ Type "90" and press Enter.

36➜ Click along the constrained line to create a new vertex.

36➜ Finish the sketch.

TIP: The angle box contains the current angle when you right-clicked. You can use this feature to measure an existing angle!

Notice how this time the new segment is at an angle of 90 degrees to the previous segment, rather than being simply vertical. Another command, Segment Deflection, can use ANY existing segment as the reference direction, not just the previous one in the sketch. To make this one work, right-click the existing segment to deflect from.

37➜ Start a new sketch and enter several vertices.

37➜ Right-click one segment of the *last line feature* (not the sketch), and choose Segment Deflection (Fig. 11.21).

37➜ Type "90" and press Enter.

37➜ Click along the constrained line to enter the vertex.

37➜ Finish the sketch.

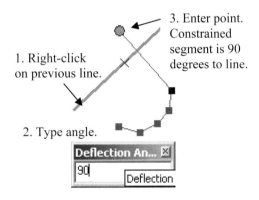

Fig. 11.21. Segment deflection

This time the new segment is perpendicular to the segment clicked in the previous line.

As a final example of using the Sketch menu, we will learn to make lines be perpendicular or parallel to an existing line. Just as with Segment Deflection, you must right-click ON the reference line when bringing up the Sketch menu.

38➔ Select and delete all the previous lines.

38➔ Start and finish a straight line segment with two vertices.

39➔ Start a new sketch by entering one vertex.

39➔ Right-click the straight line segment, and choose Parallel. The segment is constrained parallel to the first line. Click along the line to enter a vertex.

39➔ Right-click the first straight line again, and this time choose Perpendicular. Click along the constrained line to enter a vertex.

39➔ Experiment by adding a few more parallel and perpendicular lines until you are comfortable with the process.

39➔ Finish the sketch.

Creating polygons

As our final task before we start actually editing some layers, we need to practice editing polygons, first as separate entities, and then as adjacent features.

40➔ Select and delete the lines just created.

40➔ Set the target layer to practicepo.

The first polygon will be by itself. To create a first polygon, or one that is not connected to another polygon, use the Create New Feature task and the Pencil tool.

40➔ Make sure the task is set to Create New Feature, and click the Pencil tool.

40➔ Start a polygon by adding several vertices. Notice how the shape remains closed, with the last vertex always connected to the first.

40➔ Double-click the last vertex to finish the sketch.

40➔ Try adding several more polygons that don't touch each other, for practice.

Next we will create some adjacent polygons. It is very important to use the AutoComplete Polygon task to do this, so that the polygons have a coincident boundary. Vertex snapping also helps match the polygon vertices.

41➔ Open the Snapping window, and turn off all snapping. Then turn on vertex snapping for practicepo. Close the Snapping window.

42➔ Change the Task bar to AutoComplete Polygon.

42➔ Make sure the Pencil tool is clicked.

42➔ Click INSIDE an existing polygon to start the sketch (Fig. 11.22).

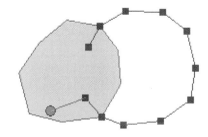

Fig. 11.22. Adding an adjacent polygon with the AutoComplete task

42➔ Add additional vertices to define the adjacent polygon.

42➔ Add the last vertex INSIDE the existing polygon.

42➔ Double-click to finish the sketch.

TIP: When using AutoComplete, always start and end well inside an existing polygon. The dangles will automatically be deleted. If the starting point does not fall inside an existing polygon, ArcMap won't be able to create the polygon and you'll have to start again. Alternatively, use vertex or edge snapping to ensure that the new polygon meets the old one.

You can also add polygons that start and end inside two different features.

43➔ Begin a new sketch inside one of the two adjacent polygons, and finish it inside the other polygon. Finish the sketch (Fig. 11.23a).

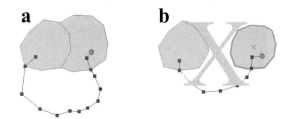

Note that when completing polygons, the sketch and the adjacent polygons cannot have any gaps, such as they do in Figure 11.23b.

Fig. 11.23. Starting and ending a sketch inside two polygons. (a) Works. (b) Does not work because the sketch does not fully enclose an area.

Splitting provides another way to create adjacent polygons with coincident boundaries. In this task the Pencil tool cuts a polygon in pieces.

44➔ Select one of the polygons (Fig. 11.24).

44➔ Change the task to Cut Polygon Features.

44➔ Click the Pencil tool.

44➔ Start a sketch OUTSIDE the selected polygon, and add vertices of a line that goes through the polygon.

44➔ Add the last vertex OUTSIDE the polygon, and double-click to finish the sketch.

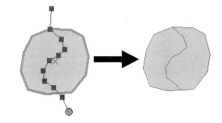

Fig. 11.24. Cutting a polygon

Cutting works on multiple polygons, too.

45➔ Check that the two polygons you just created are still selected. If not, select them. (Fig. 11.25).

45➔ Sketch another line that goes through both.

45➔ Finish the sketch.

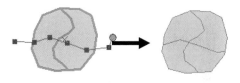

Fig. 11.25. Cutting multiple polygons

Now you have all the basic skills to begin doing some real editing. Adding to the neighborhood layers in this map document will use many of the techniques we have learned. First, let's do a little cleanup work.

46➜ Choose Editor > Stop Editing from the Editor toolbar. Choose No when prompted to save the changes.

46➜ Remove practiceln and practicepo from the map document.

46➜ Zoom back out to the full extent of the Streets layer.

46➜ Save the map document.

Adding buildings

Your task is to continue developing the database started for this small area of Rapid City. You will digitize the buildings, parcels, and water connections. We will begin with the buildings.

47➜ Zoom into Block 1 (upper left) and examine the different features present in this neighborhood survey. Pay particular attention to the buildings.

4. Do the buildings need to be snapped to any other features? _____

One feature of these buildings is readily visible—most of the corners are square. We can use one of the snapping options to make this requirement easy and exact.

47➜ Pan to Block 3, below Block 1, so that the entire block is visible.

47➜ Turn on the Orthophoto layer to show the houses. (They are somewhat fuzzy, but clear enough to digitize approximately.)

47➜ Choose Editor > Start Editing from the Editor toolbar.

47➜ Make sure that the target layer is Buildings. Set the task to Create New Feature.

47➜ Open the Snapping window. Check the box labeled Perpendicular to Sketch. Then close the window.

This last step ensures that all the building corners will be square. Now we are ready to start.

48➜ Start with the clearest square building, second from the left on the top row.

48➜ Click the Pencil tool, and click on the upper left corner to start the sketch (Fig. 11.26).

For these buildings, we want all the lines to go north-south or east-west. So we will constrain the angle of the first segment, the west wall of this building, to be 90 degrees.

48➜ Right-click off the sketch and choose Direction.

48➜ Type "90" in the box and press Enter.

Fig. 11.26. Adding the first house

48➔ Add the second vertex at the lower left corner of the house.

Now that we have the first segment established, the remaining segments can easily be made perpendicular using the snapping.

48➔ Add the remaining two vertices and double-click to finish the sketch (Fig. 11.26).

Now that we have one vertical line, we can use it to make all the others vertical or horizontal. Using Perpendicular or Parallel is easier and faster than typing an angle each time.

49➔ Start the next sketch by clicking on the upper left corner of the next house to the left.

49➔ Right-click on the west wall of the house you previously digitized, and choose Parallel from the context menu.

49➔ Enter the second vertex of the current sketch along the vertically constrained line.

49➔ Add the rest of the vertices except for the last one. Then right-click off the sketch and choose Square and Finish, to ensure that the last corner is a right angle.

49➔ Finish creating the remaining houses in this block. Get the actual shape as best you can—don't make them all boring boxes (Fig. 11.27).

49➔ Choose Editor > Save Edits from the Editor menu.

Fig. 11.27. Follow the house shape as best you can.

Next we will add the sewer connections to the fronts of the houses. We don't know exactly where they go, so we will just enter a point on the front edge of the house, but not in the driveway. We'll use snapping to make sure that the connections fall exactly on the house edge.

50➔ Open the Snapping window and turn off the perpendicular snapping.

50➔ Turn on edge snapping for Buildings.

50➔ Leave the Snapping window open, but position it out of the way.

50➔ Change the target layer to Connects.

50➔ Make sure the task is Create New Feature, and click the Pencil tool.

50➔ Add a water connection point to each house.

50➔ Save the edits. Close the Snapping Environment window.

Creating parcels

Next we will create the parcel polygons. Of course, we can't tell exactly from the picture where each property line falls, but for practice an approximation will suffice. In a real application we would need to be more exact. Parcels are adjacent polygons, so we must ensure that all of them have coincident boundaries.

First, though, we need to add a field to contain the parcel number. We must stop editing to do this task.

51➔ Choose Editor > Stop Editing. Say Yes if prompted to save the edits.

51➔ Open the attribute table for the Parcels layer.

51➜ Add a new field called "Parcel_ID". Make it a text field, and give it a length of 8 characters.

51➜ Close the table.

51➜ Start another editing session by choosing Editor > Start Editing.

Now let's begin adding parcels. We will outline the block boundary, and then cut up the large block into smaller parcels. Snapping ensures that the block boundary matches the pink street boundary, to avoid gaps between different layers in the data set.

52➜ Set the target layer to Parcels.

52➜ Make sure that the task is set to Create New Feature.

52➜ Make sure the boundary of the entire block is visible in the map.

52➜ Open the Snapping window and turn on vertex snapping for Streets and for Parcels. Leave the Snapping window open but position it out of the way.

52➜ Choose the Pencil tool and sketch the block boundary, snapping to each corner of the pink Streets layer. Finish the sketch.

52➜ Turn off the Streets layer.

Our next step is to cut this large polygon up into the smaller parcels. The polygon to cut must be selected first.

53➜ Make sure that vertex snapping is still on for Parcels, and also turn on edge snapping for Parcels.

53➜ Make sure that the block boundary is still selected. If not, select it.

53➜ Change the task to Cut Polygon Features.

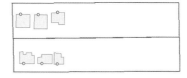

Fig. 11.28. Cut the block in two

53➜ Choose the Pencil tool.

53➜ Starting on or outside the boundary, digitize a roughly horizontal line that marks the boundary between the upper and lower houses (Fig. 11.28). Follow the white line on the photo.

53➜ Finish the sketch outside the boundary on the other side. The polygon will split into two sections.

Now create a roughly vertical line to delineate the first two parcels on the left from the two selected polygons. Change direction by doing a "dogleg" along the centerline. The edge snapping will make sure the direction change matches the existing line.

54➜ Turn off the Orthophoto to see the houses better.

54➜ Make sure both upper and lower sections are still selected.

54➜ Add a vertex on the upper polygon, approximately between the first two houses (number 1 in Figure 11.29).

54➜ Add a second vertex (2) on the horizontal dividing line.

54➜ Add a third vertex (3), still on the horizontal line, but a little to the east to align the final segment between the houses.

54➜ Add a final vertex (4) on the bottom line.

54➜ Double-click to finish the sketch.

55➜ Finish creating the parcels for the rest of this block, using the same technique.

55➜ Save the edits.

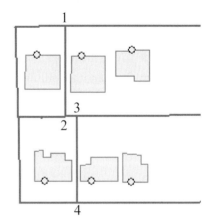

Fig. 11.29. Cutting polygons. Enter the vertices in the order shown.

TIP: Use Delete Vertex to correct any mistakes while entering vertices.

Using the Attribute Editor

We created all the parcels at once because the editing was easier that way. Now we need to go back and add a parcel number for each one. The Attribute Editor facilitates this task.

56➜ Select the parcel in the upper left corner of the block.

56➜ Click the Attribute Editor button.

56➜ Position the Attribute Editor window so that you can see all the parcels.

56➜ Click the Parcel_ID field on the right side of the window, and type in the parcel number "3-129876" (Fig. 11.30). The 3- represents the block number.

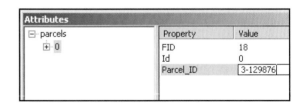

57➜ Select the next parcel on the right.

57➜ Type in the parcel number "3-129877".

57➜ Continue adding the parcel numbers, going clockwise around the block, and increasing by 1 each time.

Fig. 11.30. Adding parcel IDs with the Attribute Editor

TIP: You can highlight the first part of the number, **3-1298,** copy it using Ctrl-C, and paste it into each field (Ctrl-V). Press Enter for each record before going to the next.

Now all of these parcels belong to Block 3, so we will set the ID field to the block number. We can do this for all the parcels at once.

58➔ Choose Selection > Set Selectable Layers from the main menu bar, and make Parcels the only selectable layer.

58➔ Use the Select tool on the Editor toolbar to draw a box around and select all the parcels in Block 3.

58➔ Notice all the parcels are now shown in the Attribute Editor, with a value of 0.

58➔ Click the entry "parcels" at the top of the list of records (Fig. 11.31). The attributes on the right go blank.

58➔ Click on the Id field on the right side, and type "3". Press Enter. Notice that all the values on the left also change to 3.

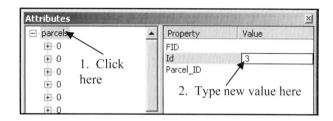

Fig. 11.31. Changing all of the parcel IDs at once

The block ID may not be the most useful value to display in the attribute editor. We can easily change what is shown in the list.

59➔ Open the Properties window for the Parcels layer.

59➔ Click the Fields tab.

59➔ Change the primary display field to Parcel_ID.

59➔ Click OK. Select all the parcels again, so that the parcel IDs are now displayed in the Attribute Editor.

TIP: You need to reselect the parcels in order to have the field change take effect. If any of the Parcel_ID values are blank, the FID will be displayed instead.

Now the parcel IDs are shown in the Attribute Editor. It is easy to find out which parcels match which records.

60➔ Click on a parcel record in the list on the left. The corresponding parcel on the map will briefly flash.

60➔ Right-click a record in the list to get a context menu (Fig. 11.32).

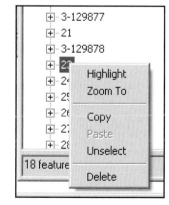

The context menu can highlight a feature (flash it), zoom to it, copy it, unselect it (it will disappear from the Attribute Editor), or delete it from the file.

60➔ When finished with the Attribute Editor, close it.

60➔ Stop Editing and save the edits.

60➔ Save the map document.

➔ Close ArcMap.

Fig. 11.32. Examining a record in the Attribute Editor

This is the end of the tutorial.

Exercises

1. Add the buildings for Block 8 to the buildings layer in the mgisdata\Rapidcity\eastpat folder.

2. Add water connectors to each building.

3. Digitize the parcels for Block 8, making sure that they all have coincident boundaries.

4. Enter the ID and Parcel_IDs for the parcels in Block 8. They will all begin with 8-, but you can make up the rest of the number. Have them increase by 1 as you go around the block clockwise.

5. Create a water mains layer and include a text attribute field for the type (Main or Lateral). Add water mains along the streets of Block 8 and set their attribute type to Main.

6. Add water laterals to the water mains layer in Block 8, connecting each house to the water mains. Set their attribute type to Lateral.

7. Create a layout showing all the elements you digitized in Block 8. Label each parcel with the parcel number. Show the water mains with a different symbol than the laterals. For clarity, do not include the orthophoto in your layout (or include it but make it about 70% transparent). Print the layout.

Challenge Problem

Open ex_11b.mxd in the mgisdata\MapDocuments folder. It shows a portion of an image of agricultural land in eastern South Dakota. You will create and digitize some layers based on this photo. Put the layers in the folder mgisdata\Sdakota\farm (the orthophoto is already there). Be sure to use appropriate snapping to ensure topological integrity.

1. Create three new shapefiles: roads (line), rivers (line), and crops (polygon). Import the coordinate system from the orthophoto.
2. Digitize the roads as single lines.
3. Digitize the rivers, showing both shores. (Hint 1: Read about streaming in the help documentation and try using it. Hint 2: Digitize one shore and trace the other.)
4. Create crop polygons. Use your best judgment here to find homogeneous areas to classify as separate fields. The field boundaries next to rivers should be coincident with the river shores. You don't have to do EVERY field, but make at least 25 different fields.
5. Create an attribute field for the crops called Crop, and assign a crop to each field (e.g., corn, soybeans, wheat, fallow). Just make them up.
6. Create a layout showing your work. Include labels for the crops. Print the layout.

Skills Reference

Begin an editing session

1. Open ArcMap and add the data layers you wish to edit. For best results, add only as many layers as needed.

2. If the Editor toolbar is not already displayed, click the Edit button on the main menu bar to launch the Editor menu.

3. Choose Editor > Start Editing from the Editor toolbar.

4. If data from more than one folder or geodatabase appears in the map document, you will be prompted to choose which one to open for editing.

You are now ready to begin editing.

Setting snapping options

1. Choose Editor > Options from the Editor toolbar.

2. Click the General tab (Fig. 11.33).

3. Choose the units of the snap tolerance.

4. Enter the snap tolerance value.

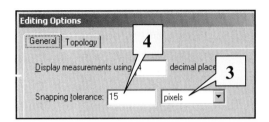

Fig. 11.33. Setting the snapping tolerance

TIP: Pixel units usually work best, because then the snap distance on the screen remains constant regardless of scale.

---

5. Choose Editor > Snapping from the Editor toolbar.

6. Check the boxes for any layers you want to snap to according to the type of snap you want: vertex, edge, or end (Fig. 11.34).

7. Click and drag a layer up or down the list to change the snapping priority. Layers in the top of the list will be snapped to first.

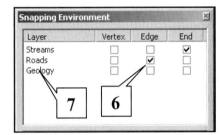

Fig. 11.34. Turning on snapping for layers

TIP: Keep the Snapping Environment window open while editing, and turn snapping on and off as needed while editing features.

Selecting features for editing

All of these techniques respond to the options used for regular interactive selection, including setting the selectable layers, changing the selection method, or changing the selection options, as described in Chapter 6. Features may also be selected for editing using the Select by Attributes and Select by Location.

Selecting features with the Select tool

1. Click the Select tool on the Editor toolbar.

2. Click the feature to be selected, OR click and hold down the mouse button to drag a box around the features to be selected.

3. To select multiple features, hold down the Shift key and click on each feature in turn.

Selecting features with a line

1. Choose Select Features Using a Line from the Task bar (Fig. 11.35).

2. Choose a sketch tool.

3. Click on the map to begin drawing a line; double-click to end the line. All features that intersect the line will be selected.

Selecting features with a polygon

1. Choose Select Features Using an Area from the Task bar.

2. Click the Pencil tool.

3. Click on the map to begin drawing a figure; double-click to end it. All features that intersect the area will be selected.

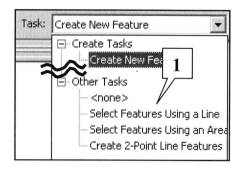

Fig. 11.35. Using the Task bar for selecting features

Moving features

1. Select the feature(s) to be moved using the Select tool on the Editor toolbar.

2. Place the cursor over the objects until crosshairs appear.

3. Click and drag the feature to a new location.

Rotating features

1. Select the feature(s) to be rotated using the Select tool on the Editor toolbar, or by using another selection method (Fig. 11.36).

2. Click the Rotate tool on the Editor toolbar.

3. Click and draw anywhere on the screen to rotate the features about their center of rotation (marked with an X).

4. To change the center of rotation, place the cursor over the X. It will change to a crosshairs symbol. Click and drag the center to a new location.

5. Click and drag anywhere outside the features to rotate them about the new center.

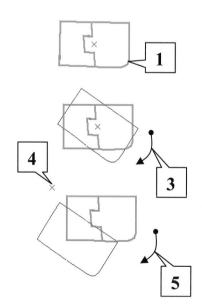

Fig. 11.36. Rotating features

Deleting features

1. Select the feature(s) to be deleted using the Select tool on the Editor toolbar, or use another selection method.

2. Make sure the Select tool on the Editor toolbar is clicked.

3. Press the Delete key to delete the features.

Creating new features

1. In the Editor toolbar, set the target layer to which the new feature will belong (Fig. 11.37).

2. In the Editor toolbar, choose Create New Feature from the Task bar.

3. Choose a sketch tool.

4. Sketch the feature by clicking on its vertices or by using one of the special techniques described in the steps that follow.

You may use the context menus or switch between different sketch tools while creating the vertices of a sketch, as described in the next two sections.

5. To delete or edit the vertex of a sketch, right-click on a vertex to bring up the vertex context menu (Fig. 11.38).

6. To add a vertex to a sketch, right-click on a segment of the sketch between two vertices.

7. To bring up the Sketch menu, right-click the screen at a location off the sketch.

8. To start again, right-click to raise one of the menus and choose Delete Sketch.

9. When the sketch is complete, right-click it and choose Finish Sketch OR simply double-click on the last vertex.

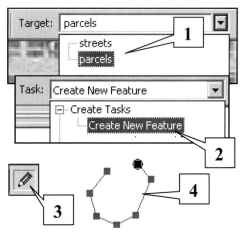

Fig. 11.37. Creating a new polygon

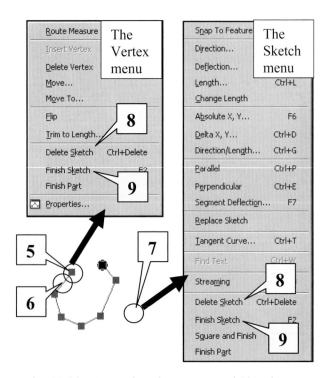

Fig. 11.38. Accessing the Vertex and Sketch menus while creating a sketch

Using sketch context menus

The following functions can be accessed during a sketch by right-clicking anywhere off the sketch.

Entering absolute *x-y* locations

1. Right-click off the sketch to open the Sketch menu.

2. Choose Absolute X,Y from the menu.

3. Enter the *x-y* coordinates and press Enter (Fig. 11.39). The coordinates must be in the same coordinate system as the data frame.

Fig. 11.39 Absolute X,Y

4. Finish the sketch, or switch to another sketch tool and keep sketching.

Using offsets from a previous location

1. Enter at least one vertex of the sketch.

2. Right-click off the sketch to open the Sketch menu.

3. Choose Delta X,Y from the menu.

Fig. 11.40. Delta X,Y

4. Enter the distance change in the *x* and *y* directions, in map units (Fig. 11.40). Press Enter.

5. Finish the sketch, or switch to another sketch tool and keep sketching.

Creating a segment at a specified angle

1. Enter at least one vertex of the sketch.

2. Open the Sketch menu and choose Direction.

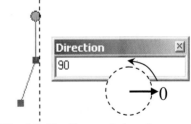

3. Type in the desired angle in degrees, and press Enter (Fig. 11.41).

4. The segment is now constrained at an angle of 90 degrees. Click at the desired distance to enter the point.

Fig. 11.41. Constraining the segment angle

5. Finish the sketch, or switch to another sketch tool and keep sketching.

TIP: The angle box contains the angle that was established when you right-clicked. You can use this feature to measure an existing angle.

Creating a segment of specific length

1. Enter at least one vertex of the sketch.

2. Open the Sketch menu and choose Length.

3. Type the length in the box, in map units, and press Enter (Fig. 11.42).

4. The segment is constrained to a set length. Click at the desired location to create the next vertex.

5. Finish the sketch, or switch to another sketch tool and keep sketching.

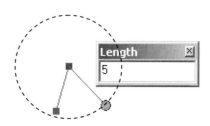

Fig. 11.42. Constraining the sketch segment to a set length

Creating a segment of set angle and length

1. Enter at least one vertex of the sketch.

2. Open the Sketch menu and choose Direction/Length.

3. Type in the desired angle and length and press Enter (Fig. 11.43).

4. Finish the sketch, or switch to another sketch tool and keep sketching.

Fig. 11.43. Constraining the angle and the length of a segment

Creating a deflected segment

Use this option to create a new segment at a specified angle from the last segment in the sketch.

1. Enter at least one segment of the sketch.

2. Open the Sketch menu and choose Deflection.

3. Enter the angle in degrees that the new segment will be deflected from the current segment (Fig. 11.44). Press Enter.

4. The new segment is now constrained along a line. Click at the desired distance to enter the new vertex.

5. Finish the sketch, or switch to another sketch tool and keep sketching.

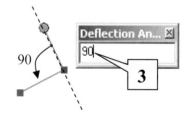

Fig. 11.44. Deflecting a vertex from the previous segment

Deflecting a segment from a feature

Use this option to create a new segment at a specified angle to an existing segment.

1. Enter at least one segment (two vertices) of the sketch.

2. Right-click *on an existing feature segment* (the side of this building) to open the context menu, and choose Deflect Segment.

3. Type in the angle of deflection from the existing segment, and press Enter (Fig. 11.45).

4. The new segment is now constrained at the specified angle (45 degrees) to the existing segment (the south side of the building). Click at the desired length to create the new vertex.

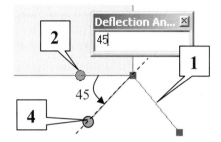

Fig. 11.45. Deflecting at an angle to an existing segment

5. Finish the sketch, or switch to another sketch tool and keep sketching.

Creating parallel or perpendicular segments

1. Enter at least one vertex of the sketch (Fig. 11.46).

2. Right-click *on an existing feature segment* (the side of this building) to open the context menu, and choose Parallel or Perpendicular.

3. The new segment will be constrained in the chosen direction. Click to enter the new vertex at the desired location.

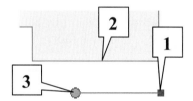

Fig. 11.46. Creating a segment parallel to an existing feature

4. Finish the sketch, or switch to another sketch tool and keep sketching.

Creating multipart features

Multipart features include more than one contiguous feature. This technique is used to create a single feature from multiple poly areas (such as Hawaii), or multiple segments forming a single street, or multiple points forming a single feature.

1. Sketch the first part of the feature using any of the sketch tools and context menus (Fig. 11.47).

2. Right-click on or off the sketch and choose Finish Part.

3. Add as many parts as desired, ending each with a Finish Part.

4. On the last part, right-click on or off the final sketch and choose Finish Sketch.

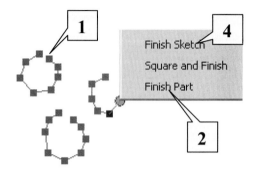

Fig. 11.47. Creating multipart features

Creating features with square corners

1. Set the target to a line or polygon layer.

2. Choose Editor > Snapping from the Editor menu.

3. Check the box labeled Perpendicular to Sketch (Fig. 11.48).

4. Click to add vertices. Each segment will be constrained to be perpendicular to the last segment.

5. To add the last vertex, right-click *off* the sketch and choose Square and Finish from the Sketch menu.

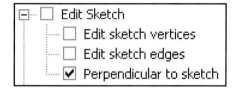

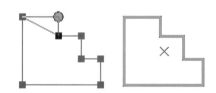

Fig. 11.48. Creating figures with square corners

Creating adjacent polygons

You can create adjacent polygons with coincident boundaries in two different ways: by using the AutoComplete task, or by creating a large polygon boundary and cutting it up into sections.

Using the AutoComplete task

1. Set the target layer.

2. Set the task to Create New Feature (Fig. 11.49).

3. Use one or more sketch tools to create the first polygon.

4. Change the task to AutoComplete Polygon.

5. Sketch the next polygon. Be sure to start and end inside the adjacent polygon, or inside two adjacent polygons.

6. Finish the sketch to create the new polygon.

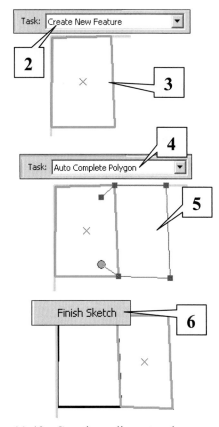

Fig. 11.49. Creating adjacent polygons with the AutoComplete task

Using the Cut Polygon Features task

1. Set the target layer.

2. Use the Select tool on the Editor toolbar to select the polygon(s) to be cut (Fig. 11.50).

3. Set the task to Cut Polygon Features.

4. Sketch a line that will form the boundary between the new polygons. Important: The line must start and end **outside** of the existing polygons.

5. Finish the sketch. The new polygons will be created.

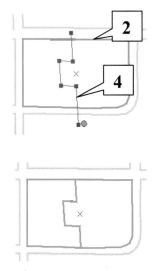

Fig. 11.50. Cutting a polygon

417

Using the Attribute Editor

1. Select the feature(s) to edit.

2. Click the Attribute Editor button on the Editor toolbar.

To edit a single feature

3. Click the record to edit in the list on the left side (Fig. 11.51). Its attributes will be shown on the right.

4. Click on the attribute field to edit, and type in the new value. Press Enter.

5. Keep editing different fields until you are satisfied.

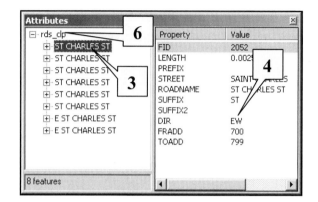

Fig. 11.51. The Attribute Editor

To edit a field for all the features

6. Click the layer name at the top of the list. The fields on the right will go blank.

7. Click the field to edit, and type in the new value. The new value will replace the existing values for each record in the selected set.

Managing records

1. Clicking an entry in the list causes the feature to flash briefly on the map.

2. Right-click one entry in the list to bring up a context menu (Fig. 11.52).

3. Highlight causes the feature to flash. Unselect removes the feature from the selected set (and from the Attribute Editor). Delete deletes the feature from the layer.

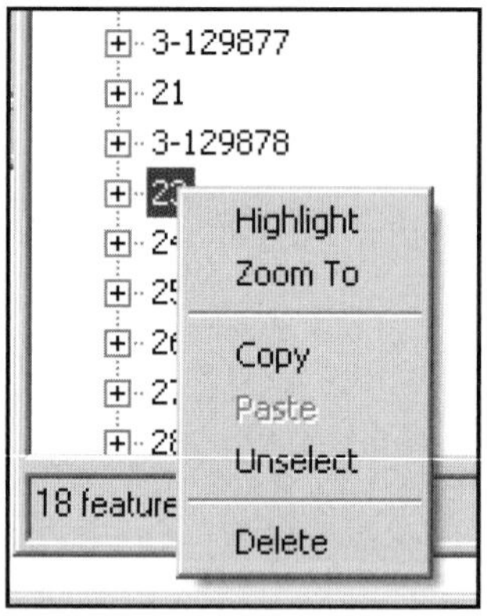

Changing the display field in the Attribute Editor

1. Double-click the layer in the Table of Contents to open its Properties window.

Fig. 11.52. Examining a record in the Attribute Editor

2. Click the Fields tab.

3. Change the primary display field to the desired attribute.

4. The change will be applied to the Editor when you make the next selection.

Saving changes during an edit session

1. Choose Editor > Save Edits from the Editor toolbar.

2. Continue editing.

Stopping an edit session

1. Choose Editor > Stop Editing from the Editor toolbar.

2. When prompted, indicate whether to save any current changes or discard them.

Switching to another folder or geodatabase

1. Choose Editor> Stop Editing from the Editor toolbar.

2. Choose whether to save the changes from the current session.

3. Make sure that the layers from the new folder to be edited appear in the Table of Contents.

4. Choose Editor > Start Editing from the Editor toolbar.

5. Select the new folder or geodatabase to edit.

Chapter 12. More Editing Techniques

Mastering the Concepts

Objectives

➢ Modifying shapes of existing features

➢ Advanced techniques for editing lines and polygons

➢ Combining features with merge, union, intersect, and clip

➢ Buffering features to create lines and polygons

➢ Editing shared lines and polygon boundaries

Concepts

Chapter 11 presented some basic and advanced techniques for creating new features in a data layer. In this chapter we examine additional ways to form and modify features. First we will examine the functions of the different types of sketch tools, look at ways to modify and reshape features, combine features together, and create new features by buffering old ones. Finally, we will discover how to easily edit features which share a common boundary.

Using different sketch tools

So far we've used only the Pencil tool to enter vertices of features. However, several different sketch tools can be used to create specialized vertices with specific characteristics. If the black arrow on the current sketch tool is clicked, the menu expands to show many different tools (Fig. 12.1). More tools have been added with later releases of ArcMap.

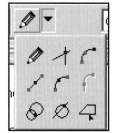

Fig. 12.1. The nine sketch tools

The **Distance-Distance tool** finds locations that fall a specified distance from two points. For example, imagine that the owner of a house left a note that the septic tank lies 50 feet from the NE corner of the house, and 32 feet from the NE corner of the tool shed. The tool could use this information to pinpoint the two possible locations of the tank when it becomes necessary to dig it up for repairs (Fig. 12.2a).

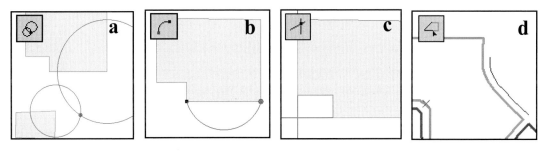

Fig. 12.2. Using some of the sketch tools. (a) Distance-Distance tool. (b) Curve tool. (c) Intersection tool. (d) Trace tool.

The **Arc tool** creates true parametric arcs of set radius. The user enters the start point, the point which the curve passes through, and the endpoint, as shown by the patio in Figure 12.2b.

The **Intersection tool** finds the intersection point of two lines, such as the location where the two sides of a parcel would intersect if the cutout did not exist (Fig. 12.2c).

The **Trace tool** creates new features which follow along existing features, either directly on top of them or with a specified offset. For example, the second side of a street could be created from the first side, with the exact width in between (Fig. 12.2d).

Additional sketch tools perform other specialized functions. The **Midpoint tool** will enter a new point midway between two points entered by the user. The **Endpoint Arc** tool enters a different kind of curve than the Arc tool. The user enters the two endpoints of the arc and then types in the radius of the arc. The **Tangent** tool adds a new sketch segment that is tangent to the previously sketched segment. The **Direction-Distance** tool allows the user to create a new vertex using a bearing direction from a known point plus a distance from another point.

Changing existing features

Once a polygon or line has been created, it can be changed by eliminating unwanted vertices, adding vertices, or moving existing vertices. Other modifications include flipping lines, trimming them, extending them, and performing a variety of other editing functions.

Modifying features

The Task bar contains an option for modifying existing features by adding or deleting vertices, or moving the existing vertices. The Modify Feature task causes the sketch of the feature to appear. When the edits are complete, the feature is updated to the current form of the sketch. Figure 12.3 shows how an original circular polygon is modified to a keyhole shape by moving vertices in the sketch.

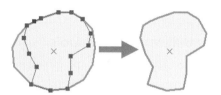

Fig. 12.3. Modifying a feature

Reshaping features

The reshaping task reenters a feature or a portion of a feature, using a new sketch to define the revised shape. The sketch must start and finish on the original feature, so it is helpful to have vertex or edge snapping on. When the sketch is finished, the original feature is modified to follow the sketch. Figure 12.4 shows the process of reshaping parts of lines and polygons.

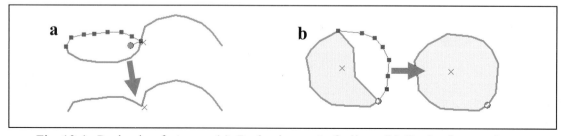

Fig. 12.4. Reshaping features. (a) Reshaping part of a line. (b) Reshaping a polygon.

Flipping lines

As we found in the chapter on geocoding, lines can have a direction. Lines begin at the "from" node and end at the "to" node. Normally the direction of a line generates little concern, and either direction works just as well. In some instances, however, the direction matters. In geocoding, it matters because it defines which way the addresses increase along the street. A streams layer used for network analysis uses the line direction to encode the flow of water. A similar situation would hold for constructing a network of water pipes or sewers, in which the direction of flow must be constrained and recorded.

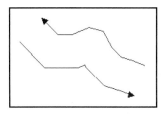

The direction of lines may be determined by drawing them with arrow-ended symbols. Once drawn, you can see the direction of flow and flip the direction if required (Fig. 12.5).

Fig. 12.5. A flipped line

Combining features

Chapter 8 introduced overlay techniques such as intersection, union, and merging. In the Geoprocessing Wizard, these functions are applied to entire shapefiles or feature classes. These functions are also available during editing and can be applied to individual features on an interactive basis. The functions perform similar operations: intersect still finds the area common to both inputs, and union includes all areas from both inputs. However, the attributes are treated differently. Instead of combining all the attribute fields from both inputs, the output feature simply has the same attribute fields as the target layer.

When using these combinations, the user should be sensitive to how any attributes of the features are handled. In cases of a merge, union, or intersect, the resulting features will be given the attributes of one of the original features. In a geodatabase, the attributes of the feature selected FIRST will be copied to the output feature. For shapefiles and coverages, the feature with the lowest feature-id will be copied. If the feature attributes matter, then the user must pay attention to which attributes are being copied and must correct any attributes that were not copied as desired.

Merge

A merge takes one or more features and combines them into a single feature (Fig. 12.6). If the two features are adjacent, then the boundaries between them are removed. If the two features are separate, then a multipart feature will be created. This function might be used frequently to update a parcels map when an owner has purchased two adjacent parcels and combined them into one.

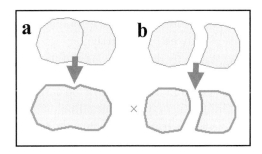

Union

A union performs the same operation as a merge, except that the original two features remain unchanged and a new feature is created in addition.

Fig. 12.6. (a) Merging adjacent polygons to create a new feature. (b) Merging separate polygons to form a multipart feature.

Intersection

An intersection creates a new feature from an area common to both original features (Fig. 12.7). A new feature is created and the original ones are maintained. If two features are selected and an intersection is performed, the resulting new polygon will consist only of the areas shared by the original polygons. The new feature is created in addition to the original features, which are not changed or deleted. This function might be used to identify repeat infestations of pine bark beetle attacks. Two features representing infestations in two different years could be intersected to reveal the area attacked twice.

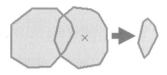

Fig. 12.7. Intersection of two polygons to create a third polygon

Clip

A clip behaves in a cookie cutter fashion. If one feature lies over another, the underlying feature will be cut along the boundaries of the overlying feature (Fig. 12.8a). Two options may be specified: preserving the area common to both features (Fig. 12.8b), or discarding the area common to both features and retaining what is outside the clip polygon (Fig. 12.8c). In either case the feature used for clipping is retained unchanged, and the feature underneath is modified. As shown in Figure 12.8c, clipping is one way to create a "donut" polygon. Clipping is useful anytime a polygon must have internal boundaries, such as a residential area with a park in the middle.

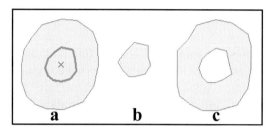

Fig. 12.8. Clipping. (a) Ready to clip the outer polygon with the selected polygon. (b) Option to preserve the common area. (c) Option to discard the common area.

Buffering features

A buffer delineates the area within a specified distance of a feature and can be created from points, lines, or polygons (Fig. 12.9a-c). The output buffers may be lines or polygons, as determined by the feature type of the target layer. The buffer distance must be specified in map units. In the case of buffering polygons, a negative distance may be used to reduce the size of the feature (Fig. 12.9d). Buffers are useful for tasks such as identifying setbacks from parcels, finding drug-free zones around schools, or creating road widths from a set of centerlines.

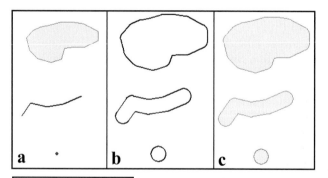

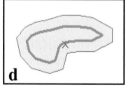

Fig. 12.9. Buffering. (a) The original features. (b) Buffer output as lines. (c) Buffer output as polygons. (d) A negative buffer.

Topology and shared features

Shared features have boundaries or nodes in common with each other. When editing them, it proves very convenient to be able to edit the shared boundary or node, rather than editing each individual piece separately. ArcMap has a special tool which allows editing of shared features.

Consider the problem of changing a boundary between two polygons. Editing the first polygon leaves a gap between the two, so that the edit must be repeated on the second polygon. Snapping can help ensure that the boundaries match, but it quickly becomes tedious to edit every boundary twice. The Topology Edit tool provides a better way.

Topology refers to spatial relationships between features. Shapefiles contain only simple features and are incapable of storing topological information. Coverages store topology as an integral part of the data structures for lines, polygons, and multifeatures. Geodatabases can store several types of topology, including network topology and planar topology. Only feature classes inside a feature dataset can participate in a topologic association. Chapter 14 demonstrates some uses of a network topology.

Planar topology is used to specify the spatial relationships allowed in and between feature classes. A topology in a geodatabase is composed of rules. For example, the Must Not Overlap rule stipulates that within a single layer one polygon cannot extend onto another, even a little bit. This situation would clearly be an error in the case of parcels, for the same bit of land cannot belong to two different parcels. Other rules specify relationships between layers, such as the rule Must Cover Each Other. This rule describes the relationship between counties and states, in which every bit of the state area must belong to a county, and no part of a county should lie outside the state. In other words, the rule helps enforce that shared boundaries of features between layers are identical, with no gaps or overlaps between them.

Figure 12.10 shows typical topological errors occurring between different layers. The boundary of the Pine Ridge Indian reservation (stippled area) should match the boundary between Shannon and Bennett counties, but it does not. Neither does it properly match the South Dakota-Nebraska boundary. Such errors can cause problems during analysis, such as the generation of sliver polygons during an intersection or union, as described in Chapter 8.

Creating planar topology makes it easier to locate and eliminate boundary errors by using special Topology tools while editing. This approach is especially helpful when errors are too small to see at normal viewing scales—in Figure 12.10 the county boundaries appear to match the state boundaries, but might not actually do so. Such errors typically occur as a result of combining data sets from different sources, or from files created by people who do not understand or use snapping very well. Users interested in topology should read the appropriate sections in *Editing with ArcMap*, one of the digital books that comes with ArcGIS.

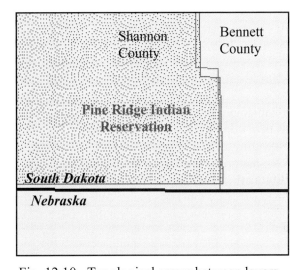

Fig. 12.10. Topological errors between layers

425

Building planar topology for a feature dataset and using it during editing requires an ArcEditor or ArcInfo license. However, users with ArcView licenses, or who are editing shapefiles rather than feature datasets, can use a function called **map topology** to edit features with shared edges or vertices. Map topology creates temporary relationships between features so that they can be edited together. Its purpose is to preserve existing coincident boundaries, however, not to fix problems like the ones shown in Figure 12.10.

The user creates map topology during editing by selecting which feature classes will participate in it. The map topology is created on-the-fly for the set of features currently in the data view. In a topological association, the endpoints of lines are called nodes and lines or polygon boundaries are called edges. The user selects the node or edge to be edited using the Topology Edit tool, and then selects one of the topology tasks from the Task bar.

The Topology Edit tool is similar to the normal Select tool, except that its purpose is to select shared boundaries in preparation for editing. When using the Topology Edit tool to select a feature, or part of a feature, all other features sharing that boundary are also affected by the edits.

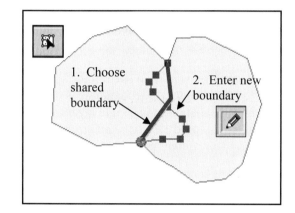

Figure 12.11 shows the use of the Topology Edit tool and the Reshape Edge task to change the boundary between two polygons. First, the shared boundary is selected using the Topology Edit tool. The selection color of the Topology Edit tool is purple, to distinguish the selection from those made with the Select tool. The Sketch tool is then used to draw the new boundary between the polygons. When the sketch is finished, the new boundary replaces the old one, and the change is applied to both polygons.

Fig. 12.11. Reshaping a polygon boundary with the Topology Edit tool and the Sketch tool

The Topology Edit tool is also convenient when working with connected line features, such as roads. If one road node is moved, the roads attached to the node move also to maintain their connection. In Figure 12.12, the road node was selected with the Topology Edit tool and then moved downward. The attached vertices of the other lines are also moved when the sketch is finished.

The Topology Edit tool can in many cases be used in place of the Select tool when editing. It can be applied to moving features, reshaping them, modifying them, and so on. However, the map topology must be created for the layers being edited before it can be used.

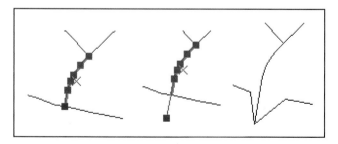

Fig. 12.12. When editing shared nodes, the node is moved, and the attached roads follow.

Summary

➢ The Sketch tool may take five different forms for creating vertices.

➢ Modifying and reshaping are two ways to change how existing features look without having to re-create them.

➢ Features may be combined to create new ones using the merge, union, intersect, and clip functions. Care must be taken to ensure that attributes are correctly copied during these transactions.

➢ New polygon or line features may be created by buffering existing points, lines, or polygons.

➢ Planar topology establishes rules about the spatial relationships within and between layers, and can be used to help locate and eliminate errors. Building and editing with planar topology requires an ArcEditor or ArcInfo license.

➢ Temporary topologic relationships, called map topology, may be used when editing with an ArcView license, allowing features which share vertices or boundaries to be edited simultaneously with the Topology Edit tool.

Chapter Review Questions

You may need to consult the Skills Reference section to answer some of these questions.

1. What is the difference between modifying a feature and reshaping a feature?

2. What is the difference between a union and a merge?

3. What determines the direction of a newly created line?

4. What are two ways to create donut polygons?

5. Examine the drawing. In (a), the polygons were created one after the other using Create New Feature. In (b), the first was created as a donut and then filled using the Trace tool. They look identical, but they are not. What is the difference? (**Hint:** What is the area of the outer polygon in each case?)

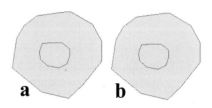

6. How do you establish whether the output feature of a buffer command is a line or a polygon?

7. What are the two types of output features possible when you divide a line?

8. Which sketch tool is used when filling a hole in a polygon?

9. If you trim a line and the wrong end gets deleted, how do you fix it?

10. Explain the differences between map topology and planar topology in a geodatabase.

Mastering the Skills

Teaching Tutorial

The following examples provide step-by-step instructions for doing basic tasks and solving basic problems in ArcGIS. The steps you need to do are highlighted with an arrow ➔; follow them carefully. Click on the video number in the VideoIndex to view a demonstration of the steps.

➔ Start ArcMap and open the ex_12.mxd document in the mgisdata\MapDocuments folder.

➔ Add the practice layers from last chapter, practicept, practiceln, and practicepo.

➔ Use Save As to rename the document, and remember to save frequently as you work.

Using other sketch tools

First we will practice using some of the different sketch tools. Figure 12.13 shows the block numbers from the last lesson as a reminder to help navigate to the right places during this tutorial.

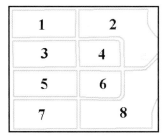

1➔ Choose Start Editing from the Editor toolbar.

1➔ Set the target layer to practiceln, and the task to Create New Feature.

1➔ Zoom into Block 1, showing the buildings and parcels.

1➔ Zoom into the third house from the left, in the lower row (Fig. 12.14).

Fig. 12.13. Block numbers

Imagine drawing a patio nestled in the cutout on the NE side of the house. The patio is flush with the edges of the house. You can use the Intersection tool to precisely lay out this patio, and you will need to switch back and forth between two sketch tools for this task. Using snapping ensures that the patio edges exactly meet the house corners.

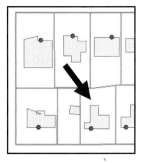

1➔ Choose Editor > Snapping from the Editor toolbar.

1➔ Turn on Vertex snapping for the buildings layer by checking its box.

Fig. 12.14. Zoom to this parcel

TIP: The videos demonstrate opening/closing the Snapping window each time it is used. For easier control, dock the Snapping window in the Table of Contents and keep it open during editing. Either method is acceptable; choose which you prefer. (Adjust the size of the Table of Contents and Snapping windows by placing the cursor over the boundary between them until a double arrow symbol appears, and dragging the boundary to the desired position.)

2➜ Make sure that the target layer is practiceln, the task is Create New Feature, and the Pencil tool is clicked.

2➜ Place the cursor very near the corner of the building marked 1 (Fig. 12.15). Move the cursor back and forth, and notice how it "snaps" to the corner of the building when you get close.

2➜ Click to add the first vertex at point 1.

For the next vertex, we want it to lie at the intersection of the two building walls enclosing the patio. We will use the Intersection tool for this.

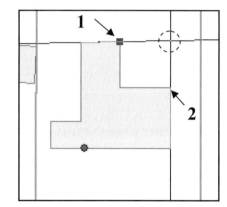

 2➜ Switch to the Intersection tool using the drop-down box next to the Pencil tool.

2➜ Place the cursor over the east wall of the house. A thin black line will appear, indicating the direction of the wall. When the correct line appears, click. The line remains.

2➜ Place the cursor over the north wall. Its line should intersect the first one. Click again.

Fig. 12.15. Finding the intersection of the east and north walls

The point is automatically placed at the intersection when you click the second line. Finally, we need to add the third point at 2.

2➜ Switch back to the Pencil tool.

2➜ Move the cursor toward point 2 until it snaps to the corner of the building.

2➜ Double-click to enter the third vertex and complete the sketch (Fig. 12.16).

➜ Make a couple more patios on other houses for practice.

Fig. 12.16. The patio

Next we'll use the Arc tool to make a curved patio.

 3➜ Select and delete the first patio.

3➜ Switch to the Arc tool.

3➜ Start the patio again at point 1 of the house (Fig. 12.17).

3➜ Click another point at the approximate intersection of the north and east walls. The arc will pass through this point.

3➜ Double-click on point 2 to add the third vertex and finish the sketch (Fig. 12.17).

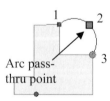

Fig. 12.17. Sketching an arc

Finally, let's see how the Distance-Distance tool works. This one is used to find the intersection points of two circles of specified radius. Imagine that you are digitizing the locations of neighborhood electric boxes. The interns locate these in the field by parcel number, and then by measuring the distance to the closest house and lot corner. You use this information to digitize the location of the box. In this particular lot, the intern's field notebook locates it 12 meters from the NE corner of the house and 9 meters from the NE corner of the lot.

1. Which kind(s) of snapping should be on for this task? _____

> 4➔ Pan or zoom to the parcel in the NE corner of Block 1.
>
> 4➔ Turn on vertex snapping for the buildings layer and the parcels layer.
>
> 4➔ Change the target layer to connects, so you can enter a point rather than a line.

> 5➔ Click the Distance-Distance tool.
>
> 5➔ Click on the NE corner of the house to specify the first distance (Fig. 12.18).
>
> 5➔ Notice the circle start to expand as you move away from the click point. The size of the circle is reported in the lower left of the ArcMap window.
>
> 5➔ To specify an exact distance, press the D key. Type "12" in the box and press Enter.
>
> 5➔ Click the NE corner of the parcel to start the next distance.
>
> 5➔ Press the D key. Type "9" in the box and press Enter.

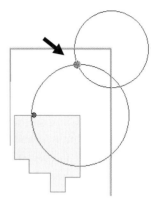

Fig. 12.18. Distance-Distance intersection

Two intersection points are defined where the circles cross. As you move the cursor on top of the intersections, a bubble appears to indicate where the next vertex would go. One of the intersections is in the street, and so it is obviously incorrect.

> 5➔ Click the bubble at the intersection inside the parcel to add the point.

The other sketch tool is called the Trace tool. We can use it to create a new feature that exactly matches an existing one, or is offset from it.

> 6➔ Turn off all the layers except streets and practiceln.
>
> 6➔ Zoom to the extent of the streets layer, then zoom into Block 3.
>
> 6➔ Select the outline of Block 3.
>
> 6➔ Choose practiceln as the target layer.

> 7➔ Click the Trace tool.
>
> 7➔ Click on one corner of the block to start the trace.
>
> 7➔ Move the cursor around, tracing the outline of the block. A thin black line will show the trace.
>
> 7➔ When you get back to the starting point, click again. The sketch appears.
>
> 7➔ Right-click on a part of the sketch, and choose Finish Sketch.

TIP: Sometimes the trace stops following the mouse because it has reached the "end" of the feature being traced. The end is determined by the way the feature was digitized. To continue tracing, click the last end of the trace to add a vertex, and click again at the same spot to start the trace again. Continue until the entire feature has been traced.

To trace alongside a feature, specify an offset value. This function would be useful for entering water mains which follow along a street but are offset from it. Let's assume that the water mains in this part of town are constructed such that they lie 5 meters from the edge of the street to the north, or 5 meters from the edge of the street to the west. We will use the tracing function to precisely locate the water mains.

8➜ Zoom to the extent of streets, and select the top four blocks.

8➜ Choose the Trace tool.

8➜ Press the O key to set the offset. Type "-5" in the Offset box and press Enter.

8➜ Start tracing along Block 2, moving from east to west (Fig. 12.19).

8➜ When you reach the end of the block, the trace will stop. Click to enter a vertex.

8➜ Click on the east end of the next block to continue tracing along Block 4.

8➜ Click to add a vertex at the end of Block 4, then click the start of Block 3 to continue tracing.

8➜ At the end of the last block, double-click to finish the sketch.

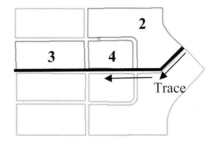

Fig. 12.19. Tracing a water main

If you made a mistake in the last set of directions, simply delete the sketch or the water main and start again.

TIP: If the water main is being traced on the wrong side of the street line, press O again to change the offset, and make it positive instead of negative. Any time a trace is on the wrong side, simply switch the sign of the offset value.

Changing polygons

Next, we will go back and modify some of the parcels we created in the last chapter. Begin by entering the parcel boundaries of an elementary school.

9➜ Turn on the Orthophoto, the buildings, the streets, and the parcels.

9➜ Pan/zoom into the open block to the west of Blocks 1 and 2, which has a star-like school building in the center (Fig. 12.20).

9➜ Set the target layer to parcels, and the task to Create New Feature.

Fig. 12.20. Zoom to the area inside the dashed line

9➜ Using the Pencil tool, sketch a polygon around this entire block, bounded by the street edges. Finish the polygon.

The side of the street is rather difficult to see on the west edge of the parcel. Let's zoom in and try to do a better job.

10➔ Zoom in as close as possible to the west edge of the new parcel, but still showing the entire street.

10➔ Turn on vertex snapping for parcels.

10➔ Change the task to Reshape Feature.

10➔ Select the parcel, if necessary.

10➔ Switch to the Pencil tool.

11➔ Click on the parcel polygon at the north end of the bounding street.

11➔ Click new vertices along the street, defining a sketch.

11➔ At the south end of the street, double-click one last vertex on the corner of the parcel to finish the sketch and create the new edge.

➔ Experiment until the line is satisfactory. Try starting and stopping the Reshape task in different locations along the polygon.

TIP: It is usually helpful to use vertex or edge snapping when reshaping features.

That is how to reshape a standalone feature, but shared editing works best when reshaping a boundary between adjacent polygons.

12➔ Click the blue Back arrow to return to the previous extent.

12➔ Turn on edge snapping for the parcels layer.

12➔ Change the task to Cut Polygon Features, and cut the block in two with a straight north-south line to the west of the school.

Now let's redraw this straight line to more closely match the real parcel boundary. In order to edit it as a shared boundary, however, you must first create a map topology.

13➔ Choose Editor > More Editing Tools > Topology from the Editor toolbar. The Topology toolbar will appear.

13➔ Click on the Map Topology icon on the Topology toolbar.

13➔ Check the box to have the parcels layer participate in the map topology. Leave the cluster tolerance at the default value. Click OK.

14➔ Change the task to Reshape Edge under the topology Tasks group in the Task bar.

14➔ Turn on vertex snapping for parcels.

14➔ Click the Topology Edit tool on the Topology toolbar.

14➔ Click on the north-south line in the middle of the block. It will turn purple to indicate it is a shared boundary (Fig. 12.21).

14➔ Click the Pencil tool.

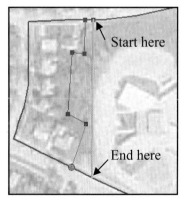

Fig. 12.21. Reshape the shared boundary

14➔ Click the north end of the boundary to start reshaping. Be sure to click ON the end.

14➔ Enter new vertices that go behind the houses, as shown in Figure 12.21.

14➔ Click again on the south end of the shared boundary to end the sketch. Be sure to click ON the end.

14➔ Right-click and choose Finish Sketch.

Next we will learn how to modify a sketch instead of reshaping it.

15➔ Change the target layer to buildings.

15➔ Change the task to Create New Feature and click the Pencil tool.

15➔ Create a building polygon for the large building in the center of the block. Be sloppy about it.

15➔ Change the buildings symbol to a hollow red line to show the photo underneath.

15➔ Zoom into the building in order to see it as clearly as possible.

As intended, the building does not exactly match the photo. Let's improve the fit.

16➔ The building should still be selected; if not, select it.

16➔ Change the task to Modify Feature. The building sketch appears.

16➔ Make sure the Select tool is clicked on the Editor toolbar.

16➔ Move the cursor over a vertex of the building. When it changes to a crosshair, click and drag the vertex to a new location.

16➔ Right-click another vertex, and choose Delete Vertex from the Sketch menu.

TIP: For best results, the Select tool should be active while modifying the feature.

16➔ Right-click in the middle of the segment, and choose Insert Vertex to replace the vertex just deleted.

16➔ Move the vertex back to its original location.

16➔ Continue moving, adding, or deleting vertices until you are satisfied with the shape of the building.

16➔ Right-click on the sketch and choose Finish Sketch.

Notice that the task changes back to Create New Feature (or possibly something else). Now clean up and prepare for the next task, changing line features.

17➔ Choose Save Edits.

17➔ Change the building symbol back to the solid beige color.

17➔ Zoom to the extent of the roadlines layer, and turn it on.

Changing lines

The roadlines layer represents the centerlines of the roads. They were exported from the rds_clp layer from the Chapter 10, and they come complete with names and address ranges. However, it is clear that they do not match well with the orthophoto. Editing these roads will provide a better match with the photo and the rest of the data.

First, we will move all of the roads as a group to get the best average fit. Examine the roads and note the direction and distance of offset in different locations of the map. Note that the offset is generally to the west and slightly south of the true location. We will move all the streets to a better position, being careful not to move any other features.

18➜ Choose Selection > Set Selectable Layers from the main menu bar, and make roadlines the only selectable layer.

18➜ Change the target layer to roadlines.

18➜ Click the Select tool on the Editor toolbar, and draw a box around ALL of the roads. Don't miss any!

19➜ Zoom into the lower left corner of the roadlines layer.

19➜ Click on the Select tool.

19➜ Click and drag the lower left road features to the right and slightly up, so that the corner falls in the exact middle of the road in the photo (Fig. 12.22). Move it several times, if need be, to get as close as possible.

19➜ Zoom back to the extent of the roadlines layer again.

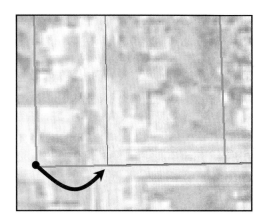

Fig. 12.22. Move the road corner into the center of the road in the photo

Already the road locations look much better, but some areas still need adjustment, such as the lower right near the corner of Block 8.

20➜ Clear the selected features so that nothing is selected.

20➜ Zoom into the roads extending down from the lower right corner of Block 8.

20➜ Turn off snapping for parcels, and turn on end snapping for roadlines.

The view should look similar to the picture in Figure 12.23. The following directions refer to the street segments according to the lettering scheme indicated. Since these features must connect at their ends, always use the Topology Edit tool during these edits. First, however, we must add the roadlines to the map topology.

21➜ Click the Map Topology button on the Topology toolbar. Uncheck parcels and check roadlines. Click OK.

21➜ Choose the Topology Edit tool.

21➜ Select the node at the intersection of streets A, B, and C by holding down the N key on the keyboard and clicking at the intersection. (The N key ensures that only nodes can be selected.)

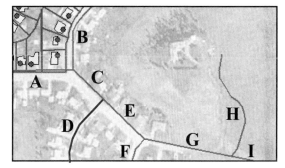

Fig. 12.23. The vicinity of East Oakland St.

21➜ Click and drag the node to the intersection point of the streets in the Orthophoto.

21➜ Select the node at the other end of segment C and move it to the intersection on the Orthophoto of segments D and C/E.

Street C moves to its new location. Because you used the Topology Edit tool, the streets connected to C moved also, so already this section looks much better.

21➜ Repeat the previous steps with the lower end of Street E to match it to the intersection in the photo.

Examining Street B up close, notice that its last segment flattens out and does not follow the photo street as well as it could. Use Reshape to fix it.

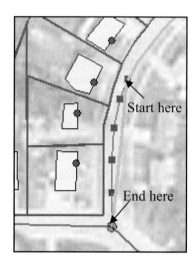

22➜ Turn on edge snapping for roadlines.

22➜ Zoom into the lower end of Street B where the flat section is (Fig. 12.24).

23➜ Click Street B with the Topology Edit tool.

23➜ Choose Reshape Edge from the Topology Tasks in the Task bar.

23➜ Click the Pencil tool.

23➜ Click above the flat section to start the reshaping sketch.

23➜ Enter new vertices along the street in the photo.

Fig. 12.24. Reshaping Street B

23➜ Double-click at the intersection of Streets A, B, and C to end the sketch.

TIP: If the sketch does not result in a changed shape, try starting the reshape sketch from the other end, and be sure to cross over the original feature at the other end.

Next, Street F has a bit of a kink where it runs into Streets E and G.

24➜ Zoom to the previous extent and zoom into the intersection of F, E, and G.

24➜ Change the task to Modify Edge in the Topology Tasks.

24➜ Select Street F with the Topology Edit tool.

24➜ Add, delete, and move vertices until the kink is gone. (You cannot move the end vertex because it is shared by other features.)

24➜ Right-click the sketch and choose Finish Sketch.

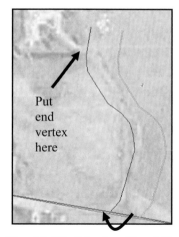

Finally, Street H is way off (Fig. 12.25). We'll begin by moving the end nodes to the correct locations. Then we'll use Modify Edge to correct the ending location, and finally use Reshape to enter the street correctly.

Fig. 12.25. Moving Street H

25➔ Zoom to the previous extent. Change the task to Create New Feature.

25➔ Click off the roads to clear the selected feature. Then use the Topology Edit tool to select the node at the south end of Street H (Fig. 12.25).

25➔ Click and drag the Street H node until it is in the correct location.

25➔ Change the task to Modify Edge.

25➔ Select Street H with the Topology Edit tool.

2. Which direction does this line run, from north to south or from south to north? _____ How can you tell? _____

25➔ Move the top vertex of the sketch to the location shown in Figure 12.25, near the end of the large white shape.

25➔ Right-click the sketch and finish it.

TIP: To have attached features proportionally stretched when moving nodes, check the box on the General tab of the Editing > Options menu. This option often provides better adjustment of features.

26➔ Select Street H with the Topology Edit tool, if necessary.

26➔ Make sure that only end snapping for roadlines is on.

26➔ Change the task to Reshape Edge.

26➔ Click the Pencil tool.

26➔ Click on the bottom end of the street to start the reshape sketch, making sure you snap to the endpoint.

26➔ Enter the vertices along the road visible in the photo.

26➔ Double-click on the end of the road to finish the sketch, making sure you snap.

Now this section of the neighborhood looks in good shape. Before we leave it, however, let's experiment with some line directions.

27➔ Click the line symbol of the roadlines layer in the Table of Contents to open the Symbol Selector.

27➔ Scroll down to the bottom of the Symbols window, and choose the symbol called Arrow at End. Increase the line width to 2 points. Click OK.

27➔ Choose the Select tool and click someplace off all roads, to clear the selection.

Examine the directions of the roads. Let us pretend that Street H accidentally changed direction because we reshaped it in the wrong direction. We need to flip it back to its original orientation, from south to north.

28➔ Select Street H with the regular Select tool.

28➔ Change the task to Modify Feature.

28➔ Right-click one of the vertices and choose Flip from the context menu.

28➔ Right-click the sketch again and choose Finish Sketch.

28➔ Clear the selection by clicking off the line with the Select tool.

28➔ Save all edits.

The line has been flipped and now travels from south to north.

More techniques for editing lines

In this next section we will experiment with some additional ways to work with line features.

29➔ Zoom to the extent of the roadlines layer.

29➔ Pan the map to the right to show the empty region west of the roadlines layer.

29➔ Turn off the Orthophoto layer.

We will use some commands to create a road layer from scratch, using geometric measurements. Before beginning, take a look at Figure 12.26 to see the final result.

30➔ Change the target layer to practiceln, and change the task to Create New Feature.

30➔ Click the Pencil tool.

30➔ Right-click on the screen and choose Absolute X,Y.

30➔ Type the values "641883" and "4879835" in the boxes and press Enter. The first vertex is entered. You may need to pan or zoom to see the first vertex.

TIP: If for some reason you cannot find the first vertex on the screen, right-click to delete the sketch, and start again by simply entering an arbitrary point.

31➔ Right-click the screen again and choose Direction/Length from the menu.

31➔ Type the values "90" and "700" and press Enter.

31➔ Right-click the sketch and finish it.

Next we'll copy this street to make several blocks to the east.

32➔ The new line should still be selected. If not, select it.

32➔ Choose Editor > Copy Parallel from the Editor menu.

32➔ Type "200" for the distance and press Enter.

32➔ Repeat the Copy Parallel to create lines at 400, 600, 800, and 1000 meters. Adjust the Pan/Zoom if needed so that all are visible.

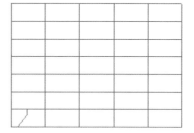

Fig. 12.26. The road segments will eventually look like this.

Next we create the cross streets.

33➔ Turn on end snapping for practiceln.

33➔ Make sure Create New Feature is the task and that the Pencil tool is clicked.

33➔ Create a horizontal line across the bottom, making sure to snap it to the end of each vertical line.

To ensure that the streets connect to each other, we will divide the vertical lines into equal pieces.

34➔ Set practiceln to be the only selectable layer.

34➔ Select the first vertical line to the west.

34➔ Choose Editor > Divide from the Editor toolbar.

34➔ Choose to place points separated by 100 meters. The resulting line segments remain selected.

The undivided line remains also, so get rid of it.

34➔ Choose the Select tool and click on the line. If the *entire* line changes to the selection color, then you have selected the correct line. Click Delete.

34➔ The line segments should remain behind. Click part of the line. A short segment will be highlighted.

➔ Repeat step 34 to divide each of the other vertical lines. Make sure to delete the original long line in each case.

Well, that was a little bit of a nuisance. For the horizontal lines, let's divide BEFORE we copy, so we won't need to divide the copies also.

35➔ Select the horizontal line and divide it into 200 meter lengths.

35➔ While the segments of the divided line are *still selected,* choose Copy Parallel.

35➔ Copy the lines to locations 100, 200, 300, …, 700 meters above the original line.

TIP: If the first line ends up below the original instead of above, Undo it and try again, this time entering the opposite sign of the first number, e.g., –100 instead of 100.

All this time, the original full-length line is still present. We need to delete it, leaving only the short segments.

36➔ With the Select tool, click off all lines to clear the selection, then click the lowest horizontal line, making sure the entire line is highlighted at once.

36➔ Press the Delete key to get rid of the line.

36➔ Save your edits.

Now we have a basic block structure of some streets, but we need to add a couple of small changes. First, there is a small alley to add in the lower left block (Fig. 12.26). We can't just draw it in because we must make sure that the street splits where the drive starts. The drive begins at 37 meters from the west end of the block.

37➔ Select the lower left horizontal line segment, the bottom of the lower left block.

37➔ Choose Editor > Split from the Editor toolbar.

37➔ Enter the distance "37" and choose OK. Watch as it creates the split. The short end should be created on the left (west).

TIP: If the split occurs at the wrong end (east), then Undo and try it again, this time choosing "From End Point of Line" as the Orientation in the Split window.

Now that we have the split at the correct location, we can snap to it and create the alley.

38➔ Make sure that end snapping is turned on for practiceln, and all other snapping is off. We want to be sure we are right on the end.

38➔ Make sure the task is Create New Feature, and click the Pencil tool.

38➔ Move the cursor to find the location where the line was split: the cursor will snap to it. Enter the first vertex.

38➔ Right-click off the sketch and choose Direction/Length. Enter "50" for the direction and "97" for the length and press Enter.

38➔ Turn on edge snapping for practiceln.

38➔ Right-click one of the vertical lines and choose Parallel.

38➔ Enter the last vertex where the constrained line crosses the next street up. Make sure it snaps to the street edge. Finish the sketch.

Now we still have to create a split where the alley meets the street above. We don't know the exact length of the split, but we can snap to the end of the alley.

39➔ Turn off edge snapping for practiceln, but keep end snapping.

39➔ Select the horizontal line that meets the alley at the top.

39➔ Click the Split tool on the Editor toolbar.

39➔ Click on the place where the alley meets the road, making sure to snap to its end. The road will split (Fig. 12.27).

39➔ Save all edits.

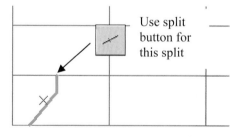

Use split button for this split

Fig. 12.27. Adding the alley

More techniques for polygons

In this next section we'll take a look at some more ways to work with polygon features, including mirroring them and combining them. We'll also learn to make donut polygons. Most of these functions are accessed from the main Editor toolbar under the Editor menu button.

40➔ Turn off the practiceln layer.

40➔ Turn off all snapping.

40➔ Change the target layer to practicepo, and turn it on.

First we'll learn how to mirror a polygon.

41➜ Make sure the task is Create New Feature, and the Pencil tool is selected.

41➜ Create an irregular polygon with an asymmetric shape such as the one shown in Figure 12.28.

41➜ Make sure the polygon is selected.

42➜ Choose the Mirror Features task from the Editor toolbar.

42➜ Make sure the Pencil tool is selected.

42➜ Enter the start and end vertices of a line which defines the mirror plane (i.e., the line across which the figure is mirrored). The feature is mirrored as soon as you enter the second point of the line.

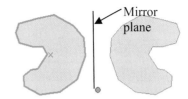

Fig. 12.28. Mirroring a feature

Now we will experiment with combining features.

43➜ Set practicepo to be the only selectable layer.

43➜ Select the two mirrored polygons with the Select tool.

43➜ Choose Editor > Merge from the Editor toolbar.

43➜ Select the first polygon in the list and watch it flash. The other will flash if the second polygon is selected. Select the first and click OK (Fig. 12.29).

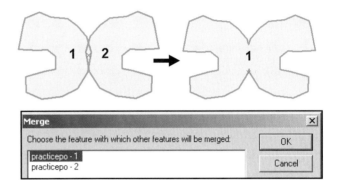

Fig. 12.29. The merged polygon is given the attribute values of the polygon selected in the Merge window.

TIP: The attribute values of the merged polygon will become the same as the attributes of the selected polygon in the window. For example, if the selected polygon is Residential land use, and the unselected one is Commercial, the resulting merged polygon will have the attribute Residential. Note that the polygon list shows the primary display field for the layer, making it easier to decide which polygon to use.

No change may be visible, but the multiple polygons are now in fact a single feature.

43➜ Click off the polygons with the Select tool to clear the selection.

43➜ Now click on ONE of the polygons.

43➜ Open the practicepo attribute table and notice that it contains only the single feature (unless you have other polygons from a previous session, in which case clicking Show Selected will demonstrate that these two shapes form one multipart polygon).

Both polygons are highlighted when you click one, showing that they are a single multipart feature. Moreover, the table contains only a single record for both.

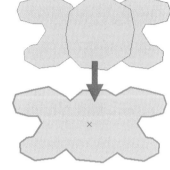

Fig. 12.30. Merging a new polygon with the merged mirror polygons

44➜ Resize the table and place it off to the side so you can see it and the polygons at the same time.

44➜ Set the task back to Create New Feature and click the Pencil tool.

44➜ Create a new polygon such that it overlaps both parts of the merged polygon (Fig. 12.30).

44➜ Select all of the features with the Select tool.

44➜ Choose Editor > Merge from the Editor toolbar.

The boundaries disappear and now all the features form a single merged feature.

3. How many polygons are selected now? _____

Now let's examine the difference between a merge and a union.

45➜ Choose Edit > Undo Merge from the main menu bar.

45➜ Select all the features again.

45➜ Choose Editor > Union from the Editor toolbar.

At first sight, the result appears the same. However, it is not.

45➜ Click on the selected union polygon and drag it off to the side.

The original polygons are still present underneath. Union, then, creates a brand-new feature and leaves the input features as they were. Next we will see how Intersect works.

46➜ Delete the union feature.

46➜ Choose the remaining polygons (there are two, the merged mirror features and the one on top) (Fig. 12.31).

46➜ Choose Editor > Intersect from the Editor toolbar.

46➜ Click on one of the selected intersection areas and drag them below the original features.

4. Are the two areas separate polygons, or are they part of a multipart polygon? _____ How can you tell? _____

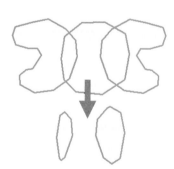

Fig. 12.31. Intersecting polygons

Finally, let us examine the clip function.

47➔ Delete the intersection result.

47➔ Select the single polygon area overlapping the merged mirror polygon (Fig. 12.32).

47➔ Choose Editor > Clip from the Editor toolbar.

47➔ Choose to discard the area that intersects and click OK.

47➔ Click and drag the clip polygon out of the way to see the result.

Now let's try clipping the other way.

48➔ Choose Edit > Undo from the main menu bar twice, to undo the move and undo the clip.

48➔ Clip again, but this time choose to Preserve the area that intersects.

48➔ Click and move the clip polygon out of the way.

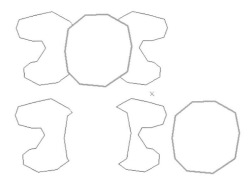

Fig. 12.32. Clipping with the circle polygon

Notice that the result is very similar to the Intersect done earlier. The only difference is that Intersect creates a new feature, but Clip with Preserve modifies the original features.

49➔ Select and delete all of the polygons in the view.

49➔ Close the attribute table.

Creating donut polygons

In this section we will experiment with two ways to create donut (or Swiss cheese) polygons. The first method uses multipart polygon creation.

50➔ Make sure the Create New Feature and the Pencil tool are selected.

50➔ Begin a sketch and create a large circular polygon area. Don't finish it yet (Fig. 12.33).

50➔ After adding the last vertex, right-click and choose Finish Part.

50➔ Click inside the first sketch, and create another ring sketch outlining the inside of the donut.

50➔ After completing the second ring, right-click and choose Finish Sketch.

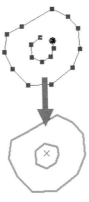

Fig. 12.33. Creating a multipart donut polygon

Although we only created one inside hole, we could have created more by choosing Finish Part after each one until the last.

51➜ Create a "Swiss cheese" polygon with several holes inside (Fig. 12.34).

The second method works to add a hole after a polygon is already finished.

Fig. 12.34. Swiss cheese polygon

52➜ Create a new whole polygon and finish it.

52➜ Create another smaller polygon inside it, and finish it.

52➜ Select the smaller polygon.

52➜ Choose Editor > Clip, and choose to discard the area that intersects.

52➜ Delete the clip polygon to see the result.

Now for an interesting effect, let's try clipping with a donut polygon.

53➜ Delete all the polygons except the original donut.

53➜ Create a new polygon next to the donut, and larger than it.

53➜ Select and move the donut completely inside the larger polygon.

53➜ Choose Clip and discard the area that intersects.

53➜ Move the selected clip polygon out of the way to see the result.

5. What happened? _____

6. Does the result contain one polygon, or more than one? _____

54➜ Delete all of the polygons.

Finally, let's see how to fill a hole with another polygon. Imagine digitizing a floodway zoning polygon, and in the middle it contains a park.

54➜ Create a polygon sketch representing the flood zone, approximately as shown in Figure 12.35. Finish the part.

54➜ Create another part of the polygon inside the flood zone, representing a rectangular park. Finish the sketch.

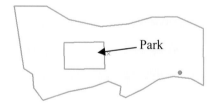

Fig. 12.35. A park inside the floodway

55➜ Choose the Trace tool.

55➜ Click on a location on the park boundary to start the trace.

55➜ Trace around the hole.

55➜ When back at the starting place, click to create the sketch.

55➜ Right-click and choose Finish Sketch.

56➜ Delete the floodway and park.

56➜ Save all edits.

Buffering

Buffers create new features around existing ones. Recall the set of streets we created in practiceln. Suppose we want to create street edges from these lines. Assuming that the average road width in this part of town is 15 meters, we will buffer half that distance, or 7.5 meters (Fig. 12.36).

57➔ Turn on the practiceln layer again and zoom to its extent.

57➔ Make practiceln a selectable layer, keeping practicepo selectable also.

57➔ Make sure that practicepo is the target layer.

57➔ Delete any polygons from practicepo that might still be in the view.

Fig. 12.36. Streets created by buffering the road lines

58➔ Use the Select tool to click and drag a box around all of the road segments to select them.

58➔ Choose Editor > Buffer from the Editor toolbar.

58➔ Type the value "7.5" in the distance box and press Enter.

TIP: Sometimes when buffering you may get an error message stating that the units or measures are out of bounds. If this happens, save any important edits, stop editing, save the map document, and close ArcMap. Start ArcMap again and reopen the map document. This routine usually fixes the problem.

Since each segment is buffered individually, the buffers intersect at the corners. Open corners are more attractive and realistic and can be created by merging the buffers (Fig. 12.36).

59➔ Make sure the buffers are still selected.

59➔ Choose Editor > Merge from the Editor toolbar. Select any of the polygons in the list and click OK.

59➔ Save all edits.

The interior polygon boundaries are removed, leaving an open network of streets.

Negative buffers can create lines showing easements. We will create a five-meter easement from the parcel boundary around the school we digitized earlier in this lesson (Fig. 12.37). In this case we are creating a line buffer from a polygon.

60➔ Zoom to the extent of the roadlines layer.

60➔ Turn off the roadlines layer.

60➔ Add parcels to the set of selectable layers.

60➔ Locate the large school parcel we did earlier, and zoom into it.

60➔ Change the target layer to practiceln, so the buffers will be lines.

Now we are ready to buffer. Since the easement falls INSIDE the parcel, we will need to use a negative buffer distance. We also use the full value, since only one side of the buffer is created when the distance is negative.

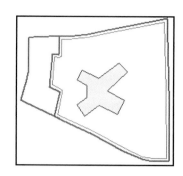

61➔ Select the school parcel (Fig. 12.37).

61➔ Choose Editor > Buffer from the Editor toolbar.

61➔ Enter the distance value "-5".

Our easement is created.

Fig. 12.37. Creating an easement

62➔ Stop editing and save all edits.

62➔ Remove practiceln and practicepo from the map document.

62➔ Save the map document. You'll want it later for the exercises.

You now know many different ways to work with features while you are editing. As a final parting shot, you are reminded to always have a backup copy of the layer you're working with, and save often!

This is the end of the tutorial.

➔ Exit ArcMap.

Exercises

For the first six questions, use your practicepo or practiceln shapefile to create the shape shown. Take care to make the feature parts the same size and/or equidistant if they look that way in the example. Also list the commands/tools you used to create it. **Capture** your figures.

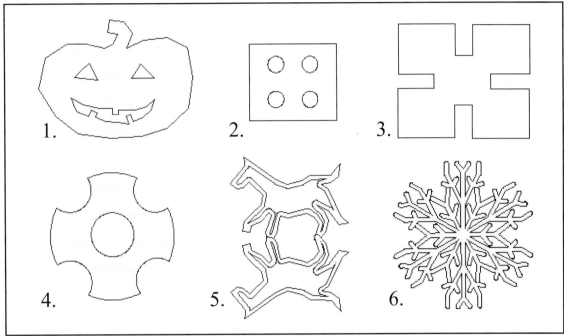

7. Finish adjusting the roads in the roadline shapefile to match the Orthophoto.

8. Use the adjusted road centerlines to create street boundaries as polygons in a new shapefile. Assume that the roads are 10 meters wide.

9. Create a layout showing your adjusted roads and road boundaries.

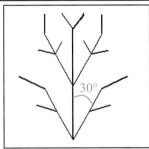

Hint on #6: Create half of this first.

Challenge Problem

Design and construct a system of water mains for the neighborhood included by the roadlines. Be sure that you establish a realistic flow of water through the network by having the lines in the correct direction. Every building must have reasonably close access to water. Create a layout showing your work, complete with arrows on the lines. Put the orthophoto (60% transparent) in the background.

Skills Reference

Using the sketch tools

The following techniques may be used to create single point locations or to place the next vertex of a line or polygon. **Note that in each case below, you have already selected the target layer and the task.** Each sketch tool can be used at the beginning, middle, or end of a sketch. You can also use multiple tools in succession during a single sketch.

Finding intersections

Imagine starting a new street edge at the intersection of two adjacent edges.

1. Click the small black arrow on the current sketch tool and choose the Intersection tool.

2. As you move the tool over a line on the screen, its extension will be indicated. Click the first line on the south side (1 in Fig. 12.38).

3. Click the second line on the west side. At this point the point of intersection will be visible where the extensions of the two lines cross.

4. Click near the intersection to enter the point or vertex.

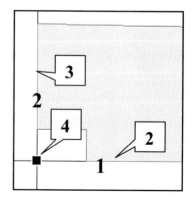

Fig. 12.38. Placing a vertex at the intersection of two lines

5. Finish the sketch, or switch to another sketch tool and keep sketching.

Using distance-distance intersection

1. Click the small black arrow on the current sketch tool and choose the Distance-Distance tool.

2. Click at the center of the first distance circle and move the mouse out to expand the circle (Fig. 12.39).

3. Press the D key to bring up the Distance window, and type in the exact distance, in map units. Press Enter to close the window.

4. Click at the center of the second circle, and type D to enter a distance for it. Press Enter to close the window.

5. Move the cursor over one of the two intersection points (a blue circle will appear). Click to create the point or vertex.

6. Finish the sketch, or switch to another sketch tool and keep sketching.

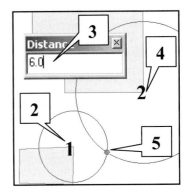

Fig. 12.39. Finding the intersection point of two distance circles

Creating a parametric arc

1. Click the small black arrow on the current sketch tool and choose the Arc tool.

2. Click the starting point of the curve (Fig. 12.40).

3. Click the point that the curve will pass through.

4. Click the endpoint of the curve.

5. Continue adding pass-through and endpoints to create multiple curves, if desired.

6. Finish the sketch, or switch to another sketch tool and keep sketching.

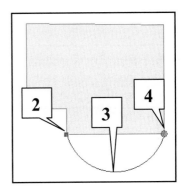

Fig. 12.40. Sketching a parametric curve

Tracing along features

1. Click the Select tool and select the feature to be traced (Fig. 12.41).

2. Click the small black arrow on the current sketch tool and choose the Trace tool.

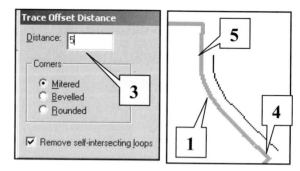

3. To offset from the original feature, press the letter O key on the keyboard to specify an offset distance in map units. Type the distance into the box and press Enter. (If the trace appears on the wrong side, press O again and make the distance negative.) If desired, also change the style of the corners.

Fig. 12.41. Creating a sketch by tracing

4. Click on the feature to begin tracing, and move the mouse along the feature to create the trace.

5. Click again to finish the trace and create the sketch.

6. Finish the sketch, or switch to another sketch tool and keep sketching.

TIP: Features have ending locations according to how they were digitized. A trace stops when it reaches the end of the feature. To continue tracing, click at the stopping point to end the current trace, then click again at the same point to begin a new trace.

Using the Endpoint Arc tool

With this tool the user specifies the beginning, end, and radius of an arc.

1. Click on the Endpoint Arc tool.

2. Enter the first point of the arc (Fig. 12.42).

3. Enter the second point of the arc.

4. Move the mouse to the desired size.

5. To specify an exact radius, press the R key, type in the radius, and press Enter.

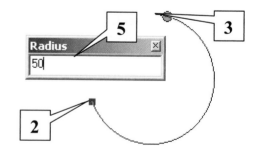

Fig. 12.42. Using the Endpoint Arc tool

Using the Midpoint tool

The Midpoint tool is useful for entering a vertex at the halfway point of a specified distance. Imagine that you wish to split an existing parcel exactly in half.

1. Turn on vertex snapping for the parcel layer to ensure that the exact midpoint is found.

2. Select the parcel polygon to split and set the edit task to Cut Polygon Features (Fig. 12.43).

3. Choose the Midpoint tool.

4. Click on the location representing the start of the measuring line to be halved.

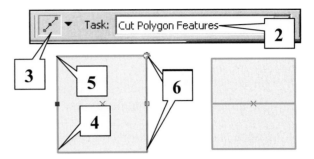

Fig. 12.43. Using the Midpoint tool to split a parcel in half.

5. Click on the other end of the measuring line. A point is entered at the midpoint of the measuring line.

6. Click to start and end another measuring line on the other side of the parcel, thereby entering a second vertex.

7. Right-click and choose Finish Sketch.

Using the Tangent Curve tool

This tool creates an arc that is tangential to the previous sketch segment, and is useful for creating smooth curves for roads or other features. At least one sketch segment must be present for the tool to be active.

1. Use the Pencil or another sketch tool to create one or more sketch segments (Fig. 12.44).

2. Click the Tangent Curve tool.

3. Move the mouse to the desired endpoint of the tangential curve and click to create the curve.

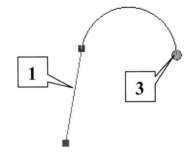

Fig. 12.44. Creating a tangential curve

Using the Direction-Distance tool

This tool creates a new vertex using a bearing direction from a known point plus a distance from another point. Imagine digitizing a service road that stops 20 feet from a well.

1. Use another tool to enter the first part of the line (Fig. 12.45).

2. Choose the Direction-Distance tool.

3. Click on the well to establish the rotation point of the direction segment. A temporary line appears.

4. Rotate the line to the desired direction and click, or press the D key to type a specific direction angle and press Enter.

5. Click on the well to define the center of the circle.

6. Move the mouse to the desired radius and click, or press R to type a specific radius and press Enter.

7. Click on one of the two possible intersections to create the vertex.

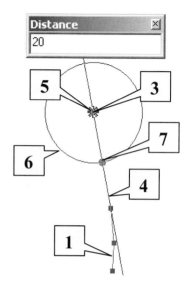

Fig. 12.45. Using the Direction-Distance tool

Modifying a feature

1. Choose Modify Feature from the Task bar in the Editor menu.

2. Select a feature to be modified with the Select tool. The sketch of the feature will appear (Fig. 12.46).

3. To delete a vertex, place the cursor on the vertex, right-click, and choose Delete Vertex from the Sketch menu.

4. To add a vertex, place the cursor over the spot to add the vertex, right-click, and choose Add Vertex.

5. To move a vertex, place the cursor on top of the vertex to be moved, so that a crosshair symbol appears. Click and drag the vertex to a new location.

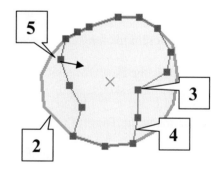

Fig. 12.46. Modifying a feature

6. When done modifying the vertices, right-click ON the sketch and choose Finish Sketch from the context menu.

TIP: For best results, the Select tool should be active while modifying the feature.

Reshaping a feature

1. Choose Reshape Feature from the Task bar in the Editor menu.

2. Select a feature to reshape.

3. Click on the Pencil tool.

4. Click ON the feature at the start of the section to be reshaped (Fig. 12.47).

5. Enter vertices defining the new shape. Use the other sketch tools and context menus as needed.

6. Double-click ON the feature at the end of the section being reshaped, OR right-click and choose Finish Sketch.

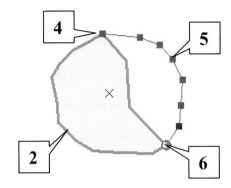

Fig. 12.47. Reshaping part of a polygon boundary

TIP: It is very helpful to use snapping when reshaping features. If the start/end points of the new sketch do not fall exactly on the feature, the reshaping will not be performed.

Flipping a line

1. Choose Modify Feature from the Task bar on the Editor toolbar.

2. Click the Select tool and select the line to be flipped. The line's sketch will appear. The red vertex shows the END of the line.

3. Right-click ON the sketch and choose Flip from the Sketch menu.

4. To flip another line, select it and repeat step 3.

5. When done, either right-click the sketch to finish it or clear the selection by clicking on the screen.

Displaying line directions

1. Click the line layer's symbol in the Table of Contents.

2. In the Symbol Selector, scroll to the bottom of the list to find the arrow symbols.

3. Choose the symbol with an arrow at the start of the line or at the end of the line (Fig. 12.48).

4. Change the color or width as desired. Click OK.

Fig. 12.48. Arrows in the Symbol Selector for showing line directions

Mirroring features

1. Choose Mirror Feature from the Task bar on the Editor toolbar.

2. Select the feature to be mirrored (Fig. 12.49).

3. Click the Pencil tool.

4. Click on the two endpoints of a line which defines the mirror plane (i.e., the line across which the feature is mirrored).

5. After the second click, the mirror feature will be created.

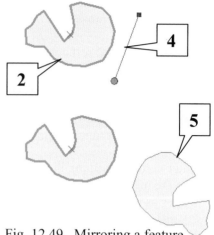

Fig. 12.49. Mirroring a feature

Dividing a line

This task makes it possible to place points along a line, either at equal intervals defined by the user, or evenly spaced along the available distance. If the target layer contains points, then point features along the line will be placed in the target. If the target layer is a line layer, then equal line segments will be created and placed in the target as lines.

1. Check the target layer type. If it is lines, then line segments will be created. If it is points, then points will be created.

2. Select the line along which to place the points (Fig. 12.50).

3. Choose Editor > Divide from the Editor toolbar.

4. Choose to place a set number of points evenly along the line, or to separate points by a set distance. Click OK.

5. The points will be placed in the point target layer.

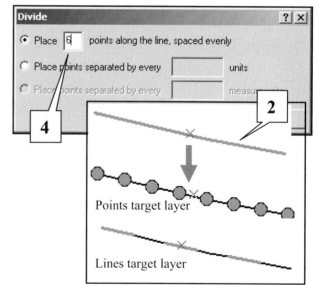

Fig. 12.50. Placing points along a line

TIP: If the target layer is the same as the line being divided, then the line segments will be placed as selected features on top of the original line. The segments may then be moved to a new location, or the original line underneath may be selected and deleted.

Splitting lines exactly

You can split lines by clicking the Split button and pointing at the location to split, as we learned in Chapter 11. This other method can split a line at a specified distance from the start or endpoint. (See "Flipping lines" to learn how to display the direction of lines.)

1. Select a line to split (Fig. 12.51) and choose Editor > Split.

2. Fill the button to choose distance or percentage measurement.

3. Enter the distance or percentage.

4. Choose the direction of the split (from start or end).

5. Click OK. The line will be split in two, with both pieces remaining selected.

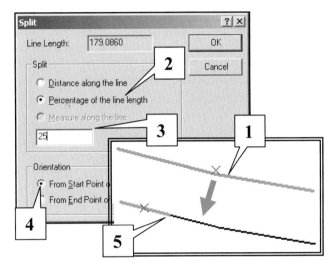

Fig. 12.51. Splitting a line at an exact location along it

Moving a feature an exact distance

1. Select the feature to be moved (Fig. 12.52).

2. Choose Editor > Move from the Editor toolbar.

3. Type the change in *x*- and *y*-coordinates as an offset from the current location and press Enter.

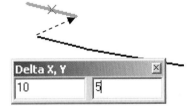

Fig. 12.52. Moving a line an exact offset

Using Copy Parallel

This function copies a feature and places it parallel to the original at a specified distance.

1. Select the feature to copy (Fig. 12.53).

2. Choose Editor > Copy Parallel from the Editor toolbar.

3. Type the distance offset.

4. Change corner style if desired.

5. Click OK or Enter to create the copy.

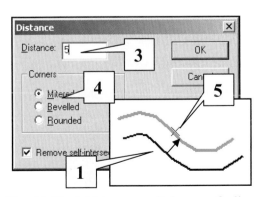

Fig. 12.53. Making a parallel copy of a line

Trimming lines

1. Choose Extend/Trim Features from the Task bar in the Editor menu (Fig. 12.54a).

2. Select the line to be trimmed.

3. Click the Pencil tool.

4. Enter two endpoints of a line crossing the feature and finish the sketch.

5. The feature will be trimmed where it crosses the line.

TIP: If the wrong side of the line disappears, choose Edit > Undo from the main menu bar (or press Ctrl-Z) to restore the original line. Then repeat step 4 but enter the line in the opposite direction.

Extending lines

1. Choose Extend/Trim Features from the Task bar of the Editor toolbar.

2. Select the line to extend (Fig. 12.54b).

3. Click the Pencil tool.

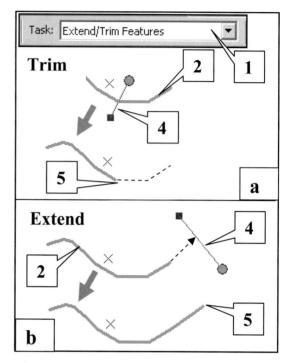

Fig. 12.54. Trimming or extending lines

4. Enter two endpoints of a line defining the stopping point of the extension.

5. The feature will be extended to meet the line.

Merging features

1. Select at least two features to be merged (Fig. 12.55).

2. Choose Editor > Merge from the Editor toolbar.

3. Select the feature that the others will be merged into. The merged feature will be given the attributes of the selected feature. The value shown in the window is the primary display field for the layer.

4. Click OK.

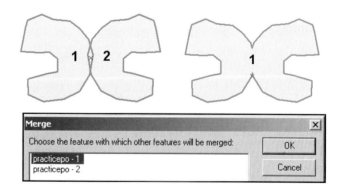

Fig. 12.55. Merging features

Union of features

1. Select at least two features to union (Fig. 12.56).

2. Choose Editor > Union from the Editor toolbar.

3. The union feature remains selected and on top of the original figures. Move or delete one set.

Fig. 12.56. Union of features

Intersection of features

1. Select at least two features to intersect (Fig. 12.57).

2. Choose Editor > Intersect from the Editor toolbar.

3. The new feature remains selected on top of the original polygons. Move or delete one or both sets.

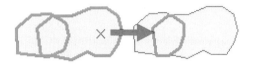

Fig. 12.57. Intersecting two polygons

Clipping features

1. Select a polygon that lies at least partially on top of another polygon (Fig. 12.58).

2. Choose Editor > Clip from the Editor toolbar.

3. If desired, apply a buffer to the clipping polygon by entering the value in map units.

4. Choose whether to preserve the overlapping area, or discard the overlapping area.

5. Click OK.

6. The clipping polygon will remain on top of the original feature. You may move or delete it.

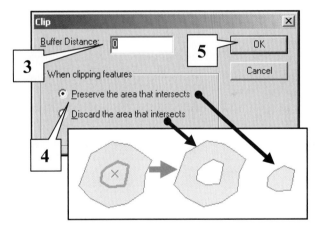

Fig. 12.58. Clipping with a polygon

Creating donut polygons

A donut polygon contains an empty hole in the middle. There are two ways to create them.

Creating multipart donuts

1. Set the task to Create New Feature, and choose one of the sketch tools.

2. Enter one ring of the donut polygon feature. On the last vertex, right-click and choose Finish Part from the context menu (Fig. 12.59).

3. Enter the next ring of the donut polygon feature, and choose Finish Part.

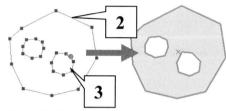

Fig. 12.59. Creating a donut or "Swiss cheese" polygon

4. Keep adding parts as long as needed. When the last part is done, choose Finish Sketch instead of Finish Part.

Using Clip to create donuts

1. Set the task to Create New Feature, and choose the Pencil tool.

2. Sketch the exterior ring of the donut polygon and finish the sketch (Fig. 12.60).

3. Sketch the interior ring of the donut polygon and finish the sketch.

4. Select the interior ring, if necessary, and choose Editor > Clip from the Editor menu.

5. Choose the option to discard the area that intersects. Click OK.

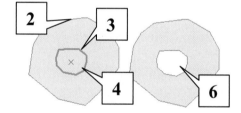

Fig. 12.60. Creating a donut by clipping

6. The clip polygon remains selected. You may keep it or delete it.

To fill a hole with another polygon

1. Set the task to Create New Feature, and select the polygon around the hole.

2. Choose the Trace tool.

3. Click on the boundary of the hole to begin tracing.

4. Trace all the way around the hole. Click again to end the trace.

5. Right-click the sketch and choose Finish Sketch.

Buffering features

1. Make sure the target layer is a polygon or line layer. If it is line, then a buffer line will be created. If it is polygons, then buffer polygons will result.

2. Select the feature (s) to be buffered. It may be a point, a line, or a polygon (Fig. 12.61).

3. Choose Editor > Buffer from the Editor toolbar.

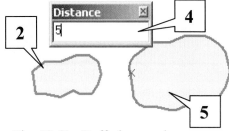

Fig. 12.61. Buffering a polygon

4. Type the buffer distance in map units and press Enter.

5. The new buffer will remain selected on top of the original feature. Move the buffer or delete the original, if needed.

Editing shared features

To edit features with shared nodes or boundaries, you must first build a map topology. Then shared features are selected using the Topology Edit tool. The selected feature will appear purple. Once selected, the shared feature may be modified, reshaped, or moved. All features sharing that vertex or boundary will be affected by the edits. The following examples represent two common tasks associated with shared editing, but there are many more. Consult the Help files or digital books that come with the software for more information about map topology and editing. Users with ArcEditor or ArcInfo licenses have even more topological editing tools available.

Creating map topology

1. Choose Editor > More Editing Tools > Topology from the Editor toolbar. The Topology toolbar will appear.

2. Click on the Map Topology icon on the Topology toolbar.

3. Check the boxes for the layers to participate in the map topology (Fig. 12.62).

4. You should normally leave the cluster tolerance at the default value. Click OK.

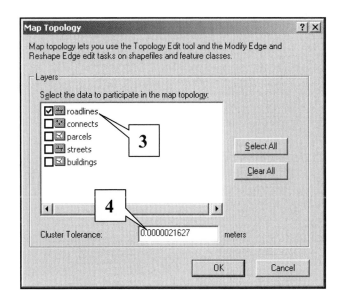

Fig. 12.62. Creating map topology

Reshaping a common boundary

1. Turn on vertex or edge snapping, and click on the Topology Edit tool.

2. Click the shared boundary to edit. It will be selected in a purple color to show it is shared (Fig. 12.63).

3. Choose Reshape Edge from the Topology Tasks group in the Task bar on the Editor toolbar.

4. Click the Pencil tool.

5. Click the beginning of the shared boundary to start entering the sketch.

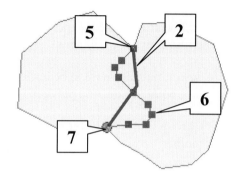

Fig. 12.63. Reshaping a shared boundary

6. Enter vertices defining the new shape of the boundary.

7. End the sketch on the other end of the shared boundary, and double-click to finish it.

8. The boundary will change to the shape of the sketch.

Moving shared nodes

1. Click the Topology Edit tool.

2. Hold down the N key and click on a node at the intersection of the lines to select it. The N key ensures that only a node is selected (Fig. 12.64).

3. Click and drag on the node to move it to its new location.

4. As you move the node, the movement of the shared features will be shown.

5. When you release the mouse button, the node and its attached lines will go to their new places.

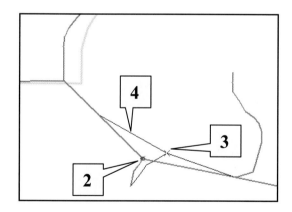

Fig. 12.64. Moving shared line features

TIP: To have attached features proportionally stretched when moving nodes, check the box on the General tab of the Editing > Options menu. This option often provides better adjustment of features.

Chapter 13. Working with Geodatabases

Mastering the Concepts

Objectives

➢ Understanding the geodatabase model

➢ Creating a geodatabase

➢ Importing layers into a geodatabase

➢ Setting up a feature dataset and importing layers

➢ Setting up and using attribute domains

➢ Using split and merge policies

Concepts

As we have learned, the ArcGIS software has a long history of development, and coverages were the first data model used. Later, shapefiles were developed for use in the ArcView package. The geodatabase model has arrived with the release of ArcGIS 8 (Fig. 13.1). The new model offers several advantages over both coverages and shapefiles. Like coverages, it can store topological relationships between features for added functionality in modeling real-world phenomena, but the geodatabase is simpler in construction and more robust in general usage. Like shapefiles, the features of a geodatabase are simply constructed and difficult to corrupt, but it provides many more benefits than shapefiles. Another advantage of geodatabases over shapefiles includes automatic tracking of lengths and areas for features. Geodatabases also provide behavior and validation rules which control how features behave, how they can connect, and what attributes are allowed. A user can also set up planar topology for features and construct topological rules, such as specifying that a state boundary must always match county boundaries, or that counties must never cross a state boundary.

About geodatabases

The geodatabase model is constructed on the architecture of standard database systems and can be implemented with several commercially available database software packages. Two types of geodatabases may be created. The personal geodatabase uses Microsoft™ Access® technology and requires no additional licensing. An organizational database uses a large database management system (DBMS) such as Oracle® to store information, and a software package called the Arc Spatial Database Engine (ArcSDE) to translate geographic data into a consistent and customized format in the large database.

Organizational databases have distinct advantages when the spatial data sets are large and/or are accessed by many different people. For example, such a system might be developed for the planning department of a large city, which may have dozens of people in several departments needing access to the information, both for reading and for editing. A DBMS/ArcSDE setup allows many people simultaneous access to data sets, provides security and permission handling, and manages cases of multiple simultaneous editing of the data. It also makes it possible to set up version tracking for added data security and protection. However, the additional work in setting up a DBMS only makes sense when these issues are present. In this chapter we will work exclusively with personal geodatabases, but most of the commands and functions are the same for the organizational version.

A geodatabase may contain a variety of objects (Fig. 13.1). Feature classes, or collections of similar objects in a spatial data layer, are used to store the actual spatial information. For all practical purposes, a feature class can be conceived as a shapefile inside a geodatabase. Geodatabases may also contain tables, layers, relationships, geometric networks, and feature datasets. A feature dataset is a collection of related feature classes with the same coordinate system. The Badlands database in Figure 13.1 contains three feature datasets: Infrastructure, Transportation, and Wildlife. The Wildlife feature dataset has been expanded to show the feature classes: home ranges of various species that live in the park. The geodatabase also contains four standalone feature classes: boundary, sacred_sites, streams, and wetlands. In this case the feature classes all share the same coordinate system, but they do not have to.

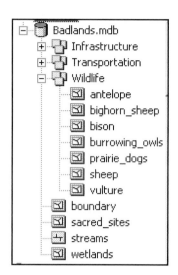

Fig. 13.1. A geodatabase with feature classes and feature datasets

---

TIP: Designing and creating good geodatabases involves many issues far beyond the scope of this book. Users planning to work with geodatabases should read more about them, particularly in the online-documentation book *Building a Geodatabase.*

Creating geodatabases

Geodatabases must be created as empty shells to which layers are later added. Layers may be loaded from a variety of sources. Coverages and shapefiles may be imported into the geodatabase, either as standalone feature classes or as part of a feature dataset. INFO and dBase tables may also be loaded. Users may also create empty feature classes and use editing to add features, much as we did for shapefiles in Chapters 11 and 12. Starting with Version 9, rasters can also be stored inside a geodatabase. Storing a photograph tied to a specific ground location is one potential use of this new capability.

Geodatabase feature classes can also be created any time a spatial operation is performed in ArcMap, such as exporting data, creating a spatial join, or buffering the features in a shapefile. In all of these cases the option to store the output as a geodatabase feature class may be chosen, instead of creating a shapefile as we have usually done in this book.

Because they are based on a new data model, many databases are currently developed by importing shapefiles or coverages of existing data. This process is usually straightforward. The entire data set, including features, attribute tables, and coordinate systems, is transferred into the database. In the case of coverages, the attribute fields must be converted from the coverage field types to the geodatabase field types. This conversion is handled automatically, although the user can customize the interpretation of certain fields if desired. The user can also request a coordinate system re-projection as part of the transfer process.

Unlike layer files, which store only a link to the source data, the geodatabase stores the feature data itself. Importing a shapefile creates a copy of the features inside the database, leaving the original features unchanged. The user then has two copies of the data.

Creating feature datasets

A feature dataset is a collection of feature classes that are related to each other in some way (Fig. 13.2). For example, a transportation feature dataset might contain feature classes for roads, railways, airways, and interstates. These features could be used to construct a geometric network, so that the movement of people or goods along the network could be modeled. Similar models could be constructed for utilities, using feature classes representing water mains, water sources, sewer lines, and treatment plants. Feature datasets may also contain planar topological relationships between the feature classes, allowing for easier identification and correction of errors such as gaps or overlaps between polygons. (See Chapter 12 for a description of planar topology.)

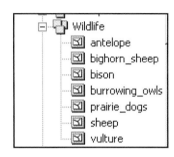

Fig. 13.2. A feature dataset contains related layers with a common coordinate system.

The feature classes in a feature dataset must all share the same spatial reference, that is, they must have the same coordinate system. A feature dataset is also created to have a maximum extent. Once the feature dataset is created, features outside the extent cannot be added to the feature dataset. Thus care must be used in setting up a feature dataset to ensure that its spatial extent is large enough to encompass all the features it might eventually contain. Nothing is more frustrating than to have loaded eight layers into a feature dataset and discover that the ninth layer extends slightly beyond the maximum extent and cannot be loaded in. At that point a new feature dataset would need to be created!

A feature dataset must be created with a defined coordinate system and extent. Although the coordinate system and extent may be specified manually, it is easier and more reliable to specify them from an existing feature class. The following general procedure is recommended for setting up a feature dataset:

➢ Collect the feature classes that will be placed in the feature dataset, or at least as many as are currently available.

➢ Determine the maximum possible spatial extent of features in the dataset. For example, if setting up a feature dataset for a state, the state boundary might be the maximum extent.

➢ Create a boundary shapefile in the desired coordinate system of the feature dataset. The boundary should be slightly outside of the possible maximum extent, rather than exactly on it. In the case of the state boundary, a buffer zone of perhaps 5 to 10 miles should be added to the extent.

➢ In ArcCatalog, create the new feature dataset. When specifying the coordinate system, import the CS information from the boundary shapefile prepared in the previous step.

Using the boundary ensures that the coordinate system is correct, and the *x-y* extent of the feature dataset is sufficiently large.

➢ Load the desired feature classes into the feature dataset from shapefiles or coverages or other geodatabases. If the feature classes have a coordinate system different from that of the feature dataset, the features will automatically be projected to the dataset CS during the conversion.

Although it may seem odd to buffer the feature dataset boundary, in fact it is very good practice. Since different data layers often have slight offsets from each other, it may well happen that some rivers in one feature class extend a few meters across the feature dataset boundary. Even this slight mismatch might be enough to prevent loading the rivers in the feature dataset. Unfortunately, features falling outside the boundary are not clipped to end at the boundary, leaving the portion that falls inside. Instead, the entire feature is omitted from the feature class. Thus it is safer to take the maximum extent and add a margin for safety. Figure 13.3 shows a boundary layer created for South Dakota by buffering the state boundary. It is far better to use the buffered version rather than the state boundary itself, in case any features in the dataset extend even a couple feet past the state line!

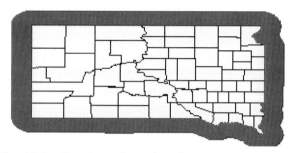

Fig. 13.3. Creating a boundary feature class that goes beyond any likely feature in the feature dataset will ensure trouble-free loading of layers later.

If the user attempts to load features into a feature dataset and they extend beyond its extent, a warning will appear in the geoprocessing box (Fig. 13.4). Notice that the import function is still considered successful even if some features fail to load, and users may not be aware of the problem if the importation is being conducted in unsupervised batch mode, or if the dialog box is set to close automatically upon exiting. Caution is advised.

Using default values

Another advantage of using geodatabases lies in setting up default values for attributes. For example, if a user is digitizing roads, residential streets (LOCAL) are usually the most common

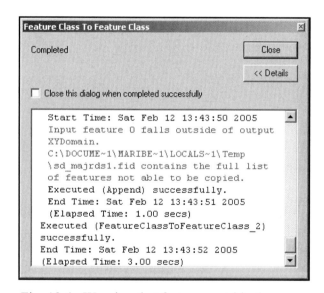

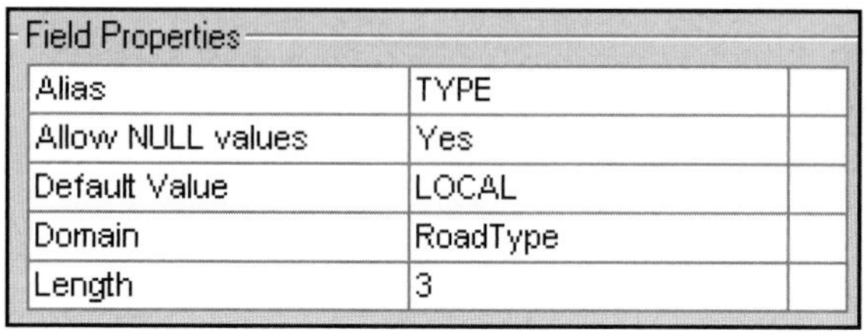

Fig. 13.4. Warning that features outside the feature dataset extent have failed to load

Field Properties		
Alias	TYPE	
Allow NULL values	Yes	
Default Value	LOCAL	
Domain	RoadType	
Length	3	

Fig. 13.5. Setting up default values for attributes can save much time during data entry.

type and typically have two lanes and a speed limit of 25 mph. The user can set default values for each of these fields (Fig. 13.5). Then when each road is added, the road type, lanes, and speed limit are automatically set to LOCAL, 2, and 25, respectively. The information only needs editing if the new road differs from the default. A user could even set the defaults one way, digitize all the local roads, and then close the layer, change the defaults, open the layer again and digitize the next group. This approach saves lots of editing time and helps reduce attribute errors.

Setting up domains

One great advantage of geodatabases includes using **domains** for attributes. An attribute domain constrains the values that may be entered for a particular attribute. Such constraints help ensure correct data entry and keep out incorrect or extraneous values. For example, imagine an attribute field named PipeDiam, which contains the diameter of water pipes. Water pipes don't come in an infinite variety of sizes; in fact, a town might only use 1-inch, 3-inch, 6-inch, or 12-inch pipes. Setting up a domain for the PipeDiam field prevents any user from mistakenly entering a 4-inch or 5-inch pipe. These constraints can enforce data entry rules in situations when more than one person is entering data and not everyone knows or can remember all the rules.

Domains come in two types. The first type is a **range domain,** which specifies the lowest and highest possible values but lets the values take any number in between. Range domains are only applicable to numeric data. For example, a range domain could be applied to a student Grade Point Average (GPA) field. A GPA ranges between 0 and 4.0. A range domain would allow any value between 0 and 4.0, but would exclude negative values or values greater than 4.0. The second type is a **coded domain,** which allows only certain values taken from a list. The PipeDiam domain mentioned previously is a coded domain because it allows only four specific values from a list: 1, 3, 6, or 12.

Domains are created and maintained in ArcCatalog as a property of the geodatabase, rather than as a property of a single feature class or attribute field. Therefore, a domain can be used more than once, and it can be used in multiple feature classes. For example, a range domain called Percent, which allows values from 0 to 100, could be used in many feature classes any time a percentage value is being stored in an attribute.

Figure 13.6 shows how the properties of a percentage domain would be set up in ArcCatalog. The field type is specified as short integer. The domain type is range, and the minimum value of 0 and maximimum value of 100 is entered. A domain can also have split and merge policies assigned to it, which we will cover in the following section.

Field Type	Short Integer
Domain Type	Range
Minimum value	0
Maximum value	100
Split policy	Default Value
Merge policy	Default Value

Fig. 13.6. Properties of a range domain

The field type of the domain must always match the field type of the attribute to which it will be assigned. Thus, a short integer domain can only be used with a short integer attribute field, and not with a long integer or float field. Users must pay careful attention to the fields for which the domain is intended, in order to select the appropriate field type for the domain.

An example of the PipeDiam domain is shown in Figure 13.7. The field type is short integer, and the domain type is set to coded values. The codes themselves are typed in below. The code is the value actually stored in the attribute field. The description appears in legends and tables during an ArcMap session, providing easily understood information to the people working with the attributes. However, the descriptions do need to fit well inside the menus and legends, so it is best to keep them reasonably short.

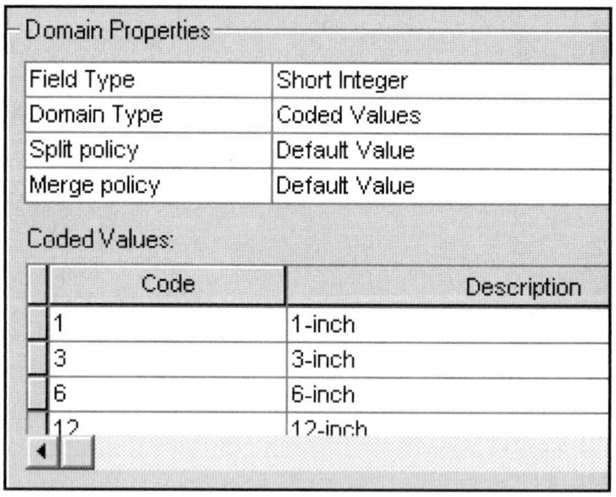
Domain Properties

Field Type	Short Integer
Domain Type	Coded Values
Split policy	Default Value
Merge policy	Default Value

Coded Values:

Code	Description
1	1-inch
3	3-inch
6	6-inch
12	12-inch

Fig. 13.7. Properties of a coded domain

Coded domains also provide a valuable way to combine the ease of numeric codes with the luxury of understandable text information for people. Land use planning, for example, often uses numeric codes to indicate various zoning types (such as 32 = Residential and 45 = Commercial). Numeric codes are advantageous because they are less susceptible to typing errors and take less space to store. However, it is difficult for people to remember the meanings of many codes. With a coded domain, the zoning codes may be stored as integers, but the information presented to the user takes the form of the textual description (Fig. 13.8).

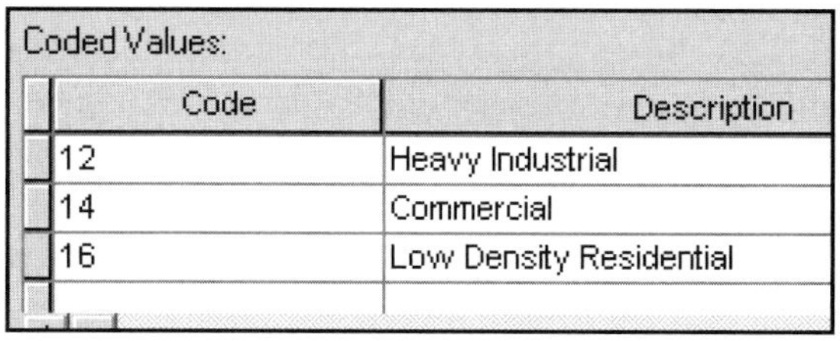
Coded Values:

Code	Description
12	Heavy Industrial
14	Commercial
16	Low Density Residential

Fig. 13.8. Using domains to link numeric land use codes to easily interpreted descriptions

Split and merge policies

Domains offer a way to control the updating of attributes during a split or a merge operation. The default policy of making copies or entering blanks may be overridden by alternate and more useful actions. Split and merge policies are associated with each attribute domain. Setting up these policies correctly can save much time during editing.

The merge policy

If we examine what happens during a merge, the usefulness of the merge policy immediately becomes apparent. Imagine two parcels that have been combined into one. Each parcel has its own attributes. After merging, there will only be one parcel. Which attributes will it have? By default, in coverages or shapefiles the merged parcel will contain the attributes of the first feature found in the database, hardly a consistent and effective method. In geodatabases, it will contain the attributes of the first feature selected before merging. The merge policy overrides this haphazard method by specifying a different action for each field.

Three different policies may be assigned to fields: Default, Weighted Average, or Sum. If Default is chosen, the merged feature attribute will contain the default value established for that field. A weighted average assigns the new value based on the relative areas of the two input features, and the sum assigns the new value as the sum of the values in the two input features. Let us examine the attribute fields in Figure 13.9 one by one.

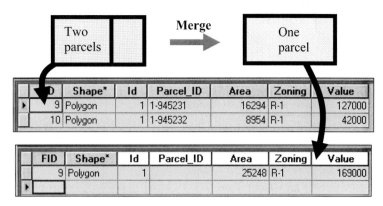

Fig. 13.9. Attributes from two parcels are updated according to the Merge policy when they are combined into one.

The Zoning is a text field. By far the majority of parcels in most towns are residential, so in setting up the database it might make sense to have the default value be R-1 for zoning. This default is usually set at the time the attribute field is created. If the merge policy is set to default, then after the merge the new polygon will contain the default value for that field, R-1. Of course it may be incorrect in some cases, but if well chosen a default will be right more times than it is wrong and will save editing time. However, some managers prefer generating a blank field so that they can manually ensure that the parcel gets the right value, rather than risk automatic assignment of a wrong value.

TIP: A user can cause a field to receive a blank during a split or merge by setting the policy to Default and leaving the default value blank.

The Area field is numeric and contains the area of the parcel in square feet. This field is an obvious candidate for the Sum policy. After the merge, the new feature will be assigned an area equal to the sum of the areas in the original polygons. The Value field would also logically be assigned Sum for the merge policy. A weighted average is similarly applied, except that the values are averaged instead of summed.

The Parcel_ID is not a numeric quantity and so cannot qualify for Sum or Weighted Average. Nor is it practical to assign a domain code for every single unique ID. This attribute cannot effectively be assigned a domain. Instead, the normal rules for splitting and merging features without domains would apply. Notice that the Parcel_ID field is blank, signaling to the operator that a new code must be assigned.

The split policy

The split policy is analogous to the merge policy, but it is applied when a single feature is split in two. The policy choices include Default, Duplicate, and Geometry Ratio. In the case of Default, the policy works the same as for merging. If Duplicate is used, then both new features retain the value of the original feature. If Geometry Ratio is used, then the new value is assigned based on the relative size of the original and split features.

Let us examine an example of splitting a parcel using the policies (Fig. 13.10). The Parcel_ID field had no domain and followed the standard rule of being duplicated in the resulting polygons. At least one of the new parcels will need a new number. The Area and Value fields used the Geometry Ratio policy. The proportion of areas in the two new polygons was 60-40; thus the smaller parcel was assigned 40% of the original area and value, and the larger parcel was assigned 60% of the original area and value. Finally, the Default policy was applied to the Zoning field, again assigning the most common value of R-1.

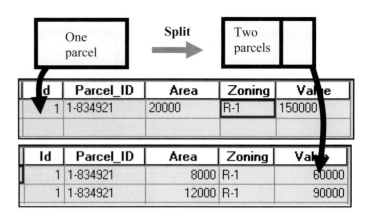

Fig. 13.10. The attributes of two parcels created by splitting one parcel according to the split policy

In deciding whether to use split and merge policies, one must consider the number of attributes involved and the amount of editing that a data layer receives. Setting up and applying domains to attributes takes some time and thought. Unless the layers are being frequently edited, then the effort may not be worth it. When appropriately used, however, domains can save a substantial amount of time and prevent many incorrect data entries.

About subtypes

Subtypes are another way to facilitate the entry and validation of attribute data. Subtypes may be set up for a particular feature class. Every feature in the layer then must belong to one of the subtypes. For example, the roads layer of a city might be constrained to contain only certain types of roads (Fig. 13.11). One field is chosen as the subtype attribute, and the allowed types are set up as a coded domain. The allowed road types might be alleys, streets, connectors, highways, and interstates. Then every road in the layer must belong to one of these types.

So far this setup sounds similar to a coded domain. However, subtypes provide added functionality. First, when a data layer is loaded into a map, the subtypes are automatically displayed with different symbols. Second, each subtype can have its own default value, instead of having a single default for the attribute field.

In the case of the roads, each road type might have default values for certain attributes, such as number of lanes. An alley typically has one lane, a local street has two lanes, a connector has four lanes, and an interstate has four lanes. Similar attributes might be determined for speed limits; 15 for alleys, 25 for streets, 35 for connectors, 65 for highways, and 75 for interstates. When a

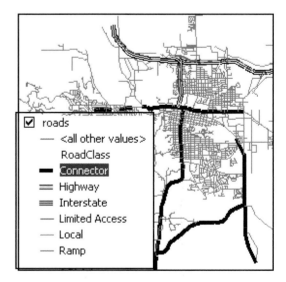

Fig. 13.11. Subtypes are formal classes of features within a layer.

subtype is established, default values can be specified for each of the types. These values are then automatically entered in the attribute table and new features are created.

During an editing session, each subtype appears as a separate entity in the target list. Thus to add local streets, the user would set the target to the Local subtype and begin adding streets. As they were digitized, each street would immediately receive the Local default values: two lanes, 25 mph speed limit, and so on. To enter a connector, the user would change the target layer to the Connector subtype, and subsequently added streets would receive the default values for connectors. Subtypes thus streamline both data entry and the assignment of attributes to features. Of course, occasionally features may deviate from the default values, such as a connector having six lanes, but such exceptions may be corrected as needed.

TIP: Subtypes can be created for geodatabase feature classes using ArcEditor or ArcInfo. ArcView can view subtypes, but it cannot create or edit them. Furthermore, *layers containing subtypes cannot be edited in ArcView,* even if the user doesn't want to change the subtypes. The layer will simply not be available for editing of any kind.

We will not study subtypes further in this chapter, but if you have an ArcEditor or ArcInfo license you may want to read more about them in the online documentation *Editing in ArcMap,* and experiment with them on your own.

Summary

> A geodatabase is a new data model developed for ArcGIS. It uses commercially available standard database formats and technology.

> Geodatabases offer significant advantages over shapefiles and coverages, including simpler and more robust implementation, the storing of topology, geometric networks, and behavior and validation rules.

> Geodatabases are created as empty containers and then filled with feature classes, feature datasets, layers, tables, relationships, and other objects.

> Feature datasets contain related feature classes with the same coordinate system. Caution must be used in setting up the extent to ensure that all features fit inside the specified envelope. Features outside the envelope, even slightly, cannot be loaded into the dataset.

> Domains define allowed values for attributes in geodatabases. Domains may cover ranges of numerical values, or they may have specific coded text or numeric values.

> Domains also have split and merge policies, which define the assignment of attributes when features are merged or split.

> Subtypes allow features in a layer to be classified into specifically defined groups. Each group may have its own set of default attribute values.

> Subtypes may be set up with ArcEditor or ArcInfo, but they can only be viewed with ArcView.

TIP: Designing and creating good geodatabases involves many issues far beyond the scope of this book. Users planning to work with geodatabases should read more about them, particularly in the online-documentation book *Building a Geodatabase*.

TIP: Subtypes can be created for geodatabase feature classes using ArcEditor or ArcInfo. ArcView can view subtypes, but it cannot create or edit them. Furthermore, *layers containing subtypes cannot be edited in ArcView at all, for any reason.* Don't create subtypes if someone must be able to edit the layer with only an ArcView license.

TIP: If the import layer has a different coordinate system than the feature dataset, then the features will automatically be reprojected as they are loaded.

TIP: Domain policies and codes may be edited after the domain is created. However, the field type and domain type, once set, cannot be altered. To change them you must delete the domain and create a new one with the desired properties.

TIP: ALWAYS, ALWAYS click Apply after EACH domain is added, before adding the next one. If you made an error in one domain, but added more before clicking Apply, then one or more of the domains may be added incorrectly, and it may not be clear which ones have problems. You may end up deleting them all and starting again if you fail to follow this rule.

Chapter Review Questions

You may need to consult the Skills Reference section to answer some of these questions.

1. List three advantages of the geodatabase model.

2. On what type of software architecture is the geodatabase model based?

3. How does a personal geodatabase differ from an organizational geodatabase?

4. What is the difference between a feature class and a feature dataset?

5. Why is the spatial extent of a feature dataset a special concern?

6. Can you have a geodatabase which contains one feature dataset in UTM Zone 13 and another feature dataset in South Dakota State Plane?

7. You have a forest feature class with polygons showing individual stands of trees. It contains the three attribute fields: TreeSpecies, CanopyCover%, and Acres. TreeSpecies contains the dominant type of tree (e.g., ponderosa pine, aspen, etc.). Canopy cover gives the percentage of the stand covered by tree crowns. The Acres field contains the area of the stand. For each of these attributes, list the most appropriate split policy and merge policy.

8. For each of the following zoning attributes, state whether it would be suitable for a domain. If yes, then say whether a coded or range domain would be more appropriate. In each case, explain your reasoning. List: Zoning, StreetAddress, Value, PercentImperviousArea.

9. Domains are set up as properties of a geodatabase rather than as properties of feature classes. Why is this arrangement an advantage?

10. How does one establish a default value for an attribute in a feature class? In a shapefile?

Mastering the Skills

Teaching Tutorial

The following examples provide step-by-step instructions for doing basic tasks and solving basic problems in ArcGIS. The steps you need to do are highlighted with an arrow ➔; follow them carefully. Click on the video number in the VideoIndex to view a demonstration of the steps.

In this examples section we will build a geodatabase of Rapid City, using the existing data layers in the mgisdata\Rapidcity folder.

1➔ Start ArcCatalog and navigate to the mgisdata\Rapidcity folder.

1➔ Right-click the Rapidcity folder and choose New > Personal Geodatabase.

1➔ Type in "rapidcity" as the name of the geodatabase and press Enter.

Adding coverages to a geodatabase

We will begin by bringing in two coverages, the city boundary and the land use data.

2➔ Right-click the new rapidcity geodatabase and choose Import > Feature Class (single) tool (Fig. 13.12).

2➔ Click the Browse button for the Input Features, navigate to the mgisdata\Rapidcity folder, and double-click the landuse coverage folder to see its feature classes.

2➔ Select the polygon feature class and click Add to close the menu.

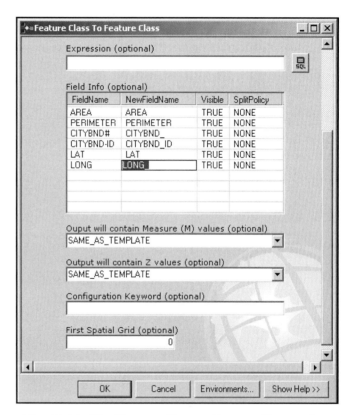

Fig. 13.12. The Feature Class to Feature Class tool

TIP: Recall that coverages can have multiple feature classes, unlike shapefiles. When importing from a coverage, which feature class to import must be specified.

2➜ Enter "landuse" for the Output Feature Class Name. Leave the rest of the options at their defaults, and click OK.

2➜ Click the + button next to the rapidcity geodatabase to expand its contents and click the Preview tab to see the polygons.

Everything looks quite in order. Next we import the citybnd coverage.

3➜ Right-click the rapidcity geodatabase and choose Import > Feature Class (single).

3➜ Set the input feature class to the polygons of citybnd in the Rapidcity folder.

3➜ Set the output name to "citybnd".

3➜ Scroll down to the lower half of the tool and examine the other input boxes (Fig. 13.12).

Although we are not going to change any settings, let's at least look at the options.

 4➜ Click the SQL button next to the box marked Expression. This Query Builder window, familiar from other chapters, allows the user to import only the records meeting certain criteria, such as only the residential parcels in a landuse feature class. Click Cancel to close the window.

4➜ The Field Info section lists the attributes in the feature class, showing the way the attributes will appear in the new feature class.

This section controls how the item names are translated. Names that are permissible in coverages often must be modified to follow geodatabase naming conventions. This process happens automatically but can be customized if needed. The first column shows the original name and the second column shows the output name (Fig. 13.13). You can edit the output names by clicking in the boxes, and you can make one or more fields invisible. The final column is not used in this tool.

Field Info (optional)			
FieldName	NewFieldName	Visible	SplitPolicy
AREA	AREA	TRUE	NONE
PERIMETER	PERIMETER	TRUE	NONE
CITYBND#	CITYBND_	TRUE	NONE
CITYBND-ID	CITYBND_ID	TRUE	NONE
LAT	LAT	TRUE	NONE
LONG	LONG_	TRUE	NONE

Fig. 13.13. Mapping item names from coverage format to geodatabase format

4➜ The M and Z inputs permit the user to set up M (measurement) or Z (height) values in the new feature class. M values are used to measure distances along a line, such as mile markers on a highway. Z values store elevations.

4➜ The First Spatial Grid box sets the spatial reference grid index used to speed drawing and searching for features. The default is set based on the extent of the data set. Don't change this value unless you fully understand how to customize indexes.

5➜ Click the Environments button. It opens the Environment Settings for the geoprocessing environment. You might use this menu under General Settings to specify an output coordinate system so that the new feature class is projected before it is placed in the geodatabase. Click Cancel to close the window.

5➔ When done examining the input boxes, click OK to import the citybnd feature class.

Often users have multiple feature classes to import into the same geodatabase. Another tool allows the imports to be set up and then run without supervision.

6➔ Right-click the rapidcity geodatabase and choose Import > Feature Class (multiple).

6➔ Click the Browse button for the Input Features and select the polygon feature class of the wshds coverage. When you click Add it is placed in the list of feature classes to process.

6➔ Click the Browse button again and select the rcwgeology2 shapefile to import.

6➔ Click OK to begin processing the feature classes.

6➔ Expand the geodatabase and examine the new feature classes.

Notice then when using the tool for importing multiple feature classes, you cannot set options individually for each one as you could with the single input tool.

Finally, let's create a new empty feature class to which we will later add water mains.

7➔ Right-click the geodatabase and choose New > Feature Class.

7➔ Enter the feature class name "waterlines".

7➔ Set the type to the first option, containing simple features.

7➔ Click Next.

7➔ Set the Configuration Keyword option to Default.

7➔ Click Next.

In the Fields window we first need to set the feature class type and the coordinate system.

8➔ Click the Shape field to select it.

8➔ Click in the Geometry Type box and change it from Polygon to Line.

8➔ Click the ellipses down by the Spatial Reference property of the field.

8➔ Choose Import, and select the landuse feature class in the geodatabase as the basis for importing the coordinate system. Click Add and OK.

Field Name	Data Type
OBJECTID	Object ID
SHAPE	Geometry
Linetype	Text
Diameter	Short Integer
Capacity	Short Integer

Fig. 13.14. Entering the fields for the water lines

Next we add more fields for the water lines: Linetype, Diameter, and Capacity (Fig. 13.14).

9➔ Click in the next empty box under SHAPE, enter the Field Name as "Linetype", change the Data Type to Text, and set the Length to 8 in the Field Properties.

9➔ Click the next empty box and create a short integer "Diameter" field. Do not change the field properties.

9➔ Create a short integer "Capacity" field. Do not change the field properties.

9➜ Click Finish to create the feature class.

9➜ Verify that the new feature class has been added to the geodatabase.

Using attribute defaults

We can streamline our data entry of the water lines by using default values. We will begin by entering water mains. Assume that the mains are made of 12-inch pipe and have a rating of 200 gallons per minute (gpm). Our first task involves setting these default values for the attributes.

10➜ In ArcCatalog, locate the waterlines feature class in the geodatabase.

10➜ Right-click the waterlines feature class and choose Properties.

10➜ Click the Fields tab.

10➜ Click on the Linetype field to select it.

10➜ Type "MAIN" in the Default Value box (Fig. 13.15).

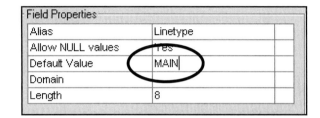

Fig. 13.15. Entering the default value MAIN for the attribute Linetype

10➜ Click the Diameter field and set the default value to 12.

10➜ Click the Capacity field and set the default value to 200.

10➜ Click Apply to establish the values and OK to close the window.

Now we will use these defaults as we digitize some water mains.

➜ Close ArcCatalog and start ArcMap with the document ex_13.mxd in the mgisdata\MapDocuments folder.

➜Use Save As to rename the document, and remember to save frequently as you work.

11➜ Add the new waterlines feature class to the map.

11➜ Change the waterlines symbol to a thick line that contrasts well with the orthophoto.

12➜ Open the Editor toolbar, if necessary, and choose to Start Editing.

12➜ Make sure the target layer is waterlines, the task is Create New Feature, and the Pencil tool is selected.

12➜ Open the Attribute Editor box. It will be blank right now because nothing is selected. Move it so that it does not block the view of the map.

One of the sources of water in Rapid City is a water gallery and small treatment site next to Rapid Creek, as marked on the map. Begin digitizing water mains from this point.

13➜ Click on the water gallery marker to start a water main. Click on the road immediately to the left of the gallery, and continue digitizing south down the road to the edge of the map. Double-click to finish the sketch.

13➜ Examine the attributes of the new feature in the Attribute Editor.

Notice that the defaults of MAIN, 12, and 200 are automatically entered in the fields. Just imagine how much time this will save. If you happen to enter a different type of water line, simply change the default values in the Attribute Editor before going on.

14➔ Turn on edge snapping.

14➔ Add several east-west water mains going along those streets.

14➔ Stop editing and save the edits.

Now, let us suppose that we have finished entering water mains for a while, and want to create laterals. We can change the defaults to make it easier to enter these also.

15➔ Remove the waterlines layer from the map. (We can't make changes to the feature class when it is open in ArcMap.)

15➔ In ArcCatalog, open the properties for the waterlines feature class.

15➔ Click the Fields tab.

15➔ Click the Linetype field and change the default value to LATERAL.

15➔ Click the Diameter field and change the default value to 3.

15➔ Click the Capacity field and change the default value to 10.

15➔ Click Apply and then click OK to close the window.

15➔ Close ArcCatalog so it cannot interfere with editing.

16➔ Add the waterlines layer back to the map document.

16➔ Start editing.

16➔ Open the Attribute Editor.

16➔ Choose the Pencil tool and digitize several new water lines coming off the main ones (Fig. 13.16).

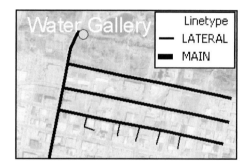

Fig. 13.16. Creating different types of water lines with default attributes

Notice that each of the new lines has the new default values of LATERAL, 3, and 10.

16➔ Stop editing and save the edits.

➔ Exit ArcMap. You don't need to save the map document.

With ArcEditor or ArcInfo, you could have created subtypes for the waterlines layer. The laterals and mains would be two different subtypes. Using subtypes, you don't have to stop editing and change the defaults. Each subtype is a separate target layer and has its own defaults.

Creating a feature dataset

Our next step in building this geodatabase will be to create a feature dataset called Commerce, which contains gas stations and the restaurants we geocoded in Chapter 10. However, recall that once the *x-y* extent of a feature dataset is established, features even slightly outside that extent cannot be loaded in. Therefore, we want to create a boundary layer that will serve as the maximum possible extent of any feature we might eventually put in the feature dataset. For this purpose we will take the largest of our Rapid City coverages and buffer it for extra safety.

17➜ Open ArcCatalog and examine the metadata for several of the layers in the mgisdata\Rapidcity folder.

1. What appears to be the coordinate system of these data layers? _____

We have no reason to change the coordinate system being used, so that will be the coordinate system of our feature dataset. Let us find out which is the largest extent.

➜ Start ArcMap with a new empty map.

18➜ Select the Rapidcity folder in ArcCatalog and change the view mode to Contents.

18➜ Hold down the Ctrl key and click on each of the coverages and shapefiles in the mgisdata\Rapidcity folder so they are all selected, EXCEPT for rcw_oops. Don't include the rasters, tables, or layer files.

18➜ Click on one of the highlighted layers and drag them all to the ArcMap window.

18➜ Zoom to the full extent of the data.

When creating a feature dataset, the *x-y* extent and coordinate system must be explicitly set. Once set, they cannot be changed. Therefore it is critical to properly set them in the first try. The easiest way to do this is to create a shapefile with the desired maximum possible extent, and use it as a template when setting the feature dataset coordinate system. To do this, we must know the maximum possible extent of our data sets, and build a little margin for error.

19➜ Open ArcToolbox and select the Data Management > Feature class > Create Feature Class tool.

19➜ Set the output location to migsdata/Rapidcity (you'll need to move up one folder so you can actually select the Rapidcity folder itself).

19➜ Set the output feature class name to "databound.shp". Make sure the geometry type is set to POLYGON.

19➜ Click the button to set the spatial reference. Set it to NAD 1927 UTM Zone 13N, matching most of the layers in this folder.

19➜ Click OK to create the feature class. Close ArcToolbox when it is finished.

20➜ Open the databound shapefile properties and examine the coordinate system, paying special attention to the extent values.

Notice the strange extent value, #QNAN0. These values mean that the extent for this shapefile is currently undefined, because it does not yet contain any features. Next we will create a boundary for the city that defines its maximum extent.

20➜ Close the databound properties window.

20➜ Now, open the data frame Properties and verify that the coordinate system is UTM Zone 13N NAD 1927. If it is not, change it. Close the window.

20➜ Open the Editor toolbar, if necessary, and choose Start Editing. Ignore the coordinate system warning, and click the button to Start Editing.

21➜ Click the Fixed Zoom Out tool once or twice to make a little room.

21➔ Make sure the target layer is databound, set the edit task to Create New Feature, and click the Pencil tool.

21➔ Create a polygon that covers the entire data frame window area, making sure all of the Rapid City layers lie inside it.

21➔ Stop editing, and choose to save your edits.

21➔ Examine the databound shapefile properties again, and verify that the coordinate system extent now has valid units.

This new shapefile will serve as a template for several feature datasets. Using it will ensure that all the features can be loaded without incident.

➔ Exit ArcMap. You do not need to save your changes.

Now that we have our template shapefile, we are ready to create the feature dataset, assigning it the same coordinate system and extent as the databound layer.

22➔ In ArcCatalog, right-click the rapidcity geodatabase and choose New > Feature Dataset.

22➔ Type the name "Commerce" for the new feature dataset.

22➔ Click the Edit button to change the coordinate system.

22➔ Click the Import button, navigate to the databound shapefile, and select it. Click Add.

22➔ Click OK and OK.

TIP: It is best to use Import to establish the coordinate system of a feature dataset, and to specify a feature class with the desired coordinate system AND a valid extent that covers the maximum possible area of your feature dataset. Using Select to set the coordinate system will not generate a valid *x-y* extent, and you won't be able to load features into the dataset.

Now we add some feature classes to the feature dataset.

23➔ Expand the Commerce feature dataset.

23➔ Right-click the Commerce feature dataset and choose Import > Feature Class (multiple).

23➔ Click the Browse button for the Input Features and locate the gas_station2 shapefile in the mgisdata\Rapidcity folder. Add it.

23➔ Locate and add the restaurants shapefile, rceats.shp, created in Chapter 10. (If you didn't create it, load the rceats.shp file from the mgisdata\MapDocuments\Results folder instead.)

23➔ Click OK to start processing.

23➔ Examine the new feature classes in the feature dataset.

TIP: If the import layer has a different coordinate system than the feature dataset, then the features will automatically be reprojected as they are loaded.

Next create another feature dataset to hold transportation features.

24➔ In ArcCatalog, right-click the geodatabase and choose New > Feature Dataset.

24➔ Name it "Transportation" and use the Edit button to import the coordinate system from databound.shp, as in step 22.

24➔ Examine the new feature dataset.

25➔ Right-click the Transportation feature dataset and choose Import > Feature Class (single).

25➔ Import the rds_clp shapefile from the mgisdata\Rapidcity folder. Name the output feature class "roads".

Adding geodatabase layers from ArcMap

You can also use ArcMap to add layers to a geodatabase. This approach is beneficial if you are trying to assemble certain features and/or change projections during the import. We will use this capability to import just the Rapid City schools from the layer containing all schools in the state.

➔ Start ArcMap with a new, empty map.

26➔ Load the rds_clp layer and the orthophoto rceast_nw.sid from the mgisdata\Rapidcity folder.

2. What is the coordinate system of the roads? _____

27➔ Add the sd_gschools layer from the mgisdata\Sdakota folder.

27➔ Rename the layer "SD Schools".

A coordinate system warning may appear when loading the schools. Let us check its coordinate system to try to identify the problem.

3. What is the coordinate system of the schools? _____

Now when ArcMap brought the data in, it automatically chose a transformation from NAD 83 to NAD 27 so it could match the schools with the other data. This transformation may not be very accurate, but we can easily check how far off it might be. The schools are large buildings compared to their surroundings and are fairly easy to spot on the orthophoto.

28➔ Turn on the orthophoto, if necessary.

28➔ Zoom into an area containing schools that are on the orthophoto.

28➔ Examine the location of the school points relative to the buildings on the orthophoto.

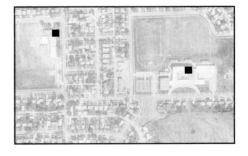

Most of the school locations are fairly close to the correct sites (Fig. 13.17). Any offsets are more than likely due to errors in locating the schools, rather than a datum shift. Thus, we can accept this

Fig. 13.17. The school points match the large buildings in the photo.

479

transformation and proceed with our purpose. The SD Schools layer contains schools for the entire state, so we want to select only the ones in Rapid City for our database.

29➔ Turn off the orthophoto.

29➔ Zoom to the extent of the rds_clp layer.

29➔ Set the Selectable Layers to SD Schools only.

29➔ Use the Select Features tool to select the schools which fall within or near the area covered by the roads layer.

Now we can export these selected features directly to the geodatabase as a standalone feature class, taking advantage of the projection capability along the way.

30➔ Right-click the SD Schools layer and choose Data > Export Data.

30➔ Choose to export the Selected Features (Fig. 13.18).

30➔ Choose to use the same coordinate system as the *data frame*.

30➔ Click the Browse button.

30➔ Use the bottom drop-down box and change Save As Type to Personal Geodatabase feature classes.

30➔ Navigate to the folder with your rapidcity geodatabase in it.

30➔ Double-click the database to open it and show the other feature classes.

30➔ Type in the feature class name "schools".

30➔ Click Save in the Saving Data window.

30➔ Click OK in the Export Data window.

30➔ Click No to add the feature class as a data layer.

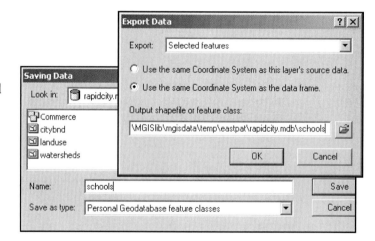

Fig. 13.18. Exporting a shapefile to a geodatabase feature class from ArcMap

31➔ Switch to ArcCatalog and examine the geodatabase to see the new layer.

31➔ Open its properties to check that the coordinate system has been changed to UTM NAD 27. Close the Properties window when you are finished.

TIP: If the schools layer does not appear, it probably means that ArcCatalog did not update the display. Close the folder containing the geodatabase, choose View > Refresh from the main menu bar, and then reopen the folder and the geodatabase. The new schools layer should now appear. If not, try exiting and restarting ArcCatalog.

➔ Close ArcMap. You do not need to save your changes.

> **TIP:** We close ArcMap because a geodatabase cannot be modified in ArcCatalog while it is open in ArcMap.

Setting up attribute domains

For the next task, we will create some attribute domains for the roads in Rapid City. Recall that using domains is one benefit of the geodatabase model over coverages and shapefiles. We will create four domains for the roads: RoadType, NumLanes, Direction, and StreetAbbrev. The road type contains a text designation such as Highway or Local, NumLanes gives the number of lanes, Direction gives one of the eight directions (N, W, SE, etc.) and StreetAbbrev indicates the street type abbreviation (RD, ST, AVE).

4. Which type of domain, range or coded, would you suggest for each of these attributes?

The first domain, RoadType, will be a coded value domain based on a text field, and will contain the values Interstate, Highway, Ramp, Connector, and Local. Note that in creating the domain, we must be careful that every road in Rapid City fits into one of these groups. Such a classification scheme is called complete. We will use a three-letter code for the actual stored value to save space but enter a longer description which will appear when we do editing. We should also consider the split and merge policies to use. If a street is split, its type will likely remain the same, so we choose Duplicate for the split policy. In the case of a merge, the street types might not match, so we should choose the Default Value policy.

32➔ In ArcCatalog, right-click the rapidcity geodatabase and choose Properties. Click the Domains tab.

32➔ Type the domain name, "RoadType", into the first box, and enter the description "Type of Road" (Fig. 13.19).

32➔ Set the Field Type to Text.

32➔ Set the Domain Type to Coded Values.

32➔ Set the split policy to Duplicate.

32➔ Set the merge policy to Default Value.

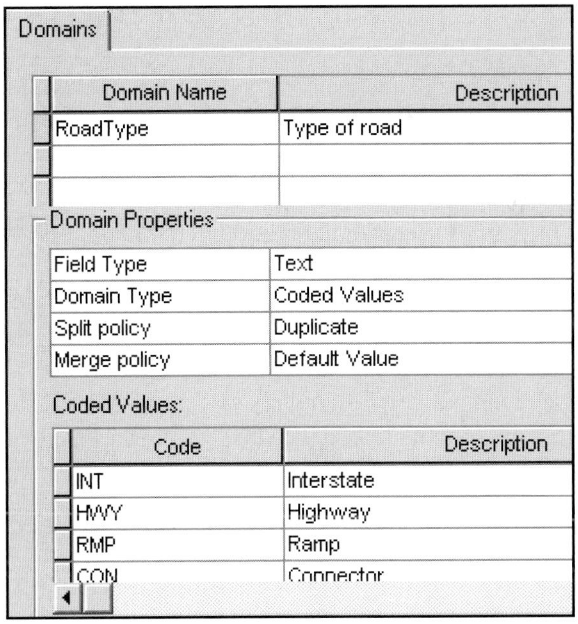

33➔ Type the three-letter codes into the Code column, and enter the description for each in the Description column. INT = Interstate, HWY = Highway, RMP = Ramp, CON = Connector, and LOC = Local.

33➔ Click Apply.

Fig. 13.19. Setting up the road type domain

TIP: ALWAYS, ALWAYS click Apply after EACH domain is added, before adding the next one. If you made an error in one domain, but added more before clicking Apply, then one or more of the domains may be added incorrectly, and it may not be clear which ones have problems. You may end up deleting them all and starting again, if you fail to follow this rule.

The second domain, NumLanes, will be a short integer range domain, and will allow values from 1 to 12. The split policy will be Duplicate, and the merge policy will be Default Value.

    34➜ Enter the domain name, "NumLanes", and the description, "Number of Lanes".

    34➜ Set the field type to Short Integer.

    34➜ Set the Domain Type to Range.

    34➜ Set the minimum value to 1 and the maximum value to 12.

    34➜ Set the split policy to Duplicate and the merge policy to Default Value.

    34➜ Click Apply. We do not need to enter any codes since this is a range domain.

Finally, we enter the third and fourth domains, Direction and StreetAbbrev. This Direction domain will be a text coded domain with values of N, S, E, W, NW, NE, SW, and SE. The descriptions will be the same as the codes. The split policy will be Duplicate and the merge policy will be Default Value. The StreetAbbrev domain will be a text coded domain containing AVE, ST, RD, DR, BLVD, CT, LN, WAY, CIR, and PL, with the full word entered as the description (Avenue, Street, etc.). The split and merge policies will be Duplicate and Default Value, as usual.

TIP: You must *always* enter a description, even if it will be the same as the code. If the description is left blank, then the values will appear blank when you try to view them or edit them in ArcMap. ArcMap always displays the descriptions, not the codes.

    35➜ Enter the Direction domain parameters and click Apply.

    36➜ Enter the StreetAbbrev domain parameters and click Apply.

TIP: Domain policies and codes may be edited after the domain is created. However, the field type and domain type, once set, cannot be altered. To change them you must delete the domain and create a new one with the desired properties. To delete a domain, select it and press the delete key.

Now that we have set up the domains, our next step is to create the attributes in the roads file itself. At the time we create them we will assign the domains and also set the default values. The default road type will be Local, the default number of lanes will be 2, and the default speed limit will be 25, since by far most of the roads in town are local residential streets.

37➜ Click OK to close the Domains window.

37➜ Expand the Transportation feature dataset if necessary, right-click the roads feature class, and choose Properties.

37➜ Click the Fields tab.

37➜ Type the new field name, "TYPE", in the first available box in the fields list (Fig. 13.20).

37➜ Click in the Data Type box and choose Text.

37➜ Type "LOC" for the default value.

37➜ Click the Domain box and choose RoadType.

37➜ Set the Length to 3.

37➜ Click Apply.

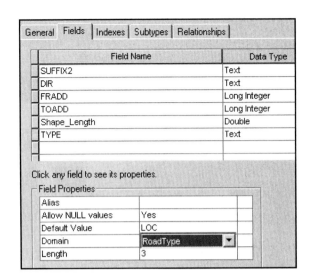

Fig. 13.20. Setting the domains and default values for a new field

Next we create an attribute for the number of lanes.

38➜ Type the new field, "LANES", in the next available box.

38➜ Set the field type to Short Integer.

38➜ Set the default value to 2.

38➜ Set the domain to NumLanes.

38➜ Click Apply.

TIP: Note that the list of domains available was different in the previous two examples. The domain field type must match the attribute field type; a short integer domain cannot be applied to a long integer attribute, for example.

Now we can assign domains and default values to some existing fields. The Direction domain actually applies to two fields: the PREFIX field and the SUFFIX2 field. One benefit of domains is that they may be applied multiple times.

39➜ Click the row for the field PREFIX to select it.

39➜ Leave the default blank.

39➜ Set the domain to Direction. Click Apply.

39➜ Select the SUFFIX2 field and set the domain to Direction. Do not set a default. Click Apply.

39➜ Select the SUFFIX field and set the domain to StreetAbbrev. Do not set a default. Click Apply.

Using domains when editing

Now we have set up our domains and are ready to see how they work.

→ Close the Properties window and exit ArcCatalog, so it won't interfere with editing the roads.

40→ Start ArcMap with a new, empty map, and load the roads layer from the rapidcity geodatabase.

40→ Open the Editor toolbar, if necessary, and choose to Start Editing.

41→ Use Select By Attributes to select all the roads where [STREET] = 'MOUNT RUSHMORE'.

41→ Close the Select By Attributes window.

42→ Click the Attribute Editor button.

42→ Click the top entry in the attribute list and examine its attributes.

First, notice that the SUFFIX field entries are listed as "Road" rather than RD, which is the code actually in the field.

42→ Click in the space next to the SUFFIX field.

A coded value domain has a drop-down list showing the available codes for that attribute (Fig. 13.21). The only way to enter a value is to choose it from the list. This restriction automatically protects the attribute from having inappropriate codes entered, such as AV instead of AVE. The same holds for the SUFFIX2, PREFIX, and TYPE fields.

42→ Click each of the other coded value fields and examine the choices.

Notice that most of the fields also include a <Null> entry. Whether or not this null choice appears depends on whether you answered "yes" or "no" to the Allow Null Values entry in the Fields properties in the previous section.

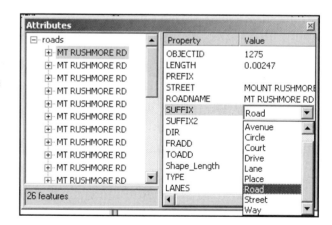

Fig. 13.21. Coded domains appear as drop-down lists when you are editing.

43→ Click on the TYPE field and choose Connector.

43→ Click on the LANES field.

Since LANES was assigned the NumLanes domain, a range domain, there is no list. Simply type the value in. No error checking is done at this point.

43→ Enter the value "4" in the LANES field.

So far, we've only entered values for the first road. Let's assign the same values to all of the Mt. Rushmore roads.

44➔ Click the row labeled roads at the top of the Attribute Editor. The attributes all go blank.

44➔ Click on the PREFIX field and choose <Null>.

44➔ Click on the SUFFIX2 field and choose <Null>.

44➔ Click on the TYPE field and choose Connector.

44➔ Click on the LANES field and enter "4".

Now let's see how to take advantage of the range domain error checking. It is done by means of a separate step called validation.

45➔ Click one of the segments of MOUNT RUSHMORE RD in the Attribute Editor.

45➔ Click the LANES field and enter the number of lanes, but pretend to make a typing mistake by entering "44" instead of "4" (Fig. 13.22).

Recall that the NumLanes domain has a range from 1 to 12. Now test for these attribute errors by using the validation function.

46➔ Choose Editor > Validate Features from the Editor toolbar.

If any errors are found, it reports the number in a dialog window, and places the incorrect records in the selected set, ready for editing. Note that only attributes with domains will be tested. Of course, these errors were entered on purpose. However, validation is very useful for catching unanticipated errors.

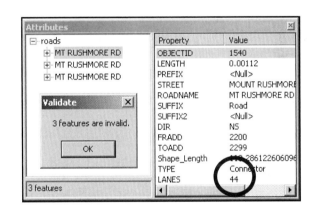

Fig. 13.22. Validating attribute entries with domains

TIP: Note that the PREFIX and SUFFIX2 fields were originally blank. If we had failed to apply the <Null> value specifically to these records, then all of the records would have shown up as errors. A blank field without a <Null> entry is considered an error.

46➔ Correct the wrong entry by setting the number of lanes back to 4.

46➔ Choose Editor > Validate Features again.

This time it reports that all features are valid.

➔ Stop editing and save your edits.

➔ Exit ArcMap.

Exploring the split and merge policies

In this section we will experiment with using the split and merge policies to help manage parcel attributes during editing. First, however, we will import some parcels into our geodatabase and add some additional domains.

47➔ Start ArcCatalog and navigate to the directory containing the rapidcity geodatabase.

47➔ Import the shapefile parcelval8.shp from the mgisdata\Rapidcity\eastpat folder into the geodatabase as a standalone feature class named "parcelval".

47➔ Preview the attribute table of parcelval feature class.

5. Other than the standard FID and Shape fields, what fields are present in this table?

_____

Recalling that the ID field contains the block number, answer the following questions.

6. Which one of these fields should not have a domain? Why not? _____

7. For each field that can have a domain, state whether a coded or range domain would be best.

_____

8. For each field that can have a domain, state which split policy would work best.

_____

9. For each field that can have a domain, state which merge policy would work best.

_____

With these answers in mind, create the domains.

48➔ Right-click the rapidcity geodatabase and choose Properties.

48➔ Enter a new domain called "ZoneCodes". Make it a text domain with coded values.

We will enter just a few zoning codes and descriptions as an example: R-1 = Low Density Residential, R-2 = Med Density Residential, R-3 = High Density Residential, and C-1 = Commercial.

48➔ Set the split policy to Duplicate.

48➔ Set the merge policy to Default Value.

48➔ Enter the zoning codes as listed above.

48➔ Click Apply.

For the ID field, instead of making a special domain just for this parcel layer, let us create a generic long integer ID field domain with a duplicate split rule, which could be used in many different circumstances.

49➜ Enter a new domain called "LongIntDup" with the description "Generic Long Integer with Dup Splits".

49➜ Make it a range domain from 0 to 999,999,999.

49➜ Set the split policy to Duplicate and the merge policy to Default Value.

49➜ Click Apply.

Since ParcelValu and Area may range quite a bit in size, there is no particular benefit to setting a maximum—parcels can get pretty large and expensive. Mainly we want the domains in order to use the split/merge policies. Both are also long integer fields, so let us simply create a single generic domain for both, called LongIntRatio, with a very wide range.

50➜ Enter a new domain called "LongIntRatio" with the description "Generic Long Integer Values with Ratio Splits".

50➜ Make it a range domain from 0 to 999,999,999.

50➜ Set the split policy to Geometry Ratio and the merge policy to Sum Values.

50➜ Click Apply.

50➜ Click OK to close the window.

Now we still need to assign the domains to the attributes in the parcelval feature class. First the ID field.

51➜ Right-click the parcelval feature class in the geodatabase, and choose Properties.

51➜ Click the Fields tab.

51➜ Select the ID field.

51➜ Set the field alias to "BlockID".

51➜ Change the default value to 0 (as a flag to identify blank records).

51➜ Set the domain to LongIntDup.

51➜ Click Apply.

Now the Parcel_ID field has no domain, but we can still set a default value.

52➜ Click the Parcel_ID field to select it.

52➜ Set the default value to 0-000000. Click Apply.

52➜ Click the Area field and set the domain to LongIntRatio. Click Apply.

53➜ Click the ParcelValu field and set the domain to LongIntRatio. Click Apply.

53➜ Select the Zoning field. Set the domain to ZoneCodes and enter the default value as R-1. Click Apply.

53➜ Click OK to close the Properties window.

Now let us see how the split and merge rules work.

➜ Close ArcCatalog.

➜ Start ArcMap with a new empty map.

54➔ Add the parcelval feature class from the geodatabase.

54➔ Label the parcel features with the Parcel-ID field in 10 point bold font.

TIP: When setting up the labels, click the Placement Properties button, click the Placement tab, and choose the Try horizontal first option. The labels will fit the polygons better.

54➔ Open the Editor menu, if need be, and choose Start Editing.

Imagine that these parcels have been approved for a new subdivision being built. During the final site preparations, however, some site difficulties have surfaced, necessitating some changes in the layout of the parcels. The first parcel listed for change is the third from the right on the bottom row, currently labeled 8-453934. This parcel must be split in half.

55➔ Click the Select tool on the Editor toolbar, and click on the parcel to be split.

55➔ Open the attribute table of parcelval, and click on the Show Selected button.

55➔ Resize and move the table to show the important fields and the map at the same time.

55➔ Examine the values in the table.

Now let's split the parcel.

56➔ Turn on edge snapping for the parcelval layer and then close the Snapping window.

56➔ Make sure that parcelval is the target layer, and change the editing task to Cut Polygon Features.

56➔ Use the Pencil tool to add a line cutting the parcel approximately in half. Double-click to finish the sketch.

56➔ Examine the updated feature attributes in the table (Fig. 13.23). Yours may differ slightly.

Notice that two polygons are now selected. The Parcel_ID field had no domain and followed the ordinary split rule of copying the old value to both new features. We set a Duplicate rule for the BlockID, so the value 8 appears in both new polygons. The Area and ParcelValu fields, with a Geometry Ratio rule, have been divided between the two polygons based on their respective areas. The Zoning value has defaulted to Low Density in both. All of these new fields are correct, except that a new parcel-ID must be added.

Id	Parcel_ID	Area	Zoning	ParcelValu
8	8-453934	17438	Low Densit	174000

Id	Parcel_ID	Area	Zoning	ParcelValu
8	8-453934	8302	Low Densit	82843
8	8-453934	9136	Low Densit	91157

Fig. 13.23. The parcel attributes before and after a split.

Add the next highest value available in that block, 8-453949.

57➔ Click in the Parcel_ID field in the bottom record of the table and change the value to 8-453949.

The next change includes merging the third and fourth parcels from the left in the top row, 8-453914 and 8-453915.

58➜ Move the table so that you can see these two parcels.

58➜ Select the two parcels to be merged.

58➜ Briefly examine their attributes in the table.

Id	Parcel_ID	Area	Zoning	ParcelValu
8	8-453914	6244	Low Densit	115000
8	8-453915	7486	Low Densit	121000

Id	Parcel_ID	Area	Zoning	ParcelValu
8	8-453914	13730	Low Densit	236000

Fig. 13.24. Parcel attributes following a merge

59➜ Choose Editor > Merge from the Editor toolbar. If asked to choose which to be merged, pick the first one. Its attributes will be used for the new polygon if no rule is available for a field.

59➜ Examine the attributes again (Fig. 13.24).

Notice that the parcel_ID was copied from the first polygon in the merge list, number 8-453914. The area and parcel values were summed, and the zoning defaulted again to Low Density.

TIP: If the Area and Value numbers were not summed, go back and check the domains. If they were correct, simply move on. The author could not get these to sum correctly for this particular file, for some unknown reason. Hopefully, yours will work.

➜ Split or merge a few more parcels for practice, if you wish.

➜ When finished, stop editing and save the edits.

This is the end of the tutorial.

➜ Exit ArcMap. You do not need to save the changes to the map document.

Exercises

1. Create a new geodatabase to contain the mgisdata\Sdakota data.

2. Add a Water feature dataset to the geodatabase for water features. Use State Plane North NAD 1983 for the coordinate system, and remember to make the extent large enough. Put the hydrology layers from the Sdakota folder including the rivers, lakes, and watersheds in the Water dataset.

3. Put the remaining layers in the mgisdata\Sdakota folder in the geodatabase as standalone feature classes. Use State Plane North NAD 1983 again for the coordinate system. (**Hint:** Use the Project tool, or the Batch Project tool.)

4. **Capture** the layout of the geodatabase in the ArcCatalog window to turn in.

5. Using the restaurants layer in the rapidcity database, add two attributes for the type of restaurant (franchise or local) and the number of tables. Create attribute domains for the fields. **Capture** the windows showing the properties for each domain.

6. Enter the information for each restaurant. (A franchise includes a national chain, such as Burger King, or Kentucky Fried Chicken. All others are considered local. Simply make up some values for the number of tables.)

7. Create a layout showing the type of restaurant, labeled with the number of tables.

TIP: The author has encountered an intermittent problem in ArcGIS 9.0 when importing/exporting shapefiles with one coordinate system to a feature class with a different coordinate system—the coordinate system conversion does not take place, leaving the new feature class with the original CS. This behavior was observed when exporting data from ArcMap using the same coordinate system as the data frame, and when using the Batch Project tool, and appears to be a program bug. Check your output coordinate systems carefully during these exercises. If a problem occurs, use the Project tool to transfer the data (i.e., project the shapefile and save the output in the geodatabase).

Challenge Problem

Create a geodatabase with at least four feature classes and at least one feature dataset. The data you use may come from any source but should comprise a logical collection of related data. **Capture** the layout of your geodatabase in ArcCatalog, and create a map layout showing your dataset.

Skills Reference

Creating a geodatabase

1. In ArcCatalog, navigate to the folder in which the geodatabase will be created.

2. Right-click the folder and choose New > Personal Geodatabase from the menu.

3. The geodatabase appears in the display window to the right.

4. Type in the name of the new geodatabase and press Enter.

Deleting a geodatabase

1. In ArcCatalog, navigate to the folder containing the geodatabase.

2. Right-click the geodatabase and choose Delete.

3. Say Yes when prompted to confirm deleting the geodatabase.

Deleting feature classes

1. In ArcCatalog, navigate to the folder containing the geodatabase and expand it.

2. Right-click the feature class to be removed and choose Delete.

3. Click Yes when prompted to confirm deleting the feature class.

Deleting feature datasets

1. In ArcCatalog, navigate to the folder containing the geodatabase and expand it.

2. Right-click the feature dataset to be removed and choose Delete.

3. Click Yes when prompted to confirm deleting the feature dataset.

Creating a feature dataset

1. Create a boundary shapefile or feature class that has the same coordinate system and data extent planned for the feature dataset.

2. In ArcCatalog, navigate to the geodatabase to contain the feature dataset.

3. Right-click the geodatabase and choose New > Feature Dataset.

4. Enter the dataset name (Fig. 13.25).

5. Click Edit to change the coordinate system.

6. Click the Import button in the next window.

7. Navigate to the folder containing the boundary data layer and select it. Click Add.

8. Click OK to finish setting the coordinate system.

9. Click OK to finish creating the feature dataset.

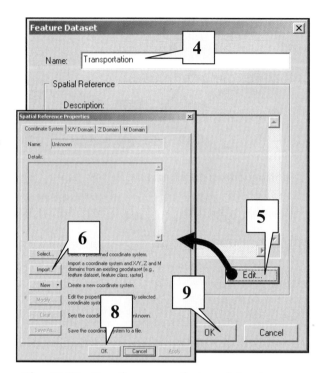

Fig. 13.25. Creating a new feature dataset

10. Import the boundary data layer as the first feature class in the new feature dataset.

TIP: If the import layer has a different coordinate system than the feature dataset, then the features will automatically be reprojected as they are loaded.

Creating a feature class

1. In ArcCatalog, right-click the geodatabase or feature dataset that will contain the feature class, and choose New > Feature Class.

2. Enter the name of the feature class (Fig. 13.26).

3. Fill the button for the feature class to contain simple features.

4. Click Next.

5. Set the Configuration Keyword setting to Default.

6. Click Next.

7. Click the Shape field to select it.

8. Click in the Geometry Type box and choose the desired type from the drop-down list (point, line, etc.).

If you are adding to a feature dataset, the spatial reference will already be defined. If it is a standalone feature class, you must define the spatial reference.

9. Click the ellipses in the Spatial Reference box to access the Spatial Reference window. Use the button to define, modify, select, or import a coordinate system.

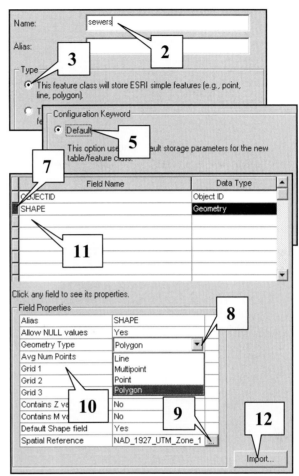

Fig. 13.26. Creating a new feature class

10. Change other properties of the shape field, if desired.

11. To add new attribute fields, click in the first available box in the Field Name list. Type in the name, set the Data Type, and change the field properties as needed.

12. To add fields from another feature class or shapefile, click the Import button. Locate the feature class and select it. All fields in the layer will be added to the new one.

13. Click Finish to create the feature class.

Importing layers into a geodatabase

The following directions apply to importing feature classes as standalone classes, or as parts of a feature dataset.

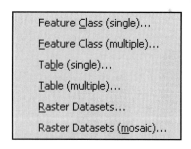

1. In ArcCatalog, navigate to the folder containing the geodatabase.

2. Right-click the geodatabase or feature dataset, choose Import, and select an import option based on the type of data to bring in (Fig. 13.27).

The following directions assume that you have chosen the Feature Class (single) tool. The other choices will be similar.

Fig. 13.27. Options for importing layers

3. Click the Browse button and navigate to the directory containing the shapefile to import (Fig. 13.28). Select it and click Add.

4. Enter the name of the feature class in the geodatabase.

5. Enter an SQL expression to bring only selected records into the geodatabase (optional).

6. Examine the fields list, and rename any of the NewFieldNames if desired.

7. Click the Environments button to set the geoprocessing environment defaults, such as the coordinate system. Use this button to project the data to a new coordinate system as it is imported.

8. Click OK to start importing the shapefile.

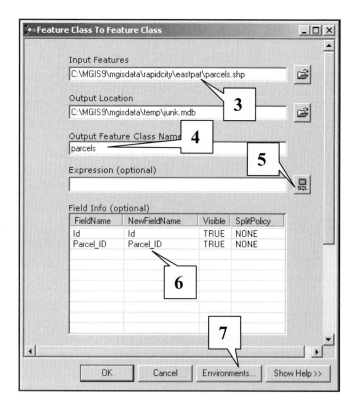

Fig. 13.28. Importing a shapefile to a geodatabase

TIP: To import several feature classes at the same time, choose Import > Feature Class (multiple). The options to rename the feature class, perform selections, and edit field names are not accessible with the multiple feature class tool, however.

Creating attribute domains

Attribute domains are properties of the geodatabase and are set at that level.

1. In ArcCatalog, navigate to the folder containing the geodatabase.

2. Right-click the geodatabase and choose Properties. The Domains tab is the only one.

3. Click in the first available Domain Name box and enter the name of the domain (Fig. 13.29).

4. Enter a longer description of the domain for documentation purposes.

5. Set the attribute type (Short Integer, Long Integer, Text, etc.). The domain can only be applied to attributes of the same type.

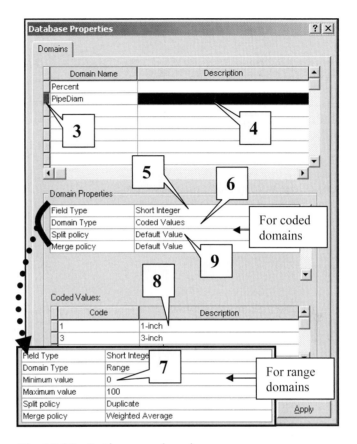

Fig. 13.29. Setting up a domain

6. Click in the Domain Type box and choose a Range or a Coded Value domain.

7. If it is a Range domain, enter the minimum and maximum range values.

8. If it is a Coded Value domain, enter each value code and description in the lowest set of boxes. The descriptions should not be too long (< 15 characters is probably best).

9. Click in the Split and Merge policy boxes and choose the type of policy desired.

10. Click Apply to create the domain. It will be checked for errors and added if OK.

11. To enter additional domains, repeat steps 3 through 10, or click OK if finished.

12. **To delete a domain,** click on the grey area to the left of its name to select it, and press the Delete key.

TIP: Domain policies and codes may be edited after the domain is created. However, the field type and domain type, once set, cannot be altered. To change them you must delete the domain and create a new one with the desired properties.

Setting default values or domains for attributes

1. In ArcCatalog, right-click the feature class containing the attribute and choose Properties.

2. Click the Fields tab (Fig. 13.30).

3. Select the field by clicking on the grey area to the left of its name. Its properties appear underneath.

4. Click the box for Null values and choose Yes to allow null values or No to disallow them.

5. Click in the Default Value box and type in the default value.

6. To assign a domain to an attribute, click in the Domain box and select a Domain from the list.

7. If desired, enter default values for additional items by repeating steps 3 through 5.

8. Click Apply to effect the changes without closing the window, or click OK to effect the changes and close the window.

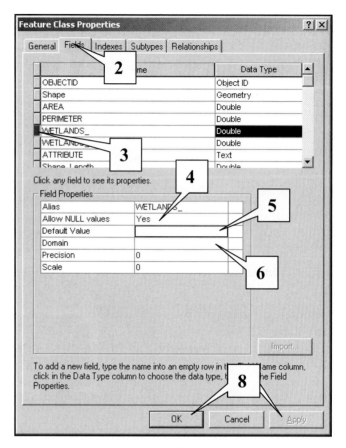

Fig. 13.30. Setting default values for an attribute

Chapter 14. Analyzing Networks

Mastering the Concepts

Objectives

> ➢ Learning the function and terminology of simple networks

> ➢ Understanding different types of networks

> ➢ Performing tracing analysis on networks

> ➢ Understanding how networks are constructed

Concepts

About networks

Networks consist of a system of paths traveled by a variety of things, such as traffic, water, sewage, or electricity. Common examples of networks include roads, utility lines, airline routes, and streams. Modeling the behavior of a "commodity" flowing through the network can help answer questions such as: How long would it take a chemical spill in this watershed to reach the city water supply? If this transformer blows, which parts of the city will be out of power? If a problem develops with a sewer line, which shutoff valve can we use to halt flow during repairs, while affecting the fewest possible people?

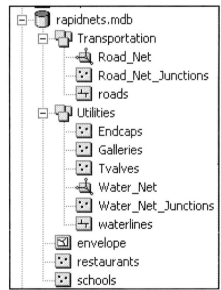

Fig. 14.1. Networks are built from feature classes in feature datasets.

Geodatabases contain a special data model developed to answer exactly these kinds of questions. Networks can be built from existing simple feature classes. A feature class containing roads, for example, can be used to create a transportation network. Multiple layers can be combined to form more complex networks. A water supply network might include feature classes representing pumping stations, pipes, valves, and meters.

Networks may only be created in feature datasets. All feature classes participating in a network must belong to the same feature dataset; however, a feature dataset with a network may also contain other feature classes which do not participate in the network. A Transportation feature dataset containing roads and rail lines could have a roads network that did not include the rails.

Figure 14.1 shows a geodatabase containing two networks. The Transportation feature dataset contains a single line feature class, roads, plus two feature classes that comprise the network: Road_Net and Road_Net_Junctions. The Utilities feature dataset contains four feature classes plus the Water_Net and Water_Net_Junctions. It is not obvious from looking at the geodatabase which feature classes in the feature dataset participate in the network; however, this information can be obtained by examining the properties of the network in ArcCatalog.

All feature classes participating in a network must take the role of either an **edge** or a **junction** (Fig. 14.2). An edge feature represents a path along the network. Examples of edges include roads, pipelines, or electric cables. Edges must come from a line feature class. A junction represents a point in a network where edges meet, or where an object such as a valve exists. A street intersection is represented by a junction. T-valves or straight-line connectors would constitute junctions in a pipeline network. Junctions can come from two sources: they can be automatically constructed during network building wherever edges meet, or they can be specific objects such as valves which come from a point feature class.

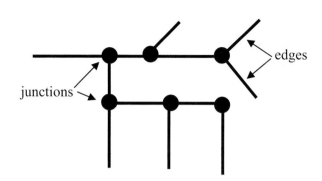

Fig. 14.2. A network is composed of edges and junctions.

Network features can take one of two states, enabled or disabled. An enabled edge or junction allows material to flow through it. A disabled edge or junction blocks travel, whether of water, electricity, or cars. The feature state is initially assigned during construction of the network; however, it may be temporarily modified during analysis, such as disabling a street edge to represent road construction that blocks travel. The states of the network features are stored in the attribute table of the feature class in a field named Enable.

Network storage in a data model has two aspects (Fig. 14.3). The **geometric network** consists of the actual points and lines in the feature classes which participate in the network. These features are stored as the familiar feature classes containing points and lines. Certain fields are added to the attribute tables of the feature classes to help track the participation of each feature in the network. The **logical network** includes a set of tables which store information about the construction and operation of the network elements. Users do not usually have direct access to the logical network. The files are created at the time the network is built from the feature classes. The geometric network and the logical network are interconnected, and changes in the geometric network require changes to the logical network. During editing, the relationships between the geometric and logical networks are maintained automatically.

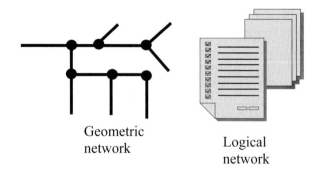

Geometric network

Logical network

Fig. 14.3. The geometric network consists of actual features in a feature class. The logical network contains the relationships, connections, and behavior as tabular data.

TIP: In order to build networks, the user must have an ArcEditor or ArcInfo license. ArcView cannot create or edit networks, but it can perform analysis on networks that already exist.

Types of networks

Networks fall into one of two basic categories: transportation networks or utility networks. The material in a transportation-type network (cars) has few constraints on travel; the drivers go where they choose (Fig. 14.4). The vehicles can usually travel in either direction along an edge and through intersections represented by junctions.

Utility-type networks have established directions of flow determined by the topology of the connections and the location of sources or sinks (Fig. 14.5). A **source** point provides the material to the network and "pushes" the material away from itself. A power plant would be a source for an electric grid; a water treatment plant might act as a source for a water line network. A **sink** represents a location where the material is used or leaves the network. Sinks draw the material toward themselves. Electric meters or water meters on a building might be considered sinks, as would a sewage treatment facility discharging to a lake.

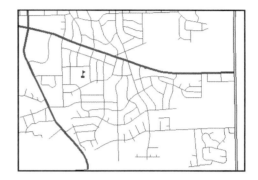

Fig. 14.4. In a transportation network, the drivers choose the way to travel.

Transportation-type networks and utility-type networks have different types of analysis functions available for use in modeling flow. A transportation network could be used to solve problems such as

➢ What is the best path to travel to 16 delivery locations?

➢ What is the likely service area of a fire station based on travel time?

➢ What is the shortest path from point A to point B?

Utility-type networks might be used to address other kinds of problems.

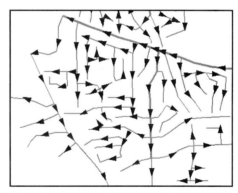

Fig. 14.5. In a utility network, the flow of the material is determined by the network configuration.

➢ If this valve fails, which customers will be affected?

➢ If I have to close this pipe for repairs, can I reroute water through another path to minimize service disruption?

➢ How will contamination at one location propagate through the network?

➢ Which sewer lines serve only residential customers?

Utility network flow must be established before the network can be used for analysis. You establish flow by specifying the sources or sinks in the network and solving for the flow direction. A utility network edge may have three possible states. It may be unassigned, if network flow has not yet been established. It may have a determinate flow, meaning that a single flow direction was successfully established. Or it may have an indeterminate flow, meaning that the flow direction is ambiguous and cannot be uniquely determined. The third case generally results from a topological error in the network or from an error in specifying the sources and sinks.

Network analysis

Network analysis problems are tackled using programs called **solvers**. Many types of solvers are possible, and they can range from very simple to very complex. ArcMap comes with some simple solvers called **traces,** which follow network connections (Fig. 14.6). More solvers will be added to future versions. More advanced and complex network analysis tools are also available in the ArcToolbox for users with an ArcInfo license. Knowledgeable users may employ the special customization languages provided with ArcGIS to write new solvers, although such tasks are beyond the general user. ArcMap solvers require input in four forms: the network itself, weights of elements along the network, the locations of interest, and any barriers to flow.

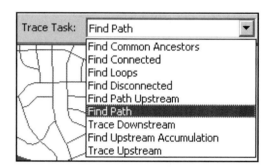

Fig. 14.6. Tracing solvers in ArcMap

Network weights are stored for each edge or junction as attributes and specify the "cost" of traversing the network. Common edge weights include distance along the edge, or travel time as dictated by the street length and speed limit. Common junction weights might include a loss of pressure at a pipe T-junction, or a waiting time imposed at a traffic light.

Flags pinpoint locations of interest, such as the starting and end points of a journey, the delivery stops on a route, or a location from which to trace up or downstream. Flags may be placed along edges or at junctions (Fig. 14.7).

Barriers represent temporary blockages or outages in the network, such as a damaged bridge, a closed valve, or a broken power line. As with flags, barriers may be placed on either edges or junctions (Fig. 14.7). When placed on a feature, barriers temporarily override the normal "Enabled" state of the feature, as stored in the layer's attribute file.

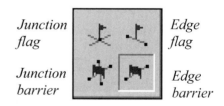

Junction flag *Edge flag*

Junction barrier *Edge barrier*

Fig. 14.7. Flags are used to mark locations of interest. Barriers stop flow through a junction or along an edge.

When placing flags and junctions before performing an analysis, the user must ensure that the types match—that all flags and barriers are at junctions, or that all of them lie on edges. The solvers do not always work properly on a combination of junction and edge flags.

The output of a network tracer usually takes the form of a set of edges and junctions along the traced path. Network analysis offers several options concerning the form of this output. The path solution may be returned either as a graphic, or as a set of selected features. If returning a selection, one may choose to return only edges, only junctions, or both.

The Utility Network Analyst toolbar

ArcMap contains a special toolbar used to direct and control the analysis of networks (Fig. 14.8). The main features of this toolbar are as follows:

➤ The Network drop-down list controls which network is being analyzed. One network may be analyzed at a time. Networks must be added to the map document in order to be used.

➤ The Flow menu controls the display of arrows and other symbols representing flow directions in a utility network. The user can turn the symbols on and off and change the symbols being used.

➤ The Establish Flow button next to the Flow menu is used during an editing session to establish the initial flow direction in a network, and it requires an ArcEdit or ArcInfo license. ArcView can perform analysis on networks for which this task is already complete.

➤ The Analysis menu manages the analysis by placing flags or barriers, clearing both flags and barriers, or clearing previous results in preparation for a new analysis. The Trace Task menu specifies which kind of analysis will be performed.

➤ The drop-down flag tool controls which type of flag or barrier to add to the map.

➤ Finally, the Solve button must be clicked to actually perform the trace. All of the trace tasks require that at least one flag be added to the map before any analysis can occur. If no flags are present, the Solve button will be dimmed.

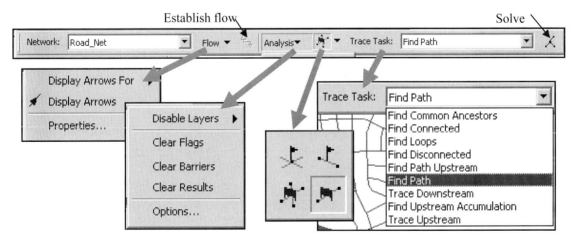

Fig. 14.8. The Utility Network Analyst toolbar contains all the functions needed to analyze networks in ArcMap.

The overall process of analyzing a network includes specifying the network to be used, choosing the tracing task, adding flags or barriers, and solving. Once a solution is obtained, it may be cleared before going on.

The Analysis > Options menu choice opens a window with four tabs. These options control the analysis by specifying which features will be traced, whether the features will be weighted, and the format of the results. The output of a trace may take the form of a drawing or a set of selected features (Fig. 14.9). The features returned can include all features matching the criteria of the trace task, or simply those stopping the trace. The user can also choose whether to return edges or junctions or both.

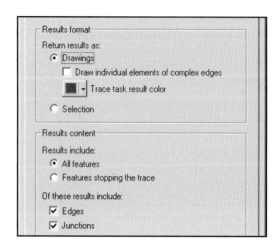

Fig. 14.9. Choosing the format of the output results

Generic trace solvers

The solvers in this section may be applied to either utility-type or transportation-type networks. They carry out their work without regard to the direction of flow through the edges and junctions.

Finding paths

This solver traces the least-cost path between two or more flags placed on the network. It can find a way to travel from one location to another, or to plan a path to most efficiently visit a series of locations, such as delivery stops. When finding paths, the solver keeps track of the "cost" of traveling the path. By default the cost is the number of edges traversed. Figure 14.10a shows the lowest-cost path, in terms of the number of edges traversed, for driving from a student's house to school. The flags are shown as green boxes at the locations of interest; in this case edge flags were used.

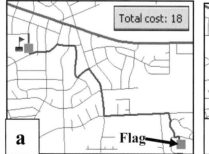 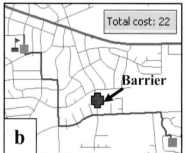

Fig. 14.10. Finding simple paths. (a) Finding the shortest path between home and school, with cost in edges. (b) Placing a barrier (construction) requires finding a new, higher-cost route.

The user can modify the solver behavior by specifying barriers to travel, such as road construction. Figure 14.10b shows the new path found when a barrier is placed along the original one. The cost has increased to 22 edges, but it is still the least-cost path given the presence of the barrier.

Users can also specify a weight, such as distance or travel time. Without weights, the solver simply counts the number of edges traversed. However, the lowest number of edges does not

always mean the shortest path; in some cases the shortest path could have more edges. Finding the shortest path requires a distance weight. In Figure 14.11a, the distance along the edges has been chosen as the weight. A slightly different path results than when simply counting edges. The cost shows up in the distance units—in this case, meters.

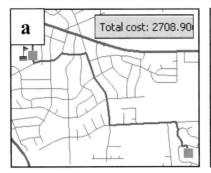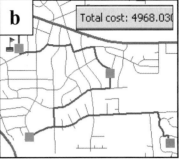

Fig. 14.11. Finding least distance and multiple stop paths. (a) Using distance as a weight results in a different path, with cost in meters. (b) Finding the best path to pick up the child's friends on the way to school.

Finally, the user can specify multiple flags in order to find the least-cost route for making several stops in a particular sequence. Figure 14.11b shows a possible path to pick up a child's friends and take them to school, also. This particular Find Path tracer simply visits the stops in the order given. Other solvers, not currently available in ArcMap, must be used to solve the classic "traveling salesman" problem: given a set of locations to visit, devise the most efficient sequence and path to follow.

Find Connected/Find Disconnected

The Find Connected solver traces along the features and highlights the ones that are connected to the flag(s) placed on the network. Such a solver might be used to find all of the water lines that are supplied by a single water intake gallery along a river (Fig. 14.12). Such information might be useful in gauging the impact of a chemical spill near the intake, and determining which customers must be warned to purchase drinking water until the contamination is cleaned up.

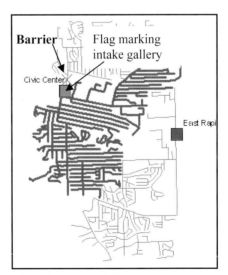

Fig. 14.12. Find Connected returns all parts of the network connected to one or more points of interest.

Barriers placed at specific locations affect the results of this solver. Any feature with a barrier between it and the flag location will be considered disconnected from the network (Fig. 14.12).

Find Disconnected performs the opposite function of Find Connected; it locates portions of the network that can't be reached from the flagged locations. This solver is particularly useful for diagnosing problems when constructing networks. Sometimes a failure to snap two line ends together, or neglecting to split a line at an intersection, will result in part of the network being disconnected from another section. The solver easily finds these problem areas so that they can be fixed.

The Find Connected/Disconnected solvers are most useful for utility networks, but they can also be applied to transportation networks. A road network is designed to facilitate connecting all locations together for easy travel. Thus, a Find Connected usually results in the selection of the entire network (not a very useful result most of the time). However, one possible useful application might include finding which parts of the city might be seriously affected by construction placed at a critical location. In addition, Find Disconnected does help in locating network errors, as previously noted.

Finding loops

A loop is a section of a network that is closed upon itself and forms a continuous loop or ring. In a transportation network, loops are generally the rule, as most streets eventually connect to other streets and can be traversed in a circular fashion. This solver finds little utility in transportation networks, therefore. In utility networks, as a general rule, loops indicate a topological error, rather than a legitimate situation. A water pipe that forms a loop would cause problems when the emerging flow reentered the loop. The Find Loops tracer then, has its primary use as a tool for detecting and fixing loops. A flag must be specified as the place to begin tracing, and the solver finds all loops that lie downstream of the flag.

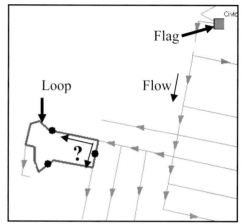

Fig. 14.13. A loop in a utility network is a topological error. Flow direction cannot be determined in a loop.

Figure 14.13 shows an example of a loop in a utility network. The flag has been placed at the upper right of the picture. The arrows indicate the direction of flow along the network. The loop found by the tracer is marked with small circles which indicate that flow in the loop is indeterminate—because of the loop the flow direction cannot be established from the two possible solutions. To fix the loop, one end must be disconnected from the circle, or an edge or junction must be disabled.

Utility trace solvers

The solvers in this section all share the constraint that they be used only in utility networks. They rely on an established flow direction in order to carry out the analysis.

Find Path Upstream

This solver works similarly to the Find Path solver previously discussed, except that movement along the edges is always constrained to be in the upstream direction. The tool can locate a path between a single location and its source (Fig. 14.14). If more than one flag has been specified, the paths of each flag are traced upwards.

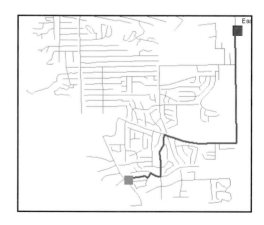

Fig. 14.14. Find Path Upstream traces the path of flow from a location to its source.

Find Upstream Accumulation

This solver is a variation of the Find Path Upstream tool, except that it keeps track of the weights on each edge and junction and applies them to determine a total accumulation of resources along the path.

Trace Upstream/Downstream

Trace Upstream and Trace Downstream provide a way to follow the flow from a specified location. Trace Upstream is essentially the same solver as Find Path Upstream. The downstream trace finds all the edges which lie downstream of a flagged location or locations. Such a solver might be used to find the service areas that would be disrupted by a downed power line or a water line break (Fig. 14.15).

Fig. 14.15. Trace Downstream could find the service area disrupted by a broken pipe at the flag location.

Find Common Ancestors

This solver takes a group of flags and finds any upstream edges and junctions which are shared by all the individuals in the group. The tool can be used by utility companies to help narrow down the possible failures in the system based on reports by customers.

For example, suppose that a utility company receives 50 calls in one evening from customers complaining about a loss of water pressure. By entering the locations of these reports as flags, the company can narrow down the possible locations of the suspected break or leak by finding the set of pipes that are common to all the calls. The break will most likely be found near the lower end of the common ancestors (Fig. 14.16).

These solvers are only a few of the possible types of analysis which could be performed on a network. Additional solvers will be released in future versions of ArcMap.

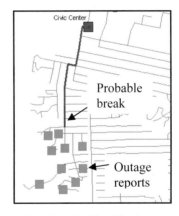

Fig. 14.16. The Find Common Ancestors solver can help locate service failures.

Building networks

Building valid networks can require much ingenuity and attention to detail, especially if the network is to be used for sophisticated modeling. The planning stage is very important, and the builder must consider a number of issues in designing the network. These issues include using simple or complex edges, assigning weights, using domains or subtypes, and establishing connectivity rules.

As previously mentioned, an ArcEditor or an ArcInfo license is required in order to create networks. We will not further address building networks in this chapter, but the next section introduces some basic design issues which can affect network analysis.

Simple edges or complex edges?

Network edges can occur in two different styles, as simple edges or complex edges (Fig. 14.17). A simple edge can only have junctions at its endpoints. A complex edge can have one or more junctions in its middle.

Complex edges offer advantages when keeping track of long features. For example, a water main going down a street has junctions composed of T-valves which branch off to laterals carrying water to each individual house. Using simple edges, the main would have to be broken into multiple pieces between each T-valve. However, for purposes of queries and maintenance, the water main is more logically and easily treated as a single feature. When clicking on the water main, it would be better to select the entire feature, not just one tiny segment between two laterals. By using a complex edge network, the water main can remain a single feature, while allowing the junctions of the laterals to connect to it.

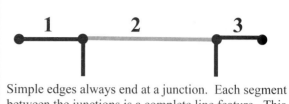

Simple edges always end at a junction. Each segment between the junctions is a complete line feature. This water main has three segments with two laterals joining it.

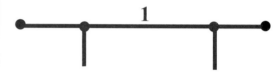

Complex edges may have junctions in the middle of the line feature. This water main is a single feature, with two laterals joining it in the middle.

Fig. 14.17. Networks with simple edges versus networks with complex edges

Junctions also can be implemented as simple or complex features. A complex junction can be built from a set of simple junctions and edges. For example, a dam in a stream network might be represented as a complex junction, which itself is composed of edges and junctions representing flow gates, generators, turbines, and more. The geodatabase model comes already equipped with simple junctions, simple edges, and complex edges. Complex junctions must be created using advanced modeling tools, and their construction is beyond the scope of this book.

Network weights

When creating a network, the user can include certain attributes which establish weights for each edge and junction. Such weights allow for more realistic modeling of flow. For example, in analyzing the most rapid street route from the fire station to a fire, the shortest route is not necessarily the fastest. A longer route with less traffic, fewer traffic lights, and a higher speed limit might serve better than the shortest route. A more realistic travel network could include a travel time attribute for the road features, and point junctions representing traffic lights with different average delays.

When creating the network, these attributes would be specified as weights, to take into account costs associated with traversing the edges or passing through the junctions. Weights can only be developed during network creation. To add additional weights to a network it must be deleted and re-created.

Domains and subtypes

Domains and subtypes are often used in conjunction with networks because they provide ways to validate and enforce correct attributes. Subtypes are particularly valuable because they create distinct classes of features with specific attributes, such as a water main always being composed of a 12-inch pipe, and connecting to another main through a 12-inch coupler. The two utility networks used in this chapter both contain subtypes.

Connectivity rules

While building a network, domains and subtype rules can enforce certain relationships between network features. For example, a connectivity rule can dictate that a standard T-valve must always connect three pipes of equal size. A lateral T-valve carrying water from a main to a house must have two large pipes and one small pipe. These rules can help users correctly enter and maintain networks.

The network model provides a powerful environment for designing, creating, editing, and maintaining information about networked systems such as roads and utility lines. Many local planning departments and utilities companies are already using systems like this to simplify maintenance, keep up-to-date records, and help provide their constituents and customers with better service.

Summary

➤ Networks formed by streets and utility lines may be stored and analyzed using the network modeling tools.

➤ Networks are stored in feature datasets, and they may include some or all of the feature classes in the dataset.

➤ Network topology, connectivity, and relationships are stored in data tables and collectively represent the logical network. The actual line and point features represent the geometric network.

➤ Network topology is created in ArcCatalog. Once established, it is updated and maintained during editing. Building networks requires an ArcEditor or ArcInfo license.

➤ Transportation-style networks are distinguished by the fact that the material can independently determine how it will flow. In a utility-type network, the direction of flow is determined by the network topology and the location of sources and sinks.

➤ A variety of programs called solvers may be applied to analyzing flow through a network. ArcMap uses a family of solvers called traces.

➤ Network analysis uses three inputs: the network itself, flags indicating points of interest such as delivery locations, and barriers which prevent flow.

➤ Networks may be composed of simple edges, which always end at junctions. Alternatively, they may contain complex edges, in which junctions can occur in the middle of the feature.

➤ Network weights are stored as attributes of edges. Weights can be used during analysis to decide on a path, such as choosing the shortest path or the shortest travel time. They can also be used to accumulate the costs of moving along the network.

TIP: Weights are set up at the time the network is created. If no weights are listed in the drop-down boxes, then the network contains no weights, and none can be used.

TIP: Some tracing tools will not work if the flags or barriers are of different types. Try to always use edge flags and barriers (or junction flags and barriers) together for an analysis.

Chapter Review Questions

You may need to consult the Skills Reference section to answer some of these questions.

1. What kinds of feature classes can be used to create networks?

2. Explain the difference between the geometric network and the logical network.

3. What characteristic distinguishes a transportation-type network from a utility-type network?

4. Networks are composed of what two spatial elements? What is the role of each?

5. What are sources and sinks? In what type of network are they found?

6. What purposes do flags and barriers serve?

7. Explain what loops are, and whether they are desirable.

8. By default, does the Find Path solver always find the shortest path? Explain.

9. What is the purpose of using network weights?

10. How do simple edges and complex edges differ?

Mastering the Skills

Teaching Tutorial

The following examples provide step-by-step instructions for doing basic tasks and solving basic problems in ArcGIS. The steps you need to do are highlighted with an arrow ➔; follow them carefully. Click on the video number in the VideoIndex to view a demonstration of the steps.

1➔ Start ArcCatalog and navigate to the mgisdata\Rapidcity folder.

1➔ Expand the rapidnets geodatabase.

1➔ Expand the Transportation feature dataset.

1➔ Right-click the Road_Net network and choose Properties.

1. What, if any, weights have been created for this network?_____

1➔ Close the Transportation network properties.

1➔ Open the properties for the Utilities network.

2. Which feature classes participate in the network, and what roles do they play?

_____

3. What weights does the Utilities network have? _____

1➔ Close the Utilities network properties.

➔ Close ArcCatalog and start ArcMap with the ex_14.mxd map document in mgisdata\MapDocuments.

➔ Use Save As to rename the document, and remember to save frequently as you work.

Finding paths

We will begin by finding some simple paths through this transportation network.

2➔ If the Utility Network Analyst toolbar is not open, right-click on the menu area on top of the window and turn it on.

2➔ Make sure that the Network is set to Road_Net in the Utility Network Analyst toolbar.

2➔ Set the trace task to Find Path.

2➔ Click on the Junction Flag tool.

2➔ Click on the ends of two outlying roads on opposite ends of town as shown in Figure 14.18.

2➔ Click the Solve button.

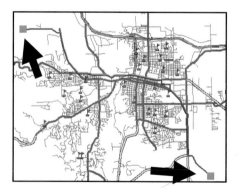

Fig. 14.18. Finding a path between two flags

The path solution appears as a bold red line connecting the two flags. This path meets the criteria that it follows the minimum number of edges through the network. Since this is a transportation network, traffic can travel in either direction on all the streets. Now let's try another one.

3➔ Choose Analysis > Clear Flags from the Utility Network Analyst menu.

3➔ Choose Analysis > Clear Results from the Utility Network Analyst menu.

3➔ Click the Junction Flag tool and add two flags at any locations you choose.

TIP: If the flag does not snap to the right location, clear the flags and start again. Zooming in will make it easier to snap to the right spot.

3➔ Click Solve.

4➔ Clear the flags and the results again using the Analysis menu.

4➔ Choose the Edge Flag tool and try adding a pair of edge flags. Then solve for the path again.

4➔ Clear the flags and the results.

TIP: Flags and barriers will be snapped to the closest feature within the snap tolerance set from the General tab in the Analysis > Options menu.

Now let's explore the use of barriers to help find the best way to take a child to school. Imagine that you live on Harney Dr. and your child goes to Southwest Middle School.

5➔ Choose View > Bookmarks > School Trips 1 from the main menu bar.

5➔ Click the Find tool.

5➔ Type "Harney" as the text to search for, and set the search layer to Roads. Click Find.

5➔ Hold down the Ctrl key and click on each Harney Dr. entry in turn to select them all. Don't select Harney Pl. or the others.

5➔ Right-click on one of the selected entries and choose Select Feature(s). Harney Dr. will be selected.

5➔ Use the Find tool to locate and select the text "Southwest" in the Schools layer.

5➔ Close the Find window.

6➔ Zoom into the extent of Harney Dr. to more easily mark the right spot with the flag.

6➔ Click the Edge Flag tool.

6➔ Place a flag near the middle of Harney Dr., and use the Zoom to Previous Extent button to return to the original extent.

7➔ Place another edge flag next to the two school symbols, using zooming as necessary.

7➔ Zoom back to see both flags, and make sure the trace task is set to Find Path.

7➔ Click Solve.

Well, that was probably the answer you expected. Perhaps you are wondering, however, how far it is from home to school.

8➔ Click the Solve button again without moving the mouse off it, and this time keep an eye on the lower left corner of the map window to see the cost.

4.   What is the cost of this trip? What units is it in? _____

Obviously it would be more interesting to see the distance. To find distances, you must use a weight.

8➔ Choose Analysis > Options from the Utility Network Analyst menu bar.

8➔ Click the Weights tab.

8➔ Set both of the along and against weights to Distance. Click OK.

8➔ Clear the results (but not the flags).

8➔ Click Solve again, and watch for the cost.

5.   What is the cost, and what are the units? _____

Now on the first day of school, you discover that an enormous traffic jam builds up at the intersection just south of the school. You wonder whether there is an alternate route through the tangle of residential streets north of the school, and what the street names are. First, though, let's turn on Map Tips.

9➔ Right-click the Roads layer and choose Properties.

9➔ Click the Display tab. Check the Show Map Tips box.

9➔ Click the Fields tab and set the primary display field to ROADNAME. Click OK.

With Map Tips on, use the cursor to hover over a street to find out its name. Also, the streets will be labeled if you are zoomed in to scales greater than 1:20,000.

10➔ Clear the results (but not the flags).

10➔ Zoom into the middle school to see it better.

 10➔ Click the Junction Barrier tool and place a barrier at the intersection of Park Dr. and Corral Dr.

10➔ Zoom to the previous extent when done.

10➔ Click Solve.

6.   What is the distance of this alternate route? _____

To make it easier to find the way tomorrow morning, make a list of the streets involved in this path. One way to get it is to return the path as a selection instead of as a drawing.

11➔ Choose Selection > Interactive Selection Method from the main toolbar and make sure that it is set to Create New Selection.

11➔ Choose Selection > Set Selectable Layers and make sure all layers are selectable.

12➔ Choose Analysis > Options from the Utility Network Analyst toolbar.

12➔ Click the Results tab.

12➔ Change the results format to Selection, and uncheck the box for returning junctions. The edges will contain the street names. Click OK.

12➔ Clear the results (but not the flags or barriers).

12➔ Solve again.

TIP: If no features are selected when solving a trace, check that the selection method is set to Create New Selection and that the network is one of the selectable layers.

13➔ Open the Roads attribute table.

13➔ Click the button to show only the selected records.

7. How many (different) street names are included on this path? _____

13➔ Set the results format back to Drawings in the Analysis > Options menu and check the box to return junctions once more. Click OK to close the window.

13➔ Close the attribute table.

Finally, your other child attends Meadowbrook Elementary School. Normally she walks to school, but you want to plan a route to take both children on rainy days. In addition, you have agreed to pick up her friend who lives on Player Dr. Find a new path including all of these elements. The Middle School starts earlier than the others so you plan to stop there first, then pick up the child on Player Dr., and then arrive at Meadowbrook. You need to enter the flags in the order of your planned stops (Fig. 14.19).

14➔ Use Clear Selected Features from the Selection menu to clear the trace.

14➔ Clear the barriers and the flags.

14➔ Place an edge flag on Harney Dr.

14➔ Place an edge flag next to Southwest Middle School.

15➔ Locate Player Dr. using the Find button or Map Tips and place an edge flag on it.

15➔ Place an edge flag next to Meadowbrook Elementary School (west and slightly north of Harney Dr.).

15➔ Solve for the path.

15➔ When finished, clear the results and the flags.

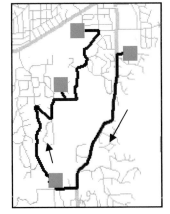

Fig. 14.19. Traveling a sequence of stops

Advanced path problems

Combinations of selections, barriers, and other techniques can tackle more advanced path problems. To start, assume that you live on Player Dr. and work at the South Dakota School of Mines and Technology (SDSM&T). You decide to map out a way to work.

16➜ Find Player Dr. again and place an edge flag on it.

16➜ Zoom to the extent of the Roads layer.

16➜ Find SDSM&T, zoom in, and place an edge flag on E. St. Joseph St. next to it.

16➜ Zoom to the extent of the Roads layer again.

16➜ Solve for the path.

Now, imagine that a doctor has placed you on a very strict low-cholesterol diet. Driving home from work in the evening, you find it nearly impossible to drive by a restaurant without stopping for something wonderfully greasy to eat. Try to find a route to work that avoids all the restaurants in town. To begin, select all the road segments that are close to restaurants.

17➜ Choose Selection > Select by Location from the main menu bar.

17➜ Choose to select roads that are within 100 meters of a restaurant.

17➜ Close the window after completing the selection.

Next, we will set the trace analysis options so that the selected roads act as barriers to travel.

18➜ Choose Analysis > Options from the Utility Network Analyst toolbar.

18➜ Click the General tab, and fill in the button to trace only on unselected features.

18➜ Click OK.

18➜ Clear the previous trace results.

18➜ Solve for the path (Fig. 14.20).

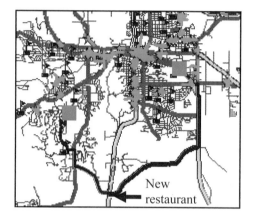

Fig. 14.20. Avoid these restaurants on the way home from work.

Unfortunately, last week a brand new greasy spoon restaurant with juicy burgers and crisp fries just opened at the intersection of Catron Blvd. and Highway 16. This new restaurant is not in the database, so enter barriers by hand.

19➜ Zoom into the intersection of Catron Blvd. and Highway 16 (see Figure 14.21).

19➜ Add two edge barriers to block off the intersection.

19➜ Zoom back to the previous extent.

19➜ Clear the previous result and solve for the path again.

Looks like you can still get to work, although the path is a little tortuous!

20➜ Clear the flags and barriers and results.

20➜ Clear the selected set of roads.

20➜ Turn off the Restaurants and Schools layers.

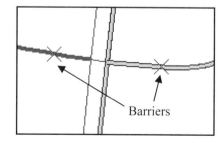

Fig. 14.21. Placing barriers on Catron Blvd.

Finding transportation network errors

Errors in topology are often created during the process of digitizing and creating networks. Once the network is built, the tracing tools can help you locate and fix these problems. The most useful tool is the Find Disconnected solver. In a road network, it should theoretically be possible to get to any road on the net. A disconnected road is usually the result of a failure to snap the end of a new road to the existing network, or failing to split a road at a junction.

21➔ Zoom to the extent of the Road Network group layer.

21➔ Choose the Junction Flag tool and click to add a flag anywhere on the network.

21➔ Set the trace task to Find Disconnected, and click the Solve button.

At first glance no trace result appears. This network is in good shape. However, look closely at the north-south extension of the yellow interstate, I-190, for a small red glint.

22➔ Zoom into this area for a closer look (Fig. 14.22).

22➔ Zoom into the triangle with the two small red spots.

22➔ Zoom into one of the spots until the red street fragment is clearly visible.

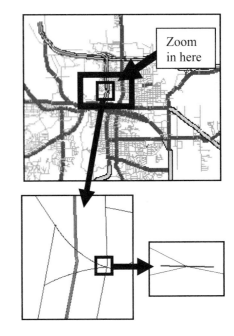

Here's the problem. A tiny sliver of road lies across the intersection, probably a digitizing error. With ArcEditor or ArcInfo, the slivers could be deleted to fix the network. For now we will simply go on; these small extra segments will not affect the behavior of the network.

22➔ Zoom back to the extent of the Road Network group layer.

We can also use the Find Loops on transportation networks.

23➔ Change the trace task to Find Loops.

23➔ Click the Solve button.

Fig. 14.22. Finding a disconnected road segment

This time, nearly the entire network is returned as the result. We can expect this, because typically road networks DO let people drive in loops. (How many times have you driven around the block?) However, the roads which are not part of the result do have a common feature—they are all "dead end" streets. Of course some of them end merely because they go outside the city limits. Generally, finding connections and loops is not nearly as useful with transportation networks as it is for utility networks.

Tracing on utility networks

Although finding driving routes is amusing, utility networks also have interesting tracers. Now we will take a look at a portion of the water utility system for east Rapid City.

24➜ Clear the flags and the results.

24➜ Turn off the Road Network group layer and collapse it using the minus box.

24➜ Turn on the Water Network group layer.

24➜ Zoom to the Water Network group layer.

24➜ Change the network to Water_Net in the Utility Network Analyst toolbar.

24➜ Examine the network.

This network contains more elements than the roads network. The water mains themselves form the edges. In addition, there are galleries, end caps on the pipes, and T-valves at the pipe junctions. All of these features participate in the network.

25➜ Right-click the Galleries layer and choose Open Attribute Table.

The attribute table contains three fields. The Name field simply indicates the name of the gallery. The Enabled field is a part of the network setup. The value True indicates that the galleries are turned on and are functioning as part of the network. The AncillaryRole field says Source, indicating that the galleries are water sources for the network. They are underground chambers that pull water in from Rapid Creek, filtering and treating it before it enters the water lines.

25➜ Close the attribute table.

26➜ Choose View > Bookmarks > Pipes from the main menu bar.

26➜ Choose Flow > Display Arrows from the Utility Network Analyst toolbar.

Arrows appear at the center of each edge, indicating the direction of flow. Here's how to edit the symbols used for portraying flow.

27➜ Choose Flow > Properties from the Utility Network Analyst toolbar.

27➜ Click the Arrow Symbol tab.

There are three different conditions of flow possible in the network. Uninitialized flow means that the flow direction has not yet been established for the network. Indeterminate flow means that flow has been initialized, but flow cannot be determined for that section. Determinate flow indicates that flow has been clearly established.

27➜ Click on each flow type to examine the symbol being used to display it.

8. What is the symbol being used for Indeterminate flow? _____

28➜ Click the symbol for Determinate flow. The Symbol Selector appears.

28➜ Change the symbol color to blue. Click Apply. The arrows in the map change color.

28➜ Change the symbol for Indeterminate flow to a 12 point pink square. Click Apply.

Just like other features in ArcMap, you can set a display scale so that the networks are not cluttered with arrows at small scales.

29➜ Click the Scale tab.

29➜ Change the settings to not display the arrows when zoomed out past 1: 30,000.

29➜ Click OK.

29➜ Zoom to the extent of the Water Network group layer to check the scaling. The arrows should disappear.

29➜ Return to the previous extent.

Notice that the arrows take a few moments to draw. This can get annoying, so we will turn them off for now.

30➜ Choose Flow > Display Arrows again to turn the symbols off.

Let us explore this network a little more using some of the tracing tools. Let's start by seeing which parts of the network are connected to each gallery.

30➜ Zoom to the extent of the Water Network group layer to show the entire network.

30➜ Choose the Junction Flag tool and click to add a flag at the Civic Center gallery.

30➜ Set the trace task to Find Connected.

30➜ Click the Solve button.

Now it is easy to see which areas of town are served by the Civic Center gallery. We can use Find Disconnected to highlight the other pipes that are not connected to this gallery.

31➜ Clear the results and change the trace task to Find Disconnected.

31➜ Click the Solve button.

Next, let's use the Find Loops tracer to see if there are any loops in this network.

32➜ Add another junction flag to the other gallery, so that both subnets are examined for loops.

32➜ Change the trace task to Find Loops.

32➜ Zoom into the single red loop result that appears (Fig. 14.23).

32➜ Choose Flow > Display Arrows from the Utility Network Analyst toolbar.

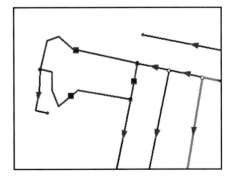

Fig. 14.23. Flow in a loop is indeterminate.

Note that all of the lines except the loop have blue determinate flow arrows. The loop has the pink square indeterminate flow symbols. The water could go around the loop in either direction. Fixing this would require editing the loop to break the connection at one point.

33➜ Clear the flags and results.

33➜ Zoom to the extent of the Water Network group layer.

One useful benefit of using network models is being able to predict what areas of town will experience service disruptions due to repairs and outages. The city has been planning to repair a leaking pipe on St. Patrick St. between 6th and 7th Ave. They will need to turn off the flow to the pipe for several hours, and they want to notify the customers who will be affected.

34➔ Turn on the Road Network group.

34➔ Use Select by Attributes to select the roads where [STREET] = 'ST PATRICK'. Click Apply rather than OK to keep the selection window open.

34➔ Change the Selection method to Add to Current Selection.

34➔ Select the roads where [STREET] = '6' OR [STREET] = '7'.

34➔ Close the Selection window.

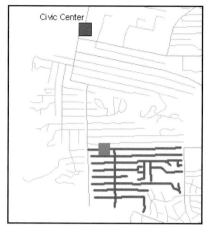

35➔ Zoom in and add an edge flag to the segment of the water line that follows St. Patrick St. between 6th and 7th. Then return to the previous extent.

35➔ Change the trace task to Trace Downstream.

35➔ Click the Solve button.

Fig. 14.24. Finding a service disruption due to repairs on St. Patrick St.

Now you know where to notify customers (Fig. 14.24). If this network were complete down to the building connections for each parcel, such as we set up in Chapter 11, then the utilities company would be able to automatically compile a mailing list to send to the affected customers.

The Trace Upstream tool performs the opposite task: finding the path that water has traveled to get to a particular location.

36➔ Clear the flags and results.

36➔ Clear the selected features.

36➔ Use the Find tool to locate and select the streets N and S Grand Vista Ct. They are on the middle of the west side of the network (Fig. 14.25).

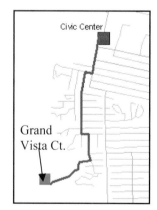

Tracing can also be done with accumulation, for example to find the total distance the water travels to get to a particular location.

37➔ Zoom into the location of these two streets.

37➔ Place a junction flag on the end of the water main serving these streets, and return to the previous extent.

37➔ Change the task to Find Upstream Accumulation.

37➔ Click the Solve button and watch for the cost in the lower left corner of the window.

Fig. 14.25. How water gets to Grand Vista Ct.

9. What is the cost, in edges, to get from the Civic Center to Grand Vista Ct.? _____

38➔ Choose Analysis > Options from the Utility Network Analyst toolbar.

38➜ Click the Weights tab, and set both the along and against weights to Distance. Click OK.

38➜ Click the Solve button again.

10. How far does the water travel to get there? _____

There is no downstream equivalent to tracing with accumulation. However, we can use the selection result option to select the downstream features from a point and determine the total length of pipe beyond that point.

39➜ Clear the flags and results.

39➜ Place a junction flag at the East Rapid gallery.

39➜ Choose Analysis > Options from the Utility Network Analyst menu.

39➜ Click the Results tab and change the result format to Selection. Click OK.

39➜ Change the trace task to Trace Downstream and Solve.

Now that the water lines are selected it is easy to get their total length.

40➜ Open the attribute table for the Waterlines layer.

40➜ Right-click the Shape_Length field and choose Statistics.

11. What is the total length of pipe served by this gallery, in kilometers? _____

40➜ Close the Statistics window and the attribute table.

40➜ Clear the selected set.

This is the end of the tutorial.

➜ Close ArcMap. You do not need to save your changes.

Exercises

1. Find the shortest path from Hidden Timbers Dr. to Stevens High School. **Capture** a map showing the route.

2. What is the distance covered by the route in Exercise 1?

Range Rd. by West Jr.

3. What is the shortest alternate route if you want to avoid the before-school congestion on Park Dr. by Southwest Middle School? **Capture** the map. How long is the new route?

4. The cafeteria at Stevens High School makes daily lunches for several elementary schools in town. They are delivered in this order: Pinedale, South Canyon, Upper Rapid, Canyon Lake, Meadowbrook, and Cleghorn. Find the most efficient route to make the deliveries. **Capture** the map.

5. You are trucking a load of Canadian hemp from a distributorship on Olde Orchard Road to a cowboy rope-making factory on Stacy St. Due to a bizarre technicality in the drug laws, you can be arrested if the truck is stopped within 250 meters of a school. Find a path to the destination that avoids this risk. **Capture** the map.

6. A construction worker accidentally breaks a water line at the intersection of E. St. Patrick St. and Ivy Ave. Create a map showing the area of disrupted service. **Capture** the map.

7. Of the two water mains leaving the Civic Center gallery, which serves a greater length of pipes? What is the total pipe length for each main (north and south)?

8. What is the total length of water *mains* served by the East Rapid gallery versus the Civic Center gallery? (Do not include laterals or other water line types.)

9. The water main that ends at 5th St. and Cathedral Dr./Fairmont Blvd. passes through how many T-valves?

10. Joe did Exercise 9 with Selection as the result format, and got 26 *selected* features. Mary did the same problem and got 51 *selected* features. Explain why they got different numbers. What is the cost in each case?

Challenge Problem

Find a route that goes from Alta Vista Dr. to the east end of E. Knollwood St., which travels ONLY on Local streets. (**Hint:** You cannot ignore the role of junctions in this problem.) **Capture** your map.

Skills Reference

Examining network properties

1. In ArcCatalog, right-click a network in the catalog list and choose Properties.

The General tab reports which feature classes participate in the network (Fig. 14.26).

The Connectivity tab shows any rules which have been established regarding how the network elements can connect (e.g., a 3-inch pipe can connect to a 6-inch pipe only through a coupler junction feature). Connectivity rules are not required but are provided to help maintain network integrity.

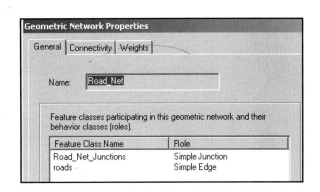

Fig. 14.26. General properties of a network

The Weights tab lists any weights which have been set up for the network. Weights are created at the time the network is built, but they may be used during an analysis if desired.

Performing a trace

1. In ArcMap, click the Add Data button and choose a *Name_*Net layer (such as Road_Net) in the geodatabase containing the network. The network and its participating feature classes will be added to the map.

2. Right-click in the menu area and turn on the Utility Network Analyst toolbar (Fig. 14.27).

3. Select the network to be analyzed, if more than one is available.

4. Choose the Trace Task to be performed.

5. Click the Flag/Barrier tool and add desired flags and barriers at edges or junctions. For reliable results, do not mix edge and junction markers; use either but not both.

6. Click the Solve button to perform the trace.

Fig. 14.27. Performing a simple tracing task

Setting general tracing options

1. Choose Analysis > Options from the Utility Network Analyst toolbar.

2. Click the General tab (Fig. 14.28).

3. Choose which features to trace. You can choose to trace only on selected features, only on unselected features, or on all features.

4. Choose whether to include edges with indeterminate or uninitialized flow when tracing.

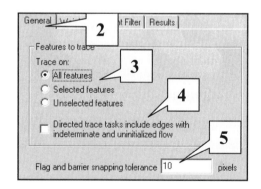

Fig. 14.28. Setting general options

5. Choose the snapping tolerance to be applied when adding flags and barriers. If the tool clicks a location within this distance from an edge or junction feature, the flag or barrier will automatically snap to the feature.

6. Click OK to apply the options.

Changing the results format

1. Choose Analysis > Options from the Utility Network Analyst toolbar.

2. Click the Results tab (Fig. 14.29).

3. Choose whether the trace result will take the form of a drawing or a selected set of features.

4. If choosing to return a drawing, change the color of the trace if desired.

5. Choose whether the result will include all features included in the trace, or only the features that are stopping the trace.

6. Choose whether the trace result will include edges, junctions, or both. By default both are returned.

7. Click OK to apply the changes.

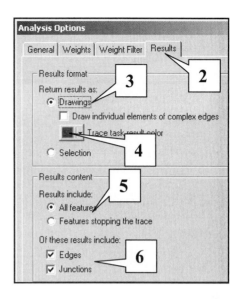

Fig. 14.29. Changing the output format

Using weights when tracing

1. Choose Analysis > Options from the Utility Network Analyst toolbar.

2. Click the Weights tab (Fig. 14.30).

3. To use weights on junctions, choose the name of the weight desired.

4. To use weights on edges, specify the name of the weight to be applied in the same direction that the line was digitized.

5. If desired, specify a weight to be applied against the digitized direction.

6. Click OK to apply the changes.

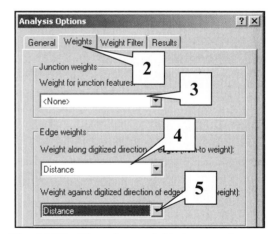

Fig. 14.30. Choosing weights to use when tracing

TIP: Weights are set up at the time the network is created. If no weights are listed in the drop-down boxes, then the network contains no weights, and none can be used.

TIP: To accumulate distances or other weights regardless of the digitized direction, specify a weight in both directions (along and against).

Using weight filters

Filtering weights means to apply only weights that fall within a specified range. For example, a junction weight called StopTime might specify an average waiting time at intersections with stop lights or stop signs. You may only be interested in adding up the total time spent at traffic lights and ignoring the wait at stop signs, which is usually 10 seconds or less. A weight filter could screen out waits less than 10 seconds.

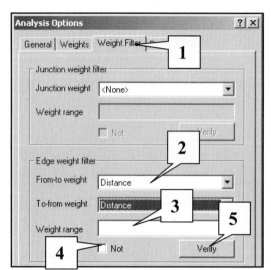

1. Click the Weight Filter tab in the Analysis > Options window (Fig. 14.31).

2. Choose the edge or junction weights.

3. Enter the weight range as min – max.

4. Click the Not button to use the weights NOT in the specified range.

5. Click Verify to check that the syntax is correct.

6. Click OK or Apply.

Fig. 14.31. Using weight filters

Viewing flow directions

1. Choose Flow > Display Arrows from the Utility Network Analyst toolbar.

Changing flow symbols

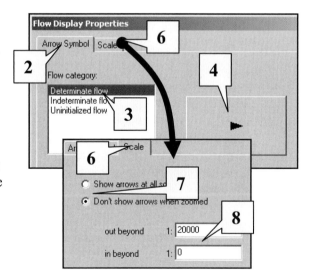

2. Choose Flow > Properties from the Utility Network Analyst toolbar and click the Arrow Symbol tab (Fig. 14.32).

3. Select the flow category to see the symbol used for that category (Determinate, Indeterminate, or Uninitialized flow). The symbol for that category will be displayed to the right.

4. To change the symbol, click on it. The Symbol Selector appears. Choose the new symbol and set its size and color properties and click OK.

5. Click OK or Apply.

Fig. 14.32. Setting the flow symbols and display scale

Setting the display scale

Use this feature to make symbols appear only at certain scales, to avoid clutter in the map.

6. Click the Scale tab (Fig. 14.32).

7. Choose to always show arrows, or to show arrows when zoomed in or out beyond a specified scale.

8. Enter the scale.

9. Click OK or Apply.

Adding and clearing flags and barriers

1. Click the Flag tool menu drop-down button on the Utility Network Analyst toolbar (Fig. 14.33), and select the type of flag or barrier to add (edge or junction).

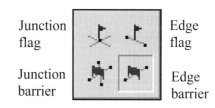

Fig. 14.33. Types of flags and barriers

2. Click on a junction to add a junction flag or barrier. Click along an edge to add an edge flag or barrier.

3. To clear flags or barriers, choose Analysis > Clear Flags or Analysis > Clear Barriers from the Utility Network Analyst toolbar. All flags or barriers are cleared together.

TIP: Flags and barriers will be snapped to the closest feature within the snap tolerance set from the General tab in the Analysis > Options menu.

TIP: Some tracing tools will not work if the flags or barriers are of different types. Try to always use edge flags and barriers (or junction flags and barriers) together for an analysis.

Finding a path

By default, the trace will return the path which traverses the fewest number of edges. This path will not necessarily be the shortest path.

1. Choose the network to trace on.

2. Set the trace task to Find Path.

3. Choose the Junction Flag or Edge Flag tool.

4. Click on the starting location to enter a flag.

5. Click on the end location to enter another flag.

 6. Click the Solve button.

> Total cost: 18

> **TIP:** To see the cost (number of edges in the path), look in the lower left corner of the ArcMap window, before the mouse moves off the Solve button. The cost will vanish once the mouse is moved. Click Solve again to bring it back.

> **TIP:** Enter more than two flags to visit a series of locations. The stops will be visited in the order in which they are entered.

Finding the shortest path

To do this, the network must have a Distance weight based on the length of the edges. (It may be called something other than Distance.)

1. Choose the network to trace on.

2. Set the trace task to Find Path.

3. Choose the Junction Flag or the Edge Flag tool. Enter the starting, end, and any intermediate flag locations.

4. Choose Analysis > Options from the Utility Network Analyst toolbar and click the Weights tab.

5. Set both the along and against edge weights to Distance (or whatever the weight name is). Click OK.

6. Click the Solve button.

> Total cost: 4968.

> **TIP:** The distance along the path will be displayed in the lower left corner of the ArcMap window, in map units. It will vanish when the mouse is moved. Click Solve again to make it reappear.

Finding connections and loops

These trace tools can be used to find features connected to a certain location, those features disconnected from a certain location, and those features that form loops downstream of a certain location.

1. Choose the network to trace on.

2. Set the trace task to Find Connected, Find Disconnected, or Find Loops.

3. Click the Edge or Junction Flag tool and enter the location from which to start tracing. More than one location flag may be entered, in which case tracing will commence from each location separately.

4. Click the Solve button.

Tracing upstream/downstream

These tracers work only on utility-type networks.

1. Choose the network to trace on.

2. Set the trace task to Trace Upstream or Trace Downstream.

3. Click the Edge or Junction Flag tool and enter a flag at the location to begin the trace. If more than one flag is entered, tracing will begin at each location separately.

4. Click the Solve button.

Tracing with accumulation

1. Choose the network to trace on.

2. Set the trace task to Find Upstream Accumulation.

3. Choose Analysis > Options and click the Weights tab to specify the weights to be accumulated during tracing.

4. Click the Edge or Junction Flag tool and enter a flag at the location to begin the trace. If more than one flag is entered, tracing will begin at each location separately.

5. Click the Solve button.

6. Read the accumulation value in the lower right corner of the ArcMap window, in whatever units the weights are in.

Total cost: 4968.

TIP: The accumulation value will vanish when the mouse is moved off the Solve button. To bring it back, click Solve again.

Chapter 15. Raster Analysis

Mastering the Concepts

Objectives

> Understanding the differences between the raster and vector data models and the benefits and drawbacks of each

> Knowing basic facts about the storage and use of the grid data model

> Getting familiar with basic raster analysis functions

> Using Boolean map algebra to perform overlay analysis with rasters

> Controlling the analysis environment when using Spatial Analyst

> Using Spatial Analyst for raster analysis

NOTE: This chapter requires Spatial Analyst, an optional extension to ArcGIS. If you do not have it installed on your computer, you will not be able to perform the functions described.

Concepts

The availability of two different data models, raster and vector, provides added flexibility to options for data storage and analysis. (Readers may wish to review the section of Chapter 1 on raster and vector data models.) Neither data model is intrinsically superior. They both have areas in which they excel and areas in which they are at a disadvantage. The GIS professional who has a grasp of both tools holds the keys to developing the most efficient and accurate analysis.

Raster versus vector models

Recall that spatial data comes in two forms, discrete data and continuous data (Fig. 1.4). Discrete data takes on relatively few values and generally forms distinct regions on the map. Soil types and land use zones provide good examples of discrete data. Sometimes discrete data is characterized as being either present or absent (such as a road, which is present in some areas but absent in most areas). Continuous data vary smoothly across a range of values, forming a surface. Elevation and precipitation are examples of continuous data. We say that data are *spatially* discrete when the values stay the same for an area, and then abruptly change to another value, as the geology areas do in Figure 15.1. Points, lines, and polygons constitute spatially discrete features by their very nature. The vector data model can ONLY store spatially discrete data. With raster data, however, we can store spatially continuous data which varies across the surface at very small scales. Elevation and precipitation are two phenomena that vary continuously across a surface and are best stored as rasters.

The benefits of the vector data model are many. First, it can store discrete features such as roads and area boundaries with a high degree of precision. The *x-y* coordinates are typically stored with up to six decimal places, such that the precision of storage far outranks our ability to map things accurately. Second, the linked attribute table provides great flexibility in the number and type of attributes which can be stored about each feature. Third, the vector model is ideally suited to mapmaking because of the high precision and detail of features which can be obtained. Vector is often a more compact way of storing data, typically requiring a tenth of the space of a raster with similar information. Projection of vectors from one coordinate system to another is faster, so much so that only recent versions of GIS have supported on-the-fly raster projection. Finally, the vector model is ideally suited to certain types of analysis problems, such as determining perimeters and areas, detecting adjacency of features, and modeling flow through networks.

However, the vector model has some drawbacks. First, it is poorly adapted to storing continuous surfaces such as elevation. Contour lines (as on topographic maps) have been used for many years to display surfaces, but calculating derived information from contours, such as slope, flow direction, and aspect, is difficult. Information that changes quickly and discontinuously over a surface, such as slope or aspect, is also difficult to portray using a vector model. Finally, some types of analysis, such as overlay and buffering, are cumbersome and time-consuming to perform with vectors.

The raster model mitigates some of the drawbacks of vectors. It is ideally suited to storing continuous and rapidly changing discontinuous information, because each cell can have a value completely different from its neighbors. Overlay and buffering functions are simple and rapid to perform. One benefit of rasters is access to an analysis method called map algebra, in which maps can be used as part of algebraic expressions.

The drawbacks of rasters lie chiefly in two areas. First, they suffer from trade-offs between precision and storage space to a greater extent than vectors do. The precision of a raster is determined by its cell size: if the cell size is 30 m, then any object or boundary can only be located to within 30 meters. Although making the cells smaller increases the precision, the storage space required increases as the square of the accuracy. Examine the three maps in Figure 15.1. Figure 15.1a shows a portion of a geology map stored as a vector file, and the boundaries are clear and precise. Figure 15.1b shows the same file stored as a raster with a cell size of 200 meters, and the chunky boundaries due to the low resolution are clearly visible. Figure 15.1c is a raster with a cell size of 50 meters. Although the map in Figure 15.1c has smoother boundaries and is four times as accurate (50 versus 200), it comes at a cost of 16 times the storage space! The 200-meter raster has 141 rows and 102 columns, or a total of 14,382 cells. The 50-meter raster has $562 \times 408 = 229,296$ cells! The increase in size also causes an increase in processing time when performing analysis functions.

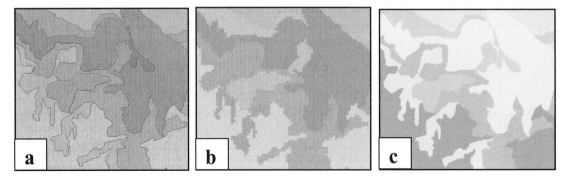

Fig. 15.1. Effect of resolution on raster precision. (a) A shapefile of geology. (b) A raster created from (a) with a cell size of 200 meters. Note the chunky boundaries. (c) The raster created with a resolution of 50 meters.

When high precision is required, the vector model is clearly superior. For example, in storing a database of land use parcels, one would like sub-meter accuracy—a homeowner would be very upset if you moved his land boundary over by a meter! Yet to store an entire 10 km × 10 km city with half- or tenth-meter accuracy in a raster data set would require *billions* of cells.

The second major drawback of rasters concerns their ability to store attributes. A raster file is an array of cells with numeric values, and each cell has only one value. To store both geology and infiltration values for an area requires storing two separate grids. Vector files, by contrast, can store hundreds of attribute values for each spatial feature, and can handle text data more efficiently.

This chapter provides an introduction to some of the basic concepts and analysis functions of raster analysis, but the topic spans a much greater breadth of theory and technique than we have time to explore here. The capabilities of raster analysis can be quite complex and sophisticated, and very powerful. The serious student of raster analysis will find plenty of additional material in the software documentation for Spatial Analyst and ArcInfo GRID.

About rasters

The grid format

Spatial Analyst can read many types of raster data formats, but it does its analysis using grids. A grid is a specific data format used by Spatial Analyst, as well as the ArcInfo GRID program from which Spatial Analyst was derived. This data format is unique to ArcGIS. Other rasters, such as images, TIFFs, SID files, and JPEG files, must be converted to grid format before they can be analyzed. All raster output generated by Spatial Analyst also takes the form of grids.

Raster grids come in two forms. Integer grids store whole numbers in binary format. Floating-point grids can store decimal values, but they require four times as much space as integer grids. Thus it is foolish to use floating-point grids except when decimal values are required. For example, digital elevation models, or DEMs, store land elevations and typically have a vertical accuracy of 7 to 15 meters. It makes no sense to store a decimal fraction for these data.

When creating output, the grid type will usually be the same as the input. If you add or multiply two integer grids, the result will also be integer. If one of them is floating-point, so will the output be. Some functions always create floating-point grids. Dividing or averaging a grid will result in floating-point output. Floating-point grids may be converted to integer grids using the INT function.

TIP: Grids have much stricter naming conventions than other ArcGIS data files. Grid names must have no more than 13 characters, and they should contain only letters, numbers, or the underscore character. Pathnames to the grid directory should contain no spaces anywhere.

Grid attribute tables

Grids can have attribute tables, just as feature layers do. However, since a grid does not contain features, only cell values, the attribute tables look different, because each record contains one unique value present in the grid. The table has three fields which are always present: the ObjectID field containing a unique ID for the table rows, a Value field showing each unique cell value in the grid, and a Count field indicating how many cells contain that value (Fig. 15.2). This table indicates that the Precambrian rock unit has the largest area in the grid because more cells (64,597) have that value than any other.

The attribute table in Figure 15.2 comes from a raster that was converted from a polygon shapefile of geology using the Name field of the geologic units. As a result, the table contains the Name field with the original text values. However, the numbers in the Value field are the actual values stored as the grid, because all grid values must be numeric. The numbers were arbitrarily assigned to the different names. Thus the

ObjectID	Value	Count	Name
0	1	22651	Cenozoic
1	2	8602	Upper Mesozoic
2	3	38657	Lower Mesozoic
3	4	32924	Upper Paleozoic
4	5	1078	
5	6	27805	Madison Limestone
6	7	64597	Precambrian
7	8	24948	Lower Paleozoic

Attributes of GEOLOGY22

Fig. 15.2. The attribute table of a geology grid

Precambrian unit is represented in the grid by the number 7. Unlike vector attribute tables, no additional fields may be added to grid attribute tables.

Only integer grids containing discrete data have attribute tables. A grid of elevation, or a floating-point grid, could have nearly as many different values as cells. In such a case, storing a table in addition to the binary data itself would only increase the storage space needed for the grid, without conferring any particular benefit. Thus floating-point grids and integer grids with an excessively large number of values will not have associated attribute tables.

NoData values

Rasters have a special type of value called NoData to indicate a null value for a cell. Because rasters are rectangular in shape, the NoData value can mask areas outside a nonrectangular study area. NoData values are typically displayed with black or transparent color. In Figure 15.3, some cells at the top of the DEM are missing data and are stored as NoData values. NoData values are far more convenient than using an arbitrary number, such as 0 or –9999, to represent a missing value (although some grids may have these). Arbitrary numbers will be used in calculations and can cause strange problems, whereas NoData values can be ignored for more consistent results.

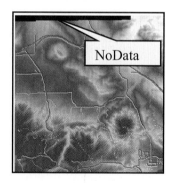

Fig. 15.3. NoData values at the top of a DEM

Coordinate systems and rasters

Just like vector layers, rasters have a defined coordinate system composed of a geographic coordinate system (GCS) with its datum, and a projection. The raster itself is georeferenced according to the *x-y* coordinates of the center of the upper left pixel. The resolution of the raster is specified using the cell size, or length of the sides of the pixels. Usually pixels are square, such that a 30-meter raster has pixels composed of 30-meter × 30-meter squares. The cell size is always specified in the coordinate system's map units, and meters or feet are typically used.

Although one can store a raster in a GCS, this practice is not common, because the inherent distortion gives an unsatisfactory view of the data. Furthermore, many raster analysis functions incorporate distance and angle formulas that will give inaccurate results when applied to a raster in a GCS or in a projection with significant distortion. Therefore, raster data are normally stored in whatever projected coordinate system gives the best and most undistorted view of the data. The main exception occurs for certain organizations, including the U.S. Geological Survey, which provide topographic data in GCS format. Such data should be projected before it is used.

Georeferencing rasters

Assigning real-world coordinates to a raster is called georeferencing. Rasters generated from maps such as land use will have the coordinate system of the source map. A raster with a real-world coordinate system is said to be **georeferenced.** Other types of raster data, particularly aerial photographs and satellite images, have distortions present in the raw data, or may not be tied to map coordinates at all. To use these images in a GIS, the image must be **rectified,** or warped to fit a map base.

This process involves finding pairs of **ground control points** which can be located on the image and on a map, such as road intersections or corners of buildings. At least four pairs of points must be chosen, but more points generally results in a more accurate transformation. Once a set of points is defined, a geometric transformation is applied to convert the image to the same coordinate system as the map.

Displaying grids

ArcMap can project rasters on the fly to match the coordinate system of the data frame. However, this process typically takes longer for a raster layer than it does for vectors, and large rasters can significantly slow the drawing speed. The results may not be particularly accurate either, because ArcMap uses a low-order polynomial transformation method to save time. Generally rasters are best stored in the same projection in which they will be frequently used. Raster projection is so slow that prior to Version 8, ArcView and ArcInfo did not even offer on-the-fly projection of rasters.

Coordinate systems and analysis

Raster functions treat layers as arrays of numbers which can be processed as a stack. To perform raster analysis, all of the input grids must have the same cell size and extent so that the array values in each stack (the cells) line up. If a command receives inputs with different cell sizes or extents, the grids will automatically be resampled to match each other. By default all grids are resampled to the intersection of all the input grid areas and the most coarse cell size present in the input grids. These defaults may be overridden. Similarly, if given input rasters with several coordinate systems or cell sizes, Spatial Analyst will automatically project the grids to match each other before analyzing them.

The resampling and projection process is invisible to the user and seems very convenient; however, it does increase the processing time. If very large grids are being resampled or projected, the user may not even be aware it is happening, yet it may overwhelm the computer and cause a system failure. Storing all grids to be analyzed together in the same coordinate system thus provides the best performance.

If multiple coordinate systems are input for analysis, then rules are applied to determine the coordinate system of the output. In general, the coordinate system of the output matches the coordinate system of the input, whether it is raster or feature data. If multiple raster coordinate systems are present, then the first one input determines the coordinate system of the output. If both raster and feature data are used as input, then the vector data are projected to match the raster coordinate system and the output matches the raster layer.

The user can override the defaults using the General tab of the Options menu under the Spatial Analyst toolbar by choosing to have all new rasters adopt the same coordinate system as the data frame. Be aware, however, that this option may significantly increase processing time if the input rasters are in a different coordinate system.

Raster analysis

Spatial Analyst offers many analysis functions which are accessed through the menus (Fig. 15.4). These same functions, and hundreds more, are available in ArcToolbox. Furthermore, any function can be used by typing its name and arguments into a window called the **Raster Calculator.** The functions fall into four main types.

1) **Local functions** are calculated on a cell-by-cell basis, such that the output depends on the result of a single cell in the grid. For example, taking the cosine of a grid only requires knowing the value of each cell one at a time. The cosine is calculated and the function moves on to the next cell.

2) **Focal functions** are based on a moving target area (window) of specific shape and size, and all the cells in the window are considered in obtaining the result. A 3 × 3 averaging filter is an example of a focal function. The average of all cells in a 3 × 3 square window is calculated, and the result is placed in the center cell of the window.

3) **Zonal functions** employ a target area called a zone, which can be of any shape and size. The function calculates a specified statistic for all cells in each zone and summarizes the results in a table. Zones are specified by an additional layer of input. For example, each parcel in a housing development has its own parcel number and would be considered a zone.

4) **Global functions** use all the cells in the grid to calculate the result at a particular location. For example, finding the easiest travel route across a surface requires knowing the values of most or all of the cells in the grid.

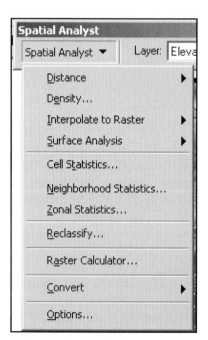

Fig. 15.4. The Spatial Analyst menu and the Spatial Analyst toolbox. The toolbox has hundreds of functions not in the menu.

Notice that each of the four types has its own entry in the Spatial Analyst toolbox. The next few sections briefly describe the functions available from the Spatial Analyst menu (Fig. 15.4). The additional functions accessible through the ArcToolbox and the Raster Calculator are described in the Help documentation for Spatial Analyst.

Distance functions

Distance functions calculate distances across grids or find the most suitable path across a surface. Weights or costs can be assigned to affect travel; for example, high slopes or dense cover increase the cost of travel. Several steps are needed to use these functions. In Figure 15.5, a DEM was used to calculate a slope grid. The slope grid was used as input to the cost-weighted distance function, to create a grid representing the cost-weighted distance from the first point, as well as a grid representing the best direction of travel from any given cell. Finally, the shortest-path function was used to create the shapefile showing the best path.

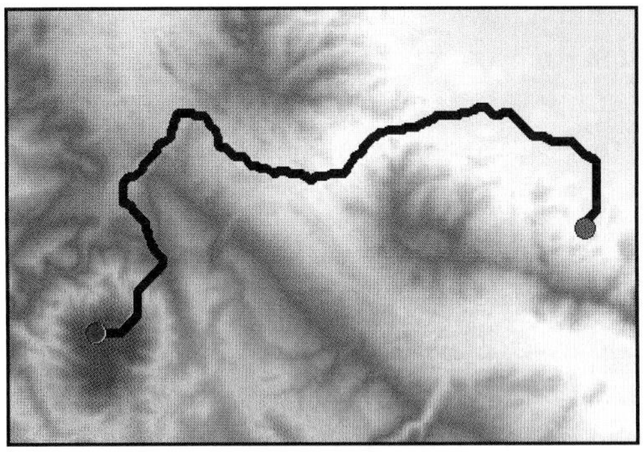

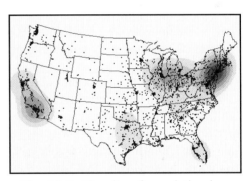

Fig. 15.5. Least-cost path between two points based on topographic slope

Density

The density function is used to create grids showing the density of features within a given radius. Figure 15.6 shows a density map of population in the United States. The darker areas represent a higher density of people and the lighter areas a lower density of people. The raster was calculated using a point file of cities (black dots) with a population attribute, and a search radius of 500 miles. Density functions have a moving window that counts the number of occurrences (people) inside the search radius and divides by the area. A large radius produces a smooth map with higher density values (in this map, 0-90 persons/ km^2); a smaller radius produces more detail and smaller density magnitudes.

Fig. 15.6. Density of population in the United States

Interpolate to raster

Interpolation takes values from points and distributes them across a grid, estimating the values at points in between the measurements. A common example involves taking precipitation measurements at weather stations and interpolating the precipitation values over an area. Figure 15.7 shows the interpolation of the total precipitation measured at the stations over the Black Hills region, with the watersheds superimposed on the image.

Three different interpolation methods are offered: inverse distance weighted, splining, and kriging. Each method uses different assumptions and different means to obtain the gridded

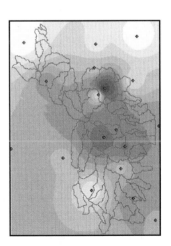

Fig. 15.7. Rainfall grid interpolated from weather station measurements. Dark colors mean more precipitation.

result. More information about these three interpolators can be found in the Help documentation for Spatial Analyst. Interpolation is a complex topic and the subject of entire textbooks. GIS professionals should attempt to become familiar with the different methods in order to understand the benefits, drawbacks, and assumptions of each one.

Although interpolated and density maps appear similar, the functions work differently. Density functions count occurrences within a radius and divide by area, yielding a density, such as people per km^2. Interpolation estimates a value for each grid cell based on its neighbors, and the rainfall grid values represent inches of rain.

Surface analysis

Surface analysis includes a set of functions for calculating properties of a surface, such as slope and aspect. Although most commonly applied to elevation data, these functions can be used on any type of continuous grid. Using these functions on a DEM could help you find an ideal location for a cabin, for example. Contouring the grid would produce a set of contour lines representing the surface. A slope map, showing the steepness of the terrain, could help identify sites flat enough to build on. An aspect map, showing the compass direction of the slope, could locate south-facing hillsides to give the cabin a warm southern exposure. A hillshade map, which represents the surface as though an illumination source were shining on it (Fig. 15.8), can help in visualizing the landforms. A viewshed analysis, which determines what parts of the surface can be seen from a specified point or points (Fig. 15.8), could help in choosing the site with the best view. Finally, the Cut/Fill function calculates the differences between two input surfaces and is useful for developing estimates of the amount of material that has been added or removed.

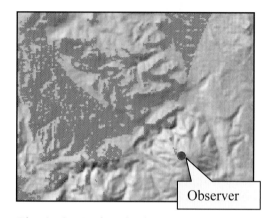

Fig. 15.8. A viewshed map showing the areas visible from a point on a hilltop. The background map is a hillshade.

Cell statistics

This function calculates cell-by-cell statistics for a stack of grids. Imagine 12 grids, each one representing precipitation during one month of 1999. Cell statistics could calculate the total precipitation for the year, the average precipitation per month, or the standard deviation of the monthly precipitation. The available statistics include minimum, maximum, range, sum, mean, standard deviation, variety, majority, minority, and median.

Neighborhood statistics

Neighborhood functions, also known as focal functions, operate on an area of specific size and shape around a cell, known as a neighborhood. A moving window, called a **kernel**, passes over each cell in turn (the target cell), calculates a statistic from all the cells in the neighborhood, and places the result in the target cell. Kernels may have several shapes, including squares, circles, rectangles, and annuli.

Consider a 3 × 3 cell averaging window, for example (Fig. 15.9). For each target cell in the grid, the algorithm finds the average value of the nine cells within the window and puts the average value in the target cell of the output grid. The window then moves to the next cell. An edge cell is still evaluated but will have fewer cells in the neighborhood, such as the cell in the upper left corner of Figure 15.9. Only four values are averaged: the target and its three neighbors. The "1" cell to its right would have the average of six cells.

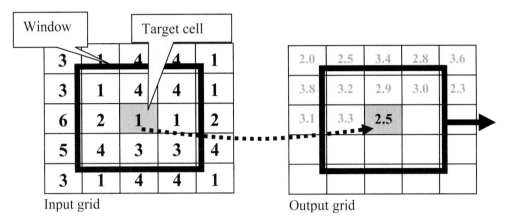

Input grid

Output grid

Fig. 15.9. A 3 × 3 averaging window averages the nine cells in the window and places the result in the target cell of the output grid.

The neighborhood can adopt a variety of sizes and shapes as specified by the user. The shapes include an annulus, a circle, a rectangle, and a wedge. The available statistical functions include: minimum, maximum, range, sum, mean, standard deviation, variety, majority, minority, and median.

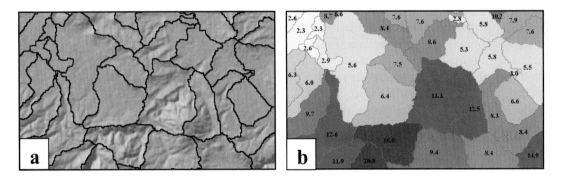

Fig. 15.10. Using zonal statistics to find the average slope of watersheds. (a) The zone grid shown over a hillshade. (b) The watersheds colored by and labeled with the average slope.

Zonal statistics

The zonal statistics function resembles the neighborhood function with two important differences. First, the neighborhoods are not a fixed size or shape, but instead are defined by another layer, called the zone layer. The zones are regions in the zone layer which share the same attribute. A zone might be a watershed, a parcel, or a soil type. Second, instead of creating an output grid, the zonal statistics function creates a table containing statistics for each zone. Figure 15.10 shows a zonal function used to find the average slope within the watersheds. In Figure 15.10a, the watersheds are displayed on top of a hillshade map, and you can clearly see which watersheds have more variable topography and therefore a higher average slope. When executing the zone function, the zone grid contains the watersheds, and a slope grid provides the values to be averaged. The analysis produces a table of statistics, which is then joined to the original watershed polygons using the zone field. From this joined table, a map showing the average slope of each watershed can be created (Fig. 15.10b). In this map, the darker colors indicate higher slopes, and the average slope values are also labeled in each watershed.

Zones need not be contiguous, however, as the watersheds are. The geology map in Figure 15.11 contains many unconnected areas with the same geologic unit. The map only has six zones, one for each color, and the four green regions comprise a single zone.

The zonal statistics function can also assign values from a surface to point data, such as attaching an elevation value from a DEM to a layer of wells. Each point should have a unique ID value that identifies it as its own zone.

Fig. 15.11. This geology map has six zones.

Reclassify

Map reclassification changes the values of a raster according to some scheme set up by the user, such as classifying a slope map into three regions of low, medium, or high slope. For a map with only a few values, such as land use codes, the user might specify a new value for each code individually. Perhaps a town has annexed a nearby planning unit that used different land use codes and must convert one set of codes to another: the code 143 becomes 347, and so on. With continuous data covering a range of values, such as slope, the values are typically classified into a discrete number of ranges. Classification can be applied using a tool similar to the one described in Chapter 4 for classifying attribute data for choropleth maps.

Raster Calculator

The Raster Calculator performs calculations on grids using a formula input by the user (Fig. 15.12). The formula is applied to each matching set of cells in the input grids to derive the cell value in the output grid. For example, adding two grids together adds the contents of each corresponding cell in the two input grids and places the sum in the output grid. Such calculations are called **map algebra.**

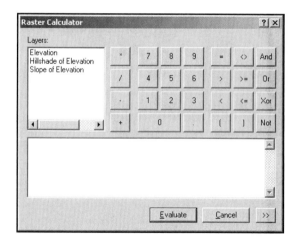

Fig. 15.12. The Raster Calculator

Imagine a grid of precipitation in inches which must be converted to centimeters. Simply enter the expression [Precip] * 2.54 to produce an output grid in cm! The formula expressions can include multiple grids, functions, and values. A model to predict erosion potential as a function of rainfall, slope, soil erodability, and vegetative cover could be implemented as a simple expression in the calculator, such as [Precip] * 2 + [Slope] * 4 / ([Erode] – [Vegcover]).

The calculator also contains logical functions that can return Boolean grids, which have values of true (1) or false (0). Figure 15.13 shows the result of entering an expression [Elevation] > 1400; the blue area in Figure 15.13 shows the cells in the elevation grid which meet this criterion. The standard Boolean operators AND, OR, XOR, and NOT provide additional support for working with Boolean grids (see Fig. 15.16).

Finally, ArcToolbox and the Raster Calculator provide access to *hundreds* of additional functions not accessible from the Spatial Analyst menu. As an example, the INT function can be used to truncate floating-point grid values to integers, and the expression INT([Ingrid] + 0.5) will round the values of a grid to the closest integer. The SETNULL function uses an expression such as Elevation > 1400 and changes those cells to NODATA while retaining the original values elsewhere. Other functions can delineate watersheds from DEM grids, perform multiple linear regression, buffer grid zones, and more.

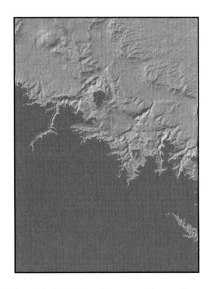

Fig. 15.13. The blue portion of this map shows the pixels where Elevation > 1400 meters.

Convert

Finally the Convert menu provides the means to convert between rasters and vectors, such as converting a polygon shapefile of geology to a grid. The user must choose *one* attribute field to populate the values of the grid, such as the geology unit name. Grids can only store numbers. Numeric fields are converted as is, but text values are assigned arbitrary numbers for storage in the grid. The original text values are placed in the grid table for reference.

You may also convert discrete grids to polygons as long as the number of polygons does not become excessive. Continuous data such as DEMs should never be converted to polygons, since each pixel is usually different from its neighbors and will become a separate polygon. If too many polygons will result from a particular conversion, the software may be unable to complete the task.

Boolean map overlay

As discussed in Chapter 8, map overlay constitutes a large proportion of the analysis carried out in a GIS. The Geoprocessing Wizard can perform intersections and unions of vector layers. The same types of analysis can be performed using raster data. In many cases, raster overlay has advantages over vector overlay. First, it can handle continuous data sets such as elevation, slope, and aspect, which cannot easily be stored as vectors. Second, the process is much faster, because corresponding cells are simply added or multiplied together, instead of splicing the existing polygons and creating new ones from them. Third, raster overlay can process more than two layers at once.

Map overlay in raster is usually accomplished using Boolean grids and the multiplication or addition operators in the Raster Calculator. A Boolean grid contains two possible values, ones or zeros. A one corresponds to a true result, whereas a zero corresponds to a false.

Imagine creating a potential habitat map for lodgepole pine trees based on two criteria: lodgepole pine tends to prefer areas above 1500 feet in elevation, with more than 15 cm precipitation. In Figure 15.14a, the grid shows areas where precipitation is >= 15 cm, created from a precipitation grid using a logical operator in the Raster Calculator. Areas where the precipitation is greater than 15 cm have the value true (1), and areas labeled false (0) have precipitation < 15 cm. Figure 15.14b resulted from a similar query to find the elevations >= 1500 meters.

To find the lodgepole habitat, we multiply the two Boolean input layers to find their intersection, (the areas where both conditions are true). In Figure 15.14c, the two input layers have been multiplied in the Raster Calculator. Areas that have a 1 in both input maps yield $1 \times 1 = 1$ (true).

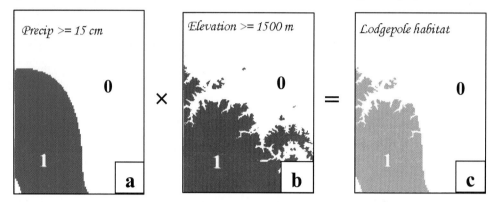

Fig. 15.14. Boolean overlay. (a) A Boolean grid in which the value 1 (true) indicates where precipitation is >= 15 cm. (b) A grid indicating elevations >= 1500 meters. (c) The result of multiplying the first two grids, showing where both conditions are true.

Thus areas that equal 1 in the output map show that both input conditions were true. If either or both of the input layers have a 0, then the output layer will also have a 0 (false). Hence the final map is the intersection of the two input maps, and it shows the potential lodgepole habitat areas.

Map *addition* provides a way to model the habitat slightly differently, to *rank* the possibility of finding lodgepole. In this model, if both elevation and precipitation are ideal, then lodgepole is most likely to occur. If only one is ideal, then it can occur but is not as likely. If neither condition is ideal, then lodgepole is very unlikely. To implement this model we add the two input maps instead of multiplying them. The final map in Figure 15.15 has three possible values: 0 if neither condition is true, 1 if either condition is true, or 2 if both conditions are true. Thus 2 indicates the highest probability of finding lodgepole, 1 a moderate probability, and 0 a very low probability.

Fig. 15.15. A Boolean addition overlay

The best model to use will depend upon which most accurately predicts the actual habitat areas. GIS modeling is always tied to knowledge of the real world. If this project were real, we would proceed to test the model predictions by comparing them to the actual distribution of lodgepole pine. If the model proved accurate, then we might use it to make predictions in other areas.

Although addition and multiplication provide one way to work with Boolean grids, there are also formal operators for use on true/false grids. Figure 15.16 shows the operators and their effects. In all cases the area inside the circles represents the "true" area of two input layers, A and B, and the blue area represents the result of the operation. The operator AND is the equivalent of the multiplication performed earlier (an intersection); the blue wedge represents the area in which both A and B are true. The operator OR performs a union. XOR stands for "exclusive or". All of the Boolean operators are available in the Raster Calculator.

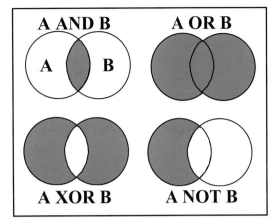

Fig. 15.16. Boolean operators

Controlling analysis options

Spatial Analyst uses several options to influence the results of an analysis. Each of these functions is set using the Spatial Analyst > Options menu.

The analysis extent

Grids do not have to have the same area in order to be analyzed together. By default, the program will automatically generate an output grid with an extent equal to the intersection of the input grids. For example, if the two overlapping grids in Figure 15.17, shown with black outlines, are multiplied together, the output grid will include only the areas common to both (the small yellow square). However, the user could alternatively specify the Union of Inputs option, which results in an output grid that covers the areas of both input grids (the large blue square). The areas outside both grids, needed to complete the rectangle, would receive NoData values.

A user could also elect to copy the area of a specific grid as the output region. Choosing Same as Layer "bhrain3" would require that all the output grids match this layer's extent. Another choice is to use the area of the current data frame as the output grid size, or to type in explicit *x-y* coordinates determining the extent. Whatever choice is made remains in effect until changed again.

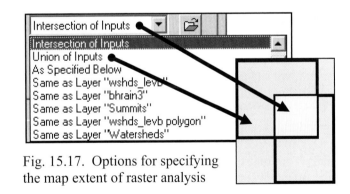

Fig. 15.17. Options for specifying the map extent of raster analysis

The analysis cell size

Spatial Analyst permits the user to specify input grids to the same function, even if the grids do not have the same cell size. However, to actually do the calculation, the grid cells must match. Spatial Analyst automatically resamples one grid to the cell size of another if they do not match.

The cell size of the output grid is determined according to the current cell size rule. By default, the output size equals the largest of the input cells. If you are adding a 50-meter soil grid to a 30-meter elevation grid, the output cell size would equal the larger of the two, 50 m. The default rule can be overridden by other options, however, as shown in Figure 15.18. The user can choose to use the smallest cell size of all the inputs, or the largest cell size. The output can be matched to the cell size of a particular grid.

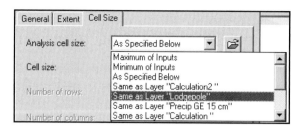

Fig. 15.18. Setting the cell size option for output grids

Or you can simply specify the cell size explicitly, such as 30 meters.

The cell size is always measured in map units, which in turn depends on the coordinate system of the raster layer. Most often the units are meters, but they can also be feet. If the grid is stored in a GCS then the units will be in decimal degrees. However, it is unusual to store grids as GCS because of the inherent distortion.

Some analysis functions always ask for a cell size. In this case, the cell size box will already be filled with a suggested value, based on the map units and area of the analysis extent. The suggested size will usually create a grid with a sufficient but not excessive number of cells. The suggested size can be changed, but making a large change in size (such as 200,000 to 50) runs the risk of creating a grid with too many or too few pixels. If the suggested size is far different than what you expect or want, then double-check the extent settings and the map units before proceeding to create the grid.

The analysis mask

An analysis mask is used to screen out areas of unwanted data. Using a mask on a grid is equivalent to clipping a feature class. The mask grid must contain NoData in the cells to be excluded in the output. The other cells to be included in the output may contain any values except NoData. When the mask grid is specified in the Spatial Analyst options, then all grids subsequently created will have NoData cells wherever the mask grid has NoData cells. In Figure 15.19, areas outside the

Mask grid   Elevation after masking

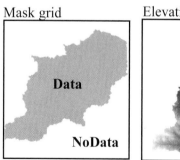 

Fig. 15.19. NoData cells in a mask grid are forced to be NoData cells in the output grid

watershed have no elevation after the mask is applied. The mask will be applied to any subsequent grid created during the analysis and remains in effect until it is changed or turned off.

The working directory

Spatial Analyst produces many new grid files. To help keep the files tidy and in one place, the user should specify a working directory. Unless you request otherwise, all grids created during an analysis session are placed there. The working directory typically defaults to the user's system temp directory, such as C:\Documents and Settings\Maribeth Price\Local Settings\Temp\, but this location is not always ideal.

Two types of output grids are created by Spatial Analyst. Often it produces a temporary grid saved in the working directory. Temporary grids have generic, automatically assigned names, like calc1, calc2, hillshade1, etc. Frequently a temporary grid is a mere stepping stone to the next analysis step and won't be needed again. Permanent grids have specific names and locations chosen by the user, with the implication that the grid is a final result that must be kept for future use. Most functions ask the user to specify whether a permanent or temporary grid should be created. A user may choose to convert a temporary grid to a permanent one at any time.

The choice of working directory will depend on whether primarily temporary or permanent grids are being created. If you are working on a project creating mostly permanent grids, then the project directory is a good choice for the working directory. If most of the output is expendable, the directory C:\temp on the computer works well. Often a special folder for grids inside the current project directory provides the best alternative; it allows permanent grids to be saved in the same area as the rest of the project data, yet temporary grids are segregated for easy removal later. Keeping temporary grids away from your permanent data provides a safety measure against inadvertent corruption of important data files.

> **TIP:** Grids should NOT be stored in folders that have pathnames containing spaces because some functions will not work properly. GIS users are well advised to avoid spaces in ALL their folder names, no matter what. You can spot an experienced GIS user by the fact that she cringes whenever she sees a space in a folder or file name (even when it's allowed).

Temporary grids tend to accumulate over time and can sequester significant space on the hard disk. Periodically, you should go into the working directory with ArcCatalog to clear out the old temporary files. Make sure that the grids are not being used in an open ArcMap session before cleaning house because otherwise you won't be able to delete them. Also be sure that these grids are not being referred to in any map document you plan to use again.

> **TIP: NEVER copy, move, rename, or delete grids using Windows Explorer.** Grids have complex storage formats, and Windows cannot manage them properly. Always use ArcCatalog to manage grids.

Spatial Analyst and ArcToolbox

All of the functions found in the Spatial Analyst menu are also found in the Spatial Analyst Tools toolbox (Fig. 15.20). The user can run them from the menu, run them from the toolbox, and employ them as tools or models in the geoprocessing environment. The toolbox includes additional toolboxes for specialized analysis, such as creating watersheds with the Hydrology tools. Dozens of additional functions can be found in the toolbox which do not appear in the Spatial Analyst menus.

Advanced users who utilize the geoprocessing environment for building models, tools, and scripts have learned that certain environment settings are applied to geoprocessing operations. Some of these settings are identical to the Spatial Analyst Options settings, including the default extent, cell size, and working directory. Users should be aware that if the toolbox is used to perform raster functions, then the geoprocessing environment settings will take precedence over the settings specified in the Spatial Analyst > Options menu. For example, if Betty sets the Spatial Analyst > Options mask grid to a watershed boundary, the mask will apply to any operation she performs with the Spatial Analyst menu. However, if she executes a tool from ArcToolbox, the mask will not be applied.

Fig. 15.20. The Spatial Analyst toolbox

Summary

➢ Rasters consist of spatial data stored as individual cells in an array. Rasters saved in the special grid format can be used for analysis in Spatial Analyst.

➢ Rasters have coordinate systems just like other spatial data. However, they are time-consuming to project accurately. Hence it is usually preferable to store rasters in the same coordinate system in which they will be used, one that best represents their features with the least distortion.

➢ Raster analysis functions fall into four categories: local functions which operate on a single cell, focal functions which act on a defined window, zonal functions which act on an irregular neighborhood, and global functions which act on the entire grid.

➢ Distance functions calculate costs and paths associated with traveling over a surface.

➢ Density functions measure the density of objects over an area.

➢ Raster interpolation estimates a surface based on measurements at isolated locations.

➢ Statistical measures can be calculated for rasters based on single cells, a specified neighborhood, or over large, irregular zones.

➢ Grids can be reclassified and given new values as needed.

➢ The Raster Calculator enables the arithmetic, logical, and functional manipulation of grids to produce new output, such as adding or multiplying grids together. Map algebra can be used in the calculator to develop advanced expressions for manipulating grids.

➢ The analysis environment controls the extent and cell size of all grids created during a Spatial Analyst session.

➢ The Analysis mask can be used to clip grids to a specific area, such as a city boundary or a watershed.

➢ The working directory should be specified by the user to keep the analysis results tidy. All temporary grids are created in it, and permanent grids will be placed there unless the user requests otherwise.

TIP: Grids have stricter naming conventions than other ArcGIS files. Grid names must have no more than 13 characters and should contain only letters, numbers, or the underscore character.

TIP: Grids should NOT be stored in folders that have pathnames containing spaces because some functions will not work.

TIP: ALWAYS use ArcCatalog to copy, move, rename, or delete grids. Grids have complex storage formats, and Windows cannot manage them properly.

TIP: The projection of large grids can be very time-consuming, so grids to be analyzed together are best kept in the same projection.

Chapter Review Questions

You may need to consult the Skills Reference section to answer some of these questions.

1. Describe the difference between the terms *raster* and *grid*.

2. Why is it best to store rasters in the same coordinate system in which you plan to use them?

3. You are adding two grids together in the Raster Calculator using the formula grid1 + grid2. Grid1 is in UTM projection, grid2 is in State Plane, and the data frame coordinate system is USA Equidistant Conic. By default, what coordinate system will the output grid have?

4. In Question 3, you could set an option to change the coordinate system of the output grid. Describe how you would change it, and what the new coordinate system of the output would be.

5. List and describe the other four analysis options that influence the output results during raster analysis.

6. Imagine that a parcels map is used as the zone layer for a zonal statistics function. Explain how the results would differ if the zone attribute used was the (1) parcel ID number or (2) the land use designation.

7. You are adding three grids with cell resolutions of 30 meters, 50 meters, and 90 meters. List the four available options which would determine the cell size of the output grid, and state what the output resolution would be in each case.

8. Why does the zonal statistics function require two input layers?

9. Imagine you have Boolean grids for 10 different criteria regarding the siting of a landfill. If you add the grids together, what is the total possible number of values in the output grid?

10. How many possible output values would there be in Question 9 if you multiplied the grids?

Mastering the Skills

Teaching Tutorial

The following examples provide step-by-step instructions for doing basic tasks and solving basic problems in ArcGIS. The steps you need to do are highlighted with an arrow ➔; follow them carefully. Click on the video number in the VideoIndex to view a demonstration of the steps.

To become acquainted with raster analysis, we will first repeat the problem we did in Chapter 8 to find the snail habitat in an area of the Black Hills (Fig. 15.21). Now we will do it in raster. Raster analysis in ArcMap is done using the Spatial Analyst extension, purchased separately. Before using it, the extension must be turned on.

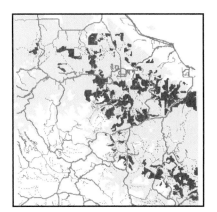

Fig. 15.21. Snail habitat determined by vector analysis

➔ Start ArcMap and open ex_15a.mxd in the MapDocuments folder.

➔ Use Save As to rename the document and remember to save frequently as you work.

1➔ Choose Tools > Extensions from the main menu bar.

1➔ Check the box to activate the Spatial Analyst extension, if it is not already checked.

1➔ Close the Extensions window.

Now that you have activated the extension, open the Spatial Analyst toolbar.

1➔ Right-click one of the menu/toolbars and choose the Spatial Analyst toolbar.

Before starting any analysis, make sure the working directory and default cell size and extent are set. To prevent new grids from getting mixed up with the Sturgis data, create a temp data folder inside the BlackHills folder.

➔ Use Windows Explorer or ArcCatalog to create a new empty folder named "temp" inside the BlackHills folder.

2➔ Choose Spatial Analyst > Options and click the General tab.

2➔ Click the Browse button to set the working directory to the new mgisdata\BlackHills\temp folder. Click OK.

2➔ Click the Cell Size tab and set the analysis cell size to Same as Layer "Elevation". Note that this value is 50 (meters).

2➔ Click the Extent tab and make sure the default extent is set to "Intersection of Inputs". Click OK.

The map document is the same one we used in Chapter 8. Let us briefly review the problem. Imagine that a rare species of snail inhabits the Black Hills. This snail prefers limey soils in the cool and dense coniferous forest, and is rarely found above an elevation of 1600 meters or below 1200 meters.

We'll use three layers, a geologic map, a vegetation map, and an elevation map, to identify the likely areas of snail habitat. The overall approach consists of creating three Boolean rasters, each one representing the ideal conditions in one of the three layers, and multiplying them together to find the area of intersection.

In Chapter 8, a polygon shapefile defined the elevation range of the snail habitat and was already created for you using a digital elevation model (DEM). Since you are now using Spatial Analyst, you can create this elevation range this time.

Performing a Boolean overlay analysis

We can use a map reclassification to choose a certain range of cells—those with elevation from 1200 to 1600 feet. This step is analogous to classifying an attribute field when creating a graduated colors map, except that the output is an entirely new raster, and the output class values are single numbers, rather than ranges. Because we are preparing a Boolean overlay analysis, we wish to create a new raster in which the cells we want (between 1200 and 1600 meters) have a value of 1, and the rest have a value of 0.

3➜ On the Spatial Analyst toolbar, set the layer to Elevation, if necessary.

3➜ Choose Spatial Analyst > Reclassify.

3➜ Click in the first old values range and type "0 – 1200". (You must include the spaces around the – symbol.) For the new values, type "0".

3➜ Click in the second old values range and type "1201 – 1600". For the new value, type "1".

3➜ Click the third row and type "1601 – 2000". For the new value, type "0".

> **TIP:** Be sure to hit the Tab key after the last new value is typed, to make sure it is actually entered. If you neglect to do this, that last category may be absent in the output raster.

3➜ Click the next row. Hold down the Ctrl key and click the remaining rows to select them all except the NoData row. Click Delete Entries.

> **TIP:** To know the highest value to type in, look at the last row with values (ending at 1956). Set the highest value just a bit higher—it doesn't matter if it extends beyond the actual range of values in the raster.

Notice that ranges can be loaded and saved for use again later. Map reclassify can also use the same classification methods described in Chapter 4 (equal interval, defined interval, etc.). To save this grid permanently you could click the Browse button by the output raster to specify a folder and file name. This time a temporary grid is fine.

3➜ Make sure the output file is temporary, and click OK.

4➜ Scroll down and find the output, named Reclass of Elevation. Rename it "Snail Elev Range".

4➜ Right-click the symbol for 0 in the new raster and change its color to No Color.

Notice that the new grid has only one value—1, and the rest of the cells are 0. Their extent matches the snail elevation habitat range used in Chapter 8. (In fact, the snail_elev.shp shapefile you used was created by taking a grid like this and converting it to a shapefile. Since we are doing the analysis in raster this time, though, we don't need to do that step.)

Another way to create a Boolean grid based on raster values is to use the Raster Calculator. Let's repeat the previous step using the alternative method. It is a little faster.

4➔ Choose Spatial Analyst > Raster Calculator.

4➔ Enter the expression [Elevation] >= 1200 And [Elevation] <= 1600 and click Evaluate.

4➔ Examine the new Calculation grid.

4➔ When finished, right-click the new Calculation grid and remove it.

Next, we will select the desired polygons from the geology and vegetation shapefiles and convert them into grids. A single field in the shapefile provides the values for the grid.

5➔ Turn off the Elevation grid and the Roads layer, and turn on the Geology layer.

5➔ Use Select By Attributes to select the Geology polygons where "NAME" = 'Madison Limestone' OR "NAME" = 'Upper Paleozoic'.

6➔ Choose Spatial Analyst > Convert > Features to Raster.

6➔ Select Geology as the input features and set the Field to UNIT, a numeric code indicating the geology unit (Fig. 15.22).

6➔ Click the Browse button and enter the folder and file name for the output grid. Call it "limegrid". Click Save and OK.

6➔ Turn off Geology and turn on limegrid.

6➔ Right-click Geology and choose Remove from the context menu.

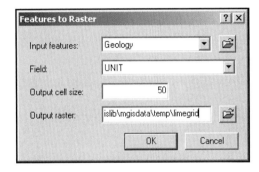

Fig. 15.22. Converting geology polygons to a raster

Notice that limegrid has two colors for units 5 and 7, the limestone units. Let's take a moment to look at the attribute table of limegrid. Grids can be stored using integer or floating-point (decimal) values. An integer grid will have an attribute table created for it, if it does not contain too many different values.

7➔ Right-click the limegrid layer and choose Open Attribute Table.

The attribute table contains two entries, one for each unique value present in limegrid (5 and 7, stored in the Value field). The Count field indicates the number of cells for each value (Fig. 15.23).

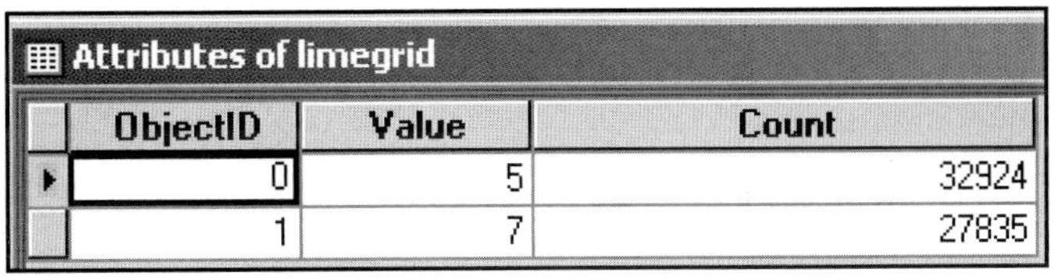

Attributes of limegrid		
ObjectID	**Value**	**Count**
0	5	32924
1	7	27835

Fig. 15.23. The limestone grid attribute table

1. What percent of the limestone in this grid consists of unit 7 (Madison Limestone)? _____

Next, we must create a Boolean grid containing only 1's and NoData, using either Reclassify or the Map Calculator. We will use the second method.

> 7➜ Close the limegrid attribute table.

> 8➜ Choose Spatial Analyst > Raster Calculator.
> 8➜ Enter the expression [limegrid] > 0 and click Evaluate.
> 8➜ The new grid is currently called Calculation; name it "Limestone".
> 8➜ Right-click and remove the limegrid layer.

Notice that we are being careful to name the output grids. Otherwise we can end up with dozens of grids with confusing names like NbrMajority of Reclass of Calculation, which makes it easy to use the wrong grid for a later step. If you are keeping a grid, name it something recognizable!

Next, we will repeat the procedure to create a vegetation Boolean raster containing the polygons of interest.

> 9➜ Turn on the Vegetation layer.
> 9➜ Use Select By Attributes on the Vegetation layer to select the polygons where "COV_TYPE" = 'TPP' AND "DENSITY96" = 'C'.
> 9➜ Close the Select By Attributes window.

> 10➜ Convert these features to a grid just as for Geology, using the COV_TYPE field. Name the raster file "Conifers".
> 10➜ Name the output layer, now appearing as Conifers – Conifers, to simply Conifers.
> 10➜ Turn off the Vegetation layer to see the raster.
> 10➜ Open the attribute table for the Conifers layer.

Notice two things about this attribute table. First, the only grid value is 1. Recall that COV_TYPE was the converted field and the selected polygons had the value TPP (Ponderosa pine). Because a grid must be numeric, the conversion automatically assigned a numeric value to the grid and placed the COV_TYPE text in the attribute table for reference. Because we only used one COV_TYPE, all the grid values are 1 already and don't need reclassification.

Now we are ready to do the overlay. We have three grids, each containing 1's where the desired condition holds, and 0 or NoData where the condition is absent. If we multiply these three grids together, any cell which has a 1 in each input grid will have a 1 in the output grid, signifying that it meets all three conditions (elevation, limestone, conifers). Use the Map Calculator to perform this simple map algebra expression.

> **TIP:** When using Spatial Analyst, you may sometimes receive a warning that the rasters are in different projections. When this happens, one of the rasters will be reprojected before analysis to ensure that they overlay properly. The projection of large grids can be very time-consuming, so grids to be analyzed together are best kept in the same projection.

11➔ Close the Conifers attribute table.

11➔ Choose Spatial Analyst > Raster Calculator.

11➔ Enter the expression Snail Elev Range * Limestone * Conifers and click Evaluate.

11➔ Turn off all layers except for Calculation.

11➔ Zoom to the extent of the Calculation layer and examine it carefully.

> **TIP:** Although *filenames* of grids must contain only 13 characters and no spaces, this rule does not apply to naming a grid layer in the Table of Contents. Give it any name you like.

Notice that the Calculation output has few zero values, as a consequence of the inputs. The Geology and Vegetation layers, converted to grids, contained NoData values where no polygons were present, whereas the Snail Elev Range layer contained either 1 or 0 everywhere. In the Raster Calculator multiplication, any cell with a NoData input was assigned NoData in the output, rather than a zero. NoData and 0 values are different, and are treated differently by different functions.

Now, in Chapter 8, the next step used an overlay to identify which primitive road sections crossed snail habitat. This analysis is not well suited to raster data. Although you could convert the roads to raster and do another overlay, the 50-meter cell size would not represent the roads well, and a smaller cell size would take longer to calculate. Instead, we will convert the snail habitat raster to polygons and overlay them with the vector roads.

However, converting the current Calculation layer will yield polygons with values of both 1 and 0. We should eliminate the 0 values before doing the conversion to polygons.

12➔ Choose Spatial Analyst > Reclassify and change the 0 values of Calculation to NoData, leaving the 1 values as 1.

12➔ Rename the result "Snail Habitat", and remove Calculation.

12➔ Choose Spatial Analyst > Convert > Raster to Features.

12➔ Set the layer to Snail Habitat, set the field to Value, and name the output layer "snailhab2.shp". Make sure the output type is Polygon.

12➔ Make sure the Generalize box is checked. This option removes extraneous vertices from the output polygons for a smoother result which uses less disk space. Click OK.

12➔ Examine the snailhab2 polygons.

At this point we could use the Intersect tool in ArcToolbox to overlay the snailhab2 polygons with the roads to get the proposed road closures just as we did in Chapter 9. Instead of repeating that step, though, we'll just go on with more raster analysis.

The last step in the snail habitat analysis included finding the habitat within 200 meters of primary and secondary roads, in order to propose thinning the trees. Buffering with rasters is based on a distance function and requires two steps, but is still faster than buffering vectors.

13➜ Turn off all the layers, and turn on the Roads layer.

13➜ Select the primary and secondary roads using the expression "TYPE" = 'P' OR "TYPE" = 'S'.

13➜ Create a layer from the selected roads and name it "Major Roads".

13➜ Turn off the Roads layer.

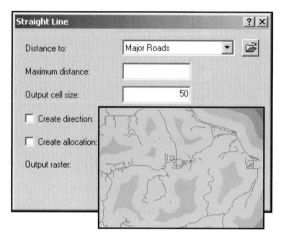

14➜ Choose Spatial Analyst > Distance > Straight Line.

14➜ Set the distance layer to Major Roads (Fig. 15.24).

14➜ Do NOT check the boxes to create direction or allocation grids.

14➜ Let it create a temporary raster. Click OK.

14➜ Name the output grid "Road Distance".

Fig. 15.24. Creating a distance grid using the Straight Line distance function

Each cell of the new grid represents the distance of that cell from the nearest road. To complete the buffering process, we will use the Raster Calculator to create a Boolean grid with a 1 within 200 meters of the roads, and a 0 elsewhere.

15➜ Choose Spatial Analyst > Raster Calculator.

15➜ Enter the expression [Road Distance] <= 200 and click Evaluate.

15➜ Name the output grid "Road Buffers" (Fig. 15.25).

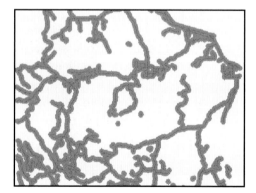

Fig. 15.25. Raster road buffers

Finally, another Boolean multiplicative overlay will yield the snail habitat inside the road buffers.

16➜ Remove the Road Distance grid.

16➜ Turn on the Snail Habitat layer and move it above the Road Buffers layer in the Table of Contents.

16➜ Turn on the Roads layer and change the colors of all three road types to medium grey.

17➜ Choose Spatial Analyst > Raster Calculator and enter the expression [Snail Habitat] * [Road Buffers]. Click Evaluate.

17➜ Name the output "Proposed Thin" and change its 0 value symbol to No Color.

17➔ Turn off all layers except Proposed Thin, Snail Habitat, and Roads.

18➔ Change the symbols so that the map shows the roads as grey, the snail habitat as green, and the proposed thin areas as purple (Fig. 15.26).

2. Using the cell size (50 meters) and attribute table of Proposed Thin, calculate the total area of proposed thinning, in square kilometers.

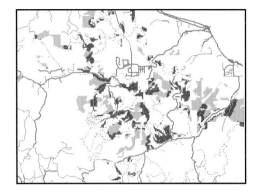

Fig. 15.26. Your map after step 18, showing proposed areas to thin timber

Well done! The next section will demonstrate some of the other capabilities of the Spatial Analyst software, starting with topographic functions.

Using topographic functions

➔ Begin a new, empty map. Choose to save your work from the previous section.

19➔ Use the Add Data button to add the raster dem2 to the map.

19➔ Set the dem2 symbology to a stretched color elevation ramp.

Note: The videos will display the dem2 grid as classified, for performance reasons, and it will look slightly different from yours.

20➔ Set the Analysis Options cell size to 50 meters so all grids default to this value. The previous Options settings were discarded when the new map was opened.

20➔ Be sure that the working directory is still set to your temp folder.

21➔ Choose Spatial Analyst > Surface Analysis > Slope.

21➔ Set the input surface to dem2, the output measurement to degrees, the Z factor (vertical exaggeration) to 1, and make it a temporary grid. Click OK (Fig. 15.27).

21➔ Turn off the dem2 grid. Rename the Slope of dem2 grid to "Slope". Examine the range of slopes present.

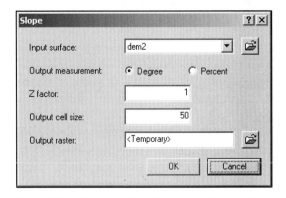

Fig. 15.27. Calculating the slope of a surface

3. What is the maximum slope of this grid?

Next we'll calculate an aspect grid. The aspect is the direction that a slope faces—in other words if you are standing on a hill looking down the steepest direction and facing south, then the hill has a southern aspect. Aspect is measured in degrees from north, with true north being 0 (and 360) degrees, east 90 degrees, south 180 degrees, and west 270 degrees.

22➔ Turn off the Slope grid.

22➔ Choose Spatial Analyst > Surface Analysis > Aspect.

22➔ Make sure the input grid is dem2, the cell size is 50, and the output is temporary.

22➔ Click OK. Rename the new result "Aspect" and examine it.

Notice that the aspect map is automatically colored and labeled according to the compass directions.

4. Which color indicates a northern aspect? _____ A southern aspect?
_____

A hillshade function creates a grid that mimics the illumination of the surface from a light source at a specified direction (azimuth) and altitude (zenith angle). The result looks similar to what a person might see from the air when flying over the surface. Note that both azimuth and altitude are measured in degrees.

23➔ Turn off the Aspect raster.

23➔ Choose Spatial Analyst > Surface Analysis > Hillshade.

23➔ Make sure the input grid is dem2, the cell size is 50 meters, the Z factor is 1, and the output is temporary. Accept the default azimuth and altitude. Click OK. (Fig. 15.28).

23➔ Rename the output "Hillshade", and examine it.

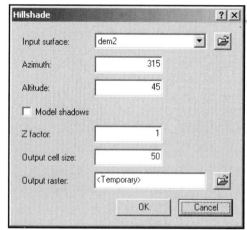

Fig. 15.28. Calculating a hillshade

Compare both the elevation and hillshade grids, and notice how the hillshade brings out fine topographic details that are difficult to see in the elevation grid. Setting a transparent layer over a hillshade looks even better.

24➔ Collapse the legends for the Aspect and Slope rasters.

24➔ Drag the dem2 layer above the Hillshade layer in the Table of Contents, and turn it on.

24➔ Open the dem2 layer properties, click the Display tab, and set the transparency to 60%. Click OK.

TIP: Try applying this display technique to the aspect map at about 80% transparency to see how the different aspect colors relate to the terrain.

The Viewshed function determines what can and cannot be seen from a point or set of points. It is used for applications such as placing fire towers for maximum viewing area, or determining whether proposed timber clear-cut areas are visible from the main roads. We will use it to determine the visibility offered by three proposed fire towers (Fig. 15.29).

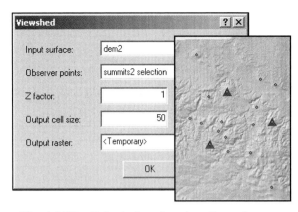

Fig. 15.29. Calculating the view from three fire lookouts

25➔ Use the Add Data button to add the summits2 shapefile from the Sturgis folder.

25➔ Use the Selection tool to select the three summits shown by pink triangles in Figure 15.29.

25➔ Create a layer from the selected summit features and name it "Towers".

25➔ Change the symbology of Towers to large pink triangles.

26➔ Choose Spatial Analyst > Surface Analysis > Viewshed.

26➔ Make sure the input surface is dem2, the observer points are set to Towers, and the Z factor is 1. Click OK.

TIP: The Viewshed function takes a few minutes, even on this small grid with only three points. Be cautious about using it on large grids or with many points. Track the progress of the analysis in the lower left corner of the ArcMap window, and if it is progressing too slowly, you can cancel the analysis.

TIP: Notice that as you accumulate grids, you have to wait while each draws. For best performance, get in the habit of turning off or removing any grids not currently in use.

27➔ Remove the summits2 layer, collapse the legend of the Hillshade layer, and turn off all layers.

Using neighborhood functions

A neighborhood function examines the area around each cell in the grid, one at a time, and calculates statistics. For example a square 3-cell-by-3-cell neighborhood with an average statistic would calculate the average of the nine cells in the neighborhood and set the value of the center cell to the average. Other statistics such as sums and standard deviations are also often used. A majority statistic determines the value which occurs most frequently. A variety statistic determines the number of values which occur in the neighborhood. The majority statistic provides a valuable tool when converting from raster to features. Imagine creating polygons showing areas with slope less than 10 degrees, as an initial step for finding some land to build a cabin on. First, reclassify the slope grid already created.

28➔ Turn off and collapse the legends of all the layers except Slope.

29➜ Choose Spatial Analyst > Reclassify.

29➜ Choose the Slope grid as the input, and create two classes: a 1 for slopes less than 10, and a 0 for slopes greater than 10.

30➜ Rename the temporary output grid Reclass of Slope to "Slope Class".

30➜ Zoom in for a closer look (Fig. 15.30).

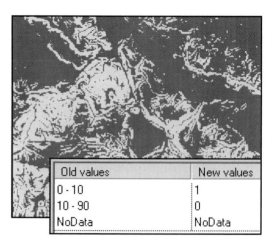

Notice that the Slope Class raster has many small specks and narrow lines. Converting this raster to features would produce an excessive number of complex polygons, and it would be unlikely to finish successfully. However, for your purposes, a tiny flat area surrounded by steep slopes is not a suitable building site, anyway. And a small bit of steepness would not necessarily disqualify an otherwise flat area. You can use the majority filter to eliminate small areas inside their opposites. If your goal is an area of mostly flat slope that is at least 250 meters square, a 5-cell-by-5-cell majority filter will locate the areas in which the slope is predominantly flat or steep and will simplify the raster.

Fig. 15.30. Reclassifying a slope grid into two classes for slope < 10 and slope > 10

31➜ Choose Spatial Analyst > Neighborhood Statistics.

31➜ Make sure that Slope Class is the input data. Choose the Majority type and set the neighborhood to a 5-cell rectangle. Make sure the cell size is 50 and the output is temporary. Click OK.

32➜ Change the symbology of the output to unique values to represent its two values, 0 and 1.

Notice the much simpler appearance of the new map (Fig. 15.31). However, there are still a few small island areas. Repeating the majority filter on the output file will clean it up a little more.

33➜ Repeat the 5 × 5 majority filter, this time using NbrMajority of Slope Class as the input, and keeping the other parameters the same.

34➜ Rename the output map "SlopeClass2".

34➜ Change the symbology of the output as you did last time to display only the two values.

34➜ Remove the NbrMajority of Slope Class layer.

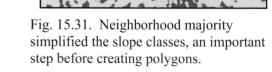

Fig. 15.31. Neighborhood majority simplified the slope classes, an important step before creating polygons.

We are almost ready to create the polygons. However, recall that the raster contains zeros and ones. To prevent the zeros from becoming polygons also, we must convert them to NoData values.

35➔ Use the Reclassify function to set the 0 values in SlopeClass2 to NoData. Set the other values to 1. Be sure to press the Tab key after entering the final value.

35➔ Name the result "SlopeClass3", and remove SlopeClass2.

35➔ Turn off all of the rasters except for SlopeClass3.

36➔ Zoom to the extent of SlopeClass3.

36➔ Use the Convert > Raster to Features function to create polygons of the high slope areas from SlopeClass3. Name the output shapefile "lowslope".

36➔ Turn off SlopeClass3 and examine the new shapefile, showing potential locations for building a cabin based on slope (Fig. 15.32).

Fig. 15.32. Polygons of low slope areas

Using zonal functions

Zonal functions are similar to neighborhood functions, except that the neighborhood is defined by another grid or feature dataset, rather than by a simple rectangle or circle. Two input grids are required: the zone grid indicating the neighborhoods to be evaluated, and the value grid containing the values used to generate the statistics. The zone layer must have a field whose values determine the zones (such as a landuse code). The output of the function is a dBase table rather than a grid. The table contains the statistics for each zone defined by the zone layer.

A common application of zonal functions occurs when hydrologists are finding watershed parameters for hydrologic runoff models, such as the average slope of a watershed. The zonal functions can provide this information easily. In this case, the watershed features constitute the zone layer, and the slope map is the values layer.

➔ Click the Open button and open ex_15b.mxd in the MapDocuments folder. Save the changes to your old map.

➔ Use Save As to rename the document, and remember to save frequently as you work.

37➔ Use Spatial Analyst > Options menu and make sure the working directory is set to your temp folder.

37➔ Make sure the default extent is "Intersection of Inputs" and the default cell size is 50 meters. Click OK.

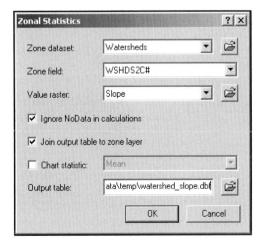

Fig. 15.33. Calculating zonal average slope for watersheds

38➔ Choose Spatial Analyst > Zonal Statistics.

38➔ Fill out the dialog box as shown in Figure 15.33, and naming the output table "watershed_slope.dbf". Click OK.

Recall that an ArcInfo coverage has a special field called the *cover#*, which contains a unique number for each feature (it is the equivalent of the FID, or feature ID). We will use this field to ensure that each watershed is its own zone. Checking the box automatically joins the output table to the Watersheds layer. We don't want a chart, so we uncheck that box.

The joined attribute table is opened automatically. Note that each watershed now has several statistics fields, MIN, MEAN, MAX, etc., indicating the slope values for that watershed. Let's create a graduated color map showing the mean slope of each watershed (Fig. 15.34).

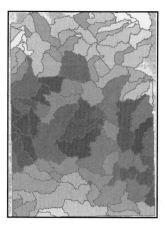

39➔ Close the watershed_slope table.

39➔ Open the properties for the Watersheds layer and click the Symbology tab. Create a graduated color map based on the MEAN field, with about 10 classes.

Notice how the high slopes in the center of the grid generate the watersheds with the highest mean slope (Fig. 15.34). Exporting the attribute table would now provide the average slope of each watershed for hydrologic modeling.

Fig. 15.34. Average slope of watersheds after step 39

5. How many watersheds have an average slope greater than 15 degrees? _____

Zone grids need not be polygons; they could also be points or lines. You could use exactly the same routine above with streams to determine the mean slope along a stream line—another important parameter for hydrologic models. When a point grid is used, the result is based on only one cell. Thus the statistics are not very interesting, but it does provide a way to assign a value from the grid to a point.

For example, the Summits layer contains points at the tops of hills, but no information is provided in the attribute table about the actual elevation. This information would be useful in planning a hiking trip. The Zonal Statistics function can add an elevation to every point.

40➔ Collapse the Watersheds legend. Turn on the Summits and Elevation layers, and turn off the others.

40➔ Open the Summits attribute table and examine the fields. Each summit has a name, but no elevation.

40➔ Close the attribute table.

41➔ Choose Spatial Analyst > Zonal Statistics, and fill out the dialog box as shown in Figure 15.35, naming the output table "summit_elev.dbf". Click OK.

41➔ Examine the summit_elev table.

Notice that all of the statistics have the same value, because the mean, min, max, sum, etc., of a single point is the same number. This number, however, is the elevation of the summit. We can now create a map with labels indicating the summit elevations.

42➔ Close the summit_elev table.

42➔ Open the Summits layer properties and click the Labels tab. Give the layer white labels in 12 point italic bold Arial font. Use any of the zone summary fields, such as MIN.

6. What is the elevation of Elkhorn Peak?

_____

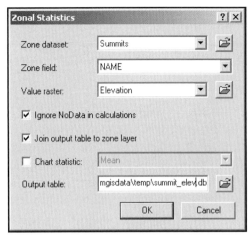

Fig. 15.35. Finding elevations of summits using Zonal Statistics

Using interpolation functions

Interpolation creates a continuous grid from data at distributed points. For example, the Black Hills has weather stations which record temperatures and precipitation. To know the precipitation at a point at some distance from a weather station requires interpolating between measurements. Many different interpolation functions have been developed; each has its own properties, advantages, and drawbacks. Unfortunately, a complete discussion of interpolation methods is beyond the scope of this book, but it is not difficult to create an interpolated map using one of the three available methods in Spatial Analyst. To demonstrate, we will create a precipitation map from the weather stations.

43➔ Activate the Black Hills data frame.

43➔ Open the Climate Stations attribute table and examine the fields. Notice that this data set contains monthly precipitation values for the year 1997, in centimeters.

43➔ Close the attribute table.

This data set covers a much larger area than the others in this chapter, and keeping the default cell size set at 50 will require an excessively large storage space for the grids. Thus, set the default cell size to a larger value for this part of the analysis.

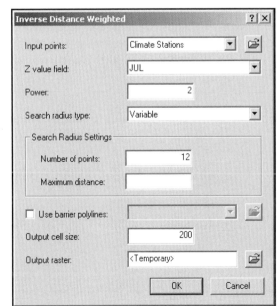

Fig. 15.36. Interpolating rainfall using the IDW method

44➔ Choose Spatial Analyst > Options and set the default cell size to 200 meters.

45➔ Choose Spatial Analyst > Interpolate to Raster > Inverse Distance Weighted.

45➔ Fill out the dialog box as shown in Figure 15.36 to create a precipitation grid for July. Set the input points to Climate Stations, the Z value field to JUL, and the cell size to 200. Take the defaults on everything else. Click OK.

45➔ Examine the new grid.

7. Do you think the precipitation grid is an integer or floating-point grid? _____ Why? _____Examine its properties to check your answer.

Now we can use this new grid to estimate the total volume of water received by each watershed during the month of July. This information could then be calibrated to stream flow measurements. The first step is to calculate the total precipitation for each cell in cubic meters. First we convert the precipitation measurements to meters.

46➔ Rename the IDW of Climate Stations grid "Precip-cm".

46➔ Choose Spatial Analyst > Raster Calculator, and enter the expression [Precip-cm] / 100. Click Evaluate.

46➔ Rename the result "Precip-m".

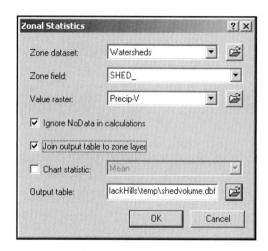

Next, we calculate the total volume of water for each cell. The area of each pixel is 200 m × 200 m = 40,000 m^2. Hence the volume of water V in each pixel is V = Precip-m * 40,000, in cubic meters.

47➔ Choose Spatial Analyst > Raster Calculator, and enter the expression [Precip-m] * 40000. Click Evaluate.

47➔ Rename the result "Precip-V".

Fig. 15.37. Summing precipitation volume over the watersheds

Finally, we use zonal statistics to sum the water volume of cells over the entire watershed. The field SHED_ contains the unique ID number for each watershed, so we use that as the zone field.

48➔ Choose Spatial Analyst > Zonal Statistics.

48➔ Fill out the dialog box as shown in Figure 15.37, naming the output file "shedvolume.dbf". Click OK.

The output table now lists the total volume of water received in July by each watershed. The values range from about 1 to 46 million cubic meters. Join this table back to the Watersheds layer to create a map, if you wish (Fig. 15.38).

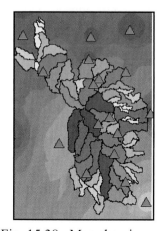

Fig. 15.38. Map showing climate stations and volume of water in each watershed

8. What is the name of the watershed which has the largest volume of water? What is the volume? _____

Using the analysis options

The analysis options provide detailed control of raster processing to get exactly the desired results. We will examine a few examples here.

Imagine that you are doing a study in a watershed, and you want only information inside the watershed to appear in maps, and that the project parameters call for a consistent cell size of 50 meters. The analysis options can be used to enforce both of these conditions.

49➔ Close the shed-precip table.

49➔ Set the selectable layers to Watersheds, and use the Selection tool to select the watershed study area (Fig. 15.39).

50➔ Convert the selected feature to a raster file, using the SHEDNAME field as the value. Make sure the cell size is still set to 200 meters. Name the output raster "studyarea".

50➔ Rename the output "Study Area".

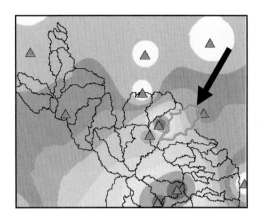

Fig. 15.39. Clipping data to this single watershed

Now take the output raster and put it in the other data frame for use in the Sturgis area.

51➔ Right-click Study Area and choose Copy.

51➔ Activate the Sturgis data frame.

51➔ Right-click the Sturgis data frame and choose Paste Layer(s).

51➔ Click the minus sign to collapse the Black Hills data frame out of the way.

Look at the data extent of the Study Area layer.

52➔ Turn off the Summits layer.

52➔ Right-click Study Area in the Sturgis frame and choose Properties. Click the Symbology tab. Set the NoData display color to a light pink.

52➔ Zoom to the extent of the Study Area grid.

Wow! This raster has a very large extent, far beyond this single watershed, because it was created using the entire Black Hills Watersheds layer. Before we go on, we will set the extent window and the cell size explicitly to prevent any unexpected results like this one. This grid takes up far more space on the disk than it needs to.

53➔ Return to the previous extent.

53➔ Choose Spatial Analyst > Options.

53➔ Click the Extent tab. Set the extent to the Same As Elevation.

53➔ Click the Cell Size tab. Set the cell size to As Specified Below. Type in "50" and press Tab to see the number of rows and columns. Click OK.

Notice that nothing has changed in the display as yet. The analysis options apply to the creation of new grids only, not the characteristics of existing grids. So now you will "clip" the Study Area grid to the smaller extent. There is no "clip" command in Spatial Analyst; rather, you set the desired extent and cell size and do a map calculation. The new output grid will be constrained by the extent and cell size set in the options.

54➔ Choose Spatial Analyst > Raster Calculator.

54➔ In the box, enter the simple expression [Study Area] and click Evaluate.

54➔ Open the Properties of the Calculation output and click the Symbology tab. Set the NoData color to pale grey to see the extent of the grid.

55➔ Right-click and Remove the previous Study Area grid.

55➔ Rename the Calculation result "Study Area".

Finally, the mask option uses a polygon layer or a raster to specify a "cookie cutter" to eliminate unwanted areas in a grid. Suppose you are doing a study of this watershed and want the grids you create to have values only inside the watershed. To do this, set the Study Area grid as an analysis mask. Any "NoData" areas in the mask will be given "NoData" values in any output grids created after the mask is set.

Masks perform a function analogous to clipping on a vector data set. To clip an existing raster, specify the mask grid and then use the Raster Calculator to create an exact replica of the grid to be clipped. The mask will screen out the values in the NoData areas, leaving only the desired portions.

56➔ Choose Spatial Analyst > Options. Click the General tab.

56➔ Set the Analysis Mask to Study Area and click OK.

56➔ Choose Spatial Analyst > Raster Calculator. Enter the simple expression [Elevation] and click Evaluate.

Notice how the study area mask has eliminated the elevation values outside the watershed (Fig. 15.40). The mask grid will remain in effect until you set it back to <None>.

9. What is the lowest and highest elevation inside this watershed?

_____

This is the end of the tutorial.

Fig. 15.40. Elevation clipped to a watershed

➔ Close ArcMap. You do not need to save your changes.

Exercises

Use the data layers in the BlackHills folder to answer the following questions.

> **TIP:** Be sure to turn off any mask grids before going on to the next problem.

1. What percentage of the Sturgis area DEM lies above 1500 meters elevation?

2. You speculate that the chance of finding new cave entrances is highest for areas of Upper Paleozoic rocks with slopes of greater than 20 degrees. Create a map showing the limestone outcrop area and the areas that meet both these criteria. **Capture** your map.

3. Create an elevation map of the Sturgis area, such that the only portions shown are inside the polygons of wshd2b. **Capture** your map.

4. Within the watersheds in wshds2b, what is the farthest you can be from a stream?
   convert streams to raster THEN do distance

5. Which of the summits is furthest from a stream? How far away is it? Which is closest, and how far away is it?

6. Create a map, inside the watersheds only, showing the areas that are closer to a stream than they are to a road. Make the areas closer to a stream blue, and the areas closer to a road grey. **Capture** your map.

7. Is there any Forest Service land (grid PVT2 = 0) more than 1 km from the roads in usfsrds2? How many sq km?

8. Which geologic unit has the highest average slope? Which has the lowest? What is the average slope for each one?

9. The geology shapefile has a field called INFILCLASS. Assume that in class 1, 10% of precipitation infiltrates, 20% infiltrates for class 2, and 30% infiltrates for class 3. Assume that all the remaining precipitation runs off. Create a raster that portrays the total runoff (in cm) for the Sturgis area in May 1997. **Capture** your map and include a legend.

10. Create an erosion index map for the Sturgis area based on the following assumptions. The basic erosion potential equals the slope divided by the infiltration class times the vegetation factor. The vegetation factor is determined from the DENSITY96 field in coverage stands2, such that 0 = 0.9, A = 0.8, B = 0.6, and C = 0.4. Finally, areas within 200 meters of a stream have their erosion index doubled. Use a Natural Breaks classification with nine classes and a yellow-brown color ramp. **Capture** your map and include a legend. (**Hint:** Use integers when classifying the vegetation and divide by 10 later.)

Challenge Problem

Imagine that you would like to build a house somewhere in the Sturgis area. Develop a map showing potential land on which to build based on the following criteria. Be sure to turn off any masks before you begin.

> ➢ Must be on private land (grid PVT2 = 1).

> ➢ Must have a slope less than 20 degrees.

> ➢ Must have a southerly aspect (between 90 and 270 degrees).

> ➢ Must be within 500 meters of an existing road (use roads2).

> ➢ Must be more than 100 meters from a stream.

> ➢ The land use category in lulc2 must be Evergreen Forest or Mixed Forest Land.

Create and capture a map showing your potential sites with the roads also shown in light grey.

What is the total area of potential land you found, in square kilometers?_____

Skills Reference

Turning on Spatial Analyst

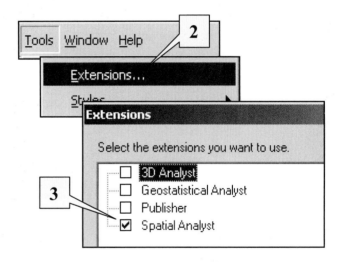

1. Open ArcMap, if necessary.

2. Choose Tools > Extensions from the main menu bar (Fig. 15.41).

3. Check the box to turn on Spatial Analyst.

4. To open the Spatial Analyst toolbar, right-click on the menu and choose the Spatial Analyst menu.

Fig. 15.41. Turning on Spatial Analyst

Managing grids

Spatial Analyst creates many grids, both temporary and permanent. Grids are a complex data format and should ONLY be copied, moved, renamed, or deleted using ArcCatalog. Using Windows Explorer may damage the grids and/or the directories containing the grids.

Copy, move, and delete grids just as you do other data sets in ArcCatalog.

TIP: Grids have much stricter naming conventions than other ArcGIS data files. Grid names must have no more than 13 characters and should contain only letters, numbers, or the underscore character. Pathnames to the grid directory should contain no spaces anywhere.

Setting the analysis environment

All settings in the analysis environment remain in effect until changed, or the session ends.

1. Choose Spatial Analyst > Options from the Spatial Analyst toolbar.

Setting the file creation options

2. Click the General tab (Fig. 15.42a).

3. Set the working directory.

4. Choose a grid as an analysis mask, if desired.

5. Set the coordinate system option desired.

6. Check the box if you want to be warned when projecting grids during analysis.

Setting the extent

7. Click the Extent tab (Fig. 15.42b).

8. Choose the analysis extent as the Intersection or Union of Inputs,

9. OR specify a specific grid or display as the extent, or enter the extent coordinates in the boxes.

10. To make the cells coincide exactly with an existing grid, choose that grid in the Snap extent to box.

Setting the cell size

11. Click the Cell Size tab (Fig. 15.42c).

12. Click the drop-down box to set the cell size method: the maximum or minimum of input grids, to a specific grid, or as specified.

13. To specify a size, type the value in the box, in map units. Press the Tab key to see the number of rows and columns that will result,

14. OR type in the number of rows and columns desired, and press the Tab key to see the cell size that results.

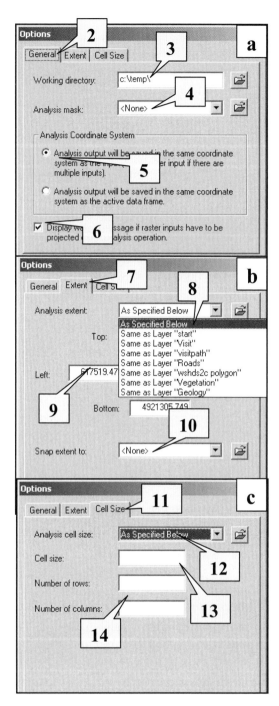

Fig. 15.42. Setting the analysis environment

Converting between grids and features

Converting features to a raster

1. Choose Spatial Analyst > Convert > Convert Features to Raster from the Spatial Analyst toolbar.

2. Choose the layer to convert from the list, or click the Browse button to locate a file on the disk (Fig. 15.43).

3. Choose the field that will become the values in the output grid.

4. Specify the cell size.

5. Specify a location and name for the output grid. Click OK.

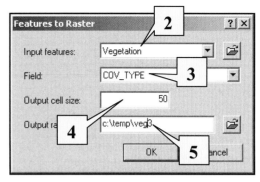

Fig. 15.43. Converting features to a raster

Converting a raster to features

1. Make sure that the raster to convert contains discrete data that will convert well to features.

2. Choose Spatial Analyst > Convert > Raster to Features from the Spatial Analyst toolbar.

3. Select the raster to convert (Fig. 15.44).

4. Choose the field of the raster (usually the Value field) to become the defining attribute for the polygons.

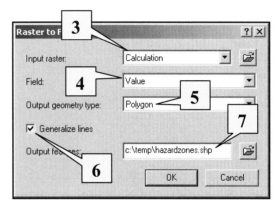

Fig. 15.44. Converting a raster to features

5. Choose the output feature type. Be sure it is appropriate for the type of raster being converted.

6. Check the box to generalize lines. You will usually want this option, which reduces the number of vertices in the output features to a manageable number.

7. Specify the location and name of the output shapefile or geodatabase feature class.

TIP: Grids have much stricter naming conventions than other ArcGIS data files. Grid names must have no more than 13 characters and should contain only letters, numbers, or the underscore character. Pathnames to the grid directory should contain no spaces anywhere.

Reclassifying a grid

1. Choose Spatial Analyst > Reclassify.

2. Choose the grid to reclassify (Fig. 15.45).

3. Choose the field in the grid to reclassify (usually Value).

If the grid has only a few values

4. Click the Unique button to see each value individually.

5. Edit the right-hand column to specify the output values.

If the grid has many values

6. Edit the ranges in the Old values column, and then edit the values in the New values column until the ranges are as desired.

7. Use the Add Entry or Delete Entries button to add or remove lines from the list of ranges.

8. Instead of entering ranges by hand, click the Classify button to open the Classify Values window, exactly like the one described in Chapter 4. Modify the classification until it is satisfactory.

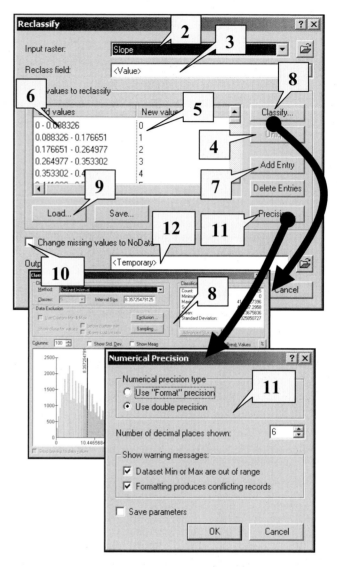

Fig. 15.45. Reclassifying a grid

9. Save or Load previously created classifications, if desired.

10. Check the box to convert missing values to NoData, if desired.

11. If the grid contains floating-point data, click the Precision key to set the format and precision of the output.

12. Specify the name and location of the output grid, or leave as is to create a temporary grid.

TIP: Grid names must have no more than 13 characters and should contain only letters, numbers, or the underscore character. Pathnames to the grid directory should contain no spaces.

Calculating cell statistics

Cell Statistics calculates a specified statistic for each cell in a stack of grids.

1. Choose Spatial Analyst > Cell Statistics from the Spatial Analyst toolbar.

2. Click the layers in the left column and click Add to add them to the stack of layers to be included in the stack (Fig. 15.46).

3. To remove a grid, click on it in the right-hand column and click Remove.

4. To add a grid from the disk, click the Browse button.

5. Specify the statistic to be calculated for the stack.

6. Specify a name and location for the output grid, or leave as is to create a temporary grid.

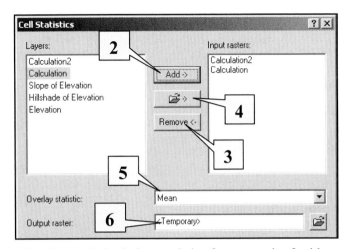

Fig. 15.46. Calculating statistics from a stack of grids

Calculating neighborhood statistics

Neighborhood Statistics utilizes a roving window centered about a target cell. It calculates a statistic for all cells in the window and assigns the value to the target cell in the output grid.

1. Choose Spatial Analyst > Neighborhood Statistics from the Spatial Analyst toolbar.

2. Choose the grid to use (Fig. 15.47).

3. Choose the grid field (usually Value).

4. Choose the statistic type.

5. Choose the neighborhood shape.

6. Specify whether the size is in cells or map units.

7. Specify the neighborhood size. The settings will differ for different shapes.

8. Specify the cell size of the output grid.

9. Specify the name and location of the output grid, or leave as is to create a temporary grid.

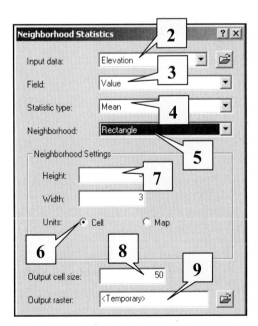

Fig. 15.47. Calculating neighborhood statistics

Calculating zonal statistics

Zonal Statistics calculates statistical measures for cells within zones. Two input grids are required: one defining the zones, and another containing the values to use in calculating the statistics. The output is a table containing a record for each zone and a field for each statistic.

1. Choose Spatial Analyst > Zonal Statistics from the Spatial Analyst toolbar.

2. Select the grid or feature layer which defines the zones (Fig. 15.48).

3. Select the field from the zone layer which defines the zones.

4. Select the grid containing the values to be calculated for the statistics.

5. Check the box to ignore NoData cells when calculating (usually the best option).

6. To use the results right away in the map, check the box to join the output table to the zone layer. You will usually want this option.

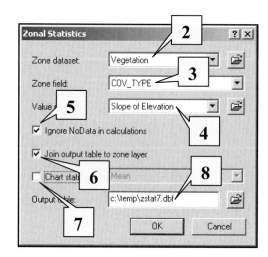

Fig. 15.48. Calculating zonal statistics

7. Check the box to create a chart showing one of the statistics.

8. Specify the name and location of the output table.

Using surface functions

The surface functions contour, slope, aspect, and hillshade all produce output grids based on an input surface and a few arguments. The general procedure is as follows:

1. Choose Spatial Analyst > Surface Analysis > *function*, where *function* is the name of the desired operation.

2. Choose the surface to use as input.

3. Specify the specific parameters for each function as described in the following paragraphs.

4. Enter the cell size for the output grid (except for contouring).

5. Specify the name and location of the output grid, or leave as is to create a temporary grid. (If you are contouring, the output will be a shapefile.)

To **Contour,** specify the base contour value and the contour interval. For example, a base of 10 and an interval of 20 will give the contours 10, 30, 50, etc. You can also specify a Z factor to multiply with the surface before contouring, for example, to create foot contours on a meter surface.

For **Slope,** enter a Z factor if desired, and specify degrees or percent for the output units.

For **Aspect,** there are no additional parameters. The resulting grid contains values between 0 and 360 degrees representing the aspect, with 0 = 360 = North. A value of -1 indicates a flat area with no aspect.

For **Hillshade,** change the default values for the azimuth and altitude of the illumination source if desired. The azimuth is given in degrees between 0 and 360 with 0 = 360 = North. The altitude is given in degrees between 0 (horizontal) and 90 (vertical). A Z factor > 1 may be specified, which will exaggerate the topographic relief. A Z factor < 1 will subdue the topographic relief. You may also check the box to model shadows. If you choose this option, cells in the shadow of another will be set to 0, so that later you can extract these to create a map of shadows. The default option calculates the illumination of all cells, including shadowed ones. The visual difference between the two options is negligible.

For **Cut/Fill**, specify the before and after surfaces. Adjust the Z factor, output cell size, and output raster if needed. The output grid will have negative values for areas that have been removed, and positive values where material has been added.

Calculating a viewshed

A viewshed shows the areas visible from a point or set of points. Please note that this function takes a substantial amount of processing time in comparison with the other surface functions. Limit the size of the input grid and the number of observation points to the lowest possible values to expedite the processing.

1. Create a feature layer which contains the observation points. If desired, this layer can contain attributes describing additional viewing factors, such as the height of the observer and the angle of view. For more information on these parameters, consult the Help document for Spatial Analyst.

2. Choose Spatial Analyst > Surface Analysis > Viewshed from the Spatial Analyst menu.

3. Specify the elevation surface used as input (Fig. 15.49).

4. Specify the point layer containing the observation points.

5. Check the box to use the curvature of the earth if desired in calculating the viewshed.

6. Enter a Z factor, if desired.

7. Enter the cell size of the output grid.

8. Enter the name and location of the output grid, or leave as is to create a temporary grid.

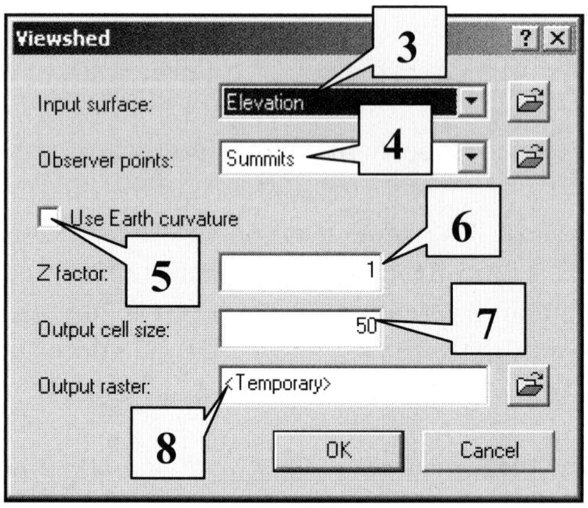

Fig. 15.49. Creating a viewshed grid

Creating a density map

1. Choose Spatial Analyst > Density from the Spatial Analyst toolbar.

2. Specify the point layer to be analyzed (Fig. 15.50).

3. Choose an attribute of the point layer, such as population, on which the density will be based. To simply count the number of points, choose <None>.

4. Choose Simple to count values in the search radius, or Kernel to weight the center values more. Kernel gives a smoother result.

5. Type in the search radius, in map units. This value specifies a circle inside which points are counted.

6. Enter the area units in which the output values will be reported, such as people/sq mile or people/acre.

7. Enter the cell size of the output grid, in map units.

8. Specify the name and location of the output grid, or leave as is to create a temporary grid.

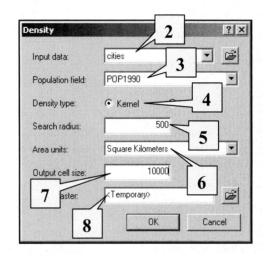

Fig. 15.50. Creating a surface density map

TIP: When calculating density maps, you must have reliable knowledge about the output extent and the map units in order to specify valid search and cell size values.

Interpolating between points

1. Choose Spatial Analyst > Interpolate to Raster > *Method* from the Spatial Analyst menu, where *Method* is Inverse Distance Weighted, Kriging, or Spline.

2. For Inverse Distance Weighted, choose the point layer containing the data to interpolate (Fig. 15.51).

3. Enter the field containing the values to be interpolated (such as precipitation).

4. Enter the power law for the distance weighting. Entering 1 will give a linear weighting, in which the influence of each point decreases linearly with distance. Entering 2 causes the influence to decrease as a square of the distance, and so on. The larger the number, the smaller the influence of each point on its surroundings.

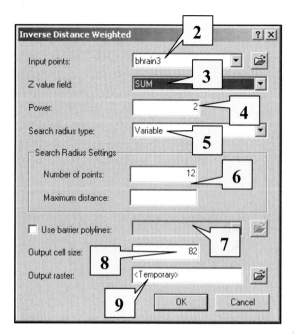

Fig. 15.51. Interpolation using IDW

5. Enter the type of search radius.

A variable radius searches for a specified number of closest points, with an optional requirement to stop searching outside a specified maximum distance. A fixed search radius uses all points within a specified distance, and it has an optional requirement to limit the points used to a specified maximum number.

6. Enter the values for the search distance and the number of points.

7. You can specify a line layer which places linear barriers between points to keep them from being used, such as using watershed boundaries to prevent interpolation from points outside the watershed.

8. Specify the cell size of the output grid.

9. Specify the name and location of the output grid, or leave as is to create a temporary grid.

Finding a least-cost path between two points

Step 1: Preparing the input layers

1. Create a shapefile containing a single point representing the start location.

2. Create a shapefile containing a single point representing the destination.

3. Create a grid representing the cost of moving across the surface.

This cost grid could be a single layer, such as slope, or it could combine rankings from several different grids. If multiple grids are used, ensure that each input is appropriately scaled to the others. You can also apply different weights when adding the grids. The final result is the cost grid.

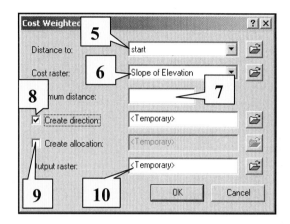

Calculating the cost distance and direction grids

Step 2. Creating the cost distance grids

4. Choose Spatial Analyst > Distance > Cost Weighted from the Spatial Analyst toolbar.

5. Enter the layer containing the start location (Fig. 15.52).

6. Specify the name of the cost grid.

7. Specify a maximum distance, if desired.

8. Check the box to create a direction grid, which is required to create the path.

9. An allocation grid is not required, but check the box if one is desired.

10. Enter the name and location of the output cost distance grid, or leave it as is for a temporary grid.

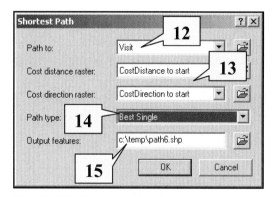

Fig. 15.52. Deriving the shortest path

Step 3. Finding the least-cost path

11. Choose Spatial Analyst > Distance > Shortest Path from the Spatial Analyst toolbar.

12. Specify the point layer containing the destination.

13. Specify the cost distance and cost direction grids created in step 2.

14. Choose to find the Best Single path.

15. Specify the name and location of the output shapefile.

Glossary

address standardization conversion of a street address into specific components such as house number, name, and type

adjust to transform the coordinates of one map layer to match those of another

alias an alternate name displayed for a field in a table which does not have to follow length and character restrictions of the actual names

analysis cell size the default cell size defined for output grids during a raster analysis session

analysis extent the default extent defined for output grids during a raster analysis session

analysis mask a raster grid applied during a raster analysis session to convert all NoData cells in the mask to NoData cells in the output grid

annotation labels created from map features and stored separately as permanent objects for detailed editing

ASCII international code used to store simple text letters, numbers, and symbols; synonymous with a "text file"

aspatial data data entries that are not tied or only incidentally tied to a location on the earth's surface

aspect the direction of steepest slope at a location on a surface, that is, the downhill direction

attribute queries pulling out specific records of information from a file based on values in the records

attribute tables tables containing rows of information for each feature in a spatial data file

attributes information about map features stored in columns of a table

automatic scaling fitting the contents of a data frame within a specified rectangle on the page

azimuth measurement of compass direction in 360 degrees with 0 being north and 180 being south

barriers objects which prevent flow or travel over a surface

batch mode automatic processing of multiple data files set up beforehand and then left to run in sequence

binary numeric data stored in base 2 as a series of ones and zeros; a computer's native data storage format

bookmarks links to a particular map area and scale for quick access

Boolean data limited to true or false values; a grid containing only true or false values; functions designed to work with them

buffers a delineated area within a specified distance of a feature; buffers may be created from labels, points, lines, or polygons

byte a unit of data storage in base 2 containing eight zeroes or ones and holding a number from 0 to 255

CAD drawings data files derived from computer-assisted drawing programs

capture using the Alt-PrntScrn keys to place a copy of the active window on the clipboard for pasting into another program

cardinality the relationship between records in two tables: one-to-one, one-to-many, many-to-one, or many-to-many

categorical data data stored as names or categories rather than as numeric quantities; examples are land use or geology

cell size the dimensions of one square of map information in a raster data set

cell statistics statistics calculated for a single map cell from a stack of grids containing the cell

central meridian the central longitude of a map projection, for which the x coordinate equals zero

chart a graph created in ArcMap from table data, such as bar charts or pie charts

chart maps a map showing several different attributes in chart form, with one chart for each feature

choropleth maps a map in which each feature such as a state is colored according to the values in a data field, such as population

class breaks break points used to classify values in a data field into a specified number of groups

classify to assign features to two or more groups based on numeric values in an attribute field

clip to remove features and portions of features which lie outside of the features of another layer

coded domain a rule which permits only certain specific values to be assigned to an attribute, such as land use codes

complex edges a network entity composed of multiple linear features but which behaves as a single linear feature in the network

conflict detection determining which labels will overlap each other when features are being labeled

conic projection a map projection derived by projecting latitude-longitude values on a paper cone covering a sphere

connection 1. A link in ArcCatalog and ArcMap which points to a folder with GIS data; serves as a shortcut to frequently used folders. 2. A link to a DBMS allowing data to be transferred.

connectivity a property of linear features when they are connected to each other via junctions

context menus computer menus that appear when certain objects on the screen are clicked, usually with the right mouse button

continuous data that take on a variety of different values and that change rapidly across a data set (elevation, for example)

contour a line indicating a constant value of a quantity on a surface, such as an elevation contour at 2000 ft

coordinate pair a single pair of x and y values indicating position in a planar coordinate system

coordinate space the range of x and y values onto which maps are plotted

coordinate system 1. A specified range and units of x-y values onto which a map is plotted. 2. The complete definition of the coordinate space used by a map layer, including the ellipsoid, datum, and projection.

coverages the spatial data format created for and used by ArcInfo

cylindrical projection a map projection derived by projecting spherical latitude-longitude values onto a cylinder wrapped about a sphere

dangle a line feature which fails to connect to another line feature, leaving a gap

data depth the number of bytes used to store a number; the more bytes used, the larger the number that can be stored

data frames containers in ArcMap which hold layers that are displayed and analyzed together; a map view

data view data frame mode optimized for the display and analysis of map data

datum a combination of an earth ellipsoid and a local reference point to reduce mapping discrepancies for a particular region

dBase a database program whose file format has been adopted for the shapefile data model and tables in ArcGIS

default automatic input values assumed by a program when no values have been given

Define Projection Wizard a set of menus to guide the user through the task of assigning a coordinate system to a spatial data layer

defined interval a classification method in which the user specifies a specific size range for all the classes

definition query setting a map layer to display only the features whose attributes meet specific criteria

deflection causing an editing sketch to branch off from a previous segment or line at a specified angle

degrees measurement units used in the spherical coordinate system; there are 360 degrees in a great circle

DEM see digital elevation model

destination table the table which receives additional data from another table during a join operation

digital elevation model (DEM) a raster array of values representing elevations at the earth's surface

digital raster graphic (DRG) a scanned image of a USGS topographic map

digitize to convert shapes on a paper map to a digital map layer by entering or tracing vertices

discrete describes data which take on a relatively small number of distinct values

dissolve to combine features together when they share the same value of an attribute

distance join combining the information from two feature tables based on the features which lie closest to each other

division units the units in which a map scale bar is measured and drawn, such as miles or kilometers

division value the length of one section of a scale bar as given in the division units, such as 100 km

divisions the number of sections given to a scale bar

domains rules which determine the values which may be entered into an attribute

dot density map a map representing attribute values by a proportional number of randomly placed dots

double-precision a numeric value stored using 16 bytes of information

DRG see digital raster graphic

dynamic labels labels determined from an attribute and placed on a map automatically each time features are drawn and redrawn

edge a line feature which participates in a geometric network

edge snapping ensuring that new features are automatically connected to the edges of existing lines or polygons

ellipsoid a spheroidal volume with unequal axes, used to approximate the shape of the earth in map projections

end snapping ensuring that new features are automatically connected to the ends of existing line features

enterprise GIS a long-term GIS project developed by a large organization and involving many people over a long period of time

equal interval a classification method in which the user specifies a number of classes which have equal size ranges

event layer a map layer of points created from a series of coordinate pairs in a table

export to create a new data file from all or a subset of features in an existing one

expression a statement containing field names, values, and/or functions used to extract records in a query or calculate values in a table

extent the range of *x-y* coordinates displayed in a map or stored in a data layer

extent rectangle the range of *x-y* coordinates occupied by the features in a data layer

false easting an arbitrary *x*-coordinate translation applied to a map projection, usually to ensure that all values are positive

false northing an arbitrary *y*-coordinate translation applied to a map projection, usually to ensure that all values are positive

feature a spatial object composed of one or more *x-y* coordinate pairs and having one or more attributes in a single record of an associated table

feature class a set of similar objects with the same attributes stored together in a spatial data file

feature datasets a set of feature classes in a geodatabase which share a common coordinate system and can participate in networks and topology

feature service an Internet map layer in which the actual features may be downloaded by the user and saved locally

feature weight the priority assigned to a layer when determining which features may be drawn on top of others

FID feature ID, a unique number assigned to every feature in a spatial data file and used for identification and tracking

field a single column of information in a data table

field definition parameters specified when creating an attribute field, including the type of data, the field length, precision, and scale

field length the maximum number of characters which can be stored in a text attribute field

filter a moving window applied to a raster data layer which calculates new values for the center pixel based on some function of the values in the window

fixed extent a constraint applied to a data frame to prevent changes in scale or the extent in layout mode

fixed scale a constraint applied to a data frame to prevent the scale of the map from being changed in layout mode

flag a marker showing a point of interest in a network, such as a start or end point for a path, or a stop along a route

flipping lines swapping the start and end nodes of a line feature so that it goes in the opposite direction

focal function a raster analysis which determines new values for a target cell based on some function derived from values in a moving window around the target

GCS see geographic coordinate system

geocoding the matching of a location stored in a table to a spatial point feature based on a reference spatial data layer; most often applies to converting addresses to locations

geodatabases a new data model developed for ArcGIS 8 which employs recent database technology for storage and implements rules and topology

geographic coordinate system (GCS) a spherical coordinate system of degrees latitude and longitude used to locate features on the earth's surface

geoid the shape of the earth as defined by mean sea level and affected by topographic and gravitational factors

geometric network a system of linear edges and point junctions used to model the flow of commodities such as traffic or utilities

geoprocessing analysis of spatial data layers such as dissolving, intersecting, and merging

georeferenced describes a spatial data layer that is tied to a specific location on the earth's surface for display with other data

global functions raster analysis functions which require knowledge of the entire grid in order to calculate new values at a target cell

graduated color map a map which divides numeric data from a polygon feature class into classes based on value and displays the classes with different colors

graduated symbol map a map which divides numeric data from a line or point feature class into classes based on value and displays the classes with different size or thickness of symbols

grid 1. Specific raster format native to GRID and Spatial Analyst, and required for raster analysis. 2. Generic term for a raster data set.

ground control points a set of points which matches easily identifiable locations on two different data layers to enable georeferencing of one layer to another

hillshade a raster which displays the brightness variation of a surface as if it were illuminated by a light source at a specified azimuth and zenith angle

histogram a graph showing the number of pixels contained for each data value in a raster

image a raster data layer, usually referring to a raster which displays brightness values as in a photograph

image service an Internet map layer available for viewing but which cannot be downloaded for local storage by the user

INFO an early database system upon which the Arc/Info software data model was based

integer a whole number without any decimal values

interpolate to calculate values at locations between known measurements; to populate a raster with values extrapolated from a known set of point values

intersect to overlay two spatial data layers and find the areas common to both, while discarding areas unique to either

interval data values which follow a regular scale but have no natural zero point, such as degrees Celsius or pH

join the temporary association of data between two tables based on a common attribute field or a location

junction a point connecting two linear edges of a geometric network

kernel a moving window of values applied to a raster to calculate new values at a target in the middle of the window. See also *filter*.

key an attribute field that is used to extract or match records in a table

label weights a priority rating assigned to values to determine which ones will be placed in case of an overlap

latitude a spherical unit measuring angular distance north or south from the equator

latitude of origin the reference latitude of a map projection where the *y* value is zero

layer a generic term representing a single file or feature class of related objects, such as roads or counties

layer file a file which stores a pointer to spatial data along with information on how to display it

layout view a mode of ArcMap used to design and create a printed map; it allows manipulation of map layers, titles, scale bars, north arrows, and more

legend a map element which displays the names and symbols used to portray layers on a map

linear regression calculation of a best-fit line showing the relationship between two variables

local functions raster analysis functions which require knowledge only of a single cell to calculate a new value

logical expression a statement composed of field names, operators, and values which specifies criteria used to select records or values from a layer or table

logical network information stored in separate tables which keep track of relationships between elements of a network and allow the network to operate

logistic regression a multilinear regression analysis in which the dependent variable takes only true or false values, or absence-presence values

longitude a spherical unit measuring angular distance east or west from the Prime Meridian

loops a connected circular path in a network in which flow is indeterminate

map algebra system which permits calculations and operations on entire raster arrays, such as adding two rasters together

map elements objects placed on a map layout, such as titles, legends, scale bars, north arrows, images, and charts

map overlay to combine two spatial data layers, either for display or to evaluate the relationships between them

map scale the ratio of feature size on a map to its size on the ground

map topology temporary spatial relationships developed between features during editing in ArcMap to facilitate editing of features with common nodes or boundaries

map units the units in which a map is stored or into which it has been projected for display (usually feet, meters, or degrees)

mask a raster layer applied during analysis to nullify unwanted cells, such as those outside a study area boundary

merge 1. To combine two or more map features into one feature. 2. To combine two or more features based on whether they share a common attribute value. 3. To combine two or more data layers into a single layer

metadata information stored about data to document its source, history, management, uses, and more

minimum candidate score the lowest score at which an address may be considered as a candidate to match an address during geocoding

minimum match score the lowest score at which a candidate will be automatically matched to an address

model a sequence of steps or calculations used to convert raw data into useful information; a scheme used to understand and predict processes in the real world based on the manipulation of data

modify an editing task which rearranges the vertices of a feature to give it a new shape

multipart feature a single feature composed of nontouching units, such as a single state feature composed of the seven Hawaiian islands

multiple attribute map a map which can display more than one attribute at a time using both color shades and marker symbols

multivariate regression an analysis used to discern predictive relationships between multiple variables

NAD see North American Datum

natural breaks a data classification scheme which divides values based on natural groupings or gaps between values

neatline a line used to enclose one or more map elements in a rectangle

neighborhood statistics calculation of a value for a target cell or feature based on surrounding values from a defined region

network weights values associated with network elements that indicate the "cost" to traverse the element

networks an association of linear edges and connecting points, used to model the flow of a commodity such as traffic or utilities

NoData a special value used to designate that a data value is absent or unknown

nodes the beginning and end points of a line feature

nominal data values which record a name or category of an object, such as a street name or land use code

normalizing data to divide the values of an attribute field by the total of the field or by the values in another field

North American Datum (NAD) a combination of a spheroid and reference point used to minimize map distortion in North America

numeric data values stored as numbers rather than as names or categories

oblique cylindrical projection a map projection in which locations on a sphere are projected to a cylinder of paper at an arbitrary angle

ordinal a data value which indicates a rank or ordering system

origin the (0,0) point of a coordinate system

orthographic projection a map projection in which locations on a sphere are projected onto a flat piece of paper which is tangent to or intersects the sphere

orthophoto an aerial photograph which has been geometrically corrected to match a map base

palette a collection of symbols that are stored and used together

pan to move the display window to another part of the map without changing the map scale

parameters specific values associated with map projections which define how it appears

parametric arc a line feature composed of a smooth curve derived from a given radius for each segment

pathname a list of the folders that must be traversed to locate a particular file, such as c:\mgisdata\usa\states.shp

pixel a square data element in a raster corresponding to one value representing conditions on the ground

planar topology an association of feature classes in a feature dataset, established by rules regarding the spatial relationships between features, such as not overlapping each other

point a one-dimensional feature defined by a single *x-y* coordinate pair

polygon a closed two-dimensional enclosed area feature defined by three or more x-y coordinate pairs

precision the number of digits allotted to store a numeric value

Prime Meridian the line of zero longitude on the earth, passing through Greenwich, England

Project Wizard a set of menus to assist the user in projecting data from one coordinate system to another

projection a mathematical transformation which converts spherical units of latitude and longitude to a planar x-y coordinate system

project-oriented GIS a GIS project with fairly limited objectives and a finite life span

proportional symbol map a map which displays attribute values with marker or line symbols proportional in size relative to the value of the feature

pyramids a set of rasters with different resolutions calculated from a raster to speed display at smaller scales

quantile map a map type in which each class has approximately the same number of features

quantities data numeric attribute data

query the extraction of records from a database according to a specified set of criteria

range domains a rule which stipulates the largest and smallest values which can be stored in a particular attribute

raster a data set composed of an array of numeric values, each of which represents a condition in a square element of ground

Raster Calculator an analysis window which applies algebraic and other operators and functions to grids to create new ones

ratio data data having a regular scale of measurement and a natural zero point, such as precipitation or population

reclassify to replace sets or ranges of values in a grid with different sets or ranges of values

rectify to rotate, resize, or warp an image to match a map base using a selected set of ground control points

reference latitude another common term for the latitude of origin

reference layer a layer containing special attributes needed for geocoding, for which a geocoding index has been built

reference scale the display scale for which a map is designed; when set, the reference scale reduces or enlarges symbols and labels whenever the user zooms out or in on a map

relate a temporary association between two tables based on a common field, whereby fields may be selected based on whether they match selected records in the other table

resolution the ground area represented by one cell value in a raster

RGB composite an image displayed by assigning one band of brightness information to each red, green, and blue color gun in a display monitor

row a horizontal line of information in a table or in a raster array

scale 1. The ratio of the size of features in a map to their size on the ground (e.g., map scale). 2. The number of decimal places allotted to an attribute field for storing numbers

scale range the range of scales for which a data layer will be displayed, set by the user to avoid clutter or the display of layers at inappropriate scales

secant conic projection a map projection in which spherical coordinates are projected onto a cone which intersects the sphere along two great circles

segment the linear entity lying between two adjacent vertices in a line or polygon feature

select to extract one or more features or records from a layer or table in preparation for another operation; to perform a query

selected set the set of features which have been extracted prior to another operation; the result of a query

shapefile the spatial data model developed for and used by ArcView 3 and later versions

shared features features which are linked to or share a boundary with other features

short integer an attribute field definition using up to five bytes to store a binary integer; can store values up to about 62,000

simple edges linear elements in a network which always end at junctions and behave as separate entities

simple join combining two layers according to common locations when a one-to-one spatial relationship exists between the layers

simple labels arbitrary text that is interactively placed on the map by the user; it remains on the map unless deleted

single symbol a map type in which every feature in a layer is displayed with the same symbol

single-precision refers to numeric values stored using eight bytes of information

sink a location on a network which pulls or consumes the commodity flowing through the network

sketch a provisional figure created during editing; when finished it becomes a feature in a data layer

sketch menu a context menu accessed by right-clicking *off* a sketch during editing

slope the drop in elevation for a specified horizontal distance, expressed as an angle or a percentage

snapping ensuring that features within a specified distance are automatically adjusted to meet at exactly the same location; avoids gaps between features

solvers programs which analyze flows or paths through a network

source 1. A location which produces or initiates the flow of a commodity through a network. 2. The spatial data file which provides the features for a map layer.

spaghetti models a model which stores spatial features as a series of x-y coordinates and does not store topological relationships between features

Spatial Analyst a program extension to ArcMap used to analyze raster data

spatial data information that is tied to a specific location on the earth's surface

spatial queries extracting records from a data layer based on its location relative to another data layer

SPCS see State Plane Coordinate System

spelling sensitivity a setting used during geocoding to control the degree to which two names must resemble each other to be considered a good match

spheroid another term for ellipsoid used by ESRI

SQL see Structured Query Language

standalone table a table of information not linked to spatial data features

standard parallels parameters of conic map projections indicating the latitudes at which the cone lies tangent or secant to the sphere

State Plane Coordinate System (SPCS) a group of projections defined for different regions of the United States and designed to minimize map distortions

stereographic projection a map projection in which locations on a sphere are projected onto a flat piece of paper which is tangent to or intersects the sphere

sticky move tolerance a distance defined such that a feature must be moved at least that far for the move to take effect

Structured Query Language (SQL) an established syntax for creating logical expressions used to extract records from a database according to specified criteria

style a collection of map symbols and colors which are stored together and used together

subdivisions the number of units into which a single division of a scale bar is split, usually appearing on the left end of the scale bar

subtypes different categories assigned to features in a map layer; each category has its own symbol and default values

suffix the part of an address which indicates the type of street, such as St, Rd, Ave, etc.

suffix direction a direction appended to the end of a street name to indicate a part of the city, such as Main St *North*

summarized join combining the records of two attribute tables with a one-to-many cardinality by assigning each output record a single value derived statistically from the many input values

surface analysis functions designed for application to rasters representing a three-dimensional surface such as elevation

symbol a shade, line, or marker with specified shape and color used to display map features

tables data stored as an array of rows and columns, with each row representing an object or feature, and each column an attribute or property of that object

tangent conic projection a map projection in which spherical coordinates are projected upon a cone which lies tangent to the sphere along one line of latitude

target layer a layer specified in the Editor toolbar into which all new features are added and all edits are applied

templates a map design which can be saved and applied to many different map documents

thematic mapping displaying the features of a spatial data layer based on values in its attribute table

thematic rasters a raster which contains categorical or nominal data values such as land use codes or soil types

themes terminology in ArcView 3 used to indicate a data layer containing similar features such as roads or states

thumbnail a small snapshot showing the appearance of a data layer and displayed in ArcCatalog to aid the user in finding data

TIN Triangulated Irregular Network; a data model for storing surfaces as triangular facets with varying orientations

topography the three-dimensional surface of the earth

topological model a data model which stores spatial relationships between features in addition to their *x-y* coordinates

topology the spatial relationships between features, such as which are connected or adjacent to each other

trace to follow an existing linear feature, as when analyzing network paths or creating a new feature which follows or is offset from an existing one

trace solvers programs which analyze network flow by tracing paths in the network

transverse cylindrical projection a map projection in which spherical coordinates are converted to locations on a cylinder tangent to the sphere along a line of longitude

union creating a new feature or set of features by combining all the areas from two input layers

unique values a map type in which each attribute value is assigned its own symbol

Universal Transverse Mercator (UTM) a family of map projections defined for 60 zones around the world and based on a transverse cylindrical projection

unprojected data a spatial data layer stored in a geographic coordinate system with units of degrees of latitude and longitude

unsigned binary numerical values stored in base 2 and constrained to be positive. A common image format.

UTM see Universal Transverse Mercator

vector model a spatial data storage method in which features are represented by one or more pairs of x-y coordinate values forming points, lines, or polygons

vertex a point at which the segments of a line or polygon feature change direction

vertex menu a short context menu accessed by clicking *on* the sketch during editing

vertex snapping ensuring that new features are automatically connected to the vertices of existing line features

viewshed the area on a three-dimensional surface which is visible from a specified point or set of points

watershed the entire upstream area contributing to the water discharge at a specified point

wizard a series of menus designed to assist the user in specifying the parameters for a particular task or operation

working directory a specified folder into which raster analysis output is placed by default

zenith angle an angular measure of the distance of an object above the horizon

zonal function raster analysis function which analyzes values from one grid for zones defined by a different grid

zonal statistics raster analysis functions which calculate statistics from one grid for zones defined by a different grid

zone the combined areas of a raster grid which share the same attribute value, such as all areas with the commercial land use code

Selected Answers

Ch. 1 Tutorial

1. 360 miles

2. 210 miles

3. Harding County

4. 7 counties

5. 331,170 people

6. 4 shapefiles, 9 coverages, 3 dbf files, and 1 info file

7. 319 records

8. Agricultural (largest and smallest)

9. Arcs, labels, polygons, and tics

10. 3; Whitewood Creek, Ellsworth Air Force Base, and Williams Pipeline Co. Disposal Pit

Ch. 1 Exercises

1. Method: Use ArcCatalog to examine the geodatabase.

2. GCS_WGS_1984, or GCS_Geographic

3. Method: Examine the fields tab of the shapefile properties.

4. The positional accuracy was originally 80 m before the arcs were generalized, but now it is Unknown.

5. Method: Preview the lake shapefile table and sort on Area.

6. Mississippi

7. Method: Preview the gas_stations table and choose Statistics.

8. dem30_fil, rceast_nw, and TM_24sep98MS

9. Method: Open the raster properties and read the General tab.

10. Vision Questing Area

Ch. 2 Tutorial

1. Lake Oahe

2. Map Tips

3. About 1:244,000

4. About 1:305,000

5. 1:300,000

6. Grandview School and River Bottom School #50

7. Rapid City

8. About 3.7 miles

9. About 6 kilometers

10. GCS_North_American_1983

Ch. 2 Exercises

1. Method: Use Identify.

2. See map.

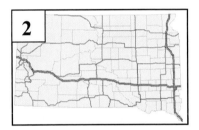

3. See map.

4. See map.

Ch. 2 Exercises, cont.

5.   See map.

6.   Method: Use Measure tool

7.   See map.

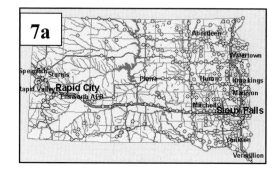

8.   See map next to question #9 in the Exercises section.

9.   See sample map in Exercises section.

10. See map.

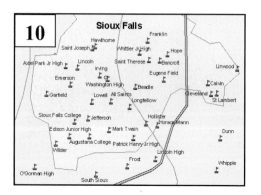

Ch. 3 Tutorial

1.   GCS_WGS_1984

2.   80° 24' 16.67"W and 25° 11' 4.3"N

3.   Decimal degrees

4.   Degrees-minutes-seconds

5.   About –5555, 1740 miles

6.   About -8952490 and 2880205 meters

7.   South America is all negative, Asia all positive.

8.   Kansas

9.   The central meridian is –96° and the latitude of origin is 39°.

10. Secant conic projection, because it has two standard parallels.

11. Same length as the equator, 25,000 miles

12. UTM Zone 11

13. Transverse Mercator (UTM) with NAD 1927, because the frame defaulted to the first data added

14. GCS_North_American_1983

Ch. 3 Exercises

1.   Method: Examine the coordinate system in the properties window or metadata.

2.   USA Contiguous Equidistant Conic. Map units are meters. Display units are miles. Units of the source data are decimal degrees.

3.   Method: Use the Measure tool

4.   Change the display units from miles to kilometers in the data frame General Properties tab. It is about 3772 kilometers.

5.   Method: Change the coordinate system to North America Lambert Conformal Conic and use the Measure tool.

6.   About 3040 miles in Mercator, about 690 miles more

Ch. 3 Execises, cont.

7. See map.

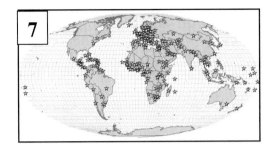

8. Area and distance are well preserved; shape and direction poorly preserved.

9. Method: Use Metadata Viewer to examine coordinate systems.

10. Rcw_oops *x-y* coordinates are in decimal degrees, but coordinate system is set to UTM Zone 13.

Ch. 4 Tutorial

1. A single symbol map

2. Facilities, Fire Fighting, Recreation, Timber, and Warning

3. North Arrows, Colors, Markers, Lines, and Fill symbols

4. Unique values map, using the FUNC_CLASS field

5. 17,942 people

6. GCS_North_American_1983

7. The min, max, and mean of band 4 are 4, 118, and 44.5

8. 240 cells

9. The min and max elevations are 872 and 1956 meters

10. Hillshd2 is a grid. The clues are: it has a unique values classification, it has an attribute table, and it has Fields and Joins & Relates tabs.

Ch. 4 Exercises

1. See map.

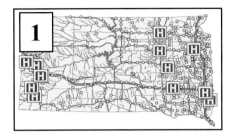

2. See map.

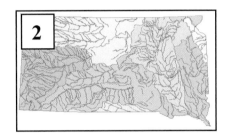

3. See legend.

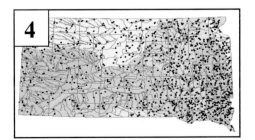

4. See map.

Ch. 4 Exercises, cont.

5. See map.

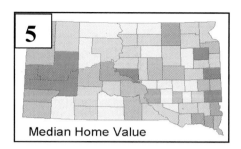

Median Home Value

6. See map.

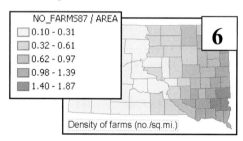

NO_FARMS87 / AREA
- 0.10 - 0.31
- 0.32 - 0.61
- 0.62 - 0.97
- 0.98 - 1.39
- 1.40 - 1.87

Density of farms (no./sq.mi.)

7. See map.

Bureau of Indian Affairs
Bureau of Land Managment
Bureau of Reclamation
Department of Defense
Forest Service
Fish and Wildlife Service
National Park Service

Federal Lands

8. See map.

9. Unit 2 has the largest area, 161.4 sq km, which is 28% .

10. See map.

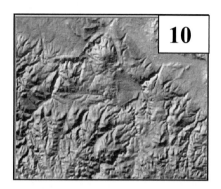

Ch. 5 Tutorial

1. California

2. Wyoming

3. 19 states

4. California, Washington, Texas, and Missouri

5. 33,603,430 people

6. 479,673 people

7. 5,397,009 people

8. 275,247,480 people

9. Yes, most states have fewer than 15 million people.

10. District of Columbia is also listed.

11. The average increased from 4,876,664 in 1990 to 5,397,009 in 2000.

12. Standalone table

13. 31 congressional districts

14. California

15. One-to-one relationship

16. One-to-many relationship

17. Joins not for one-to-many; use relate.

18. States is destination, districts is source.

19. There are 23 New England reps.

20. The Pacific subregion

Ch. 5 Exercises

1. Method: Use Select By Attributes and Statistics, then sort selected ones.

2. New York highest number; Mississippi (or D.C.) highest percentage.

3. Method: Summarize by Sub_Region and sort results.

4. Use the FIPS code.

5. See map.

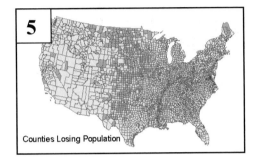

Counties Losing Population

6. Texas 63, Kansas 54, Iowa 46

7. Method: Relate US States to 106th

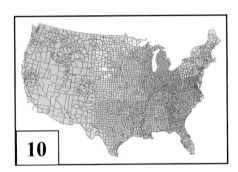

Districts and select states. Activate related records and select Democrats from those in related table.

8. 10,028,123 in state capitals; smallest 8247, largest 983,403; average 200,562.

9. Method: Add new fields and calculate percentages.

10. W N Central has 618; total population 18,910,164.

Ch. 5 Challenge

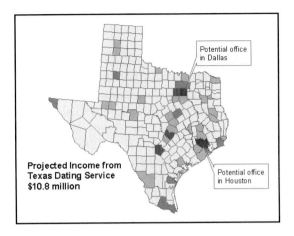

Potential office in Dallas

Potential office in Houston

Projected Income from Texas Dating Service $10.8 million

Ch. 6 Tutorial

1. All layers. This is the default.

2. 22 counties, about 223,000 people

3. 33 counties

4. Los Angeles County

5. 144 counties mostly in the Central Plains states

6. 104 counties

7. About 32% (1010 out of 3141)

8. Hughes County contains Pierre, SD.

9. About 89% (2798 out of 3149)

10. Brazos, Canadian, Pecos, Red, and Rio Grande

11. The Flathead and the Salt

12. 14 target counties containing 552,391 people

Ch. 6 Exercises

1. Method: Select by Attributes and use Statistics.

2. 566 counties out of 3141, or 18%

3. Method: Use Select by Attributes and do queries in steps.

4. 1010 out of 3141 counties, or 32%. 100,401,183 out of 275,247,480 people, or 36%.

Ch 6 Exercises, cont.

5. Method: Use Select by Location first, then Select by Attributes.

6. 109 cities, about 3.86 million people

7. Method: Select Crater Lake, then use Select By Location twice, once for volcanoes and once for interstates.

8. 91 cities. See map.

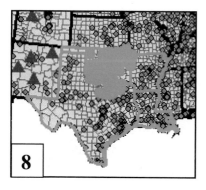

9. Method: Query for areas within 50 miles and then switch the selection.

10. Maps will vary. See picture in text for example.

Ch. 7 Tutorial

1. The FID_1 and FID_2 fields

2. Pennington 90, Custer 33, Lawrence 23

3. St. Joseph Hospital serves most counties (11), McKennan Hospital serves most people (~145,000), and Redfield State Hospital serves greatest area (12,500 sq mi).

4. The distance units are decimal degrees.

5. NAD_1983_State_Plane_South_Dakota _North

6. Meters

7. Campbell County

8. Lake Oahe received NO attributes, because it did not fall completely inside a county.

9. The Lower James, nine urban areas

10. The Lower Big Sioux, about 132,000 people

Ch. 7 Exercises

1. Method: Join the gas stations to the watersheds.

2. The Medium and Low Density residential areas contain nine gas stations.

3. Method: Join gas stations to land use.

4. The Circle S Convenient, 2143 meters away from Valley Sports Bar and Grille.

5. Method: Join gas stations to restaurants.

6. The Red Rock Canyon watershed, 503 units

7. The map should be similar to Figure 8.7.

8. Mission, SD, about 229 km

9. Method: Join towns to hospitals.

10. Day County, 16 lakes. Four largest are Day, Roberts, Hamlin, and Kingsbury.

Ch. 7 Challenge

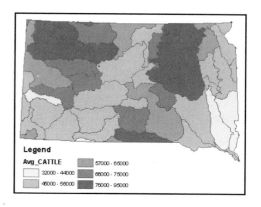

Ch. 8 Tutorial

1. ALL, NO-FID, ONLY-FID. NO-FID is best.

2. 4,859,948 sq meters

3. 31,963,508 sq meters

4. About 15%

Ch. 8 Exercises

1. See map.

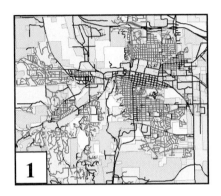

2. See map.

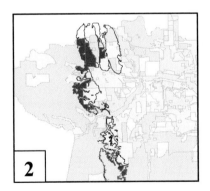

3. Method: Update the area field using the Field Calculator.

4. 5.6 sq kilometers

5. Method: Intersect streams with geology, then update the area field with the Field Calculator.

6. See map.

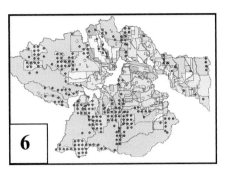

7. See map.

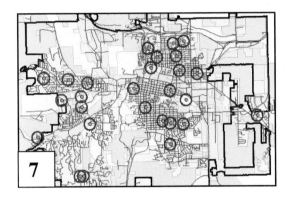

8. See map.

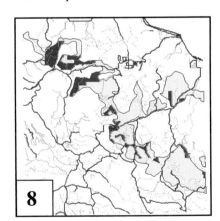

Ch. 8 Exercises, cont.

9. See map.

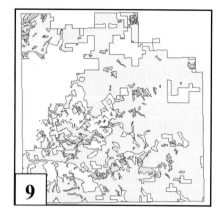

10. See map.

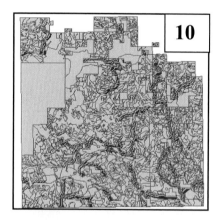

Ch. 8 Challenge

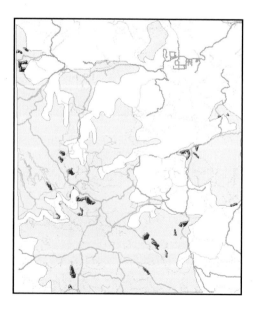

Ch. 9 Tutorial

1. 0.1 miles

2. Blue dashed line and handles around them

3. Click it with the Black Arrow tool to select it, and press Delete.

4. Only one set of bars in the graph

Ch. 9 Exercises

Answers will vary.

Ch. 10 Tutorial

1. Prefix direction = PREFIX, Prefix type = None, Street name = STREET, Street numbers = FRADD and TOADD, Street type = SUFFIX, Street direction = SUFFIX2

2. One Range

3. The ADDRESS field

4. Yes there are intersections. The connector is the & character.

5. The rds_clp shapefile in mgisdata\rapidcity

6. Shapefile

7. 74 good matches

8. Six restaurants are unmatched.

9. The problems with the addresses are:
   1301 Omaha St; prefix W omitted.
   710 Meridian Ln; Dr, not Ln.
   2250 Haines; Ave suffix omitted.

10. US Cities with State

11. CITY_NAME and STATE_NAME

12. 209 matched and 64 unmatched

13. There is a period after the state abbreviation.

14. First, geocoding can resolve ambiguities when two cities have the same name in two states. Interactive matching can pair up additional cities with slightly different names.

Ch. 10 Exercises

1. See map.

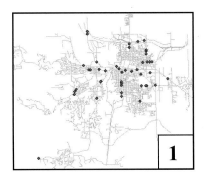

2. See map.

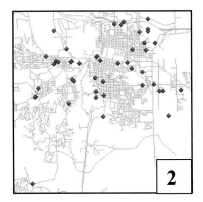

3. Method: Export the lat-lon stations and join to geocoded stations.

4. See map.

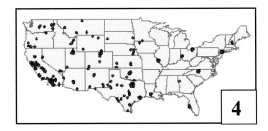

Ch. 11 Tutorial

1. UTM Zone 13 NAD 1927

2. Angle -115

3. Map units, in this case meters

4. The buildings don't need to be snapped.

Ch. 11 Exercises

Answers will vary, but should look something like this:

Ch. 11 Challenge

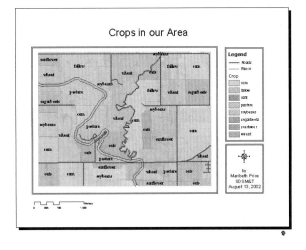

Ch. 12 Tutorial

1. Vertex snapping on for buildings and parcels

2. North to south; red (last) vertex is on southern end.

3. Still one polygon selected

4. Multipart polygon; if you select one, both are highlighted.

5. Area inside the "hole" remained.

6. Single, multipart feature.

Ch. 12 Exercises

7. Method: Use the Shared Edit tool and a variety of moving, modifying, and reshaping to match the lines to the street centers.

8. See map.

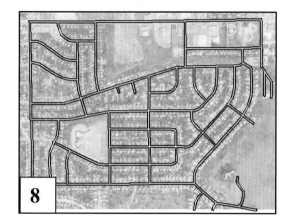

9. Answers will vary.

Ch. 12 Challenge

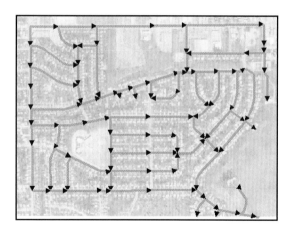

Ch. 13 Tutorial

1. NAD 1927 UTM Zone 13N

2. NAD 1927 UTM Zone 13N

3. GCS North American 1983

4. The RoadType, Direction, and StreetAbbrev would have coded domains. The NumLanes would have either a coded or a range domain.

5. Parcel_ID, Area, Zoning, and ParcelValu

6. The Parcel_ID field should not have a domain, because it would serve little purpose for such a variable field.

7. Give Zoning field a coded value domain; give Area and Parcelvalu fields range domains.

8. Give Zoning and ID fields the Duplicate policy; give Area and Parcelvalu fields the Geometry Ratio policy.

9. Give Zoning and ID fields the Default merge policy. Give Area and Parcelvalu fields the Sum merge policy.

Ch. 13 Exercises

1-4. See figure.

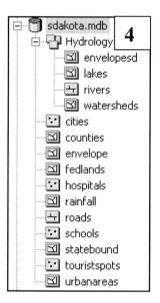

Ch 13 Exercises, cont.

5. See screen captures.

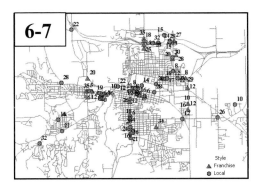

5a

5b

6-7. See layout.

6-7

Ch. 14 Tutorial

1. The Transportation network contains a distance weight.

2. Endcaps, Galleries, Tvalves, and Water_Net_Junctions are simple junctions, Waterlines are simple edges.

3. Distance and Pressure Drop

4. About 19 edges

5. About 3900 meters

6. About 4700 meters

7. 12 differently named streets

8. Black circle

9. 17 edges

10. About 3600 meters

11. About 53 kilometers

Ch. 14 Exercises

1. See map.

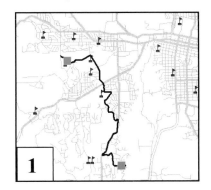

2. About 7400 meters

3. See map.

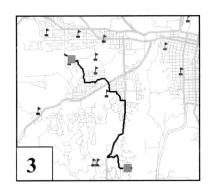

Chapter 14 Exercises, cont.

4. See map

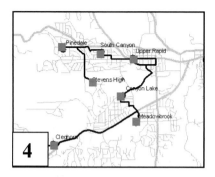

5. See map.

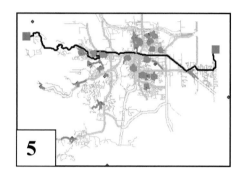

6. See map.

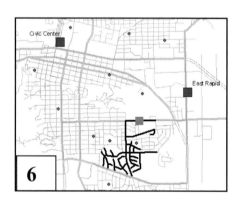

7. Method: Flag Civic Center gallery and barrier at start of north line. Trace with selection as result to get length. Move barrier to south to get north length.

8. Civic Center 17 km; East Rapid 10 km

9. Method: Find Upstream Accumulation from end and count edges (same as junctions).

10. Joe had set the analysis options to return only junctions; Mary was returning both edges and junctions. The cost in both cases is the same, 25.

Ch. 14 Challenge

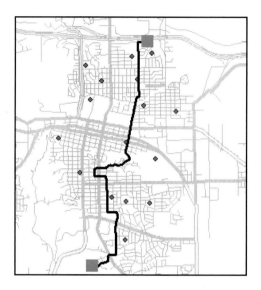

Ch. 15 Tutorial

1. About 46%

2. 2164 cells × 50 m × 50 m/ 1 million sq meter/sq km, or about 5.4 sq km

3. About 70 degrees

4. North is red, south is light blue.

5. 13 watersheds

6. 1372 meters

7. Floating-point grid; the precip values have decimals.

8. The Stockade Beaver watershed, 46.4 million cubic meters

9. Lowest 1023 meters; highest 1858 meters

Ch. 15 Exercises

1. Method: Reclassify the DEM.

2. See map.

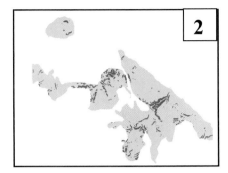

3. See map.

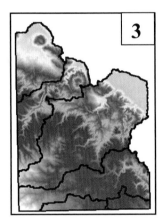

4. About 1844 meters

5. Method: Use Zonal Statistics with Summits as the value layer and the distance to streams from Exercise 3 as the value layer.

6. See map.

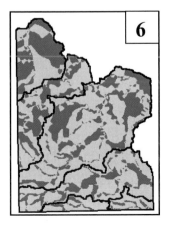

7. About 4.6 sq km

8. Madison Limestone highest average slope (14.2) ; Upper Mesozoic lowest (3.4).

9. See map.

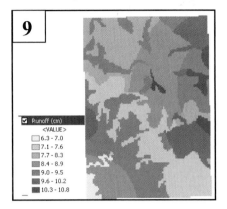

10. See map.

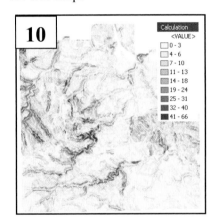

Ch. 15 Challenge

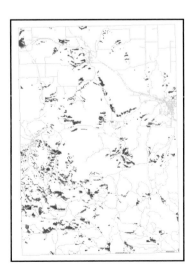

Index

A

Absolute pathnames, 64, 65
Absolute X,Y option, 413, 438
Abstract information, 50
Access format, 25, 179, 461
Accumulation (network), 505, 518, 527
Accuracy of data, 8
Active frame, 68
Add ArcIMS Server feature, 57
Add Data button (ArcMap)
    to add standalone tables, 203
    basic functions, 66
    demonstration of, 80, 93
    using data frames, 78
Add Field command, 47, 211
Addresses, geocoding, 349–53
Address locators
    selecting, 363, 376
    setup, 356–58, 362–63, 368, 375
Address standardization, 351–52, 365–66, 378
Add Text tool, 99
Add to Current Selection option, 223, 230
Add Vertex command, 398
Add XY Data function, 372–73, 379
Adjust Division Value (scale bar), 310
Adjust Number of Divisions (scale bar), 310, 318
Adjust Width (scale bar), 310
Advanced button (Join Data dialog box), 213
Advanced Options menus (graphs), 343
Advanced tab (ArcCatalog searches), 61
.aig files, 360, 366
Aliases
    defined, 178
    for duplicate field names, 264
    geocoding with, 357, 364
    for readability, 197, 204
    sizing columns with, 191
Alias tables, 357, 364
All Other values class, 156, 172
Alt-PrntScrn key combination, 55, 90, 193
AML, 20
Analysis menu, Utility Network Analyst toolbar, 501, 502
Analysis Options dialog box, 523, 524
AND operator, 219
Angle box, 413
Angles, for sketching lines, 400, 413, 414–15
Annotation
    basic functions, 73–74
    demonstration of, 84–86, 103
    reference scale, 101
Append operations, 280–81, 298
Apply, after adding individual domains, 470, 482
ArcCatalog
    assigning and modifying coordinate systems, 136–37
    basic features, 20, 32–33, 44–48
    display options, 57
    downloading Internet data, 52–53, 57
    examining coordinate systems in, 120–21, 129
    layer properties, 48–50, 60
    metadata handling, 30, 50–51
    options for starting, 56
    searching for data, 53–54, 60–61
ArcEditor, 21, 382
ArcEditor license, 34
ArcEdit program, 20, 382
ArcGIS, 19–21, 37–43
ArcGIS Desktop, 20–21
ArcInfo, 21, 382
Arc/Info, 3, 20
Arc Macro Language (AML), 20
ArcMap
    adding geodatabase layers from, 479–80
    data frame, layer, and symbol tools, 68–69, 78–80, 98–99
    display options, 67–68, 76–78, 94–96
    editing tools (*see* Editing)
    field editing with, 187
    graphical user interface, 30
    help system, 68
    identifying, finding, and measuring features, 66, 97–98
    labeling tools and concepts, 72–74, 80–88, 99–103
    major functions, 20
    map document features, 63–65
    map scale concepts, 70–71
    map types and features, 139–43
    numeric data classification, 143–46
    options for starting, 56, 92
    saving documents, 38, 93
    using projections with, 113–14, 121–25, 128–29
    Utility Network Analyst toolbar, 501–2, 522
    viewing search results in, 54, 62
    windows and menus, 66–68, 92
ArcPlot program, 20
Arc program, 20
Arc Spatial Database Engine (ArcSDE) software, 461
Arc tool, 421, 422, 430, 449
ArcToolbox
    basic features, 34
    creating practice shapefiles, 391–92
    major functions, 21
    map projection tools, 115–16
    options for starting, 56
    using, 116–17, 295
ArcView
    basic functions, 21
    editing tools, 382, 469, 470
    introduction of, 20
ArcView license, 34
Area, distortion of, 110
Area fields, 282, 291–92, 300
Arrow at End symbol, 437, 453
ASCII format, 180
Aspatial data, 1
Aspect calculations, 536, 553–54, 571
Attribute data. *See also* Queries
    assigning symbols to, 42
    in coverages, 29
    creating maps based on, 155–58, 172
    default values in geodatabases, 464–65, 475–76, 496
    defined, 21
    domains, 465–66, 481–83, 495 (*see also* Domains)
    editing in ArcMap, 387–88, 406–7, 418
    of grids, 166, 531–32, 549, 550
    labeling features with, 100
    metadata display, 50–51
    in shapefiles, 28
    showing with Identify tool, 38
    in vector data model, 24, 531
Attribute Editor
    basic features, 187, 387–88
    skills reference, 418
    tutorial exercises, 406–7
    using default values with, 475–76
Attribute indices, 48
Attribute queries, 217. *See also* Select By Attributes
Attributes button, on Editor toolbar (ArcMap), 381–82
Attribute tables. *See also* Tables
    adding and removing, 203
    appending, 280
    basic features, 177–78
    in coverages, 29
    editing in ArcMap, 392–93
    of grids, 166, 531–32, 549, 550
    joining and relating, 182–85, 196–98, 213
    map overlay results, 276, 277
    required fields, 179
    of shapefiles, 28
    viewing and editing in ArcCatalog, 47–48, 58–59
Autocomplete Polygon task, 385, 401–2, 417
Automatic scaling, 309, 330

B

Backgrounds, 339, 343
Backups, 383
Badlands National Park, 9–12, 16–17
Balance in map design, 303–4
Barriers (network)
    defined, 500
    impact on solvers, 503, 511–12
    tools, 525
Batch mode geocoding, 355, 363–64
Best-fit lines, 9
Binary format, 26, 180
Binomial analysis, 11
Black-and-white maps, 302
Black arrow (Drawing toolbar), 80, 99
Black arrow tool, 315
BLOB fields, 181
Bookmarks, 40–41, 96
Boolean grids, 539–41, 547–53
Borders, 336, 339
Border Selector button, 336
Boundaries
    coincident, 28, 385, 417
    de-emphasizing, 306
    dissolving, 281, 286
    editing shared, 425–26, 433–34, 459–60
    for feature datasets, 463–64
    mismatching, 271–72
Browse for Maps command (ArcMap), 37
Budgets, 15, 17
Buffers
    editing features with, 424, 445–46, 459
    for feature dataset boundaries, 464
    for labels, 73
    with location queries, 234, 245
    with map overlay, 279–80, 290–91, 297
    purpose of, 6
    with rasters, 552–53
Buffer tool, 297
Build command, 48
Buildings, creating with editing tools, 403–4
Burns Basin study area, 9
Bytes, 180

C

CAD drawings, 26
Calculating fields, 187–88, 212
Callouts, 72, 100
Callout Text box tool, 100
Canada Geographic Information System, 3
Canada thistle project case study, 9–12, 16–17
Capitalization in expressions, 219–20
Cardinality
    spatial joins, 250–51, 256–57
    table joins, 183–85
Cartesian coordinates, 358. *See also* Coordinate systems; *x-y* data
Case sensitivity in expressions, 219–20
Case studies, 4–7, 9–12
Categorical data, 23, 139–40
Categorical grids, 165
Categories maps, 140–41
Categories of symbols, 151
CD-ROM, 37
Cells
    effects of size on data processing, 530
    matching sizes for analysis, 541–42
    in raster model, 22
    size settings, 163, 532, 547, 566
    statistics, 536, 569
Cell Statistics function, 536, 569
Census Bureau, 3
Center of rotation, 411
Central meridians, 110, 123
Change Layout button, 332, 347
Chart maps, 142, 160–61
Chart tools, 308
Choropleth maps, 139
Clarke 1886, 107, 108
Classes, adding to maps, 87
Classification
    of labels, 72, 86–88
    numeric data, 143–46, 156, 158–59, 173
    in quantities maps, 142
    of raster data, 538, 548–49, 568
Classified method, 148
Classify Values window, 568
Clean command, 48